P9-DVL-887

POWERPLANT SECTION TEXTBOOK
BOOKS ONE THROUGHT EIGHT

An IAP, Inc.
Integrated Training Program
for the Aviation Maintenance Technician

Books 1, 4 and 5
by Dale Crane

Books 2, 3, 6, and 7
by Dale Crane and Charles Otis

Book 8
by Dale Crane, Frank Delp and Charles Otis

International Standard Book Number 0-89100-251-0
For sale by: IAP, Inc., A Hawks Industries Company
Mail To: P.O. Box 10000, Casper, WY 82602-1000
Ship To: 7383 6WN Road, Casper, WY 82604-1835
(800) 443-9250 ❖ (307) 266-3838 ❖ FAX: 307-472-5106

IAP, Inc.
7383 6WN Road, Casper, WY 82604-1835

Table of Contents

Book One   Reciprocating Engine Theory
Reciprocating Engine Maintenance and Operation
Reciprocating Engine Removal and Replacement

Book Two   Turbine Engine Theory
Turbine Engine Maintenance and Operation

Book Three   Engine Ignition Systems

Book Four   Powerplant Electrical Systems

Book Five   Powerplant Electrical Installation
Powerplant Instrument Systems
Fire Protection Systems

Book Six   Aircraft Fuel Metering Systems

Book Seven   Engine Induction Systems
Engine Cooling Systems
Engine Exhaust Systems
Engine Starting Systems

Book Eight   Engine Lubrication Systems
Propellers

Preface

Part 65 of the Federal Aviation Regulations requires that an applicant for a Mechanic Certificate with an Airframe or Powerplant rating have a general knowledge of the construction and maintenance of the aircraft covered by the rating he seeks. And each applicant must prove this knowledge by passing a written, an oral, and a practical examination, each given in five sections: General, Airframe Structures, Airframe Systems and Components, Powerplant Theory and Maintenance, and Powerplant Systems and Components.

Certificated Aviation Maintenance Technician Schools all operate under individual Federal Aviation Administration approvals, and all have in their approved curriculum enough information to satisfy the FAA that anyone completing their course will meet the knowledge requirements for certification as a mechanic.

In 1972, the Federal Aviation Administration published a series of three Advisory Circulars providing background information for mechanic certification applicants. These are entitled: *Airframe and Powerplant Mechanics General Handbook*, AC 65-9; *Airframe and Powerplant Mechanics Powerplant Handbook*, AC 65-12; and *Airframe and Powerplant Mechanics Airframe Handbook*, AC 65-15. These three handbooks were written to cover the area over which a mechanic applicant was to be tested, and they were an attempt by the FAA to coordinate the information available to the applicants with the tests that were given for certification. In 1976, the three handbooks were revised into their present A Revision format: AC 65-9A, AC 65-12A, and AC 65-15A. They have not been revised since.

Individuals attempting to certificate under FAR Part 65, as well as schools operating under FAR Part 147, have found that these handbooks actually do not follow the material used in the FAA examinations, and because of this, plus their lack of study helps, they fail to effectively provide the broad base of knowledge needed for certification and for a foundation on which to build in the aviation maintenance industry.

Realizing the need for updated information, presented in a format that is in keeping with the latest developments in educational technology, International Aviation Publishers, Inc., is offering an Integrated Training Program (ITP) that presents the basic information needed for mechanic certification, in a form that will provide a good foundation in aviation maintenance knowledge. The ITP training manuals, workbooks, study guides, and audiovisuals are designed to follow the FAA written examinations and are written to the level required for the FAR Part 147 schools.

All of this material can be used by an individual preparing for certification on his own, or by schools preparing groups of individuals for certification under FAR Part 65.

Not only has every attempt been made to meet the requirements of Part 147 in producing this series, but the authors have followed many of the recommendations of the Aviation Mechanic Occupation Study conducted by the University of California, Los Angeles, under contract from the U.S. Department of Health, Education, and Welfare.

One of the most basic premises of aviation maintenance is that all operations must be carried out according to FAA-approved data, and all information of a specific nature furnished by the manufacturer of an aircraft, powerplant, or component must be followed in detail. These training manuals are, of necessity, general in nature and cover a wide scope of aircraft, powerplants, and components. Because of this, it must be understood that they should *never* be allowed to take precedence over specific information furnished by the manufacturer.

The authors and publishers of this Integrated Training Program wish to express our appreciation to the various manufacturers, Aviation Maintenance Technician Schools, and FAA personnel who have aided our efforts to provide this coordinated training material to meet the challenge of better training for aviation maintenance technicians.

Reciprocating Engine Theory

I. INTRODUCTION

The lack of a practical propulsion system has been the limiting factor in the development of mechanical devices throughout our history. Leonardo daVinci conceived a flying machine, the aerial screw, in 1483, but with no means of propulsion it was only a dream. Explosion engines were objects of experimentation in the late 17th century, but these experiments yielded no practical results.

The first patent for a heat engine was taken out by John Barber of England in 1791; this was a turbine engine. The first piston engine that could be considered practical was that built by Etienne Lenoir in France in 1860. It used illuminating gas as its fuel.

The next major breakthrough in piston engines came in 1876 when Dr. Nikolaus Otto developed the four-stroke, five-event Otto cycle. This cycle is the principle upon which almost all of our reciprocating engines used in modern aircraft operate.

The modern reciprocating engine is pitifully inefficient, converting only about one-third of the fuel consumed into useful work, and weighing from one to two pounds for every horsepower it produces. But, in spite of this, it continues to be the most popular engine for lower powered aircraft, because it requires little of the expensive materials and technology that have made the turbine engine the standard powerplant where high power is required.

By the end of the 1800's, the basic problem of aircraft control had been identified, and the airplane designers were able to turn their attention to a means of powering the flying machines with which they were experimenting. Charles Manly, working on the first aircraft to be financed by a United States government contract, searched the United States and Europe for a suitable engine for the Langley Aerodrome, but found none.

Manly finally built a powerplant by evolving a three-cylinder rotary-radial engine from a Balzar "automobile" into five-cylinder static radial engine. The Wright brothers met with equally little success in locating an engine for their "Flyer" and had to design their own engine, which was built in their bicycle shop by the master craftsman Charles Taylor. This engine had four cylinders in-line and it lay on its side, driving the two propellers through roller chains.

The evolution of the aircraft reciprocating engine has been slow. Up until World War I, little real progress was made, but in this conflict where the airplane was first recognized as a legitimate vehicle rather than an oddity, great strides were made in engine development. Three configurations of engines emerged from this period.

The rotary-radial engine was used by all of the warring nations, producing the greatest horsepower for its weight. The cylinders of this engine mount radially around a small crankcase and rotate with the propeller. Fins cut into the outside of the cylinders provide all of the cooling needed as the engine spins around.

Fig. 1A-1 Rotary radial engines were popular on both sides during World War I. They produced more power for their weight than any other engine configuration.

2

Fig. 1A-2   The in-line engine was favored for fast airplanes during the 1930's because its small frontal area allowed good streamlining.

The in-line engine, whose cylinders are all in a line, was the second configuration, and it has a far greater reliability than the rotary-radial engine. It does have, as its greatest limitation, the problem of weight because of the long crankshaft needed for each cylinder to have its own throw.

The next logical step in engine development produced the third configuration, the V-engine. Two cylinders share a crankshaft throw and this allows a much shorter and therefore lighter crankshaft to be used.

When World War I ended, the market was flooded with thousands of usable V-type engines, primarily the 90-horsepower V-8 Curtiss OX-5, the 180 horsepower V-8 Hispano-Suiza, and the 400-horsepower Liberty V-12. While these engines were heavy and to a great extent unreliable, they were available, and for a decade they dominated the aviation powerplant scene.

In the late 1920's the Wright Aeronautical Corporation, with the encouragement of the U.S. Navy, developed a series of five-, seven-, and nine-cylinder static radial engines that revolutionized

Fig. 1A-3   V-engines offer a good compromise between weight for power ratio and small frontal area. They are difficult to cool with air, however.

aviation. The reliability of these engines was far greater than that of any other powerplant up to this time, and it made possible the long distance flights in which Lindbergh and others awoke the world to the realization that the airplane was a practical means of travel.

Fig. 1A-4 The radial engine provides the greatest power for its weight, but its large frontal area makes it difficult to streamline.

In the United States, the Wright Aeronautical Corporation and Pratt and Whitney Aircraft produced many thousands of radial engines that powered most of our military and airline airplanes until the advent of the turbine engine for these high power applications.

The radial engine has the disadvantage of a large frontal area which produces a great amount of drag, and as airplane speeds increased, the need to reduce this drag has become of increasing importance. The in-line engine has little frontal area and thus little drag, but its weight is its disadvantage. The V-engine is a good compromise between weight and frontal area, but air cooling is difficult. The horizontally opposed air-cooled engine with its cylinders in two rows, one on either side of the crankcase, has such a good combination of low weight, small frontal area, and ease of cooling that it has become the standard configuration, and today, virtually all production reciprocating engine-powered aircraft use this configuration.

Fig. 1A-5 The horizontally opposed engine is the main type of reciprocating engine being produced today. It provides the best balance between weight for horsepower and low frontal area, while being easy to cool with air.

4

II. ENERGY TRANSFORMATION

A.   Principles of Energy Transformation

The reciprocating engine installed in our modern aircraft is a form of heat engine in which chemical energy in the fuel is converted into heat energy which is in turn converted into mechanical energy.

The focus of our attention recently on the shortage of fossil fuels has caused much interest in "alternate" fuels, primarily those which can be produced in a never-ending supply from grain. There is a real probability that grain-produced alcohol will have a strong influence on aviation fuels in the near future, but today gasoline is used almost exclusively, and we will limit our study to this fuel.

The principle of energy transformation is the same whether we use energy from the sun that was stored in ages past and is available to use in petroleum, or that which was stored from the sun in the grain that has been harvested this year. The source is the same and the results are the same.

The energy in our petroleum came from our primary source, the sun, millions of years ago. Solar energy was changed into chemical energy by the photosynthesis of plant life, and during a great upheaval in the past, the plants that covered the earth and the animals that ate them were buried under millions of tons of dirt and rock, and heat and pressure changed them into the petroleum with which we are all familiar.

Crude oil is brought up from the depths of the earth and is refined into gasoline and the various forms of fuel oils and lubricating oils. The process by which this is done is discussed in detail in the section on Aircraft Fuel Metering in this Integrated Training Program.

Aviation gasoline has a nominal heat energy content of 20,000 British thermal units per pound, and one Btu is the amount of heat energy needed to raise the temperature of one pound of water one degree Fahrenheit, specifically from 59° to 60°.

There are basically two types of engines that use heat to convert chemical energy into mechanical energy, the external combustion and the internal combustion engines. A steam engine is an example of an external combustion engine. Fuel burned in a boiler heats water and changes it into steam. This steam is then carried into the engine where it either forces pistons to move and turn a crankshaft or, more effectively, spins a turbine.

External combustion engines are quite inefficient with regard to the amount of heat energy in the fuel they convert into work, but since they can operate on fuels that are available at low cost, they serve an important function in our industrial strength. Almost any fuel that can burn can be

 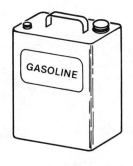 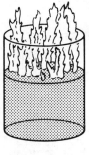 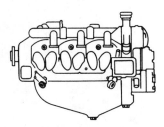

| ENERGY COMES FROM THE SUN | ENERGY IS STORED IN GASOLINE | GASOLINE BURNS TO PROVIDE HEAT | HEAT IS CONVERTED INTO MECHANICAL ENERGY IN AN ENGINE |

Fig. 2A-1   The sun is the basic source of our energy. Solar energy is converted into chemical energy and stored in the form of petroleum of which our modern aviation gasoline is a type. When gasoline is burned, the chemical energy in the gasoline is released as heat energy, and heat energy released in an internal combustion engine is converted into mechanical energy and work.

used for steam engines. Wood, coal, low-grade fuel oil, and natural or synthetic gases are all usable fuels, and today even nuclear fission is used to heat water to generate the steam.

Internal combustion engines are more selective than external combustion engines in the fuels they can burn, but even they have a relatively wide latitude. Some commercial engines burn liquefied petroleum gases such as butane or propane, and some burn natural gas. Diesel fuel and turbine fuels are used as well as many grades of gasoline, and alcohol fuels are now being studied as an alternate to petroleum for internal combustion engine fuel.

Aircraft engines are a rather specialized form of internal combustion engine because of their demand for light weight along with maximum reliability. The fuels that are allowed for use in aviation reciprocating engines must be tested under almost every conceivable condition and approved by both the engine manufacturer and the Federal Aviation Administration.

Regardless of the type of internal combustion engine, the process of releasing the energy from the fuel and converting it into work is essentially the same. Liquid fuel is measured out and converted into a fuel vapor. This vapor is mixed with the correct amount of air to provide the most combustible mixture of hydrocarbon fuel and oxygen. Then this mixture is heated until it ignites. When the mixture burns, it releases its energy and causes that portion of the air that does not enter into the combustion process to expand.

Remember that air is composed of approximately 21% oxygen and 78% nitrogen, and nitrogen is an inert gas that does not enter into the combustion process, but it expands when it is heated. The expanding gas is used to push down on a piston in a reciprocating engine or spin the turbine in a turbine engine.

All internal combustion engines, therefore, have a certain sequence of events that must take place to convert the chemical energy in the fuel into mechanical work.

1. Intake: Fuel and air must be taken into the engine.

2. Compression: The air or fuel-air mixture must be compressed.

3. Ignition: The combustible material must be ignited.

4. Power: The burning gases expand and produce work.

5. Exhaust: The burned gases must then be scavenged.

INTAKE   COMPRESSION   POWER   EXHAUST

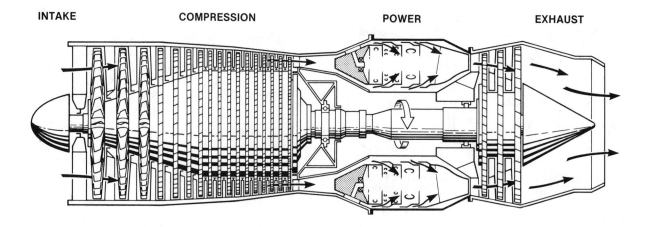

Fig. 2A-2   A turbine engine is a form of internal combustion engine that releases the energy from the fuel it burns in much the same way as energy is released in a reciprocating engine. The events in the energy-release cycle in a turbine engine are the same as those in a reciprocating engine.

B. Energy Transformation Cycles

1. Gas turbine engines

The gas turbine engine extracts energy from the fuel it burns by a constant pressure cycle. This is explained in detail in the portion of the Integrated Training Program devoted to turbine engines.

The five events we have just mentioned all occur simultaneously in a turbine engine at specific locations within the engine. The air is taken in through the inlet duct and is compressed by the spinning compressor. Downstream of the compressor, the air enters the combustion chambers where fuel is sprayed into it in the form of a fine mist, and fire that is continously burning ignites the fuel-air mixture. The burning gases expand and rush out the exhaust, passing through and spinning the turbine on the way out.

On a turboprop or turboshaft engine, the turbine extracts a great deal of energy and uses it to spin the compressor and drive the propeller or the helicopter rotors. On a pure jet engine, the compressor uses some of the energy, but as much of it as possible is used to speed up the air leaving the engine so there will be a maximum difference between the velocity of the air entering and that leaving the engine. This velocity difference is, after, all the basis of jet propulsion.

2. Two-stroke-cycle reciprocating engines

The extreme simplicity and light weight of a two-stroke-cycle engine make it useful for such applications as chain saws, lawn mowers, and other small displacement engines, and two-stroke-cycle diesel engines are used in many trucks. But for aviation applications, the inefficiency of this form of engine has prevented its widespread use.

The same five events occur in each operating cycle of this engine, and all five occur in two strokes of the piston, down and back up. In the illustration of Fig. 2A-3 we see the way this engine works. The carburetor and impeller provide the correct amount of fuel-air mixture and deliver it to the single valve in the cylinder under a positive pressure.

Since two events are occurring inside the cylinder at the same time, we will start our ex-

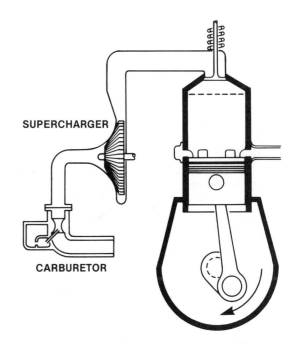

Fig. 2A-3 The two-stroke-cycle reciprocating engine has all of the events in its energy-release cycle in two strokes of the piston.

planation with the assumption that a combustible mixture has been ignited and is burning, forcing the piston down. The pressure inside the cylinder is high, but as the piston nears the bottom of its stroke, it uncovers the exhaust ports and this high-pressure air and the burned gases are exhausted into the atmosphere. As the gases leave the cylinder, the pressure drops enough for the intake air to enter the cylinder through the valve which is now open. By the time the piston passes through bottom center and moves back up enough to cover the exhaust ports, the intake valve has closed and the fresh charge of fuel and air is trapped inside the cylinder.

The piston continues to move up and compress this mixture, and near the top of the stroke, an electrical spark at the spark plug ignites the mixture. The pressure inside the cylinder reaches a maximum just about the time the piston passes through the top of its stroke and starts back down, and the cycle repeats itself.

This type of engine can be made quite simple by substituting an intake port for the intake valve and its operating mechanism, and by using a glow plug to replace the spark plug and the magneto required to produce a timed spark.

Its low efficiency has prevented the two-stroke-cycle engine finding wide application as an aircraft powerplant.

3. *Four-stroke-cycle reciprocating engines*

The Otto cycle of energy release is the most widely used operating cycle for aircraft reciprocating engines. This is classified as a constant-volume cycle because the burning fuel inside the cylinder increases the pressure with almost no change in volume. This differs from the Brayton cycle on which gas turbine engines operate, as in the Brayton cycle the volume of the heated gas increases, but there is very little increase in its pressure.

The indicator diagram of Fig. 2A-4 describes the Otto cycle. The piston moves down inside the cylinder on the *intake stroke* with the intake valve open, and the fuel-air mixture is drawn into the cylinder. This is shown as the portion of the curve between points *A* and *B*. The volume of the cylinder increases as the piston moves down and, because of the friction in the induction system, the pressure inside the cylinder drops to slightly less than atmospheric.

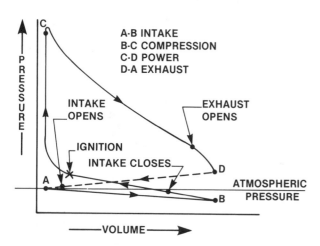

Fig. 2A-4 *An indicator diagram of the constant-volume Otto cycle of operation*

The piston starts back up for the *compression stroke*, *B* to *C*, and when it is part of the way up, the intake valve closes, trapping the fuel-air charge inside the cylinder. When the piston nears the top of its stroke, an electrical spark ignites the fuel-air mixture and as it burns in an almost

constant volume, the pressure rises rapidly. By the time the piston reaches the top of its stroke, the pressure is maximum and when the piston starts down, the pressure exerts a force on the piston, pushing it down for the *power stroke*. Near the bottom of the power stroke, when there is still a considerable amount of pressure inside the cylinder, the exhaust valve opens and the gases begin to rush out of the cylinder.

At point *D* the piston starts back up with the exhaust valve open and all of the exhaust gases are purged during the *exhaust stroke*. When the piston is near the top of the exhaust stroke, the intake valve opens. The pressure inside the cylinder is so low at this point that it does not hinder the induction of the fuel-air charge, but it does allow the intake valve to remain open long enough for a maximum amount of fuel and air to be taken into the cylinder.

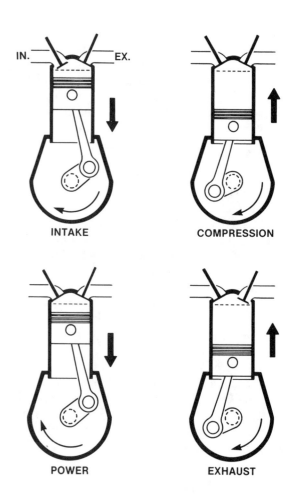

Fig. 2A-5 *The four strokes of the Otto cycle of energy release*

The timing of the opening and closing of the valves and the timing of the ignition are extremely important in a four-stroke-cycle engine. The intake, compression, power, and exhaust strokes are all considered to be the 180° movement of the crankshaft during the time the piston is moving from the top center to the bottom center or from the bottom center to the top center of its stroke. But the valves, as we have just seen, do not open or close when the piston is at either extreme of its travel; rather, they open and close at times when the pressure inside the cylinder is most conducive to getting the maximum amount of fuel-air charge into the cylinder and getting a maximum amount of the burned gases out.

In Fig. 2A-6 you will see that the piston does not move a constant amount for each degree the crankshaft rotates. When the piston is near the center of its stroke, the travel for a given amount of crankshaft rotation is the greatest, but at both the top and the bottom of the stroke the piston moves very little as the crankshaft rotates through several degrees.

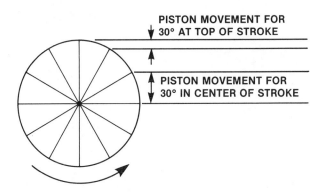

Fig. 2A-6 *Piston movement versus crankshaft travel*

In Fig. 2A-7 we have a spiral showing the timing of the opening and closing of the valves, and the location of the ignition in the cycle of events. The intake valve opens as the piston moves upward on the exhaust stroke. The actual position of the crankshaft when the valve opens depends upon the design of the engine, but it is typically about 15° of crankshaft rotation before the piston reaches top dead center. The intake valve remains open during the entire 180° of the intake stroke and until the piston is well up, somewhere around 60° degrees of crankshaft rotation, on the compression stroke.

The intake valve closes and the piston continues to move up, compressing the gases inside the cylinder until at about 30° before the piston reaches the top of the stroke, the spark from the spark plug ignites the mixture and the pressure begins to build up inside the cylinder. Both valves remain closed during the power stroke until the piston reaches a point about 60° of crankshaft rotation before bottom dead center and then the exhaust valve opens.

The exhaust valve remains open through all of the exhaust stroke and until the piston has moved down about 10° of crankshaft rotation on the intake stroke. The time during which both valves are open at the end of the exhaust stroke and the beginning of the intake stroke is called the valve overlap and is important in getting the maximum amount of fuel-air charge into the cylinder. (See Fig. 2A-7 on page 10.)

C. Work-Power Considerations

1. Work

Work is accomplished when a force moves through a specified distance, independent of time. In the English system of measurement, work is expressed in foot-pounds and in the metric system, in meter-kilograms. Work, measured in foot-pounds, is the product of the force in pounds times the distance through which the force acts, measured in feet, and it may be found by the formula:

$$\text{Work} = \text{Force} \times \text{Distance}$$

When one pound of mass is raised one foot, one foot-pound of work is accomplished.

2. Power

Time is not involved in the determination of the amount of work that has been done, but power does involve time. Power is the expression of the time rate of doing work, and it may be expressed in foot-pounds of work done in one minute or in one second.

$$\text{Power} = \frac{\text{Force} \times \text{Distance}}{\text{Time}}$$

a. Horsepower

James Watt, the inventor of the steam engine, found that an English dray horse could do,

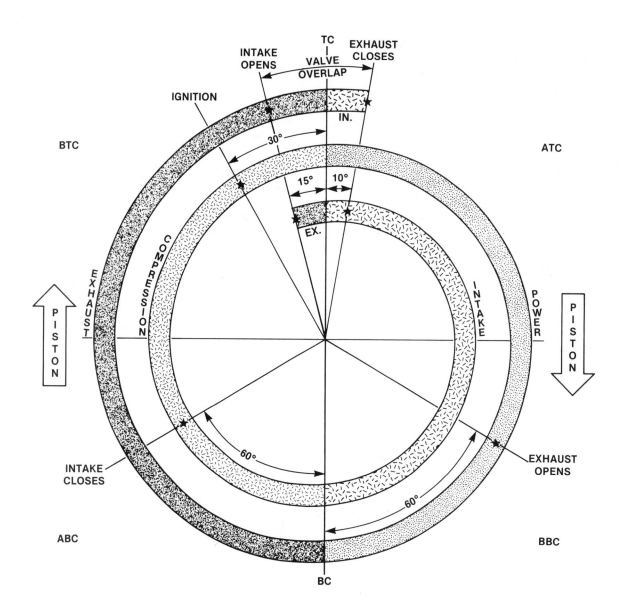

Fig. 2A-7 Valve opening versus piston position

for a reasonable period of time, about 33,000 foot-pounds of work in one minute, and from his observations came the term "horsepower," which has been standardized as 33,000 foot-pounds of work per minute, or 550 foot-pounds per second. In electrical power, there is a relationship between the watt and the horsepower. One horsepower is equal to 746 watts, or one kilowatt (1,000 watts) is equal to 1.34 horsepower.

(1) Brake horsepower

The actual amount of power delivered to the propeller shaft of a reciprocating engine is called the brake horsepower and it derives its name from the method by which it is measured. In the early days of reciprocating engine technology, power was measured by clamping a brake around the output shaft of an engine and measuring the force exerted on an arm at the speed the measurement was being taken. This measuring device was called a prony brake; thus the name brake horsepower. Today, the same type of measurement is made, but the engine drives either an electrical generator or a fluid pump. The output of the generator or pump is forced to perform work, and the amount of work done in a given period of time is used to calculate the power the engine is producing.

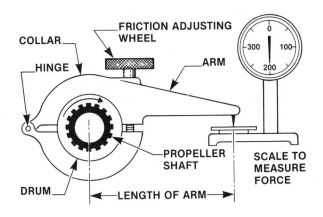

Fig. 2A-8 The basic principle of operation of a prony brake to measure the torque produced by an engine

(2) Friction horsepower

Pistons slide back and forth in the cylinders, and air is pulled into the engine and compressed. This requires power, as does the movement of all of the rotating machinery such as the gears and all of the accessories. All of the power used to drive the engine is lumped into one measurement called friction horsepower. This may be measured by driving the engine with a calibrated motor and measuring the amount of power actually needed to turn the engine at each speed.

(3) Indicated horsepower

The power delivered to the propeller plus that used to drive the engine is the total power

COURTESY OF WOOD ENGINEERING

Fig. 2A-9 Modern engines are calibrated for torque output on a dynamometer such as this one.

developed in the cylinders. This total power is called the indicated horsepower. Brake horsepower, as we have seen, may be measured by a mechanical device on the shaft, but since friction horsepower is involved, indicated horsepower cannot be measured directly. It may, however, be calculated by the formula:

$$IHP = \frac{PLANK}{33,000}$$

P stands for the indicated mean effective pressure, the IMEP, and is the average pressure inside the cylinder during the power stroke. Originally this was measured by an instrument called an indicator (thus its name), but modern technology allows us to use an electrical pressure transducer which is far more accurate and convenient to use. IMEP is expressed in pounds per square inch.

A is the area of the piston head in square inches, and since P is measured in pounds per square inch and A is in square inches, the product of P and A gives us the amount of force in pounds that acts on the piston.

L is the length of the stroke in feet. $P \times A$ gives the pounds of force acting on the piston, and L, the distance through which this force acts during each power stroke. So $P \times A \times L$ gives the number of foot-pounds of work done on each power stroke.

N is the number of power strokes per minute for each cylinder. This is found for a four-stroke-cycle engine by dividing the engine RPM by two, since there is only one power stroke for each two revolutions.

$$N = \frac{RPM}{2}$$

K is the number of cylinders in the engine.

The product of PLANK gives the number of foot-pounds of work done each minute by the engine, and since one horsepower is equal to 33,000 foot pounds of work per minute, by dividing PLANK by 33,000 we can find the indicated horsepower.

To check our understanding of this formula, let's compute the indicated horsepower for a six-cylinder aircraft engine that has a bore of five

11

inches and a stroke of five inches. It is turning at 2750 RPM and has a measured IMEP of 125 pounds per square inch.

$$P = 125 \text{ psi}$$

$$L = 5/12 \text{ feet}$$

$$A = 0.7854 \times 5^2 = 19.63 \text{ sp. in.}$$

$$N = 2750/2 = 1375$$

$$K = 6$$

$$\text{IHP} = \frac{125 \times 5/12 \times 19.63 \times 1375 \times 6}{33,000}$$

$$= 255.60 \text{ indicated horsepower}$$

b. Factors affecting engine power

(1) Thermal efficiency

The thermal efficiency of an engine is the ratio of the amount of heat energy converted into useful work, to the heat energy contained in the fuel. Simply stated, it is the ratio of power put out, to the power put in. If two engines produce the same amount of horsepower, but one burns less fuel than the other, the engine using the least fuel converts a greater portion of the available energy into useful work and therefore has the higher thermal efficiency.

Thermal efficiency may be found by the formula:

$$\text{TE} = \frac{\text{HP} \times 33,000}{F \times 20,000 \times 778}$$

The horsepower, HP, used in this formula may be either brake or indicated horsepower, depending upon the type of thermal efficiency you want. If BHP is used, the result will be brake thermal efficiency, and if IHP is used you will get indicated thermal efficiency.

The constant 33,000 is the number of foot-pounds of work per minute in one horsepower. Horsepower times 33,000 gives the foot pounds of work done in one minute. This is the output.

F is equal to the number of pounds of fuel consumed by the engine in one minute, and for aviation gasoline which nominally weighs six pounds per gallon, F is found by multiplying the gallons per hour by six and dividing this by 60. This is the same as dividing the pounds of fuel burned per hour by the number of minutes in one hour.

The constant 20,000 is the nominal heat energy content of aviation gasoline. Each pound contains 20,000 Btu of heat energy.

The constant 778 is the number of foot-pounds of work each Btu is capable of doing.

The product of $F \times 20,000 \times 778$ is the number of foot-pounds of work the fuel burned is capable of producing, and this is therefore the input.

Let's check our understanding of the thermal efficiency formula by finding the thermal efficiency of an engine that produces 150 brake horsepower while burning 14.0 gallons of aviation gasoline per hour.

$$\text{TE} = \frac{150 \times 33,000}{(14 \times 6)/60 \times 20,000 \times 778}$$

$$= 22.7\%$$

By using this formula we see that reciprocating engines are extremely inefficient with regard to the energy they use. The low percentage found in this example is for a popular engine developing its maximum rated horsepower, but even at an efficient cruise speed, these engines seldom convert more than one-third of the fuel they burn into useful work.

(2) Volumetric efficiency

Reciprocating engines are air breathing and require a maximum amount of air in the cylinder to release the most energy from the fuel. Volumetric efficiency is the ratio of the amount of air the engine takes into the cylinder to the total displacement of the piston.

For this to be meaningful, the density of the air in the cylinder must be converted to standard density by correcting the temperature and pressure of the air to the standard conditions of 59° F (or 15° C) and 14.69 psi (or 29.92 inches of mercury).

If an engine draws in a volume of charge at this density, exactly equal to the piston displacement, the volumetric efficiency of the engine will be 100%; but the volumetric efficiency of a normally aspirated engine is always less than 100% because the friction of the induction system walls and the bends in the intake tubing restrict the amount of air that can flow into the cylinder during the time the intake valve is open. Supercharged engines, on the other hand, compress the air before it enters the cylinders, and they may have volumetric efficiencies greater than 100%. Volumetric efficiency is found by the formula:

$$VE = \frac{\text{Volume of charge}}{\text{Piston displacement}}$$

Anything that decreases the mass of air entering the cylinder during the intake stroke decreases the volumetric efficiency. Some of the typical factors that affect the volumetric efficiency of a non-supercharged engine are:

Part throttle operation This restricts the amount of air that can flow into the cylinders.

Long intake pipes of small diameter The friction increases directly as the length of the intake pipes and inversely as their cross-sectional area.

Sharp bends in the induction system The air is slowed down each time it turns a corner, and the slower the air, the less will get into the cylinders.

Carburetor air temperature too high As the temperature increases, the density of the air decreases, and there will be fewer pounds of air taken into the cylinder.

Cylinder head temperature too high This also lowers the density of the air in the cylinders and provides fewer pounds of air for the same volume.

Incomplete scavenging The incoming fuel-air mixture will be diluted with exhaust gases, and there will be less fresh charge drawn into the cylinder.

Improper valve timing If the intake valve is not open long enough to draw a complete charge of fuel-air mixture into the cylinder, the volumetric efficiency will be low.

(3) Mechanical efficiency

Mechanical efficiency is the ratio of brake horsepower to indicated horsepower. It shows the percentage of power developed in the cylinders that actually reaches the propeller shaft. Aircraft engines are usually quite efficient mechanically, and it is not unusual for ninety percent of the indicated horsepower to be converted into brake horsepower.

(4) Piston displacement

Piston displacement is the product of the area of the piston, the length of the stroke, and the number of cylinders. Since the amount of work done by the expanding gases is determined by these factors, it is evident that the piston displacement is of major importance in power computations. Increasing either the bore of the cylinder or the stroke of the piston will increase the piston displacement. The formula for piston displacement is:

$$PD = A \times L \times N$$

A is the area of the piston head in square inches, and is equal to $0.7854 \times \text{Bore}^2$.

L is the length of the stroke in inches.

N is the number of cylinders.

(5) Compression ratio

The ratio of the volume of the cylinder with the piston at the bottom of the stroke to the volume with the piston at the top of its stroke is called the compression ratio of the engine, and it determines, to a great extent, the amount of heat energy in the fuel-air mixture that can be converted into useful work.

Engines with high compression ratios allow the fuel-air mixture to release its energy rapidly and produce the maximum pressure inside the cylinder just as the piston starts down on the power stroke.

The practical limit to compression ratio is determined by the fuel used. When the compressed fuel-air mixture is ignited, the flame front moves across the piston head, heating and further compressing the charge in front of it. A point is reached, called the critical pressure and tem-

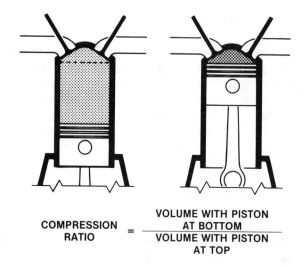

COMPRESSION RATIO = (VOLUME WITH PISTON AT BOTTOM) / (VOLUME WITH PISTON AT TOP)

Fig. 2A-10 Compression ratio is the ratio of the volume of the cylinder with the piston at the bottom of the stroke, to its volume with the piston at the top of the stroke.

perature of the mixture, where it no longer burns and releases its energy evenly, but rather explodes, releasing its energy almost instantaneously. This is called detonation and it creates such a high pressure and temperature inside the cylinder that holes may be burned in the piston heads, and connecting rods may be kinked or even cylinder heads blown off of the barrels.

Fuels with high octane ratings—that is, fuels with high critical pressure and temperature—allow high compression ratios to be used.

The pressure generated within the cylinder is important, as it must always be kept below the critical pressure of the fuel-air mixture. This cylinder pressure is determined by both the compression ratio of the engine and the pressure of the fuel-air mixture when it enters the cylinder.

The pilot has no direct way of measuring cylinder pressure, but he does have an instrument, the manifold pressure gage, that tells the pressure of the charge inside the intake system. The pressure, normally measured in inches of mercury, absolute, is related to the engine RPM to get an indication of the power being produced by the engine. For most power settings, there is more than one RPM-manifold pressure combination, and the pilot, by trying each combination can find the one that gives the most vibration-free operation for his particular aircraft.

Fig. 2A-11 The tachometer and manifold pressure gage give the pilot the information he needs to determine the amount of power the engine is developing.

(6) Ignition timing

It is of extreme importance that the maximum pressure within the cylinder be reached shortly after the piston passes its top center position and starts down. For this to occur, the mixture must be ignited quite a way before the piston reaches the top center, usually somewhere around thirty degrees of crankshaft rotation before the piston reaches top dead center on the compression stroke.

Automobile engines have a variable timing device on their distributor that changes the amount of spark advance as engine operating conditions change, but aircraft engines employ fixed timing, which is of necessity a compromise between the timing required to give good performance for takeoff and that needed for cruise.

When the engine is started, it rotates so slowly that a spark occurring at the proper advanced time would cause a serious kickback and damage the starter. To prevent this, the spark is retarded for starting, using either an impulse coupling or a set of retard breaker points to time the spark that is provided by a vibrator.

If ignition occurs too early, the engine will lose power because the maximum cylinder pressure will be reached while the piston is still moving upward and the force of the expanding gases will oppose the rotational inertia of the engine. But, on the other hand, if it occurs too late, there will be a more serious loss of power; since the cylinder volume is increasing as the gases expand, the effect of the push will be lost. Also, late ignition does not allow enough time for all of the fuel-air mixture to burn before the exhaust valve opens, and these burning gases, forced out past the exhaust valve, increase its temperature and will damage the engine by overheating.

The excess heat in the cylinder from late timing gives rise to a problem known as preignition. The burning gases leaving the cylinder may cause local overheating of valve edges or carbon particles within the cylinder. And when these particles glow from the heat, they will ignite the fuel-air mixture while it is being compressd and cause, in effect, extremely early timing. This timing is so early that the mixture will reach its critical pressure before the piston reaches top center, and detonation will occur.

Magnetos are timed to produce their spark when the piston reaches the proper position, but breaker point wear will cause the timing to drift early, and wear of the cam follower will allow the engine to rotate further before the points open, causing late timing.

(7) *Engine speed*

The amount of power produced by an aircraft engine is determined by the cylinder pressure, the area of the piston, the distance the piston moves on each stroke, and the number of times this movement occurs in one minute.

The piston area, length of the stroke, and the number of cylinders are all fixed, but the pilot has two variables with which to work, the pressure within the cylinder and the number of power strokes per minute. Engines equipped with fixed pitch propellers give little control over the number of RPM the engine develops, and the maximum RPM for the rated power is determined by the airframe manufacturer by his choice of propeller.

The load caused by the propeller limits the RPM, and so the power setting is determined solely on the basis of RPM. If the engine is equipped with a constant speed propeller, a governor will cause the propeller pitch to vary to maintain the RPM set by the pilot. Power is then determined by the relationship between the manifold pressure and the RPM.

Takeoff is made with the propeller in full low pitch, which allows the engine to reach its maximum RPM and develop its maximum power. When power is reduced, it must be done by first reducing the amount of fuel entering the cylinders by pulling back on the throttle and then reducing the RPM with the propeller pitch control.

If detonation should occur at any given power setting, the manifold pressure should be reduced and the RPM *increased*. This will distribute the energy that is being released from the fuel over more power strokes, and the cylinder pressure will be reduced enough to stop the detonation.

(8) *Specific fuel consumption*

While not actually a measure of the power itself, specific fuel consumption is an important measure for comparing the efficiencies of engines. The number of pounds of fuel burned per hour to produce one horsepower, either indicated or brake, is known as the specific fuel consumption of the engine. For most practical purposes, brake horsepower is used, and this gives us brake specific fuel consumption, BSFC.

Fig. 2A-12 illustrates the way specific fuel consumption varies with the RPM of the engine. Since brake horsepower depends upon engine speed, below 2400 RPM, the engine is not developing as much power as it is capable of for the amount of fuel it is using. Above 2400 RPM, the amount of power needed to drive the engine, the friction horsepower, increases and the brake horsepower decreases. From this curve, we can see that the most efficient operating speed for this particular engine is around 2400 RPM, at which speed the engine requires about 0.51 pound of fuel per hour for each horsepower it produces. At full throttle, because of the additional power required to turn the engine itself, the power requirements have gone up to 0.59 pound per horsepower per hour.

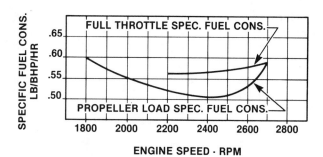

Fig. 2A-12 The relationship between brake specific fuel consumption and engine RPM

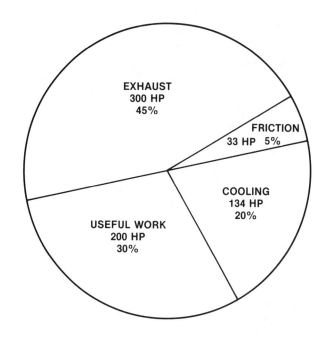

Fig. 2A-13 Typical distribution of the heat energy released in an aircraft engine

c. Distribution of power

When we considered the amount of power that is available in aviation gasoline compared to the amount of power actually delivered to the propeller shaft, we can easily see that an aircraft engine is an inefficient machine. A typical engine may develop 200 brake horsepower when burning 14 gallons of aviation gasoline per hour, but when this amount of aviation gasoline is burned in one hour it releases enough heat energy to produce 667 horsepower. About 33 horsepower is used just to turn the engine and compress the air in the cylinders, and an equivalent of about 434 horsepower is lost to the air through the cooling and exhaust systems. Anything that can be done to minimize these losses will increase the efficiency of the engine.

Even when power is delivered to the propeller shaft, we do not have what we really need because for an airplane to be pulled through the air or a helicopter lifted, we must convert the torque produced by the engine into thrust by the propeller or rotor. Torque is a force that acts perpendicular to the axis of rotation of the propeller, and thrust acts parallel to this axis. A propeller converts

about 90% of the torque it receives into thrust, and this is limited by the tip speed and by the inefficiency of the blade near the root.

d. Power curves

The engine manufacturer produces a set of power curves for each engine he builds to show the power the engine will develop for each RPM and give the specific fuel consumption for each power.

In Fig. 2A-14 we have a typical power curve for a popular four-cylinder aircraft engine. You will notice that there are two power curves and two specific fuel consumption curves that meet at the normal rated power. The upper power curve shows the maximum amount of power the engine will produce with a wide open throttle on a dynamometer. You can tell that under these conditions the engine will produce 160 horsepower at 2400 RPM and a maximum of 168 horsepower at 2700 RPM. The lower power curve is the propeller load curve and when the engine is equipped with the recommended propeller, it will produce 118 horsepower at 2400 RPM and its normal rated 168 horsepower at 2700 RPM. At 2700 RPM, the throttle will be wide open.

There are also two specific fuel consumption curves, the upper for full throttle operation and

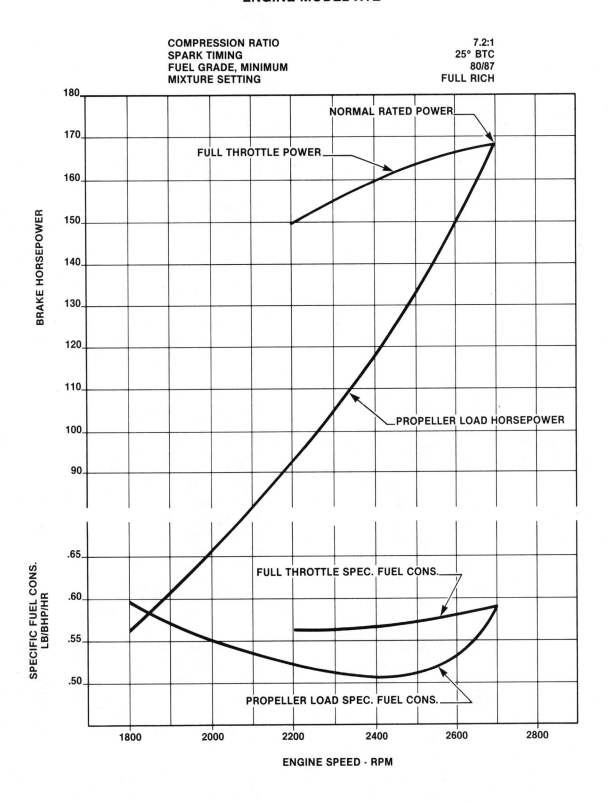

Fig. 2A-14   Power curve for a typical aircraft engine

17

the lower for the propeller load conditions. At 2400 RPM, the propeller load specific fuel consumption curve shows the BSFC to be 0.51 and since the engine is developing 118 horsepower, it is consuming 118 × 0.51 or 60.18 pounds of fuel per hour. Aviation gasoline has a nominal weight of six pounds per gallon, and so the engine is burning 10.03 gallons of gasoline per hour.

At full throttle, the engine has a BSFC of 0.59 and develops 168 horsepower. At this power it is burning 16.52 gallons of fuel per hour.

III. RECIPROCATING ENGINE REQUIREMENTS AND CONFIGURATIONS

A. Engine Requirements

The method of transforming chemical energy into mechanical power in a heat engine is basically the same for any type of heat engine, but in this text we are primarily concerned with reciprocating engines as they apply to certified aircraft, and we will find that there are some requirement that are unique to these engines.

All aircraft and aircraft engines require a series of compromises. We may trade off weight for power or flexibility for economy, but there is one thing we must never compromise, and that is safety. Since the aircraft engine is so vital a part of a transportation system it must meet and adhere to strict design, production, and maintenance requirements set by the Federal Aviation Administration.

Let's look at the basic requirement for an aircraft engine:

Reliability The ability of an engine to consistently live up to the manufacturer's specifications is one of its most important requirements. It must be designed and built in such a way that chances of it failing are minimized.

Before an engine is granted its Approved Type Certificate, its ATC, it must prove its reliability by an endurance run witnessed by the Federal Aviation Administration in which it is operated for 85 hours at maximum continuous power, 15 hours at takeoff power, and 50 hours at high cruise power. After this amount of running, the engine is disassembled and it must show no sign of abnormal wear nor any indication of impending failure.

Durability This relates to the amount of service life you should be able to expect from the engine. A measure of durability, the TBO, or time between overhauls, has been established by the engine manufacturers to give us a benchmark from which we can estimate the life expectancy of an engine. This time, given in hours, is a *recommendation* for the maximum amount of time the engine should be operated between major overhauls, and it presumes the engine is being operated in favorable conditions and is always given the maintenance recommended by the manufacturer. There is no guarantee that the engine will run for its full TBO without needing an overhaul, but it is the best measure of durability we have.

Compactness It is necessary for the engine to be as small and compact as it is possible to make it, so it can be streamlined into the structure to cause the minimum wind resistance.

Weight per horsepower This has been one of the areas in which aircraft reciprocating engines have made great strides, but also an area in which they are woefully inferior to turbine engines. Some of the first aircraft engines weighed, including all of their vital accessories, almost 16 pounds for every horsepower they produced. By the end of the 1930's the weight of the larger engines was down to about one pound per horsepower, but two to four pounds per horsepower was more typical for the smaller engines. Today, four- or six-cylinder engines that power most of the general aviation fleet weigh between one and a half and two pounds per horsepower.

The radial engine with its short crankshaft has the best weight to power ratio of any configuration, but it is seldom used on modern aircraft because its large frontal area causes so much drag that the advantage of weight saving is lost at the high airspeeds we expect from our modern airplanes.

Fuel economy The need for reliability, compactness, and low weight per horsepower has encouraged engine manufacturers to improve their designs, but the increase in the price of aviation fuel has had perhaps a greater impact on engine development than any other factor in recent years.

In the past we have been content to enrich the fuel-air mixture and use fuel to cool the engine during times of high power output. But today with fuel costing so much and very little prospect of its ever costing less, engine manufacturers are conducting basic research into ways of recovering some of the power we lose in our inefficient engines.

Cylinders are being designed to allow for higher operating temperatures, and turbochargers are installed to use part of the energy that would otherwise be wasted to compress the induction air before it enters the cylinders. Power recovery turbines that have been used only on some of the largest engines in the past are now being considered for the smaller engines, in order to use part of the exhaust gas to drive a turbine coupled to the crankshaft to help turn it.

COURTESTY OF TELEDYNE-CONTINENTAL

Fig. 3A-1 A modern six-cylinder horizontally opposed aircraft engine with a turbocharger

Freedom from vibration Airframes are built as light as modern technology will allow and with lightness we have a low resistance to vibration. It is important, therefore, that the aircraft engine be made to operate as smoothly as possible. The more cylinders the engine has, the more power impulses will overlap and the smoother the engine will operate. This consideration must be balanced against the greater efficiency of using fewer

cylinders and taking more power from each of them. To aid in damping vibrations, counterweights and dynamic dampers are installed on the crankshaft and special vibration absorbing engine mounts are used.

Operating flexibility This demands that an aircraft engine operate efficiently at both idle and cruise RPM, and it must never hesitate when full power is applied for takeoff. It must operate well at sea level conditions and be capable of being adjusted to provide the needed power at any altitude the aircraft will fly. The engine must be designed so rain, dust, sand, heat, and vibration will have a minimal damaging effect.

Cost Reasonable cost is an important criterion for all components of an aircraft. The first cost must be low enough that the engine will be able to meet the competition and be accepted by the airframe manufacturers, and the operating costs must be low enough that it will be profitable for its operators.

No single engine can be best in all areas, and so engine manufacturers must compromise their design and construction to fit the engine to its application.

B. Engine Configurations

Aircraft reciprocating engines are classified in three ways: by the arrangement of the cylinders around the crankshaft, by the method used for cooling, and by the type of lubrication system they use.

1. Cylinder arrangement

a. In-line engines

The engine used by the Wright Flyer, that got aviation started, had its four cylinders in-line on the crankcase, and the engine lay on its side. The crankshaft was long enough that each piston was connected to its own individual throw.

The in-line configuration which gained a fair amount of popularity was used up until the years immediately following World War II, and it was built in both four- and six-cylinder models. In-line engines can be mounted either upright or inverted. When upright, the cylinders are above the crankcase, which is best for the engine as the crankcase holds the oil for lubricating the connecting rod bearings. But the cylinders stick up so high they restrict the pilot's visibility. Inverted in-line engines have the advantage of the propeller shaft being high and the cylinders out of the line of the pilot's vision, but oil from the crankcase does leak past the piston rings when the engine is not operating and causes hydraulic locks in the cylinders.

In-line engines have the smallest frontal area and they were used in many racing airplanes during the 1930's. Their major drawback was the high weight for their horsepower because of their long crankshaft.

Many of the early in-line engines were water-cooled, but the Menasco and the Ranger engines, two of the last of the popular American-built engines, were air-cooled. Air scoops in the cowling brought air to one side of the engine and it flowed across the cylinders to the other side and out into the airstream. This provided far more uniform cooling than would have been obtained if the cooling air flowed over the engine from the front to the back.

The cylinders of an in-line engine start numbering at the anti-propeller end and increase toward the front.

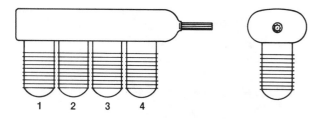

Fig. 3A-2 Cylinder configuration for an inverted in-line engine.

b. V-engines

The crankshaft is the heaviest part of an engine, and when all of the cylinders are in-line, it must be long enough for each piston to connect to its own throw. But if the cylinders are arranged in two banks with one cylinder in each bank connected to a single crankshaft throw, the engine can be made much shorter, with lighter weight,

and have only a small increase in the frontal area. This is the approach many of the early engine designers took.

The World War I surplus Curtiss OX-5 engine, the one that powered many of the airplanes the early-day barnstormers used to introduce flying to the American public, was a 90-horsepower V-8, water-cooled engine. Another World War I engine powered most of the early mail planes; it was the 400-horsepower, Liberty V-12 engine. Both of these engines were water-cooled and neither was noted for its high degree of reliability.

The reliability and low weight for its horsepower caused the radial engine to replace the V-engines until the need for a high-powered engine with a very small frontal area was brought out in the early stages of World War II. The British Rolls-Royce and the American Allison and Packard V-12 Prestone-cooled engines powered the British Spitfires and Hurricanes and the American Lightnings and Mustangs. These engines all developed in excess of 1,000 horsepower and proved excellent for their purpose.

Some of the lower powered V-engines have been air-cooled, but it is so difficult to get uniform airflow through the cylinders that almost all of the successful V-engines have been liquid-cooled.

The cylinders of a V-engine start numbering at the anti-propeller end and increase toward the propeller, and are numbered according to the position and the bank. The cylinders in the left bank of a V-12 engine, for example, as viewed from the anti-propeller end, are numbered 1L through 6L, and those in the right bank are numbered 1R through 6R.

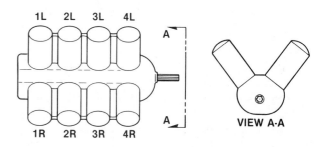

Fig. 3A-3 Cylinder configuration for an upright V-8 engine

c. Radial engines

The need for low weight has caused engine designers to concentrate their efforts on the heaviest component in the engine to decrease its weight. The crankshaft is by far the heaviest single component, and by shortening it, weight can be saved.

Before World War I, the French brought out a very successful engine configuration, the rotary-radial engine. Rather than having a crankshaft as such, a cam arrangement was fastened to the aircraft structure, and the crankcase, with seven or nine cylinders attached radially, turned around the cam. The propeller was rigidly attached to the crankcase. Pistons were connected to the cam with connecting rods, one end of which rode in grooves in the cam, and as the engine rotated, the pistons moved up and down inside the cylinders. These engines turned quite slowly, seldom more than 1200 RPM maximum, and their reliability left quite a bit to be desired.

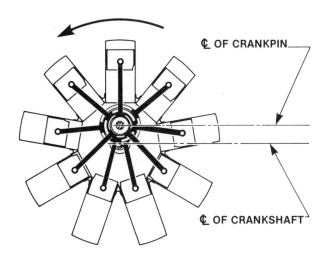

Fig. 3A-4 *The cylinders of a rotary-radial engine spin around the crankshaft which is mounted rigidly on the airframe. One end of each connecting rod rides in a groove in a cam which is offset from the center of the crankshaft.*

The logical evolution from the rotary-radial engine was the static-radial engine whose cylinders fastened radially around a small crankcase and the pistons attached to a single throw of the crankshaft through a master rod and articulated rod arrangement. It was the Wright Whirlwind

engine, a nine-cylinder 220-horsepower radial engine that was used by Charles Lindbergh in his history-making flight in 1927 that convinced the world that airplanes had become a reliable means of transportation.

Wright and Pratt and Whitney radial engines powered the majority of American fighters and bombers during World War II, and virtually all of the airliners until the turbojet engine with its extremely high power for weight completely dominated the high power aspect of aviation.

The cylinders of a radial engine start numbering with the top cylinder and continue around clockwise as viewed from the anti-propeller end. A double-row radial engine has all of the odd-numbered cylinders in the rear row and all of the even-numbered cylinders in the front row.

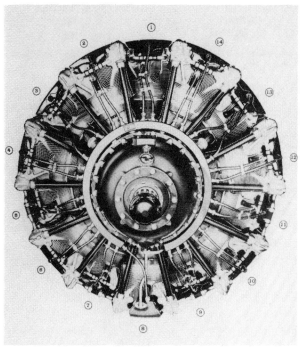

COURTESY OF PRATT & WHITNEY OF CANADA

Fig. 3A-5 *Cylinder arrangement of a 14-cylinder two-row radial engine*

d. Horizontally opposed engines

The angle between the banks of cylinders in a V-engine can cause uneven power applications for some engines, and to eliminate the roughness it causes, engine manufacturers have turned to the horizontal opposed configuration with two banks of cylinders, one on each side of the crankshaft, to get a smooth engine regardless of the number of cylinders.

It was the availability of inexpensive horizontally opposed two- and four-cylinder engines in the 37- to 40-horsepower range that allowed the

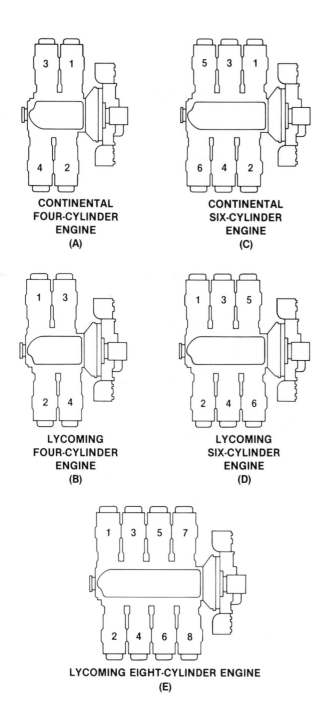

CONTINENTAL FOUR-CYLINDER ENGINE (A)

CONTINENTAL SIX-CYLINDER ENGINE (C)

LYCOMING FOUR-CYLINDER ENGINE (B)

LYCOMING SIX-CYLINDER ENGINE (D)

LYCOMING EIGHT-CYLINDER ENGINE (E)

Fig. 3A-6 Cylinder arrangement for popular horizontally opposed aircraft engines

"lightplane" to be introduced in the 1930's. These engines have a weight advantage almost as good as a V-engine, and they can be housed in a streamlined cowling with far less frontal area than a radial engine.

Horizontally opposed engines have evolved as the configuration that is most widely used, with almost all reciprocating engines now in production having four or six cylinders and made by either Teledyne-Continental or Avco-Lycoming.

Teledyne-Continental and Avco-Lycoming both make four- and six-cylinder horizontally opposed engines that operate in exactly the same way, but the cylinder numbering method used by the two manufacturers differ. Because the method of numbering differs, the firing orders differ, but the actual firing impulses are the same.

The cylinders of a horizontally opposed engine are slightly staggered, so two cylinders, one on either side of the engine, can each have a crankshaft throw of its own, and Continental engines number the cylinder at the anti-propeller end on the right side as number one, and the cylinder nearest the propeller on the left side as number four or six.

Lycoming engines also start numbering on the right side, but the cylinder nearest the propeller is number one and the one on the anti-propeller end on the left side is number four or six.

2. Cooling systems

a. Liquid cooling

We convert less than one third of the heat energy that is released inside the engine cylinder into useful work, and about 65% of the heat must be disposed of either through the exhaust system or by some form of cooling system that can remove the heat that is absorbed into the metal of the cylinders, pistons, and valves.

Many in-line and V-engines have their cylinders encased in metal jackets through which water flows to absorb heat from the cylinder wall and carry it into a radiator where it is given up to the air. Liquid cooling provides more precise temperature control than air cooling, but it requires a drag-producing radiator and the weight of the coolant, the pumps, and all of the plumbing makes this a heavy engine.

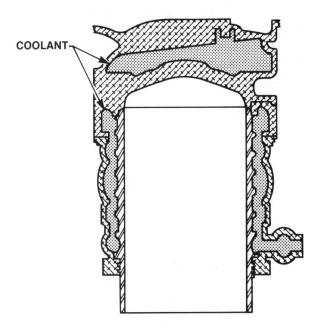

COOLANT

Fig. 3A-7 Typical liquid-cooled cylinder. The coolant is carried in a jacket around the outside of the steel cylinder sleeve. This liquid coolant absorbs heat from the cylinder and carries it outside the aircraft, where it is given up to the passing air in the radiator.

Most of the early liquid-cooled engines used water as the coolant, and at altitude the water boiled away unless the cooling system was sealed, in which case the high pressure inside the system caused leaks. The vulnerability of the liquid cooling system to damage from gunfire led the U.S. Navy, in 1928, to stop using liquid-cooled engines in any aircraft operating with the fleet, but the Army Air Corps and later the Air Force continued to use liquid-cooled engines.

In World War II, the ethylene glycol-cooled Allison and Rolls-Royce engines powered some of our fastest fighters. Ethylene glycol used as a coolant allows the engine to operate at a higher temperature and it will not freeze when the aircraft is parked outside in low temperatures. The smaller radiator required by the ethylene glycol has overcome the problem of the increased drag of the liquid-cooled engine.

b. Air cooling

Most of our modern aircraft engines have fins cast into the aluminum alloy cylinder heads and machined into the cylinder barrels. Air is taken in at the front of the cowling above the engine, and it is directed by baffles so it will flow between the fins and out the bottom of the cowling at the rear of the engine. As it passes through the fins, it picks up heat from the engine and carries it away. The amount of cooling is controlled by adjustable cowl flaps in the opening in the cowling through which the air exits.

COURTESY OF AVCO LYCOMING

Fig. 3A-8 A popular four-cylinder horizontally opposed air-cooled aircraft engine.

3. Lubrication systems

Some of the smaller two-stroke-cycle engines use fuel in which the required lubricating oil has been mixed and therefore require no separate lubricating system, but all four-stroke-cycle engines use a separate system for the oil that is used to lubricate, cool, and seal the moving parts of the engine. There are two basic types of systems used: the wet sump system in which the entire oil supply is carried inside the engine itself, similar to that used in almost all modern automobiles, and the dry sump system in which only a very small amount of oil is carried inside the engine while the majority of the oil is carried in an external tank.

Radial engines have a small crankcase and it is above many of the cylinders, so only a small amount of oil is allowed to remain inside the engine. As it is used for lubrication, the oil picks up heat and is carried out of the engine by a scavenger pump and held in an external oil tank until it is pumped back through the engine.

C. Firing Order

1. In-line engines

All four-stoke-cycle engines must fire all of the cylinders in two complete revolutions of the crankshaft, and smooth engine operation requires an even distribution of power impulses. A four-cylinder in-line engine fires either 1-3-4-2 or 1-2-4-3. These engines use a 180° crankshaft that causes pistons one and four to be at the top of the stroke when pistons two and three are at the bottom.

You will notice in Fig. 3A-9 that the distribution of the power impulses is the same with either firing order.

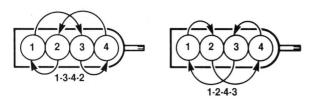

FOUR CYLINDER IN-LINE ENGINE

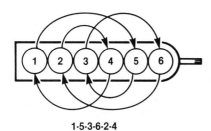

1-5-3-6-2-4

SIX CYLINDER IN-LINE ENGINE

Fig. 3A-9 Firing order of four- and six-cylinder in-line engines

Six-cylinder in-line engines use a 120° crankshaft that puts cylinders one and six at the top; then 60° of crankshaft rotation later, pistons two and five come to the top, and in another 60° of rotation, three and four are on top. These engines fire 1-5-3-6-2-4.

2. V-engines

V-engines use the same type of crankshaft as in-line engines, except that one cylinder in each bank is connected to each throw. These engines may fire either 1L-2R-5L-4R-3L-1R-6L-5R-2L-3R-4L-6R or 1L-6R-5L-2R-3L-4R-6L-1R-2L-5R-4L-3R. If you will trace these firing impulses, you will find that the left bank fires 1-5-3-6-2-4 while the right bank is firing 6-2-4-1-5-3, which is the same sequence.

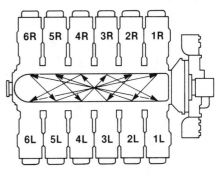

1L-6R-5L-2R-3L-4R-6L-1R-2L-5R-4L-3R

ALLISON V-12

Fig. 3A-10 Firing order for a V-12 aircraft engine

3. Radial engines

These engines fire every other cylinder in the direction of rotation. For example, a three-cylinder radial engine fires 1-3-2, and a five-cylinder model fires 1-3-5-2-4. Seven and nine-cylinder engines fire 1-3-5-7-2-4-6, and 1-3-5-7-9-2-4-6-8.

Two-row radial engines are actually two single-row engines connected together on a crankshaft that has its throws 180° apart. The rear row of cylinders are all odd numbered, and those in the front row are staggered between the ones in the back row and they all have even numbers. Cylinder number one, the top cylinder in the rear row, fires first, and then the cylinder across the engine and over one in the front row is the next to fire. In a 14-cylinder engine this is cylinder number 10, and in an 18-cylinder engine this is cylinder number 12.

The next cylinder in the rear row to fire is number five. This skips one cylinder in the same way it does in a single-row engine. The firing order for a 14-cylinder twin-row radial engine is

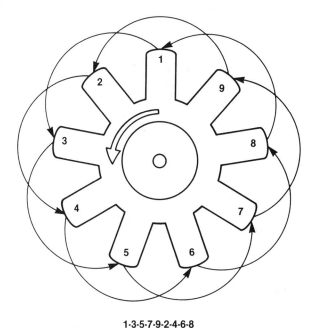

1-3-5-7-9-2-4-6-8

Fig. 3A-11 Firing order for a nine-cylinder single-row radial engine

1-10-5-14-9-4-13-8-3-12-7-2-11-6. For an 18-cylinder engine it is 1-12-5-16-9-2-13-6-17-10-3-14-7-18-11-4-15-8.

A handy way to remember the firing order of an 18-cylinder radial engine is to start with

1-12-5-16-9-2-13-6-17-10-3-14-7-18-11-4-15-8

Fig. 3A-12 Firing order for an 18-cylinder two-row radial engine

number one and add 11 when you can, and when you cannot add 11, subtract seven. Just keep all of the numbers between one and 18. You can do the same thing for a 14-cylinder engine by adding nine and subtracting five.

4. Horizontally opposed engines

The fact that Lycoming and Continental engines number their cylinders differently gives us two sets of firing orders, but if you will trace out the firing impulses, you will see that the two engines are the same.

Lycoming four-cylinder horizontally opposed engines have a firing order of 1-3-2-4, and the firing order of a four-cylinder Continental is 1-4-2-3. Both of these engines use a 180° crankshaft. In a Lycoming engine, pistons one and two are at the top of their stroke at the same time pistons three and four are at the bottom of theirs. Cylinder one fires and, as its piston is being forced inward, it rotates the crankshaft and the piston in cylinder three moves outward and compresses the fuel-air mixture. The piston in cylinder two is moving inward on the intake stroke, and the piston in four is moving outward on the exhaust stroke. If you will look at Fig. 3A-13 you can visualize that if you flip the diagram over, the firing impulses will be exactly the same as they are for the Continental engine.

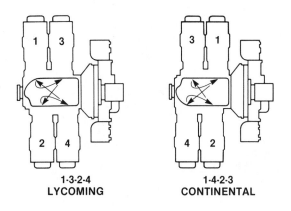

| 1-3-2-4 | 1-4-2-3 |
| LYCOMING | CONTINENTAL |

Fig. 3A-13 Firing order of four-cylinder horizontally opposed engines

The six-cylinder Continental and Lycoming engines both use sixty-degree crankshafts. This means that with each sixty degrees of crankshaft rotation, one piston is reaching the top center of its stroke. In Fig. 3A-14 we have the crankshaft and cylinder arrangement for a Lycoming engine. When the piston in cylinder one passes top center and begins its power stroke, it rotates the crankshaft and brings the piston in cylinder four up at the end of its compression stroke. Then as it moves down on its power stroke, the piston in cylinder five moves up and passes over top center to start its power stroke. This is followed by the pistons in two, three, and six. You will find the same power impulses occur in the Continental engine.

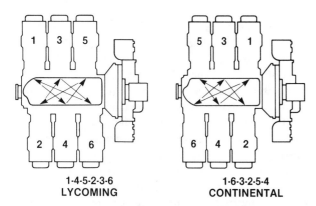

1-4-5-2-3-6
LYCOMING

1-6-3-2-5-4
CONTINENTAL

Fig. 3A-14 Firing order of six-cylinder horizontally opposed engines

The eight-cylinder Lycoming O-720 engine has a firing order of 1-5-8-3-2-6-7-4.

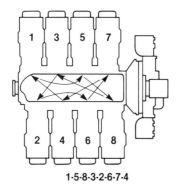

1-5-8-3-2-6-7-4

Fig. 3A-15 Firing order for an eight-cylinder horizontally opposed engine

D.   Engine Identification

Military engines used descriptive identification numbers far earlier than civilian engines, but today most manufacturers use pretty much the same system for their engine identification numbers.

The first letters designate the cylinder arrangement and the basic configuration of the engine.

O—Horizontally opposed engine

R—Radial engine

I—In-line engine

V—V-engine

T—Turbocharged

I—Fuel injected

S—Supercharged

G—Geared nose section

L—Left-hand rotation for multi-engine installation

H—Horizontal mounting for helicopters

V—Vertical mounting for helicopters

A—Modified for acrobatics

The numbers indicate the piston displacement of the engine in cubic inches.

The last letters indicate the model differences of the basic engine. Continental uses one or two letters to indicate the model, and Lycoming sometimes uses a group of letters and numbers to indicate the crankcase and crankshaft differences that make up the model variations.

An example of a couple of engine designations:

LIO-360-C

L—Left-hand rotation

I—Fuel injected

O—Horizontally opposed

360—360-cubic-inch piston displacement

C—Model C

GTSIO-520-F

G—Geared nose section

T—Turbocharged

S—Supercharged

I—Fuel injected

O—Horizontally opposed

520—520-cubic-inch piston displacement

F—Model F

IV. HORIZONTALLY OPPOSED ENGINE CONSTRUCTION

As we have already noted, there are a number of configurations of engines that have had their place in aircraft development, and each of these has had its own unique construction features. For example, the radial engine uses a crankshaft and connecting rod arrangement that is unlike that used in any other engine. Some liquid-cooled V-engines have their cylinders arranged in blocks rather than using individual cylinders.

Because the vast majority of engines used in the modern general aviation fleet are of the horizontally opposed type, we will discuss their construction first and then we will cover the differences that are found in some of the V-engines and radial engines.

A. Cylinders

Power in an engine is developed in the cylinder, in the combustion chamber where the burning and expansion of the gases takes place. The cylinder houses the piston and contains the valves through which the combustible mixture enters the cylinder and through which the exhaust gases leave.

The cylinder must be strong enough to withstand all of the internal pressures developed during engine operation, and it must be able to do this while operating at elevated temperatures. It must be lightweight so it will not create a weight penalty on the engine and it must be designed

and built so it will conduct the maximum amount of heat away from the engine. It must also be relatively simple to build, inspect, and maintain.

Some of the earliest two- and four-cylinder horizontally opposed engines had the cylinder barrels cast as part of the crankcase halves and these cylinders had removable cylinder heads, but almost all of the modern engines use individual cylinders with permanently attached heads.

Fig. 4A-1 Typical air-cooled engine cylinder

1. Cylinder barrels

A high-strength chrome molybdenum steel barrel is machined with a skirt which projects into the crankcase and a mounting flange to attach the cylinder to the crankcase. Thin cooling fins are machined onto the cylinder wall and threads are machined at the top of the barrel so it can be screwed into the head. The bore of the cylinder is normally ground with a slight choke, which means that the diameter at the top portion of the barrel is slightly smaller than the diameter of the main part of the barrel. The reason for this choke is that the upper end screws solidly into the cast aluminum cylinder head and the heat that concentrates in the head will cause the upper end to expand more than the rest of the barrel; as it expands, it straightens the bore so the diameter is uniform throughout the part of the cylinder in which the piston moves.

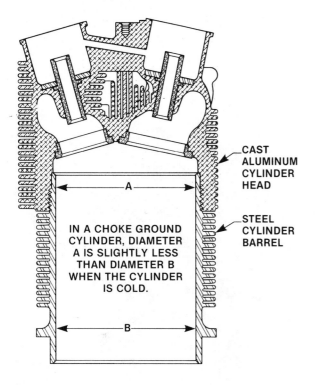

Fig. 4A-2 Construction details of an air-cooled aircraft engine cylinder

If the cylinder were ground straight rather than choked, the top would expand more than the main part of the barrel and there would be the possibility of combustion gases leaking past the rings and damaging the piston.

The piston rings continually rub up and down inside the cylinder, and to increase the life of the cylinder barrel, some cylinder walls are hardened. There are two commonly used methods of providing a hard wearing surface: chrome plating and nitriding.

a. Chrome-plated barrels

Chromium is a hard, natural element, which has a high melting point, high heat conductivity,

and a very low coefficient of friction, only about one-half that of steel. The cylinder barrel is prepared for chrome plating by grinding it to the required size and submerging it in a plating solution where a coating of chromium is electrolytically deposited on the inside of the barrel.

Chromium has a natural tendency to form surface cracks, and after the plating process is completed, current is sent through the plating solution in the reverse direction for a controlled period of time, and these tiny surface cracks open up and form a network of interconnecting cracks that hold lubricating oil on the cylinder wall.

Chromed cylinders have many advantages over both plain steel and nitrided cylinders. They are less susceptible to rusting or corrosion, both because of the natural corrosion resistance of chromium and because oil adheres to the chromed cylinder walls better than it does to plain steel. Chromium is much harder than steel, and so it wears less.

Cylinders that have been worn outside their usable limits may often be ground until the bore is round and straight and then chrome-plated back to their original size. Cylinders that have been chrome-plated are identified by a band of orange paint around their base or by a stripe of orange paint on certain of their fins.

Fig. 4A-3 A reproduction of a photomicrograph of the tiny cracks in the chrome-plated wall of an aircraft engine cylinder

b. Nitrided barrels

Case hardening is a process in which the surface of steel is changed by the infusion of some hardening agent. It differs from plating in that no material is deposited on the surface, but there is actually a change in the surface of the material itself. Nitriding does not require quenching, and so it does not warp the cylinders as other forms of case hardening might do. After the cylinder barrel has been ground to the required size and smoothness, it is placed in a furnace or retort, in an atmosphere of ammonia gas. The length of time the barrel is kept in the retort and the temperature are both carefully controlled and, in the nitriding process, the ammonia gas (NH_3) breaks down, or disassociates, into nitrogen and hydrogen.

The steel in the cylinder barrel has a small percentage of aluminum as an alloying agent, and the nitrogen combines with the aluminum to form aluminum nitrides, the hard, wear-resistant surface we want. Since nitriding is not a plating or a coating, it causes a dimensional growth of only about two to four ten thousandths of an inch, but the hardened layer varies in depth to about 0.002 inch, and gradually decreases in hardness from the surface inward, until it corresponds to that of the metal itself. After the nitriding process is completed, the cylinder walls are honed to a microsmooth finish.

One of the problems with a nitrided surface is its susceptibility to corrosion and rust. Nitrided cylinder walls must be kept covered with oil, and if an engine is left out of service for any period of time, the cylinder walls should be coated with a sticky preservative oil.

Nitrided cylinders are identified by a band of blue paint around their base or by certain fins being painted blue.

2. Cylinder heads

Most air-cooled aircraft cylinder heads are sand cast of an aluminum alloy that contains copper, nickel, and magnesium. This alloy has the desirable characteristics of relatively high strength and the ability to maintain this strength to temperatures of up to 600° F.

Fins are cast into the surface of the head to increase its cooling area, and because of the dif-

ference in temperature at various parts of the head it is necessary to provide more and deeper cooling fins on some sections of the head than on others. The exhaust valve region is the hottest and there are more fins there than anywhere else on the head.

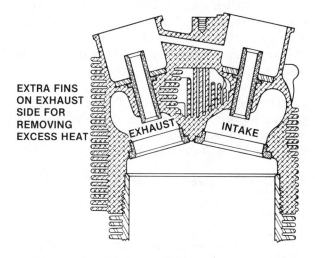

Fig. 4A-4   Cast aluminum alloy cylinder head

The holes for the spark plugs are bushed with either bronze or steel bushings that are screwed, shrunk, and pinned in place or by stainless steel Helicoil inserts. Bronze, cast iron or steel valve guides are shrunk in place, and hardened, ring-type valve seats are shrunk in the head to provide

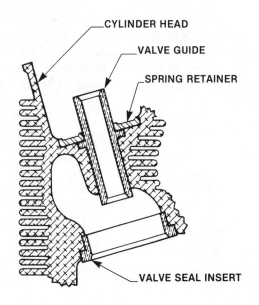

Fig. 4A-5   Valve guide and valve seat inserts in a cast aluminum alloy cylinder head

a wearing and seating surface for the intake and exhaust valves. These seats also protect the aluminum alloy head from erosion from the hot exhaust gases.

Most cylinder heads are heated to expand them, and then they are screwed onto the steel cylinder barrels which have been chilled to contract them. When the head and the barrel reach the same temperature, they fit so tightly that there is no leakage of the hot gases.

The Teledyne-Continental Tiara engine, rather than screwing the heads onto the barrels, use long through bolts to hold the heads in place as well as hold the barrels to the crankcase.

3. Valve mechanism

a. Valves

The valves in the cylinders of an aircraft engine are subject to high temperatures, corrosion, and operating stresses, and the metal used in their manufacture must be able to resist all of these attritional factors.

Because intake valves operate at lower temperatures than exhaust valves, they may be made of chrome, nickel, or tungsten steel, but the exhaust valves are usually made of some of the more exotic metals such as inconel, silicon-chromium or cobalt-chromium alloys.

The face of the valve is ground to form a seal against the valve seat in the cylinder head when the valve is closed. The face is usually ground to an angle of 30, 45 or 60 degrees, with the choice made that will give the best airflow efficiency and sealing.

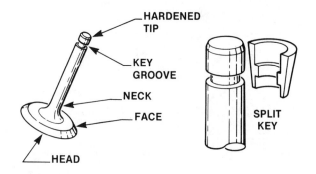

Fig. 4A-6 *Typical poppet-type valve for an aircraft engine*

Valve faces are often made more durable by the application of a material called Stellite, an alloy of cobalt and chromium. About 1/16-inch of this alloy is welded to the valve face and is then ground to the correct angle. Stellite is resistant to high temperatures and corrosion, and it also withstands the shock and wear associated with valve operation.

The valve stem is surface hardened so it will act as a pilot for the valve head as it rides up and down in the valve guide that is installed in the cylinder head for that purpose. The tip of the stem is hardened to withstand the hammering of the valve rocker arm as it opens the valve, and a groove is machined around the valve stem near the tip for the split keys which hold the valve spring retaining washer in place.

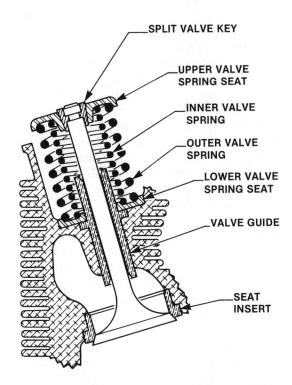

Fig. 4A-7 *Method of securing a poppet valve in an aircraft engine cylinder*

Some exhaust valve stems are hollow and are partially filled with metallic sodium. The sodium melts at approximately 208° F, and the up and down motion of the valve circulates the liquid sodium so it can carry heat from the valve head into the stem where it can be dissipated through the valve guide into the cylinder head and then out to the air through the cooling fins. In this way, the operating temperature of the valve can be reduced by as much as 300 to 400 degrees Fahrenheit.

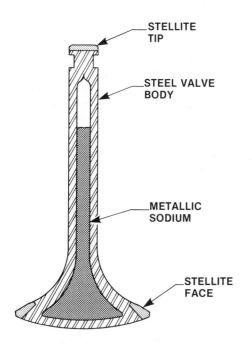

Fig. 4A-8 *Hollow sodium-filled exhaust valve*

Most intake valves have either flat or tulip-shaped, solid heads, and in some engines, the intake and exhaust valves are similar in appearance, but they are definitely not interchangeable because they are constructed of different materials.

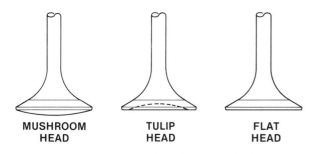

Fig. 4A-9 *Typical head shapes for poppet valves used in aircraft engines*

b. *Valve seats*

The valve operates in the hot environment of the inside of the cylinder head and is continually exposed to a pounding action. Because of this, an extremely durable seat must be installed in the soft aluminum cylinder head. Rings of aluminum, bronze or steel are machined with an outside diameter about 0.010 to 0.015 inch larger than the hole into which they are to fit, and the head is heated in an oven to around 575 to 600 degrees Fahrenheit, and the seat is put into place. This interference fit will then hold it in place when the head and seat are at the same temperature.

c. *Valve guides*

The hardened valve stem rides in the cylinder head in a bronze or cast iron valve guide that has been chilled and pressed into the heated cylinder head. The stem receives the heat from the valve head and transfers it into the guide where it is carried into the cylinder head and then dissipated into the air. Because of this heat transfer, it is important that the tolerance allowed by the manufacturer for the stem-to-guide clearance not be exceeded. This fit is achieved by reaming the guide after it has been pressed into place.

After the valve seat and guide have been properly installed in the cylinder head, a pilot is slipped through the valve guide and the seat is ground so that it is concentric with the hole in the valve guide.

d. *Valve springs and retainers*

The valves are opened by the rocker arm pushing on the valve stem tip, but they are closed by two or sometimes three coil springs. Multiple springs are used, not only because of the additional force they provide, but because of the resonance of the spring. Valve opening is done with a series of impulses, and at a certain engine RPM, these impulses occur at the resonant frequency of the spring. When this happens, the spring loses its effectiveness and allows the valve to float.

To prevent this floating, two or more springs are used. These springs are wound with different pitch, diameter, and wire size. Because of their different configuration, the springs have different resonant frequencies so the engine can operate through its full range of RPM without the valves floating.

A valve spring seat is installed around the valve stem and the two springs are slipped in place. An upper valve spring seat is put over the springs and they are compressed, and two split valve keys are installed in the grove in the valve stem. When the springs are released, the upper spring seat forces and holds the valve keys in the stem groove and the springs keep the valve tightly closed.

B. Pistons

The piston in a reciprocating engine is a cylindrical device that moves back and forth within the cylinder and acts as a movable wall for the combustion chamber. As the piston moves inward in the cylinder, it creates a low pressure and draws in the fuel and air mixture. Then as it moves outward, it compresses this charge, and when the ignition occurs, the expanding gases force the piston back inward. This force is transmitted to the crankshaft through the wrist pin and the connecting rod, and on the next outward stroke, the piston pushes the burned gases from the cylinder.

The majority of aircraft pistons are machined from aluminum alloy forgings and have grooves cut into their outer surface to receive the piston rings. The underside of the head is usually finned to aid in cooling and to provide extra strength.

Almost all aircraft pistons are of the trunk type, such as those we see in Fig. 4A-10, with the top of the head either slightly domed or flat. Many pistons have recesses machined into their heads to provide more clearance for the valves.

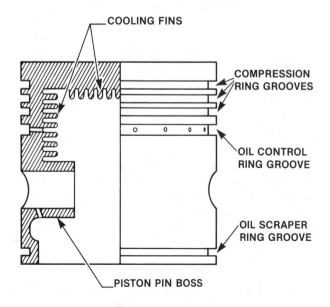

Fig. 4A-10 Typical piston for an aircraft engine

The piston pin boss is an enlarged portion of the skirt which is machined to fit the piston pin, with the additional material providing the strength needed to transmit the force of the expanding gases to the connecting rod.

The additional mass of material in the piston pin boss causes the piston to expand more in the direction of the wrist pin than it does perpendicular to the pin. Because of this non-uniform expansion, many pistons are cam-ground; that is, they are ground with a diameter several thousandths of an inch less parallel to the wrist pin than their diameter perpendicular to the pin. When the piston heats up to its operating temperature, the piston will have expanded to its required round shape.

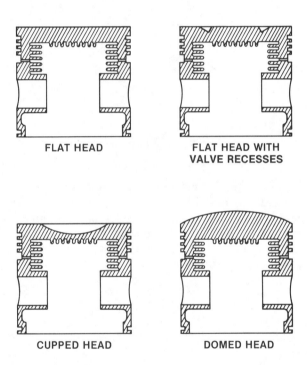

Fig. 4A-11 Head shapes for aircraft engine pistons

The compression ratio of the engine may be varied by changing pistons and this makes it vital that the proper part number be used any time a piston is replaced in an engine.

Since the piston is a reciprocating part of the engine, the inertia involved in its continual starting and stopping will produce vibration if there is any appreciable difference in the weight of the pistons. Because of this, the manufacturing tolerance usually requires that the weight of the pistons be held to within about a quarter of an ounce (seven grams) of each other.

1. Piston rings

The piston serves as a plunger to compress the fuel-air charge and transmit the heat from the burning gases into the cylinder wall, but the wide variations of piston temperature cause so much dimensional change that it is impossible to use a solid plug fitting into the cylinder. Instead, grooves are machined around the piston and it is fitted with cast iron rings expanded so they will fit against the bore of the cylinder with considerable pressure. The top two or three grooves are fitted with solid rings called compression rings, whose primary purpose is to provide a seal and prevent gases escaping around them, and to transfer heat into the cylinder wall.

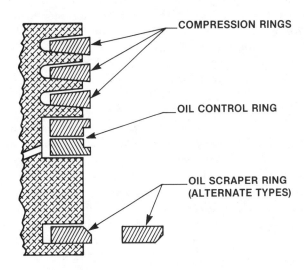

Fig. 4A-12 Typical piston rings fitted to an aircraft engine piston

Oil control rings are installed in the groove immediately below the compression rings. These rings are often constructed of two or more pieces and may have corrugated expanders in the ring groove behind them. The purpose of the oil control ring is to maintain the proper quantity of oil between the piston and the cylinder wall, and the grooves for the oil control ring often have holes drilled in them to drain some of the oil back into the crankcase.

An oil scraper, or wiper ring, may be installed at the very bottom of the piston skirt. The face of these rings is usually tapered or beveled, and they may be installed in such a way that they wipe the oil either toward or away from the piston head, depending upon the engine. The installation of these rings is critical and when installing them, you must follow the manufacturer's instructions in detail.

Piston rings are usually made of gray cast iron and may have one of several cross-sectional shapes such as those seen in Fig. 4A-12. Wedge-shaped piston rings are more or less self-cleaning, and they reduce the amount of carbon accumulation in the ring grooves.

Newly installed piston rings do not form a good seal with the cylinder wall, and they must be seated or worn in before they will stop high oil consumption. The high pressure and temperature encountered in the combustion chamber of an aircraft engine make it extremely important that only the piston ring approved by the manufacturer be used. The ring tension, the end gap clearance, and the clearance between the ring and the side of the groove must be within the tolerance specified by the engine manufacturer if you are to expect the rings to seat and seal properly.

Piston rings used in cylinders with plain steel walls may be chrome-plated for better wear characteristics, but rings used on chrome-plated or nitrided cylinder walls must be made of plain cast iron, since hardened piston rings will not seat in a hardened cylinder barrel, and if the rings do not seat, they will not stop pumping oil.

2. Piston pins

Hollow, hardened alloy steel pins fit into the hole in the piston pin boss and attach the piston to the connecting rod. Most of the piston pins, or

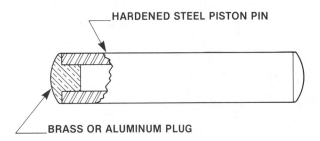

Fig. 4A-13 Hardened steel piston pin with brass plugs in its ends to prevent the pin scoring the cylinder walls

wrist pins, as they are called in modern aircraft engines, are of the full-floating type. They are free to rotate in both the piston and the connecting rod, and soft aluminum or brass plugs fit into the ends of the pin to prevent the pin scratching the cylinder wall as the piston moves in and out.

Piston pins are usually a push fit into the piston and are lubricated by oil flowing through holes drilled in the bosses of the piston.

C. Connecting Rods

The connecting rod is the link which transmits the forces from the piston to the crankshaft. It must be strong enough to remain rigid under load and yet light enough to reduce the inertia forces which are produced when the rod and piston stop, change direction, and start again at the end of each stroke.

The cylinders of a horizontally opposed engine are staggered on the crankcase so each piston can be attached to its own crankshaft throw. The end of the connecting rod that attaches to the piston is fitted with a pressed-in bushing in which the wrist pin floats, and in the big end, or the end that attaches to the crankshaft, there is a removable cap and a two-piece bearing insert. The cap is held onto the rod with bolts and locknuts, and the rod and cap assembly are paired and must always be installed as a matched set.

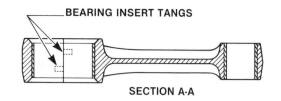

SECTION A-A

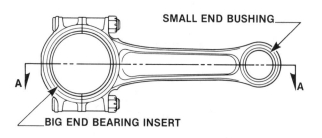

Fig. 4A-14 Connecting rod for a horizontally opposed aircraft engine

D. Crankshaft

The crankshaft is the backbone of a reciprocating engine and is subjected to most of the forces developed by the engine. Its main purpose is to transform the in-and-out motion of the piston into rotary motion to turn the propeller. The crankshaft, as the name implies, is a shaft composed of one or more cranks or throws, located at specific points along its length. These throws are formed by forging offsets into the shaft before it is machined. Since crankshafts must withstand such high stresses, they are generally forged from a very strong alloy steel such as chrome nickel molybdenum.

1. Design

Since the crankshaft is the heaviest single part of an aircraft engine, the engine's evolution has been largely dictated by the crankshaft. Radial engines use a crankshaft with one throw for each row of cylinders. V-engines have one crankshaft throw for each two cylinders, but in-line and horizontally opposed engines have one crankshaft throw for each cylinder.

In Fig. 4A-15 we have a drawing of the crankshaft of a four-cylinder horizontally opposed engine. The throws are paired, and the two end throws are 180° from the two in the middle of the shaft. This crankshaft has three main bearings, one at either end and one between the two center throws. The crankshaft is hollow and has drilled passages through the crank cheeks through which lubricating oil flows. The throws are all fitted with sludge plugs, and the oil to lubricate the connecting rod bearings flows into the space between the outside of the sludge plug and the inside of the hole in the crank throws.

Any sludge or contaminants in the oil is slung by centrifugal force to the outside of the throw, and only clean oil is allowed to flow to the bearing. When the engine is overhauled, these sludge plugs are knocked out and the sludge removed.

The hollow shaft allows oil for controlling a hydraulically operated propeller to flow into the propeller and then to drain back out as the propeller or control valve directs.

Six-cylinder horizontally opposed engines use a 60° crankshaft. This means that the throws are

60° apart. The crankshaft we have in Fig. 4A-16 is typical for six-cylinder engines. It has three pairs of throws and four main bearings. The throws for cylinders one and two are opposite each other, and those for three and four are opposite, as are those for cylinders five and six.

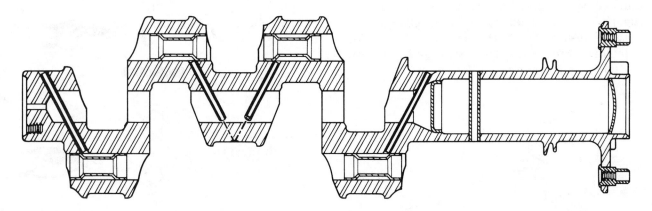

Fig. 4A-15   Crankshaft for a four-cylinder horizontally opposed aircraft engine

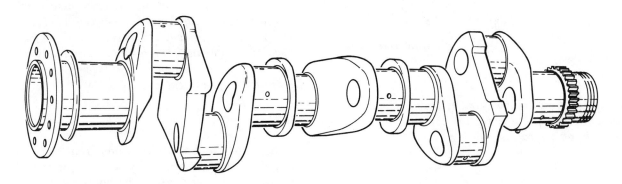

Fig. 4A-16   Crankshaft for a six-cylinder horizontally opposed aircraft engine

2. Construction

Aircraft crankshafts are forged of high-strength alloy steel. They are drilled for oil passages, and the main bearing journals and crank pins are ground and polished, and then the crankshaft is nitrided. The hollow crank pins have thin steel sludge plugs pressed into them.

3. Dynamic dampers

Crankshafts are made as light as possible in keeping with their strength requirements, and they are balanced both statically and dynamically. The force that causes the crankshaft to rotate is not a smooth push, as we would like, but it is delivered in a series of pulses or intermittent pushes. At certain RPM of the engine, the fre-

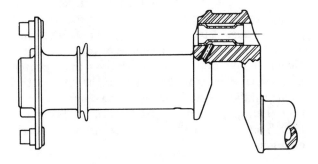

Fig. 4A-17 Sludge plugs are pressed into the hollow connecting rod journals of an aircraft engine crankshaft to collect sludge from the engine oil. This sludge is held in the recesses around the plugs until the engine is disassembled at overhaul.

quency of these pulses will be the same as the resonant frequency of the crankshaft and propeller combination, and the forces generated can be strong enough to damage the engine. This damage can be prevented by installing dynamic dampers on the crankshaft that change the natural resonance of the crankshaft-propeller combination to some value that is not excited by the power impulses.

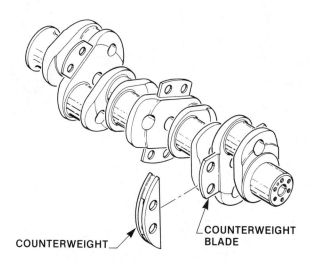

COUNTERWEIGHT

COUNTERWEIGHT BLADE

Fig. 4A-18 Movable counterweights serve as dynamic dampers to reduce the torsional vibrations in an aircraft engine.

In Fig. 4A-19 we see the way dynamic dampers work. A weight suspended from a point and free to swing back and forth is called a pendulum, and it has a natural period; that is, there is one frequency that will require a minimum amount of push to cause it to swing. This period is determined by both the length and mass of the pendulum. Now if the pendulum rod were hinged in its center and another weight installed at the hinge, the energy pulses would not cause the same swings they had previously caused, because both the length and the distribution of the mass have been changed.

Aircraft crankshafts have counterweight blades forged as an integral part of the shaft, and counterweights fit over these blades. The pins that hold the counterweights to the blades are considerably smaller than the holes in the blades, and they form a loose fit, allowing the weight to oscillate, or rock back and forth. When the engine is running, centrifugal force holds the counter-

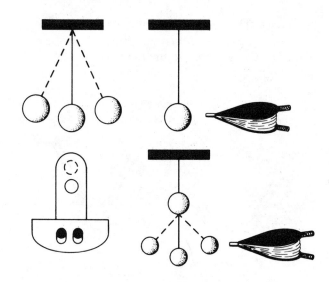

Fig. 4A-19 Operating principle of a dynamic damper. The crankshaft of an aircraft engine behaves as a pendulum and has a resonant frequency. Power impulses supplied to the pendulum at its resonant frequency cause a maximum amplitude of deflection. A weight placed on the pendulum in such a way that it is free to oscillate will change the resonant frequency of the pendulum, and the power impulses will cause less amplitude of vibrations.

weights out and the vibrations from the power impulses are absorbed by the counterweights rocking back and forth in the holes.

4. Propeller attachment

The method of attaching the propeller to the crankshaft of an aircraft engine has undergone evolutionary changes, as has almost everything else in the engine. Most of the early engines had a tapered propeller shaft, and the wooden propellers used on these engines were mounted in a steel hub that fitted over the tapered shaft and were held in place by a propeller retaining nut that was part of the hub. The retaining nut could back out against a snap ring inside the hub to pull the hub loose from the shaft to remove the propeller. A keyway cut in both the tapered shaft and inside the hub held a steel key to prevent the propeller rotating on the shaft.

As the power produced by the engines increased, so did the strength of the propeller attachment. The propeller shaft in many high-powered engines is splined, using standard SAE splines. Most splined shafts have a master spline, one that is as wide as two splines and the slot between them. This wide spline assures that the propeller will be positioned on the engine with the correct relationship between the propeller and the crankshaft counterweights. This assures the smoothest engine operation.

The propeller hub is centered on a splined shaft at the rear with a single-piece, split, bronze rear cone and at the front by a two-piece, chrome-plated steel front cone. The retaining nut fits into a groove in the front cone to allow it to be used as a puller to break the propeller loose from the rear cone when removing it from the engine.

Most modern engines have a flange forged on the front end of the crankshaft to which the propeller is attached with a series of bolts and nuts. In most installations, the nuts are in the form of bushings pressed into the flange.

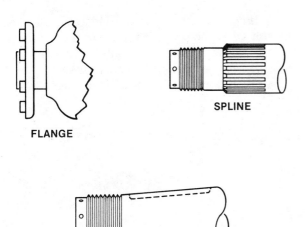

FLANGE

SPLINE

TAPER

Fig. 4A-20 Types of propeller attachment to an aircraft engine crankshaft

E. Propeller Reduction Gearing

The power produced by an aircraft engine is determined by the pressure that acts on the piston during each power stroke and the number of power strokes each minute. The faster the engine turns, the more power it produces. We are limited in an aircraft engine, though, by the tip speed of the propeller. As propeller speed increases, its efficiency drops off.

To get the needed power and at the same time maintain a reasonable propeller tip speed, most engine manufacturers have built propeller reduction gear systems into their engines. This gearing naturally uses some of the power from the engine and it adds weight, but the gain in power more than compensates for the losses and we find geared engines are used when high power output is needed.

The simplest type of reduction gearing uses a spur gear attached to the engine crankshaft, mating with a larger spur gear on the propeller shaft. This type of gearing requires a very strong crankcase to withstand the force put into it by the propeller, as it acts like a gyroscope.

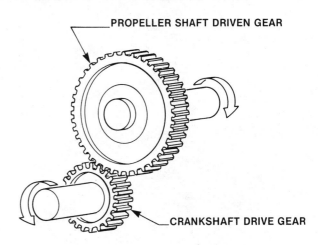

PROPELLER SHAFT DRIVEN GEAR

CRANKSHAFT DRIVE GEAR

Fig. 4A-21 External spur gear-type propeller reduction gearing system

One method of overcoming some of the problems of the simple spur gear arrangement has been the use of an internal tooth-driven gear with

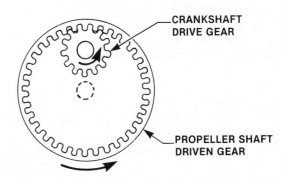

CRANKSHAFT DRIVE GEAR

PROPELLER SHAFT DRIVEN GEAR

Fig. 4A-22 Propeller reduction gearing system using an external spur gear driving an internal gear attached to the propeller shaft

an external tooth drive gear. This arrangement keeps the propeller shaft pretty much in line with the crankshaft, and it allows the propeller to turn in the same direction as the engine.

Torsional vibration problems are compounded when an engine is geared, and in addition to the counterweights on the crankshaft, some of the popular engines use a quill shaft to further reduce these vibrations. One end of the quill shaft, as we see in Fig. 4A-23, is splined into the front end of the crankshaft, and the opposite end is splined into the front end of the propeller shaft. With this arrangement, the propeller drive gear is driven through the quill shaft that flexes torsionally enough to absorb some of the shocks.

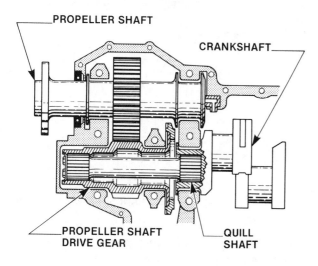

Fig. 4A-23 A quill shaft used to minimize torsional vibration between the propeller shaft and the crankshaft.

Planetary gears are used in some engines to keep the propeller shaft in line with the crankshaft and not reverse the direction of rotation between the crankshaft and the propeller, as the external-tooth spur gear system does.

One of the most efficient types of propeller reduction gear arrangements where large amounts of power must be handled is the planetary gear system. Power is transmitted from the crankshaft to the propeller with a minimum of weight and space and the direction of rotation of the propeller is the same as that of the engine. This type of reduction gear arrangement is used on some horizontally opposed engines, as well as for radial and turboprop engines.

The planetary system uses a sun gear that is rigidly attached to the nose section of the engine and the crankshaft drives a bell, or ring, gear. Turning between the bell gear and the sun gear is a series of small planetary gears that mount on a spider arrangement on which the propeller shaft is mounted. When the crankshaft turns the bell gear, it rotates the planetary gears around the sun gear and the spider drives the propeller at a reduction rate that is found by the formula:

$$\text{Gear ratio} = \frac{\text{Teeth on bell gear} + \text{Teeth on sun gear}}{\text{Teeth on bell gear}}$$

You will notice that neither the number of teeth on the planetary gears nor the number of planetary gears attached to the spider enter into the computation for gear reduction.

If there are 72 teeth on the bell gear and 36 on the sun gear, the propeller turns at a ratio of 1.5 to 1; but this is most generally spoken of as a three-to-two reduction. The crankshaft turns three revolutions for two of the propeller shaft.

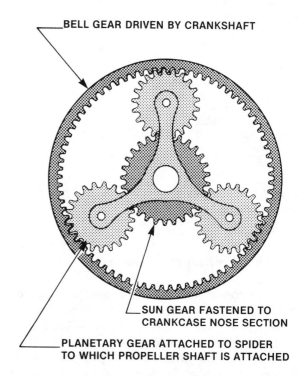

Fig. 4A-24 Planetary gear type propeller reduction gear system

F. Crankcase

The crankcase of an engine is the body that holds the engine together and it is by the crankcase that the engine mounts into the aircraft.

The crankcases of horizontally opposed engines are made of two halves of cast aluminum alloy that have been manufactured either by sand casting or by a method using permanent molds. Crankcases made by the permanent mold process are denser than those made by sand casting, and the metal is somewhat thinner.

Webs to support the bearings for the crankshaft are cast into the case, as well as the bosses for the camshaft bearings.

The crankcase halves split vertically and are held together with studs and with through bolts at the crankshaft bearings, and with smaller bolts and nuts around the outside of the case.

The camshaft rides in holes that have been line-bored through the crankcase webs, and they do not use any type of bearing insert.

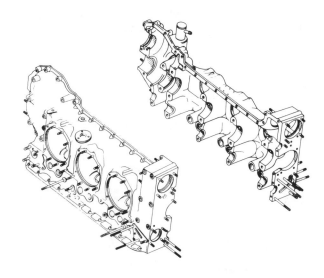

Fig. 4A-25   A typical aircraft crankcase made of cast aluminum alloy

1.   Crankshaft bearings

The crankshaft is supported in the crankcase by steel-backed, lead-alloy bearing inserts held in place by a tang on the bearing which fits into a groove in the bearing cavity; or, in some cases, by a dowel in the bearing cavity which fits through a hole in the bearing insert.

Thrust loads are carried by either a separate thrust washer or by a thrust surface on one of the main crankshaft bearings.

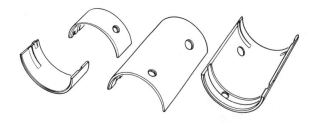

Fig. 4A-26   Typical crankshaft main bearings

2.   Oil sealing provisions

Since the supply of oil in most of the modern horizontally opposed engines is carried inside of the crankcase, provisions are made to seal the crankcase to prevent its leaking.

The two halves of the crankcase must not only be permanently sealed against oil leakage, but the seal must not interfere with the tight fit of the bearings. Most crankcase halves are sealed with a very thin coating of non-hardening gasket compound applied to the surface with a fine silk thread imbedded in the compound extending the full length of the case. When the assembled crankshaft and cam shaft are installed in the crankcase, and the halves are bolted together with the proper torque, the gasket material and the silk thread form an effective oil seal without interfering with the bearing fit.

The oil sump and rear case of the engine are sealed with gaskets made either of cork or of an asbestos impregnated sheet material coated with a light film of non-hardening gasket compound.

Cylinders are sealed to the crankcase with a cylinder base packing ring. This is a form of O-ring that fits between the cylinder base skirt and the chamfered edge of the cylinder mounting pad in the crankcase.

When a rotating shaft comes through the crankcase, an oil seal must be installed that will allow the shaft to rotate, yet not allow any oil to leak out. These seals are usually made of neoprene, leather, or some form of synthetic resin such as Teflon, and the sealing surface is held against the rotating shaft by either a coil spring or a flat leaf spring.

Fig. 4A-27 Propeller shaft oil seal prevents oil leaking from the crankcase around the rotating crankshaft.

G. Valve Operating Mechanism

1. Camshaft

A straight camshaft is installed either above or below the crankshaft and it rides in bearing surfaces that are line-bored in the crankcase. This shaft is geared to the crankshaft and turns at one-half crankshaft speed. Lobes are machined on this shaft that ride against the cam follower, or tappet body, to operate the valves. The camshaft of a six-cylinder engine has nine lobes, and the one used on a four-cylinder engine has six lobes. The reason for the difference in the number of cam lobes and the number of valves is the fact that the cylinders of a horizontally opposed engine are staggered so two or three of the lobes can operate valves in two cylinders.

2. Rocker arms

The valve, as we remember from the discussion of the cylinder, is held closed by two or more coil springs, and it is opened by the action of the rocker arm operated by the pushrod. The valve stem and springs are located in the portion of the cylinder head called the rocker box. A rocker arm, normally made of forged steel, pivots on a rocker shaft and one end presses against the valve stem tip. The other end of the rocker arm has a socket machined in it, into which the pushrod fits.

3. Pushrods

Hollow steel or aluminum alloy tubes with polished ball-type ends pressed in, transmit the movement of the cam follower to the rocker arm. The pushrods are surrounded by a thin sheet metal shroud tube, and oil for lubricating the valve mechanism flows under pressure from the engine oil gallery up through the hollow pushrod. Depending upon the engine, this oil drains out of the rocker boxes back into the crankcase through

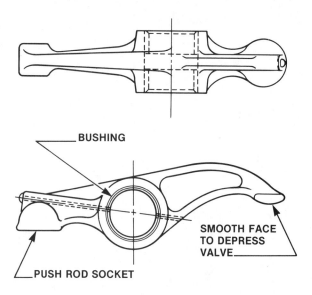

Fig. 4A-29 Rocker arm for a horizontally opposed engine using hydraulic valve lifters

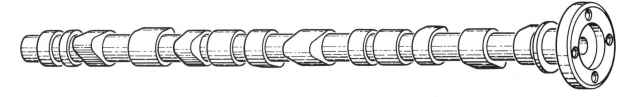

Fig. 4A-28 Camshaft for a typical six-cylinder horizontally opposed engine.

the pushrod shroud tubes or through external oil drain tubes.

4. Cam followers

The cam followers, or valve lifters, ride between the cam and the pushrod. There are two types used in horizontally opposed engines: solid lifters and hydraulically operated lifters.

a. Solid valve lifters

Some of the smaller engines use a solid valve lifter that rides in holes machined in the crankcase, with their large end riding on the lobes of the camshaft. A spherical end of the pushrod rides in a socket machined in the lifter, and holes drilled in the lifter allow oil from the oil pressure gallery of the engine to flow through the lifter and lubricate the valve mechanism in the rocker boxes.

The aluminum alloy cylinder heads expand much more than the pushrods, and as the engine gets hot, the valve mechanism in the cylinder head actually moves away from the crankcase. This causes the clearance between the rocker arm and the valve stem to open up. Engines with solid valve lifters must have some means of adjusting the clearance in the valve train, as the mechanism between the camshaft and the valve is called. To do this, the rocker arm is fitted with an adjustment screw in the end in which the pushrod rides. The engine specifications list the clearance that must be set between the rocker arm and the valve stem, and this clearance is measured with a feeler gage as the adjustment is made with the adjusting screw. After the proper clearance is set, a lock nut is tightened to prevent vibration changing it.

b. Hydraulic valve lifters

Most modern horizontally opposed engines are equipped with hydraulic valve lifters. These units are correctly called zero-lash valve lifters because they maintain zero clearance in the valve train.

The hydraulic valve lifters automatically adjust the valve clearance each time the valve is opened, and in this way, it compensates for any dimensional changes that are caused by the expansion of the cylinder head.

The hydraulic valve lifter consists of a lifter body that moves in and out in holes bored in the crankcase, and its hardened and polished face rides on the cam lobe. The toe of the cam lobes is slightly tapered so it will rotate the lifter body each time it lifts it to prevent a concentration of wear on the lifter body face.

The hydraulic plunger assembly and a pushrod socket fit inside the lifter body and are held in place with a retaining snap ring. The plunger assembly consists of a precision-ground plunger that fits into the plunger body. At the bottom of the plunger body is an oil passage into the oil reservoir in the lifter body and a check valve allows the oil to flow from the lifter body into the plunger body, but no oil can flow in the opposite direction. A coil spring holds the plunger up against the pushrod socket.

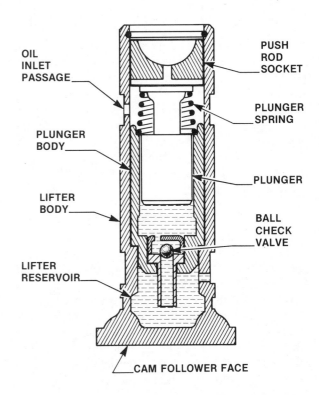

Fig. 4A-30 Hydraulic valve lifter using a ball-type check valve

When the engine is not operating, the lifters are all full of oil and the plunger springs are holding the plungers against the pushrod sockets. The plunger bodies are full of oil that is held trapped by the check valves. The plunger and plunger bodies are machined with a precision fit

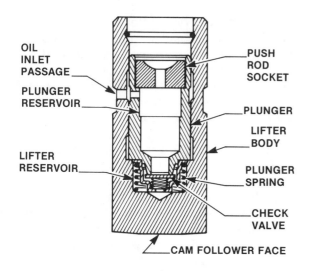

Fig. 4A-31 Hydraulic valve lifter using a disk-type check valve

so there is an accurately controlled amount of leakage that allows the lifter to leak down at a predetermined rate. When the engine is stopped with some of the valves partially open, the force of the valve springs will cause the lifters to leak down. The plunger is still full of oil, though its capacity has been decreased by the plunger moving down. As soon as the engine starts, engine lubricating oil will immediately pump the lifter back to its operating condition.

When the engine is running, oil from the pressure galleries flow into the valve lifter body through holes drilled in the pushrod socket, through the hollow pushrod, and into the rocker box to lubricate the rocker arm, valve stem, and springs. Oil also enters the lifter reservoir, and during the time the cam follower is on the base circle of the lobe, the plunger spring forces the plunger up to keep all of the clearance out of the valve train. When the plunger is pulled up, oil flows through the check valve and fills the plunger reservoir.

Now as the camshaft rotates and begins to lift the cam follower, the oil that is trapped in the plunger reservoir by the check valve lifts the plunger and opens the valve. During the time the valve is open, there is a small controlled amount of leakage between the plunger and the plunger body, and when the valve closes, the plunger spring lifts the plunger and the oil that has leaked out is replaced.

This automatic adjustment that keeps all of the clearance out of the valve train increases the life of the valve train components, increases the volumetric efficiency, and gives us a much quieter running engine.

V. RADIAL ENGINE CONSTRUCTION

From the late 1920's through World War II, the radial engine was the configuration that we could call the "typical" aircraft engine. They were available from three through 28 cylinders, and with horsepower from less than 40 up to more than 3,500. Small personal airplanes used engines with three, five, seven, or nine cylinders arranged in one row around the crankcase. Some military fighters and bombers and the airliners used nine-cylinder single-row engines or 14- or 18-cylinder two-row engines. Some of the largest airplanes used the 28-cylinder, four-row Pratt and Whitney R-4360 engine, which was the largest reciprocating engine to find popular acceptance.

Fig. 5A-1 Radial engines have an excellent weight-to-power ratio and are used where the drag produced by their large frontal area does not create a problem.

Radial engines are easy to maintain and overhaul, and their operating costs are competitively low. The reason for the decline in their popularity for high power applications has been the advent of the turbine engine with its extremely low weight for power ratio.

The large frontal area of the radial engine has caused it to be replaced in the lower power applications by the horizontally opposed engine that can be housed in a much smaller nacelle and streamlined much more easily.

Today there are no new radial engines being manufactured in the United States, but a large number of these engines are still available from World War II surplus stocks and are still used to power many agricultural aircraft and airplanes that were built in the World War II era and are still being flown in special applications where speed is not too essential.

A. Cylinders

The cylinders of a radial engine are all individual units with chrome molybdenum steel barrels onto which is screwed and shrunk a cast aluminum alloy head. In all of the important details, these cylinders are similar to those used on horizontally opposed engines.

B. Pistons

The pistons used in a radial engine are similar in their design and construction to those used in horizontally opposed engines.

C. Crankshaft

Here we find the major difference between a radial engine and any of the other configurations. The big advantage of a radial engine is its low weight for its horsepower, and this is obtained primarily by the use of a short crankshaft. Single-row radial engine crankshafts have only one throw, and those used for two-row engines have two throws.

In Fig. 5A-2 we have a typical two-piece crankshaft used in a seven- or nine-cylinder radial engine. These shafts are usually supported in ball or roller bearings at their front and rear journals and normally use ball bearings for carrying the thrust loads.

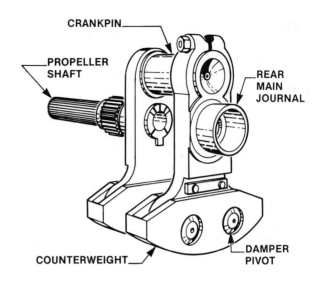

Fig. 5A-2 *Crankshaft for a single-row radial engine*

There are two types of crankshafts used in radial engines. A single-piece crankshaft requires a split master rod, and if the engine uses a master rod whose big end is in one piece, the crankshaft must be made in two pieces so it can be taken apart, the master rod slipped over the crank pin, and then the master rod reassembled.

The crankshaft used in a single-row engine has counterweights on the opposite side of the shaft from the crank pin, and usually one of these

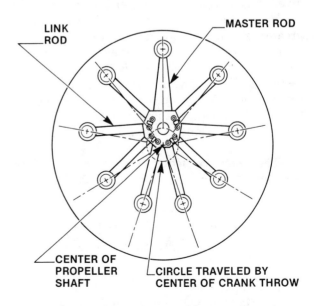

Fig. 5A-3 *Master rod and link rod arrangement for a nine-cylinder single-row radial engine*

43

counterweights is fitted with a dynamic damper similar to that described for the horizontally opposed engine. This movable weight is free to oscillate back and forth and absorb torsional vibrations in the engine.

Two-row radial engines are essentially the same as two single-row engines mounted on a single crankshaft. The crankshaft has two throws, 180 degrees apart, with one main bearing between the throws.

D. Crankcase

Radial engines use a small crankcase made up of a forged power section to which the cylinders bolt, a cast nose section which houses the propeller reduction gears, and a cast blower, or diffuser section where the fuel-air mixture from the carburetor is distributed through thin sheet metal intake pipes to all of the cylinders. The engine mounting pads are usually cast into the diffuser section. At the rear of the engine a cast aluminum accessory section houses all of the accessory drive gears, and the accessories mount on this housing.

E. Propeller Reduction Gearing

Radial engines whose propellers are geared to reduce the speed, use planetary gear systems. The type of planetary gear arrangement described for the horizontally opposed engine is used in some of the radial engines. This arrangement gives a gear reduction typical for this type of engine of 16:9 or 3:2, but some engines require a greater amount of speed reduction and these engines use a reverse planetary gear arrangement. The ring, or bell, gear is fixed in the nose section of the engine and the crankshaft drives the sun gear. The propeller shaft is fastened to the cage that holds the pinion or planetary gears that rotate between the sun and the ring gears.

The gear reduction ratio for a reverse planetary system may be found by the formula:

$$\text{Gear ratio} = \frac{\dfrac{\text{Teeth on}}{\text{sun gear}} + \dfrac{\text{Teeth on}}{\text{ring gear}}}{\text{Teeth on sun gear}}$$

If we assume a system having 72 teeth on the ring gear and 36 teeth on the sun gear, a reverse planetary system will give us a reduction ratio of 3:1.

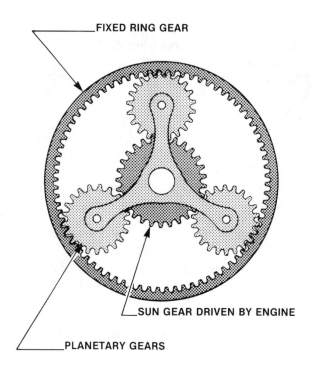

Fig. 5A-4 *Reverse planetary gear system used to provide a greater gear reduction ratio than is provided by the normal planetary gear system.*

F. Valve Operating Mechanism

It takes little imagination to realize that the valve operating mechanism for a radial engine must be totally different from that used on a horizontally opposed engine.

Rather than a camshaft, radial engines use a cam ring to operate the valve train. In Fig. 5A-5 we see a typical valve train for a single-row radial engine. The cam ring is geared to the crankshaft and the relationship between the number of lobes on the ring and the speed and direction the ring turns relative to the crankshaft is such that all of the valves open in their proper sequence in two revolutions, or 720 degrees of crankshaft rotation.

The cam followers, or tappets, in a radial engine are rollers that ride on the cam ring and as the cam ring turns, the tappet rollers ride up onto the lobes and force the valves open by the pushrods and the rocker arm. The lobes are designed with a ramp to prevent a shock to the valve mechanism as the rollers ride up the lobe.

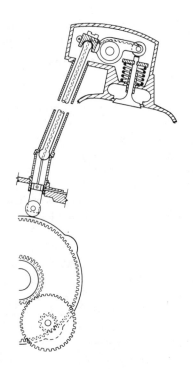

Fig. 5A-5 *Typical valve train for a radial engine*

RADIAL ENGINE CAM SPEED AND DIRECTION					
SEVEN CYLINDER			NINE CYLINDER		
NO. LOBES	SPEED	DIRECTION	NO. LOBES	SPEED	DIRECTION
4	1/8	SAME	5	1/10	SAME
3	1/6	OPPOSITE	4	1/8	OPPOSITE

Fig. 5A-6 *Relationship between the number of cam lobes, the cam speed, and the direction of cam ring travel for radial engines*

A specific clearance must be maintained between the rocker arm and the valve stem tip, a clearance large enough that the valve will never be held off of its seat, and yet small enough to minimize pounding which causes excess wear. There is so much more mass in the cylinder head than there is in the pushrod that, as the cylinder heats up, it expands and the valve clearance increases. When the cam plate is timed to the engine, the valves of one cylinder, usually number one, are adjusted to the hot, or running, clearance and the cam is timed to the crankshaft. Then all of the valve clearances are adjusted to the proper cold clearance, which is much smaller.

Many of the larger radial engines are designed with so much clearance between the cam

plate and its bearing that the cams are called "floating cams." When the valve clearance is adjusted, the spring tension on two valves across the engine must be relieved so the cam plate can move down against the cam bearing and the clearance adjustment can be kept consistent all around the engine.

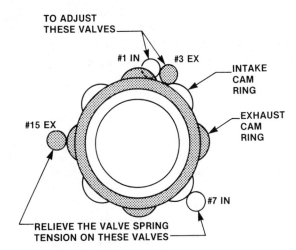

FLOATING CAM RING USED ON SOME LARGE RADIAL ENGINES

(A)

SET PISTON ON T D C	UNLOAD VALVES ON CYLINDERS		ADJUST VALVES ON CYLINDERS	
	IN	EX	IN	EX
11	7	15	1	3
4	18	8	12	14
15	11	1	5	7
8	4	12	16	18
1	15	5	9	11
12	8	16	2	4
5	1	9	13	15
16	12	2	6	8
9	5	13	17	1
2	16	6	10	12
13	9	17	3	5
6	2	10	14	16
17	13	3	7	9
10	6	14	18	2
3	17	7	11	13
14	10	18	4	6
7	3	11	15	17
18	14	4	8	10

CHART SHOWING THE VALVES THAT MUST BE UNLOADED TO ADJUST THE VALVES ON AN 18-CYLINDER RADIAL ENGINE HAVING A FLOATING CAM

(B)

Fig. 5A-7 *Valve adjustment for a radial engine with a floating cam*

VI. V-ENGINE CONSTRUCTION

V-engines differ from both radial and horizontally opposed engines in the construction of the cylinders and the connecting rods and in the method of valve actuation. Most V-engines are liquid cooled because of the difficulty in getting uniform airflow through all of the cylinders. Liquid cooling lends itself to banks of cylinders. These banks usually have steel cylinder barrels fitted into a cast aluminum coolant jacket. The cylinder heads are held onto the barrels with long through bolts from the crankcase up through the heads.

The connecting rods may be of either the master and link rod configuration similar to that used with a radial engine or in more recent engines with the fork and blade configuration. A two-piece bearing insert similar to that used in a horizontally opposed engine is fitted into the big

end of the forked rod, and the outside portion of the bearing insert shell in the opening of the fork is treated so it can act as the journal for the bearing insert in the blade rod.

The latest American V-engine to achieve poularity was the Allison V-1710 that used four valves in each cylinder. The volume of gases that must be moved in this engine would require a single intake and exhaust valve to be so large that it would likely overheat, and so rather than using one large valve of each type, these engines use two intake and two exhaust valves with much smaller heads.

These valves are actuated by an overhead camshaft mounted in the cylinder head rather than with pushrods from a camshaft in the crankcase. There is a camshaft in the heads of each bank of cylinders and the rocker arms ride directly on the lobes of the camshaft. This arrangement opens the valves with a minimum of mechanism. The rocker arms are in the shape of a Y and adjustable tappets in each arm of the Y ride on the tips of the valve stems.

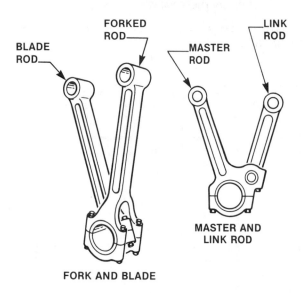

Fig. 6A-1 *Two connecting rod arrangements used on V-type aircraft engines*

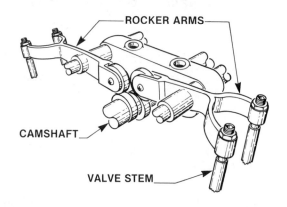

Fig. 6A-2 *Valve adjusting mechanism used for V-engines having an overhead cam*

Section 1-II

Reciprocating Engine Maintenance and Operation

I. ENGINE SERVICE LIFE

All we need to do is look back into some of the older service manuals of aircraft engines to see the tremendous strides we have made in increasing engine service life. The Curtiss OX-5 engine that introduced much of post World War I America to aviation had a service life of 50 hours, and engines built in the late 1930's and 1940's could reasonably expect a service life of between only 300 and 600 hours before major overhaul was needed. Today, with our improved materials and technology, we can expect almost any engine to run 1,000 hours before it needs a major overhaul, and most modern engines have a manufacturer's recommended time between overhaul (TBO) of between 1,400 and 1,600 hours, with many engines running up to 2,000 hours. When the decision is made on which aircraft to buy, the engine manufacturer's recommended TBO is one factor to be considered.

A. Manufacturer's Recommended Time Between Overhauls

With the costs of overhauls continually rising, economical operation of an aircraft requires that the engine must run a maximum number of hours before it requires a major overhaul. The hours specified in the manufacturer's TBO are only *recommended* hours, and there is no guarantee that the engine will actually run that long, and they most certainly do not imply that the engine will operate that long without maintenance. On the other hand, there is no implication that when the engine reaches its TBO, it must be given a major overhaul. Although some operators do specify in their operations manual that the engine be overhauled at the manufacturer's recommended TBO, for the private owner there is no such requirement.

What is actually meant by the manufacturer's recommended TBO is that if the engine is operated under average conditions, with all of the maintenance done according to the manufacturer's recommendations, the reusable parts should not have worn beyond their serviceable limits in this time. This allows major overhauls to be done economically.

When *average* operating conditions are specified, this naturally excludes such things as operating in excessively dusty or sandy conditions, operating continually at high power or exceeding the allowable RPM. Defective cooling baffles, worn air filters, improper grade of fuel, inadequate oil change intervals are only a few of the operational problems that can immeasurably shorten the time between overhauls.

The overhaul specified in the manufacturer's TBO is a major overhaul, rather than a top overhaul. A top overhaul consists only of reconditioning the cylinders and pistons, and includes removing all of the carbon and lead accumulation, grinding and reseating the valves, replacing the piston rings, and doing whatever is necessary to the cylinder walls to cause the new rings to seat properly. Because of the many variables that can damage the cylinder and piston assemblies, there is no specified time between top overhauls.

Many operators consider a top overhaul to be poor economy, but there are times when it can be wise to perform one. If an engine has run to almost its recommended TBO, it would not be economical to top-overhaul it and then, just a few operational hours later, to repeat the operation in a major overhaul. But if there is some obvious cause of cylinder damage, such as broken baffles, leaking carburetor air filter, clogged fuel injection nozzles or any other localized condition that has damaged the cylinders while there are relatively few total hours on the engine, it would be expedient to overhaul the cylinders to extend the life of the engine.

If an engine has good maintenance and has been carefully held within the operating limits allowed by the manufacturer, when it reaches the recommended TBO it may be given a careful inspection, especially noting the differential compression and the oil consumption, and a spectrometric inspection should be made of a sample of the engine oil. If everything appears to be in good operating condition, the A&P can certify the engine as airworthy and release it for another one hundred hours of operation. A continual watch is kept on the compression and oil consumption, and a careful observation is made of the contamination growth trend as shown in subsequent spectrometric oil analysis.

When an engine is operated beyond its TBO, it is a good idea to confer with the engine manufacturer and follow his recommendations to know exactly how long this extension should be allowed. The danger of catastrophic failure of the engine itself is minimal if these precautions are followed, but operating vital engine accessories beyond their normal overhaul time may be courting disaster. Check with the engine manufacturer and wisely follow his experience-based recommendations.

B. Attrition Factors

1. Wear

The economy of an aircraft engine is based not only on the number of hours allowed between overhauls, but by the number of parts that must be replaced when the engine is overhauled.

Operating too long between oil changes or with inadequate oil filtering can allow the oil to carry contaminants through the engine which act as a fine abrasive, wearing out the engine parts. A leaking air filter box or a damaged air filter can allow sharp sand crystals to enter the engine and in a short time wear away the cylinder walls and piston rings. Excessive use of carburetor heat on the ground will also allow unfiltered air to enter the engine.

Abrupt power reduction can cause the cylinders to shrink enough around a heat-expanded piston to cause piston scuffing and shorten the life of the engine.

Operating an engine with fuel containing more tetraethyl lead than the engine is designed to use will cause spark plug fouling, oil contamination, and sticking valves.

The recent increased cost of aviation gasoline has made many pilots more conscious of the mixture control than they have been in the past, and this can be a blessing as well as a curse for the engine life. Operating with too rich a mixture will cost more not only in terms of more fuel used, but the extra lead will contaminate the oil and foul the spark plugs, both of which require additional maintenance. But excessive leaning of the mixture can also be bad because of the higher temperature of the departing gases. One of the most useful instruments a pilot has at his disposal is the exhaust gas temperature gage which, if used properly, can allow the pilot to operate his engine with the most economical fuel-air mixture without the danger of burned valves or preignition. More will be said about this instrument in the portion of this Integrated Training Program dealing with fuel metering systems.

If an engine has been allowed to operate beyond its redline RPM or if the cylinder head temperature has been in excess of that allowed by the manufacturer, there is a good possibility that wear of the engine parts will be excessive. Even if the engine should be able to reach its TBO, the number of parts requiring replacement will be greater than that for a normal overhaul.

2. Corrosion

Wear affects an aircraft engine during its operating life as the moving parts rub against each other, but more damage is actually done to an aircraft engine when it is not operating than when it is used every day. This is because of corrosion.

Corrosion is an electrochemical process that forms inside an engine where dissimilar metals are in contact with each other in the presence of an electrolyte. When the engine is not being operated, moisture will condense inside the crankcase and react with the aluminum, bronze, steel, and other metals to form corrosion. If the engine is run for only short periods of time, the oil will not have a chance to heat up thoroughly, and moisture will collect in the oil and react with the sulfur to form sulfuric acid. This is an extremely potent electrolyte and will cause extensive corrosion damage inside the engine.

Many cylinders and crankshafts have their surfaces hardened by a process known as nitriding. In this process, the surface of a metal is converted into an extremely hard material that is highly resistant to wear, but it is also highly susceptible to pitting from corrosion caused by moisture. Any time an aircraft engine is allowed to remain out of service for a period of time, it should be protected from rust and corrosion by spraying the inside of the cylinders with a tenacious rust preventive oil that will cling to the cylinder walls. After the oil is sprayed in, the engine must not be turned, as the movement of the piston will wipe away the protective coating.

3. Improper maintenance

The service life of any aircraft engine may be increased by proper maintenance. Small discrepancies such as loose exhaust nuts can cause a blown gasket and a damaged mounting surface. Such a simple act as tightening a nut, if not done in time, can require the replacement of a cylinder. Cracks in the exhaust system will grow and loose induction system packings will leak and cause the engine to overheat and detonate.

Careful inspections by knowledgeable A&P's and immediate correction of any discrepancies will do more to increase the time between overhauls of an engine than anything else.

II. POWERPLANT INSPECTIONS

There is possibly no aspect of engine servicing as important as routine inspections. When each inspection is diligently carried out, impending problems may be found before they become major and, in addition to the safety of flight being increased, aircraft operation will be made more efficient.

A. Preflight Inspection

Federal Aviation Regulations require that before any pilot begin a flight, he satisfy himself that among other things the airplane is in all regards safe enough for flight. This can only be determined by carefully conducted preflight inspection, and here we are concerned with only the part of the inspection that applies to the powerplant.

Before starting an actual inspection, it must be determined that the ignition switch is in the Off position; then, if the cowling can be opened, a visual inspection should be performed on the engine. Start at the rear of the accessory section near the firewall, inspect all of the wiring and plumbing to be sure that none of it is loose, chafing against any of the engine components, or showing any sign of wear or deterioration.

COURTESY OF PIPER

Fig. 2B-1 The cowling on some aircraft may be opened for a thorough preflight inspection.

Check the oil level to be sure that it is within the operating level specified by the aircraft flight manual; it does not necessarily have to be full. Most engines have a tendency to throw oil out of the crankcase breather if the sump is completely full, and for local flights, many aircraft are operated with less than full oil.

Check all of the wire connections to the magneto to be sure there are no loose connections or chafed wires. Check all of the wiring to the generator or alternator and to the voltage regulator. If the battery and master relay are ahead of the firewall, check them as well.

Gently shake all of the wires to be sure that none of them are loose at the connection.

Check below the engine for any indication of fuel or oil leaks, and examine all of the baffles for integrity of the air seals; carefully inspect those around the cylinders for cracks or broken sheet metal that could cause local hot spots on the cylinders. Check the paint on the cylinders for any indication of discoloration which could indicate that detonation may have taken place inside the cylinders. Check the primer lines as they enter the cylinders or, if the engine is fuel injected, the injector lines where they attach to the nozzles.

Examine the pushrod tubes and rocker box covers for indication of oil leakage, and the intake pipes for any traces of dye from the fuel which might indicate a leak in the induction system.

Check the spark plug leads for their general condition and for any indication of looseness where they screw into the spark plugs. Be sure that all of the leads are secured in such a way that they cannot be burned by the exhaust stacks.

Carefully examine the exhaust connections where the pipes join the cylinders to be sure that there are no blown gaskets and that all of the nuts are in place.

Drain the main fuel strainer and examine the gasoline that came out of it. There must be no indication of water in the fuel.

Inspect the carburetor air filter for both its security of mounting and cleanness, and check as much of the induction system as is visible for traces of fuel dye stain.

Check the propeller for nicks or scratches, and all of the attaching bolts or nuts for security and for proper safety.

When the engine is run up, it should develop the required static RPM and manifold pressure, and the magneto drop should be within the limits specified by the manufacturer with equal drops on both magnetos. A magneto switch check should be performed to be sure that the switch is operating properly, and the oil and fuel pressure should be within the proper operating range. The propeller should cycle smoothly between low and high pitch, and there should be the proper drop in RPM when the carburetor heat control is pulled.

B. Fifty-Hour Inspection

Although this inspection is not required by the Federal Aviation Regulations, the engine manufacturers have proven that by observing it, the life of the engine may be increased. A typical fifty-hour inspection consists, in addition to all of the items performed on a preflight inspection, of removing the cowling and examining all of these systems:

1. Ignition system

The spark plug leads should be checked for security and for any indication of corrosion. All of the leads should be tight in both the spark plug and in the magneto distributor block, and there should be no evidence of chafing or wear. The spark plugs should be examined where they screw into the cylinder heads for any indication of leakage of the hot gases from the cylinder.

2. Fuel and induction system

Check the primer lines for indication of leaks and for security of all of the clamps. Remove and clean the fuel inlet strainers, check the mixture control and throttle linkage for travel and freedom of movement and for security. Lubricate the controls if it is necessary. Check the air intake ducts for leaks and for any indication of filter damage and for evidence of dust or other solid contaminants that may have leaked past the filter. Check the fuel pump vent lines to see if there is evidence of fuel or oil seepage which could indicate that either the fuel or oil seal is leaking.

3. Lubrication system

Drain and replace the engine oil if it is recommended at this time by the engine manufacturer. Under some conditions when a full-flow oil filter is installed, the oil drain interval may be increased, but in almost all circumstances, the filter should be replaced every fifty hours. Remove the filter if one is installed, and cut the element open so you can inspect it for any traces of metal particles which would most likely indicate an impending engine failure. Check all of the oil lines for any indication of leakage or signs of chafing.

4. Exhaust system

Check all of the flanges in the exhaust pipes where they attach to the cylinder head for evidence of leakage. If they are loose or show any signs of warpage, they must be removed and machined flat before they are reassembled. Check the entire exhaust manifold and muffler assembly for its general condition.

5. Cooling system

Check all of the cowling and baffles for any indication of damage or missing parts, and check the entire system for security.

6. Cylinders

Check the rocker box covers for indication of leaks, and replace the gasket if leaks are found. Carefully check the entire cylinder for signs of overheating which would cause the paint to burn or be discolored. This could indicate detonation and require further inspection by borescope or by removing the cylinder.

Very carefully look for any indication of discoloration or seepage between the cylinder head and the barrel and between the fins for evidence of cracks.

7. Turbocharger

Check all of the oil lines for leaks or chafing. Check all of the brackets for security and for any indication of damage or cracks. Check the waste gate for freedom of action and the alternate air door for operation and sealing.

C. One-Hundred-Hour or Annual Inspection

Perhaps the most important maintenance tool for prolonged engine life is the required one-hundred-hour or annual inspection. The actual inspections are identical, the only difference being the person authorized to perform them. A one-hundred-hour inspection is required for aircraft operating for hire, and the inspection may be performed by an A&P mechanic. All aircraft operating under FAR Part 91, Subpart C, whether they are operated for hire or not, must have an annual inspection every 12 months, and it must be performed by an A&P mechanic that holds an Inspection Authorization.

1. Preliminary paperwork

One of the most important aspects of this inspection is the paperwork involved. Before starting the actual inspection, check all of the aircraft records. Check and list all of the Airworthiness Directives against the engine and all of its components such as the carburetor, magnetos, alternator or generator, propeller, and ignition switches. Check the General Aviation Airworthiness Alerts to become familiar with the types of problems other mechanics have found with similar engines. Go through the manufacturer's service bulletins and service letters to be sure there is nothing that should be done to the engine to make it safer or more efficient.

A shop work order should be started for the inspection and the engine records examined. Be sure that all of the accessories on the engine are of the approved type and that everything that is included in the equipment list is actually on the engine. Be sure the propeller is the proper model and its blades are approved and are the ones listed as being installed. Check the total time on the engine and propeller and compare it with the list of life-limited parts to see if any of them are nearing their retirement time. See if there are any Major Repair and Alteration Forms (FAA 337) on the engine or propeller, and if there are, check the work to be sure that it is actually as is described on the form.

2. Pre-inspection run-up

After all of the paperwork has been inspected, give the engine a good pre-inspection run-up to determine its actual condition and get the oil warm and the cylinder walls well oiled. Check all of the temperatures and pressures to be sure they are within the proper operating range, and see that the engine develops its proper static RPM and that the magneto drop is within the range specified by the engine manufacturer. The drop must be the same or nearly so on both magnetos. Check for any abnormal noises in the engine and for any vibrations that are not characteristic of that engine.

The engine should respond smoothly to power changes, and when the pitch of the propeller is changed, it should be smooth and the recovery should be within the time limit allowed.

The engine should idle smoothly, and when the mixture control is pulled into the cut-off position, there should be a slight rise in RPM before the engine dies.

ANNUAL INSPECTION FORM

MAKE & MODEL_____ N_____ DATE_____

SERIAL NO._____ YEAR_____ TACH TIME_____

OWNER_____ AIRFRAME TOTAL_____

ADDRESS_____ ENGINE TOTAL_____

_____ ENGINE TSMOH_____

ITEM	MAKE	MODEL / PART NO.	SERIAL NO.	APPROVED
ENGINE				
PROP				
CARB				
MAGS: LEFT				
RIGHT				
GENERATOR / ALTERNATOR				
STARTER				
MISC. ACCES.				
ELT				
ALTIMETER				
SEAT BELTS				

II

AD NO.	DESCRIPTION	COMPLIANCE	ADDITIONAL COMPLIANCE

Fig. 2B-2   Typical annual and 100-hour inspection form for an aircraft engine (1 of 3)

III FLIGHT MANUAL REQUIRED: YES_____ NO_____

SPECIAL MARKINGS/PLACARDS REQUIRED: _____

_____

_____

_____

_____

_____

_____

IV AIRCRAFT PAPERS IN ORDER:

	YES	NO
REGISTRATION		
AIRWORTHINESS CERTIFICATE		
RADIO LICENSE		
FLIGHT MANUAL		
SPECIAL MARKINGS		

V ALTIMETER STATIC TEST [FAR 91.170] DATE OF COMPLIANCE _____

VI SPECIAL INSPECTIONS [SERVICE MANUALS, BULLETINS, INSPEC. AIDS, ETC.]

REFERENCE	INSPECTION	COMPLETED

Fig. 2B-2   Typical annual and 100-hour inspection form for an aircraft engine (2 of 3)

B. ENGINE GROUP

	L	R	100	500	INSP.
1. REMOVE ENGINE COWLS					
2. CLEAN COWLING, CHECK FOR CRACKS, MISSING FASTENERS, ETC.					
3. COMPRESSION CHECK: /80					

L. #1 #2 #3 #4 #5 #6
R. #1 #2 #3 #4 #5 #6

	L	R	100	500	INSP.
* 4. DRAIN OIL					
5. CHECK OIL SCREENS AND CLEAN					
6. REPLACE OIL FILTER ELEMENT					
7. CHECK OIL TEMP SENDER UNIT FOR LEAKS AND SECURITY					
8. CLEAN AND CHECK OIL RADIATOR FINS					
* 9. REMOVE AND FLUSH OIL RADIATOR					
10. CHECK AND CLEAN FUEL SCREENS					
11. DRAIN CARBURETOR					
*12. SERVICE FUEL INJECTOR NOZZLES					
13. CHECK FUEL SYSTEM FOR LEAKS					
14. CHECK OIL LINES FOR LEAKS AND SECURITY					
15. CHECK FUEL LINES FOR LEAKS AND SECURITY					
*16. SERVICE AIR CLEANER					
17. CHECK INDUCTION AIR AND HEAT DUCTS					
18. CHECK CONDITION OF CARB HEAT BOX					
19. CHECK MAG POINTS FOR PROPER CLEARANCE					
20. CHECK MAGS FOR OIL SEAL LEAKAGE					
21. CHECK BREAKER FELTS FOR LUBRICATION					
22. CHECK DISTRIBUTOR BLOCK FOR CRACKS, BURNED AREAS, CORROSION, HEIGHT OF CONTACT SPRINGS					
23. CHECK IGNITION HARNESS AND INSULATORS					
24. CHECK MAG TO ENGINE TIMING: LEFT_____ RIGHT_____					
*25. SERVICE OR REPLACE SPARK PLUGS					
26. CHECK CONDITION OF GENERATOR OR ALTERNATOR					
27. CHECK CONDITION OF STARTER					
28. CHECK CONDITION AND TENSION OF DRIVE BELTS					
29. CHECK HYDRAULIC PUMP AND STRAINER					
30. CHECK VACUUM PUMP AND LINES					
31. INSPECT EXHAUST STACKS, GASKETS, ETC.					
32. INSPECT MUFFLER AND SHROUDS					
33. CHECK ENGINE BAFFLES					
34. CHECK BREATHER TUBE FOR OBSTRUCTIONS, SECURITY					
35. CHECK CRANKCASE FOR LEAKS, CRACKS, ETC.					
36. CHECK ENGINE MOUNTS FOR CRACKS, LOOSE MOUNTS					
37. CHECK ENGINE MOUNT BUSHINGS					
38. CHECK FIREWALL SEALS					
39. CHECK THROTTLE, CARB HEAT, MIXTURE AND PROP GOVERNOR CONTROLS FOR TRAVEL AND OPERATING CONDITION					
40. CHECK COWL FLAP CONDITION AND OPERATION					
41. INSPECT ENGINE FOR GENERAL CONDITION, LOOSE PARTS, CHAFING, PROPER SAFETIES, PROPER INSTALLATION, ETC.					
*42. FILL ENGINE WITH OIL					
43. CLEAN ENGINE					
44. LUBRICATE ALL CONTROLS					
45. REINSTALL ENGINE COWL					

Fig. 2B-2   Typical annual and 100-hour inspection form for an aircraft engine (3 of 3)

54

Fig. 2B-3   Aircraft engine uncowled for a 100-hour or annual inspection

The alternator or generator should show a current output and all of the engine instruments should indicate that the systems are functioning properly.

If the engine is fuel injected, the injection pressure should be within the range specified in the latest manufacturer's service manuals.

After the engine has been thoroughly checked, bring the aircraft into the hangar and remove the cowling. Drain the oil while it is hot and remove one spark plug from each cylinder so you can make a differential compression test.

3.   Compression test

There are two types of compression tests that can be made, the direct test and the differential test, but the differential test is far more uniform and is the type used in almost all shops today. This type of test is not only quantitative, but if there is any trouble in the engine, this test makes it easy to locate the source of the trouble.

Fig. 2B-4   The differential compression tester is used to determine the condition of the piston rings and valves in an aircraft engine cylinder.

Turn the propeller until the piston in cylinder number one is moving outward on its compression stroke. This stroke is identified by air blowing out of the spark plug hole as the piston moves out. Screw the adapter into the open spark plug hole in cylinder number one and attach the air hose from the compression tester. Open the shut-off valve and turn the regulator in until there is between 10 and 15 psi of regulated air pressure inside the cylinder.

Now, slowly move the propeller in the direction of normal rotation to bring the piston to its top dead center position. You will know the piston is on top dead center when the pressure inside the cylinder no longer pushes back on the piston. If the piston is moved too far, the pressure will push it down and the propeller will have to be brought back until the piston is well below its top center and then started back up. The reason for this is to be sure that the rings are properly seated in their grooves so they will provide the best seal.

When you are sure that the piston is exactly on top dead center, increase the air pressure to 80 psi and read the cylinder pressure gage.

CAUTION: With 80 psi air pressure inside the cylinder, the propeller must not be bumped or the piston will move off of top center and the air will force it down, spinning the propeller fast enough to create a serious hazard.

The difference between the reading of the two gages indicates the amount of leakage in the cylinder. If the cylinder gage reads less than 60 psi, there is leakage of more than 25%, and the cause of the leakage should be found and corrected. Listen for escaping air to locate the problem. If air is heard to be coming from the crankcase breather or the oil filler, the problem is air leaking around the piston rings. If the noise is heard at the exhaust pipe, the problem is exhaust valve leakage, and if the intake valve is leaking, the sound will be heard at the carburetor air inlet.

After the test is completed, reduce the air pressure on the regulator and remove the adapter. Test each cylinder in firing order until all of the cylinders are tested.

4. Lubrication system

While the oil is hot from the pre-inspection run-up, remove the drain plug and drain the oil. Remove the oil filter and cut the element apart so you can examine the contaminants it has trapped. The presence of any metal particles in the oil is an indication of impending engine failure and their source must be found.

Install a new filter element and properly torque the filter in place. Reinstall and safety the drain plug and put the proper amount of the correct oil in the engine. Check all of the oil lines, the cooler, and the entire engine for any indication of oil leakage that might indicate a cracked crankcase or sump or a leaking oil seal or gasket.

Fig. 2B-5 The oil filter should be cut open to show if there are any metal particles or other unusual contamination in the oil.

5. Ignition system

One spark plug has already been removed from each cylinder for the compression check, and as it was removed, it should have been put into a rack in numbered holes to indicate the cylinder from which it came. Remove the other spark plugs from the cylinders and examine all of them. Much of the condition within the cylinders can be determined by studying the spark plugs. Normal operation is indicated by a spark plug having a relatively small amount of light brown or tan deposits on the nose of the center electrode insulator. If there is an excess of deposits in the firing end cavity, and if these deposits are gray and clinker-like, lead fouling is indicated which may be caused by using a fuel with a higher than recommended tetraethyl lead content. Replacing the spark plug with one having a hotter heat range may reduce the fouling, but be sure the hotter plug is specifically recommended for the engine.

A dry, black, soot-like deposit in the firing end of the spark plug indicates that the engine is operating with an excessively rich mixture, and the induction system should be carefully checked for obstructions in the filter or for a malfunctioning carburetor heat valve.

If the black deposits are oily, there is a good probability that the valve guides are worn excessively, or there may be a broken piston ring. A brown, shiny glaze on the nose insulator of the spark plug could indicate silicon contamination, and this calls for a careful inspection of the carburetor air filter for air leaks around the filter element or for holes in the element itself.

Any unfiltered air leaking into the induction system will allow sand or dust to enter the engine, and the intense heat inside the combustion chamber will turn the silicon in the sand or dust into a glass-like contaminant that, while it is an insulator at low temperature, becomes conductive at high temperature and causes the spark plug to fail to fire when maximum engine power is needed.

Any spark plug whose electrodes have worn away to one-half of their original dimensions is considered to be worn out and should be replaced. If the spark plugs are not worn out, they may be reconditioned and reinstalled. Remove all of the lead deposits with a vibrator-type cleaner, and then very lightly blast the firing end cavity with an approved abrasive. The ground electrode is then carefully moved over with a spark plug gapping tool to get the proper gap distance between the ground and the center electrodes.

Clean the inside of the terminal cavity with trichlorethylene and test the spark plug for operation. If the plug fires consistently under pressure in the tester, it may be returned to the engine. Use a new gasket and a very small amount of the thread lubricant recommended by the engine manufacturer.

All of the spark plugs should be returned to the cylinder next in firing order to the one from which they were removed, and they should be swapped bottom to top.

The threads in the spark plug bushing in the cylinder head should be clean enough that the spark plug can be screwed down against the gasket with the fingers only, and then the spark plug

is tightened to the torque recommended by the engine manufacturer, using a torque wrench of known accuracy.

Wipe the spark plug lead terminal free of all fingerprints with a rag dampened with trichlorethylene and insert the lead straight into the spark plug. Tighten the lead nut to the torque recommended by the manufacturer.

Check the magneto to engine timing, using a timing light, and inspect the condition of the breaker points and the inside of the breaker compartment for any indication of moisture or oil. Be sure the ignition switch connection, the P-lead, is tight and the wire is secure and is neither worn nor chafed.

6. *Fuel system*

Clean the main fuel screen and the screen in the carburetor or fuel injection system. Replace the cleaned screens, using a new gasket, and after testing them under pressure for any indication of leaks, safety them. Check all of the controls and lubricate them as specified by the airframe manufacturer, being sure to use only the approved lubricant.

If the engine is fuel injected, check the manufacturer's recommendations for cleaning the injector nozzles, and if they are to be cleaned, be sure to follow the recommended procedure in detail. Check the entire fuel system for any indication of dye stain that would indicate a fuel leak.

7. *Induction system*

Remove and clean or replace the induction air filter and check the entire system for any leakage or deformation. Check the carburetor heat or alternate air valve to be sure that it does not allow any unfiltered air to enter the engine. Be sure the valve and alternate air doors are fully open so the induction airflow into the engine will not be restricted.

The flexible hose that attaches the induction air valve to the hot and cold air source should be carefully inspected to be sure that there are no kinks and that the hose is in good condition with no possibility of its collapsing.

8. Exhaust system

Remove the shroud from around the muffler and check it for any indication of leakage. If there is an AD or manufacturer's service bulletin requiring pressure testing, be sure that this is complied with in detail, as even a pinhole leak can fill the cabin with deadly carbon monoxide gas. Be sure the cabin heat valve operates freely and has no obstructions in either the valve or the hose carrying heated air into the cabin. Check all of the cylinder heads at the exhaust port for indication of blown gaskets and for any indication of cracks or other leakage. Very carefully check where the exhaust gas temperature probes enter the exhaust pipes, as this is a possible source of leakage.

9. Turbocharger

If the engine is equipped with a turbocharger, be sure to follow the inspection procedure specified by the manufacturer, as each installation has peculiarities that make them different. Essentially, the exhaust portion, including the operation of the waste gate, should be checked as well as the induction air section which includes any relief valves, intercoolers, or manifold pressure sensors. The lubrication system and mounting should be carefully checked, as some of these small turbines spin in excess of 100,000 RPM while operating red hot. Extreme care should be exercised to follow all of the manufacturer's inspection recommendations in detail.

10. Cooling system

Check all of the cooling system for cracks or damage and all of the baffles and seals that direct air through the fins on the cylinders. The cowl flaps should be checked for security and for full travel, both in their open and closed position.

11. Electrical system

Here, again, systems vary between models of aircraft, and the manufacturer's recommendations must be followed in detail. The alternator and its mounting should be checked for security and for any indication of vibration-induced cracks. The voltage regulator and any relays or solenoids in the system should be checked for security and for tightness of all wires attached to them.

Follow the manufacturer's recommendation regarding inspection of brushes, bearings, and commutators of the starter and generator.

12. Accessories and controls

Check all of the air, fuel, and hydraulic pumps for indication of leakage, and for the condition of their seals. Check the vent and breather lines for security and for the proper positioning of their lower ends.

Check the instrument air system to be sure that all filters in the pump inlet or at the regulator are changed according to the manufacturer's recommendations and the oil separator shows no sign of malfunctioning.

Check the condition of the firewall to be sure that all controls, lines, and wires passing through are properly sealed and there is no corrosion or other indication of damage.

Check all controls for freedom and for proper travel. The stops on the engine component should be reached just before the stop in the cockpit. Be sure that all of the shock mounts are in good condition, and the mounting bolts are properly torqued. It is important that the electrical grounding strap between the engine and the airframe be in good condition since the engine is mounted in rubber and all of the return current from the starter must flow through this strap.

13. Propeller

The extreme stresses the propeller encounters make it important that the propeller be inspected critically. Any nicks, cracks, or scratches in the blade must be carefully stoned out, or burnished, and any questionable area must be checked with one of the approved non-destructive test methods. Be sure the blades are secure in the hubs and that there is no oil leakage. Check the spinner and spinner bulkhead for any indication of cracks or damage, and check the governor for security and for full travel of its control.

14. Post-inspection run-up and records

After the inspection has been completed, wash the engine down, recowl it, and check it out with a post-inspection run-up. When it checks out satisfactorily, complete the engine maintenance records and fill out all of the shop records.

The purpose of the one-hundred-hour or annual inspection is to determine that the engine is in the condition required for its certification, and there must be no alterations or modifications that have not been approved.

III. POWERPLANT OVERHAUL

A. Major Overhaul and Remanufacture

An overhaul of an aircraft engine consists of disassembling the engine, and cleaning, inspecting, and repairing or replacing any parts that do not meet the manufacturer's specifications.

When the manufacturer builds an engine, he establishes a set of fits and clearances between the moving parts that he considers allowable tolerances for new engine parts. Along with these limits for new parts, the manufacturer also sets serviceable limits. These limits show the maximum amount of wear a part can have and still be considered serviceable. An engine can be overhauled to either set of limits, and it is in this area that the difference in overhauls exist.

The FAA states in FAR Part 91.175 that "The owner or operator may use a new maintenance record, without previous operating history, for an aircraft engine rebuilt by the manufacturer or by an agency approved by the manufacturer." This means simply that the engine manufacturer is authorized to overhaul one of his engines and sell it with a new name plate and new logbooks, or records, stating that the engine has zero operating time. Both of the major manufacturers of general aviation reciprocating engines do this and call these engines "remanufactured," rather than overhauled.

Remanufactured engines are built up from parts of used engines that have been returned to either the manufacturer or to a facility approved by the manufacturer for this purpose. The used engines are disassembled and all of the reusable parts cleaned, inspected, and repaired if necessary. If these parts comply with the new parts fits and tolerances, they are used in the remanufacturing process.

These parts are not required to have the new parts dimensions; rather, it is the *fit* that is important. For example, if a crankshaft is in good condition except that the journals are slightly pitted, the shaft can be ground undersize and undersize bearings can be fitted into the crankcase to maintain the fit within new engine tolerances.

In the remanufacturing process, no attempt is made to keep all of the parts of a particular engine together, but all of the parts which are used must meet the standards for a new engine. After assembly, the remanufactured engine is given the same run-in as a new engine receives, and it usually gets a warranty similar to that given for a new engine.

An engine may be overhauled by a facility that is not approved by the engine manufacturer, and it may be overhauled to the same tolerances used by the manufacturer, but this engine must carry forward its previous operating time. For example, if the engine had 1,800 hours before it was overhauled, after the overhaul it must carry in its operating record that it has a total of 1,800 hours, with zero time since major overhaul. This is usually abbreviated SMOH.

An engine may be overhauled to serviceable limits standards, but because of the excessive clearances it is unlikely that it will be able to run to the recommended time between overhaul, and when it is overhauled, in all probability many parts will need to be replaced that could have been used for another period if the overhaul had been done to new engine fits.

B. Top Overhaul

The overhaul of the cylinders of an engine is called a top overhaul. The crankcase is not opened, but the cylinders are removed and everything that is normally done to them during a major overhaul is done. The valves are ground and reseated, the piston rings are replaced, and the cylinders are reinstalled and the engine given the proper run-in to reseat the piston rings.

There is no recommended time for a top overhaul because of the widely varying factors that could cause it to be needed. There is no normal reason for an engine to require a top overhaul between major overhauls. But if there is some type of localized damage to the cylinders, or if the engine has had unusual operating conditions, the cylinder wear may be such that it is economical to perform a top overhaul.

C. Major Overhaul Procedures

We will follow a typical horizontally opposed engine as it progresses through an overhaul shop for a major overhaul. In this procedure we will assume that the engine is to be kept together as a unit.

When the engine is brought into the shop for an overhaul, the work order is started and a record is made of the serial number of all of the components that are to be kept with the engine. When it is sent to the overhaul shop, the engine normally carries with it all of the inter-cylinder baffles, the carburetor or fuel injection system, the magnetos, ignition leads, and spark plugs, as well as the induction system between the carburetor and the cylinders.

The exhaust system, vacuum pump, hydraulic pump, propeller and its governor, and most of the other accessories are not considered to be part of the engine, as such, and they are usually sent to specialty shops for their overhaul. If they are overhauled by the same shop that is overhauling the engine, they are usually run through on separate paperwork.

COURTESY OF SCHNECK

Fig. 3B-1 Aircraft engine being disassembled for major overhaul

1. Disassembly

The engine is mounted on a stand and the magnetos and the fuel metering system are removed and sent to the proper department for their overhaul. The engine is completely disassembled and laid out in an orderly fashion so it can be given its preliminary inspection, and all of the parts that are normally replaced are discarded. Discarded parts normally include the crankshaft bearings, oil seals, gaskets, stressed bolts and nuts, exhaust valves, pistons and rings, and all of the ignition cables.

The condition of all the parts is noted and a record is made of any parts that are obviously damaged, so this damage will not be overlooked as the part progresses through the overhaul process.

2. Cleaning

After all of the parts have been disassembled, they are degreased by soaking or spraying them with some form of mineral spirits solvent such as varsol or Stoddard solvent. Water-mixed degreasing compounds usually contain some form of alkali which, if allowed to remain in the pores of the metal, will, when the engine is returned to service, react with the hot lubricating oil to form soap and cause the oil to foam. This will often cause failure of the lubricating system.

Degreasing removes dirt, grease, and soft carbon, but many parts will have deposits of hard carbon on their interior surfaces and all of this must be removed. An approved decarbonizing solution must be used, following the manufacturer's recommendations in detail. These solutions are usually quite active and are often heated, so care must be taken not to leave these parts in the solution longer than is absolutely necessary to loosen the carbon deposits. No magnesium parts should be put into the solution unless it is known that it will not react with the magnesium.

After the parts have been removed from the decarbonizing vat, they are thoroughly cleaned of all traces of the solution with a blast of wet steam or by brushing them with mineral spirits.

Any hard carbon that was not removed by the decarbonizing solution may be removed by dry blasting with plastic pellets or with such

Fig. 3B-2 Typical cleaning vat for washing engine parts during an overhaul

organic materials as rice, baked wheat, or crushed walnut shells. Be sure that all machined surfaces are masked off and all passages are plugged or covered. Use as low pressure as practical and blast only enough to remove the carbon.

Parts may be vapor-grit blasted, but care must be exercised that only small grit be used and adequate protection taken to prevent any of it remaining in the part after it is cleaned.

When cleaning pistons, exercise extreme care to prevent scratching their highly stressed surfaces. Do not scrape off the carbon deposits, but remove them by solvent action, or by soft-grit or vapor blasting, and especially do not use automotive-type ring groove cleaners, since it is critical that the bottom radius of the groove be maintained. Soften the carbon and remove it by drawing a hard twist cotton cord through the groove. Many aircraft pistons are cam ground— that is, they are not perfectly round, but are slightly oblong, so they will round out as they expand from the heat, and it is important that the skirt of a cam ground piston not be abraded in the cleaning process.

All bearing surfaces must be polished with crocus cloth moistened with mineral spirits, and afterward with dry crocus cloth. All passages in the crankcase must be thoroughly cleaned by flushing them with wet steam and, following this, with a spray of mineral spirits.

After all of the parts are thoroughly cleaned and the steel parts coated with a film of protective oil to prevent their rusting, they are ready to be inspected.

3. Inspection

Before starting the inspection process, an inspection and overhaul form should have been started, and must be with the engine at this point, along with the manufacturer's overhaul manual and all of the proper inspection equipment. Throughout the overhaul, the engine manufacturer's manuals, bulletins, and other service information must be available, as they are the final authority on the suitability of parts for reuse.

a. Visual inspection

Inspect all parts under a good light for any indication of surface damage such as nicks, dents, deep scratches, visible cracks, distortion, burned areas, pitting, or pickup of foreign metal. Inspect all studs for indication of possible bending, looseness, or partial backing out. All threads should be free from nicks or any other forms of damage.

The visual inspection should allow all of the parts to be separated into three groups: those which should be discarded, those which require work, and those which are apparently serviceable.

b. Fluorescent penetrant inspection

All nonferrous parts that have been determined to be apparently serviceable are further inspected by the fluorescent penetrant inspection method. This procedure checks for cracks and discontinuities that cannot be detected by the naked eye.

The most important requirement for a good fluorescent penetrant inspection is that the part being inspected be absolutely clean, with all contamination removed, not only from the surface, but especially from any cracks that may exist.

Fig. 3B-3 An aircraft engine disassembled and laid out for the initial visual inspection

For best results, the part should be vapor-degreased with hot trichlorethylene vapors, and while it is still hot, it should be immersed in the fluorescent penetrant solution and allowed to remain long enough for the solution to thoroughly soak into all cracks. After the required dwell time, as the soaking is called, the part is removed and all of the penetrant on the surface is flushed off with a soft spray of hot water, and the part baked dry in an oven. Now, put the clean, dry part into the developer tank where a fluffy talcum-like powder covers it and pulls the penetrant out of any crack that may exist in the metal.

FLUORESCENT OR DYE PENETRANT SEEPS INTO DIS-CONTINUITY

(A)

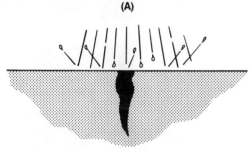

WATER SPRAY REMOVES SURFACE PENETRANT

(B)

DEVELOPER IS APPLIED, DRAWING PENETRANT TO SUR-FACE

(C)

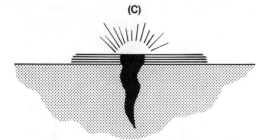

INDICATION IS VIEWED UNDER VISIBLE LIGHT OR BLACK LIGHT

(D)

Fig. 3B-4   Fluorescent penetrant inspection procedure

The part is then inspected under a black light where any crack extending to the surface will show up as a green line.

The interpretation of the results of the fluorescent penetrant inspection is its most critical aspect. Knowing exactly where to look for cracks and being able to distinguish a true crack from a false indication can best be learned by working with an experienced NDI technician.

c.   Magnetic particle inspection

Ferrous parts, those containing iron, can be inspected for hidden cracks by magnetic particle inspection. With this method, the part is magnetized and then a fluid containing particles of iron oxide treated with a fluorescent dye is flowed over the part. If there is a crack in the part, the crack will interrupt the flow of magnetic flux and produce magnetic poles on the surface of the metal. These poles will attract the iron oxide which is flowed over the part. When the part is inspected under a black light, any crack that is present will show up as a brilliant green line.

There are two ways a part can be magnetized, and the engine overhaul manual specifies the method to be used. In Fig. 3B-5 we have duplicated a page from an engine overhaul manual specifying the methods to be used. For example, the crankshaft must be magnetized both longitudinally and circularly to get an indication of all of the cracks that might appear. In Fig. 3B-6 we show the methods of getting both types of magnetization and the way each type of magnetization will show up a crack. (See Figs. 3B-5 and 3B-6 on page 64.)

When checking for cracks that run across the part, the part should be magnetized longitudinally—that is, it is placed in a coil, or solenoid— and current flowing through the coil will magnetize the part so that its poles are at its ends. Now, any crack that runs across the part will show up by attracting the oxide.

After the part is inspected for transverse cracks, as these are called, it is completely demagnetized and remagnetized by circular magnetization. This time the part is clamped tightly between the heads of the machine and current is passed through the part itself. The lines of magnetic flux radiate out from the part and will magnetize it so it has no external poles, but if a crack

PART	METHOD OF MAGNETIZATION	D.C. AMPERES	CRITICAL AREAS	POSSIBLE DEFECTS
CRANKSHAFT	CIRCULAR AND LONGITUDINAL	2500	JOURNALS, FILLETS, OIL HOLES, THRUST FLANGES, PROP FLANGE	FATIGUE CRACKS, HEAT CRACKS
CONNECTING ROD	CIRCULAR AND LONGITUDINAL	1800	ALL AREAS	FATIGUE CRACKS
CAMSHAFT	CIRCULAR AND LONGITUDINAL	1500	LOBES, JOURNALS	HEAT CRACKS
PISTON PIN	CIRCULAR AND LONGITUDINAL	1000	SHEAR PLANES, ENDS, CENTER	FATIGUE CRACKS
ROCKER ARMS	CIRCULAR AND LONGITUDINAL	800	PAD, SOCKET UNDER SIDE ARMS AND BOSS	FATIGUE CRACKS
GEARS TO 6 INCH DIAMETER	CIRCULAR OR ON CENTER CONDUCTOR	1000 TO 1500	TEETH, SPLINES, KEYWAYS	FATIGUE CRACKS
GEARS OVER 6 INCH DIAMETER	SHAFT CIRCULAR TEETH BETWEEN HEADS TWO TIMES 90°	1000 TO 1500	TEETH, SPLINES	FATIGUE CRACKS
SHAFTS	CIRCULAR AND LONGITUDINAL	1000 TO 1500	SPLINES, KEYWAYS, CHANGE OF SECTION	FATIGUE CRACKS, HEAT CRACKS
THRU BOLTS ROD BOLTS	CIRCULAR AND LONGITUDINAL	500	THREADS UNDER HEAD	FATIGUE CRACKS

Fig. 3B-5   Typical magnetic particle inspection schedule for aircraft engine parts

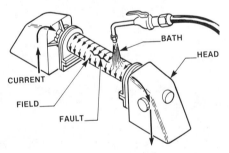

MAGNETIZING A PART BY CIRCULAR MAGNETIZATION
(A)

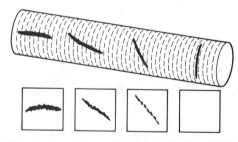

**MAGNETIC FIELDS AND APPEARANCE OF FLAWS AFTER
A PART HAS BEEN CIRCULARLY MAGNETIZED**
(B)

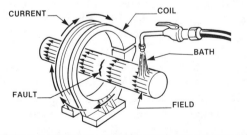

MAGNETIZING A PART BY LONGITUDINAL MAGNETIZATION
(C)

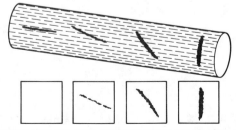

**MAGNETIC FIELDS AND APPEARANCE OF FLAWS AFTER
A PART HAS BEEN LONGITUDINALLY MAGNETIZED**
(D)

Fig. 3B-6   Methods of magnetic partical inspections

is present, it will interrupt the flux and poles will appear, attracting the oxide. Small cylindrical parts such as nuts, washers, springs, and pins may be circularly magnetized by placing them on a conductive rod which is clamped between the heads of the machine.

It is extremely important that after a piece has been magnetically inspected, every trace of the magnetism be removed. This may be done by placing the part in a coil through which alternating current is passed. Rotate the part back and forth and remove it from the field while current is flowing. A more effective method of demagnetizing is to use pulses of direct current whose polarity reverses with each pulse and whose intensity decreases with each pulse until it reaches zero. Either of these two methods disorients the magnetic fields and demagnetizes the part.

d. Dimensional inspection

Use the proper measuring tools such as micrometer calipers, telescoping gages, and dial indicators to measure the parts and determine that the fits are as specified in the engine overhaul manual. You will notice that in Fig. 3B-7 on page 66 both new parts limits and serviceable limits are given, but it is poor economy to overhaul an engine to serviceable limits, as it will run only a short while before it will wear outside of these limits. The increased clearances will accelerate wear and decrease the time between overhauls. When the engine is next overhauled, it will most likely require more parts to be replaced than should be because of these excessive clearances.

A typical measurement is that of the crankshaft clearance. In our example, new main bearing inserts are installed in the crankcase halves, and the case is assembled and torqued to the values recommended in the table of torques. Telescoping gages are adjusted to the inside diameter of the bearings and are then measured with micrometer calipers. The journals of the crankshaft are then carefully measured with the same micrometer caliper, and the difference between the two measurements is the clearance between the journal and the bearing. For a new engine, this fit is allowed to be between 0.0012L and 0.0032L. The "L" following this dimension indicates that the fit is loose, meaning that the inside diameter of the bearing is larger than the outside diameter of the crankshaft journal.

The serviceable limit for this fit is 0.005L, and a crankshaft whose fit falls outside of this limit or one with journals more than a half thousandth inch (0.0005") out of round can be ground as much as ten thousandths inch (0.010") undersize, its surface re-nitrided, and then fitted into the crankcase with undersize bearing inserts.

All of the dimensions specified in the table of limits must be measured, and if the part does not fall within the tolerances given, it must be replaced.

Some fits, such as bushings in the small end of the connecting rods, are called interference fits, and in our example are dimensioned as 0.0025T to 0.0050T. This means that the bushing must be from two and a half to five thousandths of an inch *larger* than the hole into which it fits. To press this bushing into place, you must use an arbor press and a special bushing installation drift. Other interference fits such as the valve guides and valve seats in the cylinder heads are so tight that the cylinder heads must be heated in an oven and the guides chilled with dry ice and then assembled while they have large dimensional differences because of the temperature extremes.

4. Repair

If the inspection shows that parts need repairing, they must be brought back to the standards required by the engine manufacturer before the engine is reassembled.

a. Crankcase

Crankcases are subject to such high stresses that cracks are likely to appear, and these must be repaired. Crankcases are quite expensive, and modern welding technology has made the welding of cracks an acceptable repair. It must be noted, however, that repairs of this nature must be done by either the engine manufacturer or by a facility with an FAA repair station approval for this specialized type of work.

Welding is done by one of the forms of inert gas arc welding, and after some of the weld metal has been deposited, the bead is peened to relieve stresses built up by the welding. After the complete weld is made, the repair is machined to match the rest of the surface. Not only are cracks repaired by welding, but bearings have been known to turn in their saddles, and damage the

DESCRIPTION	SERVICEABLE LIMIT	NEW PARTS	
		MIN.	MAX
First piston ring in cylinder (P/N 635814)............................Gap:	0.059	0.033	0.049
First piston ring in cylinder (P/N 639273)............................Gap:	0.074	0.048	0.064
Second piston ring in cylinder (P/N 635814)........................Gap:	0.050	0.024	0.040
Second piston ring in cylinder (P/N 639273)........................Gap:	0.069	0.043	0.059
Third piston ring in cylinder..Gap:	0.059	0.033	0.049
Fourth piston ring in cylinder..Gap:	0.050	0.024	0.040
Fifth piston ring in cylinder...Gap:	0.059	0.033	0.049
Piston pin in piston (standard or 0.005" oversize).....Diameter:	0.0013L	0.0001L	0.0007L
Piston pin in cylinder...............................End Clearance:	0.090	0.031	0.048
Piston pin in connecting rod bushing......................Diameter:	0.0040L	0.0022L	0.0026L
Bushing in connecting rod.......................................Diameter:		0.0025T	0.0050T
Connecting rod bearing on crankpin........................Diameter:	0.006 L	0.0009L	0.0034L
Connecting rod on crankpin...........................Side Clearance:	0.016	0.006	0.010
Bolt in connecting rod..Diameter:		0.0000	0.0018L
Connecting bearing and bushing twist or convergence per inch of length............................:	0.001	0.0000	0.0005
CRANKSHAFT			
Crankshaft in main bearings.....................................Diameter:	0.005 L	0.0012L	0.0032L
Propeller reduction gear shaft in bearing..................Diameter:		0.0012L	0.0032L
Propeller drive shaft in shaft...................................Diameter:		0.0012L	0.0032L
Crankpins....................................Out-of-Round:	0.0015	0.0000	0.0005
Main journals.............................Out-of-Round:	0.0015	0.0000	0.0005
Propeller drive shaft...................Out-of-Round:	0.002	0.0000	0.002
Propeller drive shaft in thrust bearing..............End Clearance:	0.020	0.006	0.0152
Crankshaft run-out at center main journals (shaft supported at thrust rear journals) full indicator reading............................:	0.015	0.000	0.015
Propeller shaft run-out at propeller flange (when supported at front and rear journals) full indicator reading............................:	0.003	0.000	0.002
Damper pin bushing in crankcheek extension..........Diameter:		0.0015T	0.003 T
Damper pin bushing in counterweight.......................Diameter:		0.0015T	0.003 T
Damper pin in counterweight..............................End Clearance:	0.040	0.001	0.023
Alternator drive gear on reduction gear.....................Diameter:		0.001 T	0.004 T
Crankshaft gear on crankshaft..................................Diameter:		0.000	0.002 T
CAMSHAFT			
Camshaft journals in crankcase................................Diameter:	0.005 L	0.001 L	0.003 L
Camshaft in crankcase................................End Clearance:	0.014	0.005	0.009
Camshaft run-out at center (shaft supported at end journals) full indicator reading............................:	0.003	0.000	0.001
Camshaft gear on camshaft flange...........................Diameter:		0.0005 T	0.0015L
Governor drive gear on camshaft...............................Diameter:	0.006 L	0.0005 L	0.002 L
CRANKCASE AND ATTACHED PARTS			
Thru bolts in crankcase...Diameter:		0.0005 T	0.0013L
Hydraulic lifter in crankcase....................................Diameter:	0.0035L	0.001 L	0.0025L
Governor drive shaft in crankcase.............................Diameter:		0.0014 L	0.0034L

Fig. 3B-7 Typical table of limits for dimensional inspection

66

Fig. 3B-8 *Crankcases may be repaired by welding and machining. When properly done, the repaired crankcase is as serviceable as a new crankcase.*

TYPICAL PART NO.	OVERSIZE ON PITCH DIA. OF COARSE THREAD (INCHES)	OPTIONAL IDENTIFICATION MARKS ON COARSE THREAD END		IDENTIFICATION COLOR CODE
		STAMPED	MACHINED	
XXXXXX	STANDARD	NONE		NONE
XXXXXXP003	.003			RED
XXXXXXP006	.006			BLUE
XXXXXXP009	.009			GREEN
XXXXXXP007	.007			BLUE
XXXXXXP012	.012			GREEN

Fig. 3B-9 *Identification of oversize studs used in aircraft engines.*

bearing cavity. These cavities may be repaired by building them up with weld material and line boring the case after all of the repairs have been made. (See Fig. 3B-8 on page 67.)

Camshafts normally run in bearings cut in the crankcase without any bearing inserts. If the clearance between the case and the camshaft is greater than is allowed, the bearings may be line-bored oversize and an oversize camshaft installed.

All of the studs must be checked for any indication of their having been loosened in operation, and any that are bent or loose must be replaced, using oversize studs if necessary. Oversize studs may be identified by one of several methods which are shown in Fig. 3B-9 on page 67.

If the threads in the case are stripped out, they may be drilled clean with a special drill, the hole tapped with a special tap, and a stainless steel Helicoil insert screwed into place. Helicoils provide new threads for the hole and standard studs may then be installed. There is no decrease in strength when this type of repair is made.

b. Crankshaft

The crankshaft is the heaviest and most highly stressed part of an aircraft engine, and there are very few repairs that can be made to it. Those repairs which can be made must be done by either the manufacturer or by a repair station that is specially approved for this type of work.

If the journals are scored or out of round, the shaft can be ground to the proper undersize. When it has been determined by magnetic particle inspection that there are no cracks in the crankshaft, and that the shaft is not bent, it is placed in a special lathe and all of the main and connecting rod journals are ground to the proper size. Special care must be taken with the radius between the bearing surface and the crank cheek because of the extremely high stresses encountered with aircraft crankshafts. After all of the journals have been ground, they are polished and the crankshaft is surface hardened by the nitriding process and the propeller flange is cadmium plated. The crankshaft is given a final magnetic particle inspection and all of the sludge plugs, counterweights, and all other removable components are reinstalled.

c. Cylinders

Bent cylinder barrel fins may be straightened and any cracked head fins may be dressed smooth, being sure that no more fin area is removed than that allowed by the manufacturer, as these fins are extremely important for proper cooling.

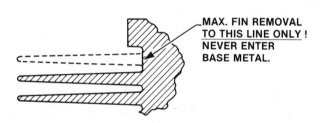

Fig. 3B-11 *When dressing a damaged fin on an aircraft cylinder head, do not remove more material than the manufacturer allows.*

If the cylinder bore is worn beyond the allowable limits, it may be restored to its original dimensions by porous chrome plating done by a repair station approved by the FAA for this work.

The valve guides may be replaced by pressing the old guides out and heating the cylinder head in an oven while chilling the new guide with dry

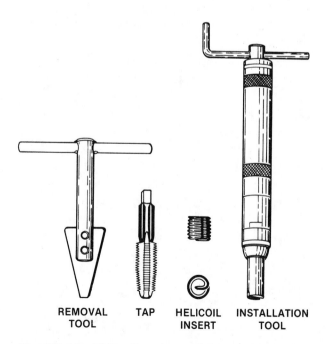

REMOVAL   TAP   HELICOIL   INSTALLATION
TOOL   INSERT   TOOL

Fig. 3B-10 *Helicoil tools used for removing and installing Helicoil inserts in an aircraft engine*

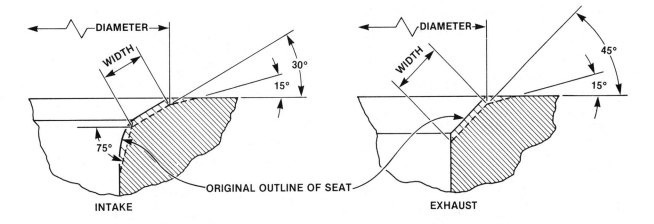

Fig. 3B-12 *Method of narrowing a valve seat after it has been ground*

ice and then pressing the guide into the head. New valve seats are installed in the same way.

When the new seats and guides have been installed, the guides are reamed to size and the seats ground, using the guide as a pilot hole. The recommendations of the engine manufacturer must be followed in detail when grinding the seats, but a typical procedure for grinding a 30° intake valve seat would be to use a 30° stone, then a 15° stone to cut the top of the seat to the diameter specified in the overhaul manual. Then use a 75° stone to narrow the seat to the specified width. (See Fig. 3B-12.)

d. Valves and valve springs

Some valves may not be reused, and since the valve is a part of the engine that is subject to extremely severe operating conditions, those that may be reused must be carefully examined for any indication of overheating that could render them unserviceable.

Any nicks or scratches in the valve stem near the spring retainer groove is cause for rejection of the valve, and any valve whose stem diameter in the center measures less than the diameter at the spring end should be rejected, as this is an indication that the valve has been stretched.

Chuck the valve in the valve grinding machine, and, using a dial indicator, check to see that the face runs true with the stem within the very small tolerance allowed, usually only about one and a half thousandth (0.0015) of an inch. Remove only enough material to clean up any wear marks or pits on the valve face, and be sure

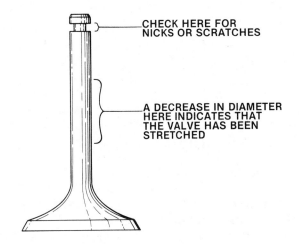

Fig. 3B-13 *Method of checking a poppet valve stem for stretch*

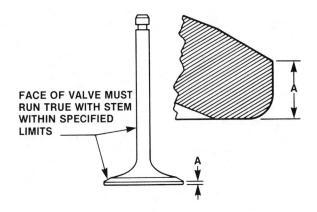

Fig. 3B-14 *It is important that the valve head thickness not be reduced more than is allowed by the engine manufacturer.*

that the surfaced valve has at least the minimum edge thickness specified in the overhaul manual.

If the overhaul manual specifies an interference fit be ground between the valve face and valve seat, the face is ground between one-half and one degree flatter than the seat so the valve will seat with a line contact on its outer edge.

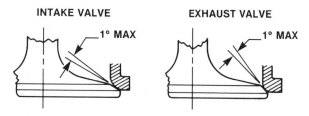

INTAKE VALVE   EXHAUST VALVE

1° MAX   1° MAX

Fig. 3B-15 *An interference fit is ground between the valve and its seat to provide a line contact seal.*

Check the valve springs with a spring tester to be sure that they require the proper force to compress them to the specified height.

e. Pistons and rings

After the pistons have been cleaned and all of the carbon removed from the ring grooves and the oil relief holes in the lower ring groove, the piston is dimensionally checked and inspected for any indication of overheating or scoring. A new set of the proper piston rings is installed after first measuring the end gap by placing the ring in the cylinder barrel, squaring it up with the piston, and using a feeler gage to measure the gap between the two ends of the ring. If the gap is correct, the rings may be installed on the piston with the part number on the ring toward the top of the piston. Use a ring expander and be careful to not scratch the piston with the ends of the ring. When you have the rings installed, check the clearance between the rings and the side of the ring groove. If tapered rings are installed, hold a straightedge against the side of the piston and measure the side clearance with a feeler gage.

It is extremely important that only piston rings approved for the engine be used and that the ring be compatible with the cylinder walls. Only cast iron rings can be used with nitrided or chrome-plated cylinders, but chrome-plated rings may be used in plain steel cylinders.

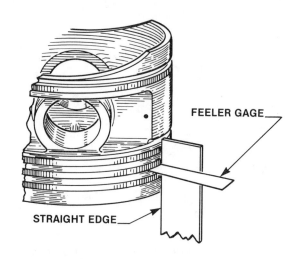

FEELER GAGE

STRAIGHT EDGE

Fig. 3B-16 *Method of checking side clearance of a tapered piston ring in the ring groove*

f. Connecting rods

After the connecting rods have passed the magnetic particle inspection and a new small-end bushing has been pressed in place and reamed to size, a new bearing insert is installed in the large end.

Arbors are installed in both ends of the connecting rod and a measurement is made, using a

Fig. 3B-17 *Method of checking a connecting rod for parallelism between its ends*

parallelism gage to check for a bent rod. This would be indicated if the two arbors are not exactly parallel.

With the arbors still installed, lay the rod across parallel blocks on a surface plate and check to see whether or not a feeler gage can be passed between the arbor and the block. If it can, it indicates that the rod is twisted. If the rod is twisted or bent beyond the limits specified, it must be replaced.

Fig. 3B-18   Method of checking a connecting rod for twist

When installing a new rod, match it to the rod on the opposite side of the engine within one-half ounce to minimize vibration.

g. Camshaft

Since the camshaft is responsible for opening the valves at the proper time, all of the lobes must be examined to see that they are not excessively worn. If any of the hydraulic valve tappet bodies were spalled or pitted, the lobe that operated that valve should be inspected for surface irregularities or feathering of the edges. If any such conditions are found, the camshaft should be rejected.

If the overhaul manual specifies that the lobes be tapered, a dimensional check should be made to indicate whether or not the amount of taper is within the limits specified.

If all of the lobes are in good condition, check the shaft for bends by supporting its end journals in V-blocks and checking the center journal for

run-out, and then check the journal diameters and compare them with the bearings in the crankcase for the proper fit.

h. Valve operating mechanism

Almost all modern horizontally opposed aircraft engines use hydraulic valve lifters to maintain zero clearance in the valve operating mechanism.

The valve tappet bodies that ride on the cam lobes are examined visually and dimensionally, and if they pass this, they are inspected by magnetic particle inspection and are then thoroughly demagnetized.

The hydraulic plunger assembly must never be magnetized as magnetization will prevent the steel check valve seating.

The plunger and cylinder are matched units and parts from one should never be interchanged with parts from another. After the part is thoroughly cleaned and visually checked for any chipped shoulders or evidence of other damage, it is given a leakage test by inserting the plunger in the cylinder and quickly depressing it. If it bounces back, it shows that the check valve is seating properly and the assembly is satisfactory

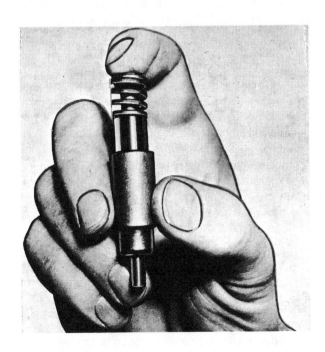

Fig. 3B-19   Method of checking a hydraulic valve lifter for proper operation

for reinstallation. If it does not bounce back, the valve is not seating and the unit must be replaced.

The pushrods must be inspected for straightness by rolling them across a flat surface plate, and the ball ends should be checked to be sure they are tight in the rod.

5. Reassembly

After all of the parts have been inspected and repaired as necesary, the engine is ready to be reassembled. The procedure here is typical, but the actual assembly is, naturally, done in strict accordance with the details in the manufacturer's overhaul manual.

a. Cylinders

Lubricate the valve stems with the recommended lubricant, and insert the valves in the valve guides. Place the cylinder over a post and hold the valves in place while you slip the valve springs and retainers over each valve stem. Using the proper spring compressor, compress the valve springs, and install the valve stem keys and any valve rotators that are used.

Install any of the intercylinder baffles or fin stabilizers that are required and slip the cylinder base seal around the skirt of the cylinder.

b. Pistons and rings

Lubricate the piston, the rings, and the wrist pin with the appropriate lubricant, and stagger the ring gaps in the way specified in the overhaul manual. Slip the wrist pin into one side of the piston and, using the proper ring compressor, compress the piston rings and slip the piston into the cylinder up to the wrist pin.

c. Crankshaft

All of the sludge plugs and expansion plugs must be installed in the reconditioned crankshaft, all of the counterweights properly assembled, and any gears that are attached to the shaft properly secured and safetied. The crankshaft can now be mounted in a buildup fixture to hold it vertically upright, while the connecting rods are lubricated and assembled to the shaft, using new bolts and nuts. The rods are now torqued and safetied. Be very sure when assembling the rods to the crank-

shaft that the numbers stamped on the rods are on the side of the crankshaft specified in the overhaul manual.

d. Crankcase

Place the crankcase halves on a flat work surface and install the main bearing inserts, being sure that the locking tangs or dowels are properly in place. Lubricate and install the hydraulic tappet bodies in one of the crankcase halves and then slip the camshaft in place after lubricating its bearing surfaces.

Now the main bearings may be lubricated and the crankshaft installed. The front oil seal is installed around the propeller shaft and a *very thin* layer of non-hardening gasket compound is applied to the outside mating surface of each half of the crankcase.

If the manufacturer recommends it, a very fine silk thread may be embedded in the gasket compound on one of the crankcase halves.

Very carefully, holding the tappet bodies in place, lower the crankcase half over the one in which the camshaft and crankshaft are installed, being very careful that the front oil seal is properly seated. Any special instructions regarding the seating of the main bearings must be observed. Install new nuts on all of the studs and through bolts and using torque hold-down plates over the cylinder pads, torque all of the fasteners to the specifications listed in the overhaul manual. Be very careful to use the exact torque and tightening sequence specified.

The crankcase may now be mounted in its vertical position in the buildup fixture, and the gears in the accessory case installed, with careful attention being paid to the timing of the cam gear to the crankshaft gear.

The torque hold-down plates may be removed and the cylinders slipped in place, installing the proper pushrods and pushrod housings with the appropriate seals.

The rocker arms and rocker shafts are lubricated and installed and the rocker shaft secured in the cylinder head.

e. Final assembly

The oil sump is installed on the crankcase and the magnetos are installed and timed. The carburetor or fuel injection system and all of the induction system is installed, as well as all of the various accessories such as the pumps, the generator or alternator, and any baffles or other devices needed for the engine to operate properly.

6. Testing

a. Test facilities

After an engine has been overhauled, it should be run in and tested either on a dynamometer or with a special test club propeller installed to absorb the power developed by the engine.

A sheet metal cooling air scoop should be installed on the engine to assure that all of the cylinders will be properly cooled, and a blast of cooling air should be directed across the generator or alternator.

A clean induction air filter of sufficient size to prevent any obstruction to the airflow must be installed and an exhaust system, either of the type used in the aircraft installation or one that is specially approved by the engine manufacturer for run-in, must be installed.

The engine must be equipped with at least the following test instruments, and because of the

Fig. 3B-20 After an aircraft engine has been overhauled, it should be given its run-in under controlled conditions.

general inaccuracy of the instruments installed in most aircraft, those on the aircraft should not be used.

An accurate tachometer
An accurate oil pressure gage
An accurate oil temperature gage
A cylinder head temperature gage that reads the temperature of the cylinder specified by the engine manufacturer
A water manometer for measuring the pressure inside the crankcase
An accurate ammeter

The test stand must be equipped with a fuel system that will deliver an adequate flow of clean fuel at the correct pressure.

b. Test preparation

If the engine is equipped with a pressure carburetor, it should sit with fuel in its fuel chambers for at least eight hours before the engine is run.

The engine must be pre-oiled. This may be done with a pressure oiler which forces oil through all of the oil passages in the engine until it runs out, indicating that all of the passages are full. But another way is quite satisfactory for some of the smaller engines: removing one spark plug in each cylinder and spinning the engine with the starter until pressure is indicated on the oil pressure gage.

c. Test run

After the engine is pre-oiled and you are sure that all of the controls are operating properly, you may start the engine and run it through the test schedule specified in the engine overhaul manual. A typical schedule is shown in Fig. 3B-21. Part of this run is an oil consumption test. After the engine has operated for a specific amount of time under controlled conditions, all of the oil is drained and weighed, and then it is put back into the engine and the oil consumption run is made. After the run is completed, the oil is again drained and weighed.

V. POWERPLANT TROUBLESHOOTING

Efficient troubleshooting is based on a systematic analysis of what is happening so you will be able to determine the cause of a malfunction. There is no magic in successful troubleshooting, but it is rather the application of logic and a thorough knowledge of the basics of engine operation.

If you are faced with a problem of deteriorating engine performance, for example, the first thing to do is to get all of the facts. Take nothing for granted, but ask questions of the pilot to determine such things as: Did the trouble come

MAJOR OVERHAUL TEST RUN		
TIME-MINUTES	RPM	T/C OUTLET PR."Hg (REFERENCE)
5	1200	
10	1500	
10	2100	
10	2600	
10	2800	
10	3000	
5	3200 ± 25 100% POWER	35.0 - 36.0
5	*3000 82.3% POWER - (280 BHP)	34.7 - 35.7
5	*2600 53.5% POWER - (182 BHP)	34.4 - 35.4
10	600 ± 25 IDLE COOLING PERIOD	
NOTE STOP ENGINE, DRAIN OIL, WEIGH OIL FOR OIL CONSUMPTION DETERMINATION AND REPLACE IN ENGINE.		

Fig. 3B-21   Typical run-in schedule for an aircraft engine to be used after a major overhaul

about suddenly or did he notice a gradual decrease in performance? Under what conditions of altitude, humidity, temperature, or power setting does this performance loss show up? Does temporarily switching to a single magneto cause any change in performance? What effect did leaning the mixture or applying carburetor heat have on the problem? Did switching from one fuel tank to another or turning on the fuel boost pump have any effect on the problem?

After getting all of the facts, the next step is to eliminate all of the areas that are not likely to cause the trouble. For example, if the magneto drop is normal, but there is a loss of power, the ignition system is more than likely *not* the problem.

We are including some typical engine problems that you are most likely to encounter, along with some of the most probable causes of the trouble. This list is definitely not all-inclusive, but it does give an idea of some of the most generally found troubles and areas that should be examined first.

ENGINE IDLES ROUGH

Problem	Fix
1. Idle mixture too rich or too lean.	1. Adjust idling mixture and idling RPM.
2. Plugged injector nozzles.	2. Remove and clean the nozzles in acetone or MEK.
3. Induction air leak.	3. Check entire induction system and listen for whistling sound when the engine is idling.
4. Cracked engine mount or defective shock mounts.	4. Repair engine mount or replace shock mounts.
5. Improperly adjusted fuel pressure.	5. Adjust it to the engine manufacturer's specifications.
6. Uneven cylinder compression.	6. Perform a differential compression check and remedy the cause of low compression.
7. Fouled spark plug or defective ignition harness.	7. Remove and check spark plugs and check the leads with a harness tester.

ENGINE IS HARD TO START

Problem	Fix
1. Too much fuel; engine is flooded.	1. Place mixture control in cut-off, open the throttle, and keep cranking.
2. Too much air.	2. Close throttle to about 1/4" from full-closed.
3. Impulse coupling not operating.	3. Turn engine slowly *with switch off*, and listen for impulse coupling snapping.
4. Fouled spark plug or defective ignition harness.	4. Remove and check spark plugs and check the leads with a harness tester.
5. Inoperative starting vibrator.	5. Check for proper voltage and listen for vibrator buzzing.
6. Improper magneto timing.	6. Check both the internal timing of the magneto and the timing of the magneto to the engine.

ENGINE WILL NOT DEVELOP RATED POWER

Problem	Fix
1. Induction system leak.	1. Check entire induction system, and listen for whistling sound when the engine is idling.
2. Exhaust system leak on turbocharged engine.	2. Check for presence of *feather* indications around exhaust system.
3. Restricted fuel-flow.	3. Check all filters in fuel system. Check for full opening of fuel selector valve. Check calibration of fuel-flow gage.
4. Restricted air inlet.	4. Clean or replace air filter. Check full travel of carburetor heat valve.
5. Improper grade of fuel.	5. Check color of fuel to be sure it conforms to that recommended for engine.
6. Carburetor controls not properly adjusted.	6. Check rigging on all controls to be sure all stops are reached.
7. Internal blockage of muffler.	7. Check the muffler for broken baffles and replace if necessary
8. Low compression.	8. Perform a differential compression check and a borescope inspection, if it is needed.

ENGINE WILL NOT DEVELOP FULL STATIC RPM

Problem	Fix
1. Restricted air inlet.	1. Clean or replace air filter. Check full travel of carburetor heat valve.
2. Propeller out of adjustment.	2. Check low-pitch angle and adjust if necessary.
3. Governor out of adjustment.	3. Adjust the governor, after making very sure that everything else is working properly first.
4. Internal blockage of muffler.	4. Check the muffler for broken baffles and replace if any are found.

LOW ENGINE OIL PRESSURE

Problem	Fix
1. Insufficient oil.	1. Carry the correct grade and amount of oil specified by the aircraft manufacturer.
2. Pressure relief valve out of adjustment.	2. Adjust to engine manufacturer's specifications.
3. Dirt or carbon under oil pressure relief valve.	3. Remove valve and clean it.
4. High oil temperature.	4. Check oil grade and quantity. Check oil cooler for air obstructions. Check thermostatic oil temperature control valve. Check for excessive blow-by past piston rings.
5. Oil pump inlet restricted.	5. Clean oil pick-up screen.

HIGH CYLINDER HEAD TEMPERATURE

Problem	Fix
1. Cooling baffles missing or broken.	1. Check all baffles and be sure they conform to those shown in the parts manual.
2. Partially plugged fuel injection nozzles.	2. Remove nozzles and clean them in acetone or MEK.
3. Induction system leak.	3. Check entire induction system, and listen for whistling sound when engine is idling.

HIGH OIL CONSUMPTION

Problem	Fix
1. Improper grade of oil.	1. Use only grade specified in the aircraft service manual.
2. New rings not properly seated.	2. Use only straight mineral oil for break-in period, unless otherwise specified by the engine manufacturer. Be sure that break-in is conducted in strict accordance with engine manufacturer's recommendations.
3. Worn or damaged piston rings or cylinder walls.	3. Perform a differential compression test, and if this indicates bad rings, pull a cylinder and verify with a dimensional check.
4. Worn valve guides.	4. Remove cylinder and dimensionally check the valve-to-guide clearance. Replace the guide if it is excessive.
5. Oil control rings improperly installed.	5. Remove cylinder and examine rings to be sure they are installed in strict accordance with the manufacturer's recommendations.

Section 1-III

Reciprocating Engine Removal and Replacement

I. REMOVAL

When it is necessary to remove an engine from an aircraft you should follow the instructions in the aircraft service manual. There are so many differences in the procedures to follow with different aircraft that we can only list the generalities here, but the things we will discuss are typical for a single-engine airplane.

Before starting to remove the engine, be sure the aircraft is placed in the hangar in a position where it will not need to be moved while the engine is out. The aircraft is balanced with the engine in place, and when this weight is removed, most nose-wheel-type airplanes are so tail heavy that the tail must be supported on a stand to prevent it dropping to the floor.

Have all of the needed hoist and slings ready and have the stands or other provisions to receive the propeller and the engine when they are removed. If either the engine or the propeller is to be shipped off to an overhaul facility, it is a good idea to have the shipping containers handy by the aircraft so they can be packed with a minimum of handling.

Turn all of the electrical switches and fuel selector valves Off and remove the cowling. Store the cowling in a location where it will not be in the way and where it will not be damaged during the time the aircraft is out of service.

As a safety precaution, disconnect the battery lead at the battery. Drain the fuel strainer and disconnect the fuel line from the carburetor or fuel injection system and drain all of the fuel

from these lines. Be sure to have a container to collect all of the fuel and oil from the lines you disconnect from the engine. While not a large quantity of fluid drains from these lines, there is enough to create a fire hazard or make a messy work area.

Drain the oil sump and oil cooler and disconnect the magneto P-leads, the primary leads that go to the ignition switch.

Remove the propeller and place it on the proper stand or in its shipping crate. When the propeller is removed, cover the crankshaft or propeller shaft flange to keep out any dirt or other contaminants.

With the major components disconnected, you are ready to disconnect the smaller fluid lines, controls, and electrical connections to the engine. The choice to disconnect these items at the engine or at the firewall depends to a large extent upon the ease with which you can get to them. Some lines can best be left on the engine until it is out of the aircraft, and others should be removed from the engine while it is still installed. Some of the typical lines and controls to be disconnected are:

Starter cable.

Cylinder head temperature thermocouple and the probes for the exhaust gas temperature gage.

Wire to the oil temperature probe or, if a direct indicating system is used, the capillary tube. Be sure that if a capillary and bulb is installed, the capillary tube is not kinked.

All of the electrical wires that connect to the alternator or generator.

All of the electrical wires to such components as landing gear warning horn switch on the throttle linkage.

Vacuum hoses to the vacuum pump.

Fuel supply line to the fuel pump.

Manifold pressure line to the intake manifold.

Fuel flow line to the fuel injection manifold valve if the engine is fuel injected.

Oil pressure gage line at the engine.

Engine primer lines.

Flexible ducts from the heater shroud to the cabin heater valve.

When you are sure that the engine is completely disconnected from the airframe, attach an engine hoist that you know to have sufficient capacity to the engine with the proper hoisting sling and lift the engine just enough to remove its weight from the engine mounts. Remove the engine mount bolts and slowly hoist the engine clear of the airframe and mount it either on the stand that will be used to carry it into the overhaul area or mount it in its shipping crate.

II. REPLACEMENT

When the engine is received from the overhaul shop, it must be built up ready for installation. It is normally easier to mount such accessories as the vacuum and hydraulic pumps, starter and alternator, and much of the wiring and lines on the engine before it is hoisted into place. Be sure that the serial numbers of all accesories are recorded before they are installed, as it is much easier to see the numbers before the components are mounted on the engine. Check to be sure that all safety wiring that is needed is properly done and is tight, with no sharp wire ends sticking out with which to cut yourself.

Hoist the engine in place just above the engine mounts and install the shock mounts and grounding strap. Lower the engine into place, and as it is being lowered, check to be sure that all of the controls and attachments are properly positioned and routed.

Install the exhaust system and all of the flexible heating ducts, and connect all of the fluid lines and electrical wiring connections. Install all of the accessories that have not been previously installed, and install the propeller and spinner, being very careful to follow the airframe manufacturer's instructions in detail on this critical installation.

Check the magneto primary wire connections for continuity, and install them in the magnetos, checking carefully that the correct P-lead goes to the magneto. After these wires are installed, it is important that the magneto switch remain in the Off position so the magnetos will both be grounded.

When connecting the controls for the engine components, be sure that the stop on the component is reached before the control in the cockpit reaches its stop. There should be a slight spring-back on each of the cockpit controls assuring that you have full travel of the control.

Put the recommended oil in the sump and install the cowling. Give the engine the ground run recommended by the manufacturer, and then uncowl it and thoroughly check it for indications of oil leaks, fuel leaks, and any loose or missing connectors. After you are sure that everything is in condition for a test flight, complete the paperwork for the installation and approve the aircraft for return to service.

It is important that the engine be run for the first few hours in the manner recommended by the engine manufacturer. Some operators have the mistaken notion that a freshly overhauled engine should be run at reduced power for the first few hours, but this has been proven false as it allows varnish to build up on the cylinder walls and will prevent the piston rings seating. Almost all engine manufacturers recommend that their engines be run for the first 25 hours or so with straight mineral oil before changing oil and putting in an ashless dispersant oil. The characteristics of these oils are discussed in some detail in the Lubrication section of this Integrated Training Program.

Book Two

I. INTRODUCTION

The progress made in flight has always been directly related to the development of the propulsion system. Around 1490 Leonardo da Vinci made sketches of an ornithopter-type flying machine that he envisioned could fly, but of course there was no propulsion system to power it. By the middle 1800's practical concepts of flight had been reached, but they were only good ideas and not practical machines because the then available powerplants were far too heavy for the amount of power they produced. When manned flight did become a reality in 1903, the engine used by the Wright brothers was extremely inefficient and heavy for the meager amount of power it produced.

Even the concept of energy exchange used by the reciprocating engine leaves much to be desired. Air and fuel must be drawn into the cylinders through an accurately timed valve, and work must be exerted by the piston to compress the fuel-air mixture, then it must be ignited by a timed high-voltage spark. Reciprocating engines are complex and heavy for the amount of power they produce. Early engines weighed more than four pounds for every horsepower they produced, and modern engines seldom weigh less than one pound per horsepower.

The very principle in which three strokes of the piston require work for every stroke that produces work limits the ultimate efficiency of the reciprocating engine. But engines using this principle have been available and they are usable for flight, so for the first four decades of flight this type of engine propelled all of our flying machines.

It was not that there were no other principles of propulsion available, because as early as the year 250 B.C. the Greek writer and mathematician Hero devised a steam engine he called the aeolipile that used jets of steam to rotate a sphere about its axis.

COURTESY OF TELEDYNE CONTINENTAL

Fig. 1A-1 Modern reciprocating engines will continue to power smaller aircraft, but they are inherently much heavier for the power they produce than a gas turbine engine.

1

Fig. 1A-2 Hero's aeolipile, conceived more than 200 years before Christ proved that power by reaction was possible.

Sir Isaac Newton applied the principle used by Hero to formulate his third law of motion that states that for every force action there is an equal and opposite force reaction. The steam within the aeolipile's sphere produced a force as it sprayed out the jets, and Newton concluded that it was a reaction from these jets that rotated the sphere. He went so far as to design a jet-propelled carriage in which a water-filled sphere was heated by a fire to produce steam which was directed out of a nozzle at the rear of the vehicle to propel it forward. Needless to say, this system never did replace the horse.

Propulsion by reaction appeared to be a good idea, but, as with many other good ideas, the hardware was lacking.

Airplanes are propelled by the reaction with the air. The reciprocating engine turns a propeller which by aerodynamic action forces a large mass of air backward, and as a result, the airplane is forced forward. In 1940 the Italian Caproni-Campini flew as a ducted fan type airplane in

which a 900 horsepower Isotta-Fraschini reciprocating engine mounted inside the fuselage drove a three-stage compressor. Fuel was injected downstream of this fan and ignited. This arrangement forced a high-velocity stream of air out behind the engine, but the weight of the engine made it an inefficient means of producing thrust. What was needed was an efficient way of producing the compressed air into which the fuel could be sprayed and burned.

Dr. Sanford Moss, in 1900, published a thesis on gas turbines which later led to the successful turbo supercharger that allowed our airplanes to reach the high altitudes needed for success in World War II.

Research done by Dr. Moss influenced Frank Whittle in the development of what became the first successful turbojet engine. To overcome the weight penalty caused by driving the compressor by a reciprocating engine as was done in the Caproni-Campini, Whittle drove a centrifugal compressor with a gas turbine in much the same way the compressor was driven in Dr. Moss's turbo supercharger.

The engine developed by Frank Whittle was a pure reaction type engine that was, in May of 1941, installed and flown in a Gloster model E28/39. The engine produced about one thousand pounds of thrust and propelled the aircraft at a speed in excess of 400 miles per hour.

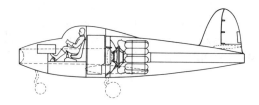

Fig. 1A-3 The Whittle turbojet engine powered the Gloster E28/39 that flew in 1941. This engine provided the breakthrough in technology that allowed the development of an entire new era in aviation.

While Whittle was developing the gas turbine engine in England, a German engineer, Hans Von Ohain, working with the Heinkle company, produced a jet engine that produced 1,100 pounds of thrust. This engine was used to power the Heinkel HE-178 that made a successful flight on August 27, 1939, to become recognized as the first practical flight by a jet propelled aircraft.

82   2

In the United States, research in jet propulsion lagged because our efforts were all being directed toward the development and production of high powered reciprocating engines. But in 1941, the General Electric Company of Schenectady, New York, was given a contract to research and develop a gas turbine engine. G.E. was chosen for this important project because of their extensive experience, both in turbines used for electrical power generation, and in the production of turbo superchargers.

The result of the contract with General Electric was the GE-1A engine, a centrifugal-compressor type engine which produced about 1,650 pounds of thrust. Two of these engines were used to power the Bell XP-59 which was first flown in 1942.

Fig. 1A-4 The Bell XP-59 which first flew in 1942 was the first American designed and built jet propelled airplane.

II. TYPES OF TURBINE ENGINES

Jet propulsion is a method of propulsion that uses the reaction produced by the acceleration of a fluid through an orifice, or nozzle, to move an object forward. In nature, the squid propels itself through the water by jet propulsion. It takes water into its body, and then, using its muscles, adds energy to the water and expels it in the form of a jet to force itself forward through the water.

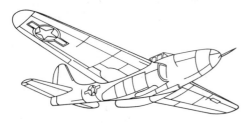

Fig. 2A-1 Many developments in technology were made by watching nature in action. The squid propels itself through the water by jet reaction in much the same way a turbojet engine propels an airplane.

A. Reaction Engines

Four types of reaction engines are used in aviation, and these are commonly lumped together and called "jet" engines. All produce thrust in the same way, by the reaction produced by accelerating a mass of air within the engine. These engines are: the rocket, the ramjet, the pulsejet, and the turbojet.

A rocket is a non-air-breathing engine that carries its oxygen within its fuel. There are two types of rockets in use; solid-propellant rockets and liquid-propellant rockets. Solid-propellant rockets use a solid fuel with an oxidizer mixed in it and formed into a special shape that allows the optimum burning rate. The propellant is formed into the rocket body and when it is ignited, it produces an extremely high velocity discharge of gas through the nozzle at the rear of the rocket body. Solid fuel rockets are used primarily as the propulsion system for some military weapons, and have been used to provide additional thrust for takeoff of heavily loaded aircraft. These booster rockets, called RATO, or rocket assisted takeoff units, attach to the aircraft structure and provide the needed boost for special-condition takeoffs.

Fig. 2A-2 RATO, or rocket assisted takeoff, devices are small, solid propellant rocket motors that may be attached to an airplane to provide additional thrust for high-altitude or overweight takeoff conditions.

Liquid-fuel rockets use a fuel and an oxidizing agent such as liquid oxygen. The two liquids are carried in tanks aboard the rocket and when they are mixed, the reaction is so violent that a tremendous amount of heat is generated and the resulting high velocity gas jet behind the rocket provides the thrust.

A ramjet engine is actually an athodyd, or Aero-Thermo Dynamic Duct. It is an air-breathing engine that depends upon a high forward velocity to compress the incoming air. Fuel is added to the air in the engine and ignited to provide the expansion and thus the acceleration of air that is needed to produce thrust. At present, ramjets are used only in weapons delivery systems for the military where is it possible to accelerate the vehicle to a high initial velocity by some means such as a rocket, then the ramjet can take over for sustained flight.

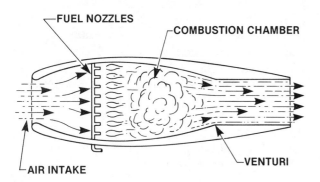

Fig. 2A-3   *The ramjet engine is able to produce thrust only after it is moving through the air at a high velocity.*

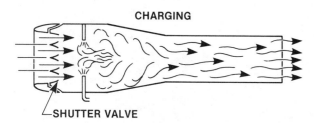

THE SHUTTER VALVES ARE OPEN AND AIR IS BEING DRAWN INTO THE COMBUSTION CHAMBER AND MIXED WITH FUEL.

(A)

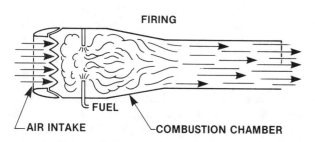

THE FUEL IS IGNITED AND BURNS. THE HEAT FROM THE BURNING FUEL EXPANDS THE AIR. THIS CLOSES THE SHUTTER VALVES AND ACCELERATES THE AIR LEAVING THE TAIL PIPE.

(B)

Fig. 2A-4   *Pulsejet engine*

Pulsejet engines are similar to ramjets except that the forward end of the duct is fitted with shutter-type valves. Air is forced into the combustion chamber where fuel is added to it and ignited. As the burning gases expand, the pressure from the expansion forces the shutters closed and the gasses exit the engine through the tail pipe. As they leave the venturi-shaped combustion chamber, they produce a low pressure that opens the shutter valves and pulls air in for another operational cycle. (See Fig. 2A-4.)

Pulsejet engines have an advantage over a ramjet because they can produce some static thrust; that is, they can produce thrust without having to be carried to a high initial velocity by some outside force. They are light in weight, simple and relatively inexpensive, but their loud operating noise, limited valve life, and relatively low speed have limited their practical use.

The gas turbine engine is by far the most practical of the reaction engines in use today. It has revolutionized aviation to the extent that today we find only a very few specialized applications for reciprocating engines in military aircraft, and almost no airliners are powered by reciprocating engines.

COURTESY OF WILLIAMS RESEARCH

Fig. 2A-5   *This small gas turbine engine takes air in through the forward fan, compresses it, adds energy from burning fuel and produces thrust as the expanded air is forced out of the tail pipe at an accelerated rate. Part of the energy released by the burning fuel is used to turn a turbine which drives the compressor.*

A gas turbine engine uses a large compressor to pull in air and compress it; then fuel is sprayed into the compressed air and ignited, and as the expanding gases leave the rear of the engine, they spin a turbine which drives the compressor. The energy added to the air by the burning fuel causes it to leave the engine at a higher velocity than it entered, and it is the difference between the exiting and the entering velocity of the air that produces thrust.

1. Gas turbine engines

The gas turbine engine has progressed from its infancy into a sophisticated family of engines, both for aviation applications and for commercial purposes as well. It is only logical that as the state of the art advances, there will be more types of gas turbine engines built to fit more applications. Today it is cost that keeps the turbine out of personal airplanes and automobiles, but cost always has a way of decreasing with increased utilization and new developments will surely bring cost into the range of practicality for many of the applications that are being missed today.

a. Turbojet engines

The turbojet engine as patented by Frank Whittle consisted essentially of an impeller-type compressor, an annular combustor, and a single-stage turbine. The modern turbojet engine is built with many variations, but the basic components are still the compressor, the combustor, and the turbine.

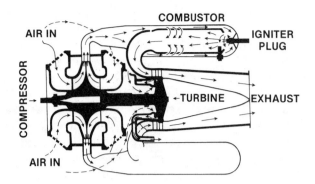

Fig. 2A-6 The Whittle engine of four decades ago had the same basic components as a modern gas turbine engine: an air inlet, a compressor, combustors or burners, and a turbine to drive the compressor.

A modern turbojet engine produces its thrust from the acceleration of the flow of hot gases. Air enters the engine inlet and flows into the compressor where its pressure is increased. Fuel is added in the combustor where it is ignited and burns, expanding the gases. As the expanded gases flow out of the engine, they pass through the turbine where part of their energy is given up to spin the turbine which drives the compressor. Energy that remains in the gases as they leave the tail pipe is in the form of velocity energy, and this produces the reaction we know as thrust. (See Fig. 2A-7.)

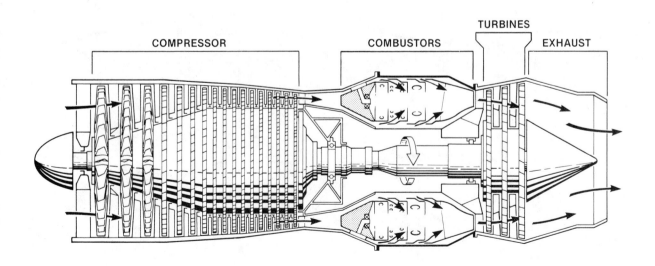

Fig. 2A-7 The modern turbojet engine consists of a turbine-driven compressor, combustors, or burners, and an aerodynamically shaped exhaust duct through which the expanded air exits at a high velocity.

5

b. Turbofan engines

A turbofan engine consists basically of a multibladed ducted propeller driven by a gas turbine. The fan has a compression ratio somewhere in the range of 2:1, or it produces a compression pressure of approximately two atmospheres. The ducted fan gives a turbofan engine cruise speed capabilities similar to those of a turbojet engine, yet its greater propulsive efficiency at low speeds gives an airplane equipped with turbofan engines far better short-field takeoff characteristics than one having turbojet engines.

The fan has a diameter and mass flow much less than that of a propeller, but it moves air from its convergent exhaust nozzle with a greater velocity. On the other hand, the fan discharge is much slower than the exhaust of a comparable turbojet engine, but it has a greater mass flow.

There are several fan configurations. The fan can be bolted directly to the compressor and rotate at the same speed, or it can be connected through a reduction gear system to the compressor. The fan on some engines is driven by a separate turbine and rotates independently of the

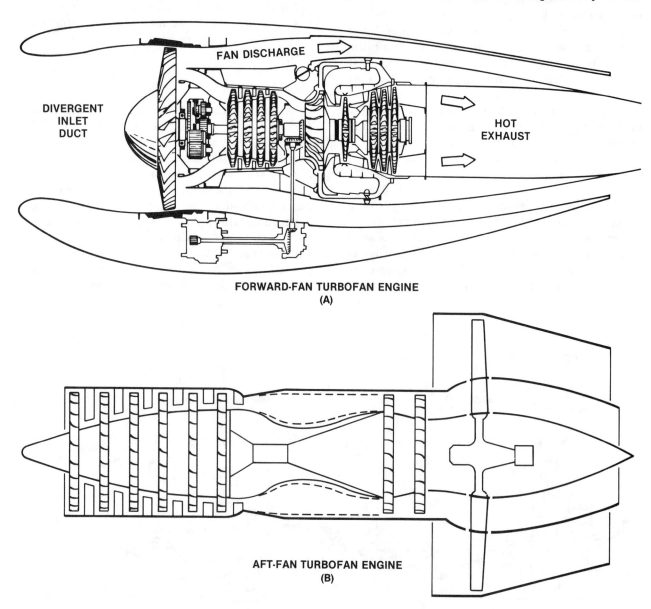

FORWARD-FAN TURBOFAN ENGINE
(A)

AFT-FAN TURBOFAN ENGINE
(B)

Fig. 2A-8 (Top) A forward-fan turbofan engine uses a relatively large diameter ducted fan for its first stage of compression. The forward fan produces thrust and provides additional air to the first stage of the low-pressure compressor. (Bottom) The aft-fan turbofan engine has its fan blades on the aft turbine.

6

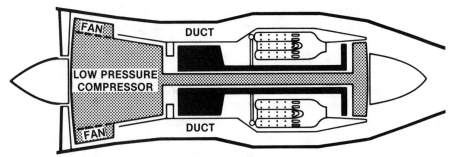

FORWARD-FAN ENGINE WITH LONG DUCT AND MIXED EXHAUST
(A)

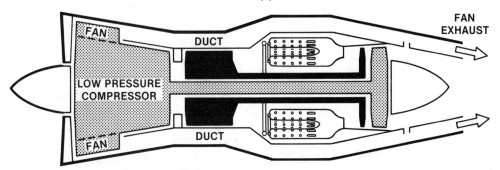

FORWARD-FAN ENGINE WITH LONG DUCT AND UNMIXED EXHAUST
(B)

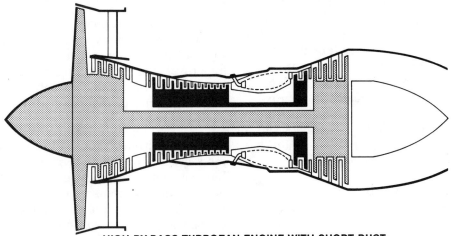

HIGH BY-PASS TURBOFAN ENGINE WITH SHORT DUCT
(C)

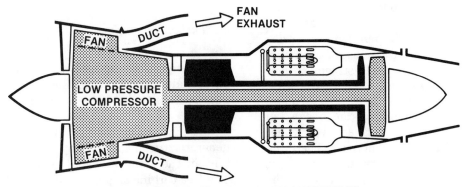

FORWARD-FAN ENGINE WITH SHORT DUCT
(D)

Fig. 2A-9 Types of turbofan engines

7

compressor, and in some engines the fan is mounted in the turbine section as an extension of the turbine wheel blades. An engine with the fan in the turbine section is called an aft-fan engine, and those with the fan in front are called forward-fan engines. The aft-fan configuration is not a popular design today because the fan does not contribute to the compressor pressure ratio.

Turbofans are usually grouped into three classifications depending upon the amount of air the bypass around the gas generator.

In a low by-pass engine, the fan and compressor section utilize approximately the same mass airflow, but the fan discharge will generally be slightly greater than that of the compressor. The fan discharge air may be ducted directly overboard from a short fan duct, or it may pass along the entire length of the engine in what is called a long fan duct. In either case the end of the duct has a converging discharge nozzle to produce a velocity increase and reactive thrust.

In a fully ducted fan engine, the hot and cold air streams generally mix before they are discharged into the atmosphere and this mixing helps attenuate, or lessen, the noise (Fig. 2A-9A). The air in the core engine is compressed, burned, and discharged in the normal manner and the thrust ratio of the two gas streams is approximately 1:1, with each stream in an engine such as the Pratt and Whitney JT8D delivering between 8,000 and 10,000 pounds of thrust, depending upon the particular model of the engine.

A medium, or intermediate, by-pass engine has an airflow by-pass ratio of between 2:1 and 3:1, and has a thrust ratio that is approximately the same as its by-pass ratio. The fan used on these engines has a larger diameter than that on a low by-pass engine of comparable power and its diameter is determined by both the by-pass ratio and the thrust output of the fan compared with the thrust obtained from the core engine. This latter ratio is often called the cold-stream to hot-stream ratio.

A high by-pass turbofan engine has a fan ratio of 4:1 or greater and has an even wider diameter fan in order to move more air. The Pratt and Whitney JT9D engine represents the current state of the art in large engine design for the jumbo-jet series airplanes. It has a by-pass ratio of 5:1 with 80% of the thrust provided by the fan and only 20% by the core engine (Fig. 2A-10).

The turbofan engine has become the most widely used engine type because it offers the best fuel economy. This economy is obtained by increasing the total mass airflow while decreasing the velocity of the hot exhaust wake. The thrust produced by the Pratt and Whitney JT3D, one of the first fan engines, is approximately 50% greater than the thrust produced by the core turbojet engine while the fuel flow remains basically the same.

B. Torque Turbine Engines

1. Turboshaft engines

A gas turbine engine that delivers power through a shaft to operate something other than a propeller is referred to as a turboshaft engine. These are widely used in such industrial applications as electrical power generating plants and surface transportation systems, while in aviation, turboshaft engines are used to power many modern helicopters.

The turboshaft power takeoff may be coupled to and driven directly by the turbine that drives the compressor, but it is more likely to be driven by a turbine of its own. Engines using a separate turbine for the power takeoff are called free-turbine engines or free power turbine-type turboshaft engines.

A free-turbine turboshaft engine has two major sections, the gas generator and the free turbine section. The function of the gas generator is to produce the required energy to drive the free turbine system, and it extracts about two thirds of the energy available from the combustion process leaving the other one third to drive the free power turbine (Fig. 2A-11).

2. Turboprop engines

A turboprop engine is similar in design to a turboshaft engine except that it uses a reduction gear system to drive the propeller.

The propeller may be driven from the gas generator turbine, or it may use its own free turbine in the same way as the turboshaft engine. The free turbine arrangement allows the propeller driving turbine to seek its optimum speed while the compressor turbine operates at the best speed for the compressor efficiency.

Fig. 2A-10 The Pratt and Whitney JT9D is a high by-pass turbofan engine used on our largest airliners.

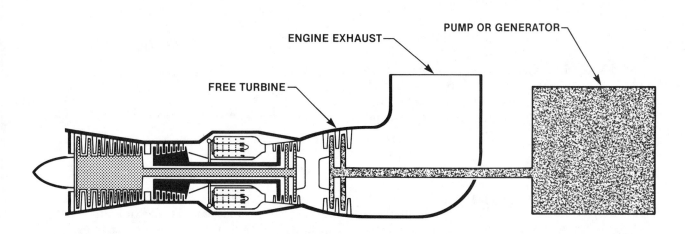

Fig. 2A-11 Industrial turbine engines are usually of the free-turbine type in which the free turbine drives an electrical generator or some form of pump.

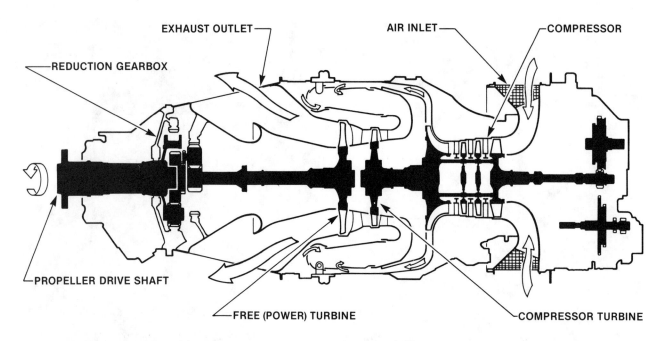

EXHAUST OUTLET

AIR INLET

COMPRESSOR

REDUCTION GEARBOX

PROPELLER DRIVE SHAFT

FREE (POWER) TURBINE

COMPRESSOR TURBINE

Fig. 2A-12   The Pratt and Whitney of Canada PT6 turboprop engine extracts most of the energy from the burning fuel with its compressor turbine and its free (power) turbine.

The basic difference between a turbojet engine and a turboprop engine, other than the reduction gearing, is that turboprop engines normally have more stages of turbines. These additional stages drive the compressor and accessories as well as the propeller. The total thrust produced by a turboprop engine is the sum of the propeller thrust and the exhaust nozzle thrust, with about 10% to 25% of the total thrust contributed by the exiting exhaust gases.

III. PRINCIPLES OF PHYSICS

Before we look at the way a turbine engine produces its thrust, let's review a few pertinent principles of physics. All of these principles are covered in the General Section of this Integrated Training Program, but here we will review those principles that are most directly related to jet propulsion.

A.   Force

Force is defined as the capacity to do work. But it may also be expressed as a vector quantity that produces acceleration in a body in the direction of force application. In turbine engine problems we often consider force to be produced by a fluid, so we can use the formula:

$$F = P \times A$$

F = Force measured in pounds

P = Pressure measured in pounds per square inch

A = Area on which the force is acting, measured in square inches.

B.   Work

Mechanical work is produced when a force acts on a body and causes it to move through a distance. In the English system of measurement, work may be found by the formula:

$$W = F \times D$$

W = Work in foot-pounds

F = Force in pounds

D = Distance through which the force acts, in feet

C.   Power

Power is the time rate of doing work and it is expressed in foot-pounds per minute, or foot-pounds per second. It is found by the formula:

$$P = \frac{F \times D}{t}$$

P = Power in foot-pounds per minute, or in foot-pounds per second

F = Force in pounds

D = Distance through which the force acts, in feet

t = Time in minutes or in seconds

Horsepower is the most commonly used term for mechanical power, and it has been standardized in the English system as 33,000 foot-pounds of work done in one minute, or 550 foot-pounds of work done in one second. Foot-pounds per minute of work can be converted into horsepower by dividing by 33,000 and foot pounds per second can be converted by dividing by 550.

D. Velocity

Velocity is similar to speed and is often expressed in the same terms such as feet per second, or miles per hour, but to be technically correct, velocity must include direction as well as magnitude. We will consider this in much more detail when we study the airflow through a compressor and a turbine where we will use vector quantities to express both the direction and the magnitude of velocity.

E. Acceleration

Acceleration is an increase in velocity with respect to time, and is found by the formula:

$$A = \frac{V_2 - V_1}{t}$$

A = Acceleration in feet per second, per second (feet per second^2)

V_1 = Initial velocity in feet per second

V_2 = Final velocity in feet per second

t = Time in seconds

Deceleration is decrease in velocity with respect to time and it is found in the same way as acceleration, but by subtracting the final velocity from the initial velocity.

F. Energy

Energy is the capacity for performing work, and in the gas turbine engine this energy produces both motion and heat. There are two forms of energy we must consider; potential and kinetic.

Potential energy exists within a body because of its configuration or its position. In a fluid, potential energy is often stored in the form of pressure.

Kinetic energy is possessed by a body because of its motion.

G. Newton's Laws as They Apply to Gas Turbine Engines

Newton's first law of motion illustrates the effects of inertia on a body. It states that a body at rest will tend to remain at rest, and a body in motion will continue in uniform motion in a straight line unless it is acted upon by some outside force.

Newton's second law deals with acceleration and is the one that explains, to a great extent, the thrust produced by a turbine engine. The acceleration produced within a mass by the addition of a given force is directly proportional to the force and is inversely proportional to the mass. A propeller accelerates a large mass of air, but it imparts a relatively small change in velocity, while the exhaust gas stream from a turbojet engine has a relatively small mass, but the acceleration that has taken place within the engine is large. Both types of acceleration produce thrust.

Newton's third law states that for every force action there is an equal and opposite force reaction. When the turbojet engine, or an engine turning a propeller accelerates a mass of air backward, an equal amount of force is produced that pushes the aircraft forward.

H. Mass and Weight

It is easy to think of mass and weight as being the same, and they are often used interchangeably, but actually there is a difference. Mass is the amount of matter in a body and it is the same at any location on the earth's surface or in space. But weight is the attraction of gravity on a mass. The mass of an object may be found by dividing the weight of the object, in pounds, by the acceleration due to gravity. On the surface of the earth the attraction between the earth and an object will cause the object to accelerate at a rate of 32.2 feet per second each second it is acted on. In a formula, this is expressed:

$$M = \frac{W}{g}$$

M = Mass of the object

W = Weight, expressed in pounds

g = Gravitational effect (32.2 feet per second^2)

I. Bernoulli's Principle

Bernoulli's principle deals with pressure and velocity in moving fluid. Pressure is changed in a gas turbine engine by adding or removing heat, by changing the number of molecules present, or by changing the volume in which the gas is contained. The first two of these changes will be mentioned later in our discussion of the compressor and combustor, but understanding the effect of changing the volume will help us better understand the energy release cycle used by the gas turbine engine.

Bernoulli's principle is based on the fact that air acts as an incompressible fluid when it flows at a subsonic rate. When a liquid or gas flows at a constant rate through a duct, the sum of the pressure (potential energy) and the velocity (kinetic energy) is constant. Another way of saying this is that if the total energy remains constant, as the pressure increases, the velocity will decrease proportionally. In other words, the velocity of the flow is inversely proportional to its pressure.

To understand Bernoulli's principle, consider that air flowing through a duct has both potential and kinetic energy. This may be also be thought of as static pressure and pressure caused by velocity. When this air flows through a section of the duct that has a divergent, or widening, shape, its kinetic energy (velocity) will decrease as the air spreads out radially, and since, by definition, the total energy remains constant, the mass flow rate of the air will be unchanged. The potential energy (pressure) must increase in proportion to the decrease in kinetic energy. If we measure the static pressure inside the duct, we will find that the pressure in the straight portion of the duct is lower than it is in the divergent portion.

If the duct converges, or becomes smaller, an airstream at a constant flow rate must speed up. The kinetic energy of the airstream increases and

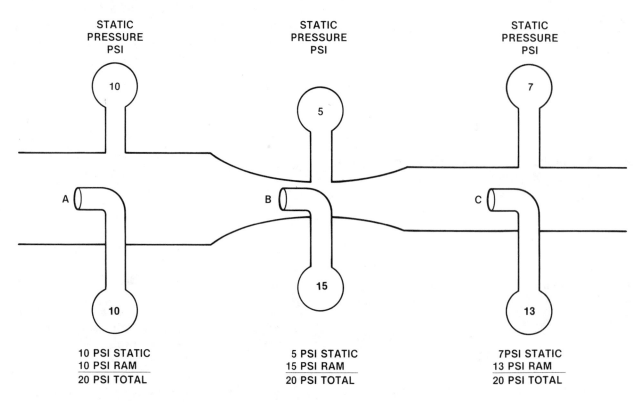

Fig. 3A-1 *According to Bernoulli's principle, if the total pressure in an airstream remains constant, an increase in the kinetic energy (velocity) of the air in the smallest portion of the tube will cause a decrease in its potential energy, or static pressure.*

its potential energy, as seen by the static pressure, decreases.

The total pressure of an airstream is the sum of the static pressure plus the ram pressure. In Fig. 3A-1 we see that the flow rate as measured by the total pressure is constant throughout the duct. At point B, the area is smallest and the velocity is greatest. The ram pressure is highest, but the static pressure is lowest. At point C, the duct has diverged, that is, has become larger, and the ram pressure has decreased. Notice that at all three locations, the total pressure is the same.

J. Diffusers and Diffusion

Throughout our study of jet engine principles, we encounter diffusers and diffusion. There are two types of diffusion we must be familiar with; subsonic and supersonic.

A subsonic diffuser in a jet engine is an air or gas passage that has a divergent, or expanding shape. When gas flows at a subsonic rate through a divergent passage, diffusion takes place as the duct progressively increases its cross-sectional area. Kinetic energy in the moving gas is converted into potential energy, or static pressure, energy.

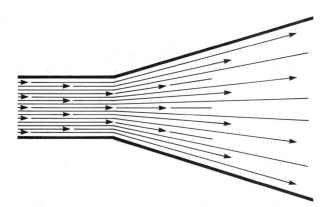

Fig. 3A-2 When gas flows at a subsonic rate through a divergent duct, some of the kinetic energy or velocity is converted into potential energy or static pressure.

Supersonic diffusion takes place in a *convergent* duct. When the air flowing at a supersonic rate passes through a converging duct, one whose cross-sectional area decreases, rather than speeding up, the air will compress or increase its static pressure and become more dense. The velocity of supersonic air passing through a con-

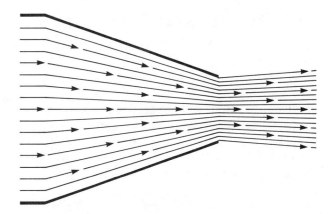

Fig. 3A-3 Supersonic airflow through a converging duct will compress rather than speed up, and its static pressure as well as its density will increase.

verging duct will decrease, and its static pressure will increase.

K. Sonic Choking

Another concept we frequently encounter is that of sonic choking. When air or gas flows at subsonic rate through a converging duct, or nozzle, its velocity increases, as we have just seen. But if it continues to speed up until it reaches a speed of Mach one, a shock wave will form that prevents further acceleration. The pressure of the air flowing through a choked nozzle will increase but since the airflow cannot speed up any more, this pressure energy cannot be converted into velocity energy, and it will remain as an increased pressure after the air passes through the nozzle. We will see this principle applied to thrust that is produced by a choked exhaust nozzle.

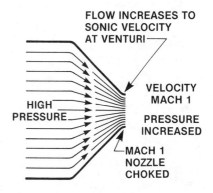

Fig. 3A-4 When subsonic air flows through a converging duct, it speeds up until it reaches Mach 1. Since it cannot speed up more, the pressure behind the choked nozzle increases.

IV. PRINCIPLES OF GAS TURBINE OPERATION

A. Energy Release Cycle

There are two cycles of energy release used in aircraft engines. Both cycles describe the sequence of events that takes place when the chemical energy in the fuel is released to become heat energy and then to become mechanical energy.

The Otto cycle describes the events that occur when energy is transformed in a reciprocating engine. This is a constant volume cycle because the energy added to the air causes almost no change in volume. All of the events in the Otto cycle (intake, compression, ignition, power, and exhaust) take place at the same location, inside the engine cylinder, but they take place at different times.

The Brayton cycle describes the events that take place in a turbine engine as the fuel releases its energy. When energy is added, the air remains at a relatively constant *pressure*, but its volume is increased which increases the velocity of the air as it leaves the engine.

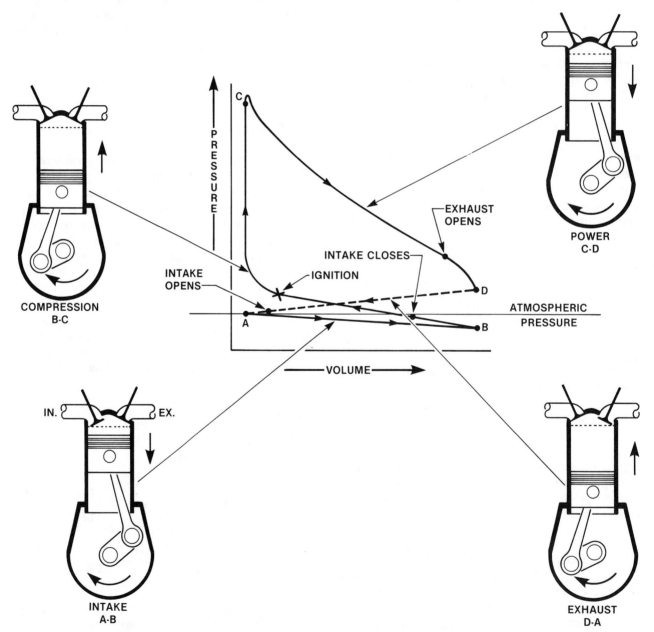

Fig. 4A-1   The Otto cycle of energy release is used in the reciprocating engine. All of the events take place in the same location, but at different times.

The four continuous events shown on the pressure-volume graph of Fig. 4A-2 are: intake, compression, expansion, and exhaust; the same events that take place in a reciprocating engine.

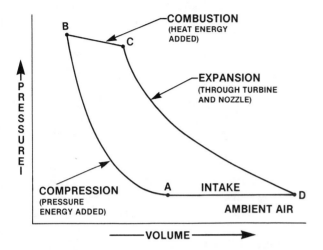

Fig. 4A-2 *The Brayton cycle of energy release is used in gas turbine engines. All of the events take place at the same time, but at different locations within the engine.*

The air entering the inlet duct of the engine is at essentially ambient pressure. The portion of the curve between A and B shows that the air pressure rises from ambient as the compressor does work on the air. It increases its pressure and decreases its volume. When energy is added to the air from the fuel burned in the combustors, the pressure remains relatively constant, but you will notice that the volume increases greatly. It is because of this characteristic that the Brayton cycle is called a constant pressure cycle.

When the heated air leaves the combustion chamber, it passes through the turbine where the pressure drops, but its volume continues to increase as we see in the section of the curve between points C and D. The burning gases have heated the air and expanded it greatly, and since there is little opposition to the flow of these expanding gases as they leave the engine, they are accelerated greatly. Some of the energy is extracted from the exiting gases by the turbine and this is used to drive the compressor and the various engine accessories (Fig. 4A-3).

After the air leaves the turbine, it passes through a converging exhaust system where the pressure continues to drop to ambient and the velocity continues to increase. In this cycle, work is accomplished by increasing the velocity of the air as it passes through the engine.

B. Thrust Calculations

Thrust is produced by a turbojet engine as air is taken into the engine and expanded by adding energy from burning fuel. The expanded gases are accelerated as they leave the engine, and it is this change in the velocity of the air between the time it enters the engine and the time it leaves that produces the thrust.

1. Gross thrust

Gross thrust, represented by the symbol F_g, is the thrust produced when the engine is not in motion. The acceleration of the gas within the engine is the difference in velocity between the air in the inlet duct, and the air as it leaves the ex-

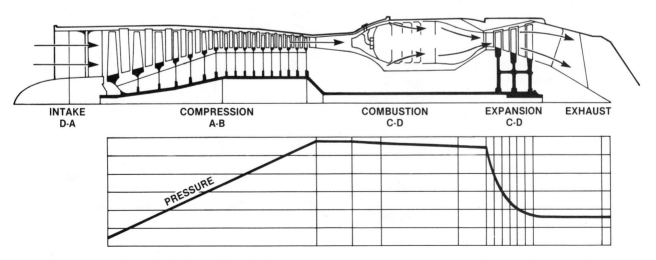

Fig. 4A-3 *Location of the energy-release events in a gas turbine engine.*

haust nozzle. We can calculate the gross thrust by the formula:

$$F_g = M_s \frac{(V_2 - V_1)}{g}$$

F_g = Gross thrust in pounds

M_s = Mass airflow in pounds per second

V_1 = Initial velocity of the air in feet per second

V_2 = Final velocity of the air in feet per second

g = Constant for acceleration due to gravity 32.2 feet per second^2.

Let's see the way this formula works: assume that there is a business jet airplane on the runway with the engine producing takeoff power. However, the airplane has not yet begun to roll. The compressor is pulling 50 pounds of air per second through the engine; this is M_s. Since the airplane is not moving, V_1 is zero, but we will assume the exhaust velocity, V_2, to be 1,300 feet per second.

$$F_g = M_s \frac{(V_2 - V_1)}{g}$$

$$= 50 \frac{(1,300 - 0)}{32.2}$$

$$= 2,018.6 \text{ pounds}$$

2. *Net thrust*

When an aircraft is flying, its inlet air has an initial momentum and the velocity change across the engine will be greatly reduced.

We can consider the same airplane whose gross thrust we have just computed as flying at 500 miles per hour (734 feet per second). Its net thrust can be found by using the same formula, only this time there is an initial velocity.

$$F_n = M_s \frac{(V_2 - V_1)}{g}$$

$$= 50 \frac{(1,300 - 734)}{32.2}$$

$$= 50 \frac{566}{32.2}$$

$$= 878.9 \text{ pounds}$$

3. *Thrust with a choked nozzle*

Some gas turbine engines are fitted with a choked exhaust nozzle. Pressure in the exhaust duct pushes the gas with such force that it reaches the speed of sound and cannot be further accelerated. The pressure at the nozzle opening does not return to ambient, but it remains somewhat higher than ambient. The pressure at the exhaust nozzle opening produces additional thrust because of the pressure differential across the area of the exhaust nozzle. When we studied Bernoulli's principle, we saw that total pressure is the sum of the static and the ram pressure, and if a body of gas is accelerated without the addition of energy, its static pressure will decrease. But even when energy is added to accelerate the gas it can only be accelerated up to the speed of sound.

There are two types of energy inside the tail pipe: energy from the flow, which in this case acts rearward, and energy from the internal pressure which acts in all directions. When the flow is unchoked, only the flow energy of the gas is significant in creating thrust because pressure energy decreases in proportion to its velocity increase through the nozzle. But if the nozzle is small enough that the flow chokes, the internal energy will build up and a second thrust factor called pressure thrust is produced in addition to the reactive thrust caused by the flowing gases. We can illustrate this using the same airplane we have previously worked with. The airplane is still flying at 500 miles per hour (734 feet per second), only now, the exhaust velocity has been increased to 1,700 feet per second and the pressure at the exhaust nozzle is 11.5 pounds per square inch, absolute. The area of the exhaust nozzle is 130 square inches and the ambient pressure at 25,000 feet, where the airplane is flying, is 5.5 psi, absolute.

The formula we will use is:

$$F_n = M_s \frac{(V_2 - V_1)}{g} + [A_j (P_j - P_{am})]$$

F_n = Net thrust in pounds

M_s = Mass airflow in pounds per second

V_2 = Final velocity of the air in feet per second

V_1 = Initial velocity of the air in feet per second

g = Constant for acceleration due to gravity (32.2 feet per second^2)

A_j = Area of exhaust nozzle in square inches

P_j = Absolute pressure at the exhaust nozzle in psia

P_{am} = Ambient air pressure in psia

$$F_n = M_s \frac{(V_2 - V_1)}{g} + [A_j (P_j - P_{am})]$$

$$= 50 \frac{(1,700 - 734)}{32.2} + [130 (11.5 - 5.5)]$$

$$= 2,280 \text{ pounds}$$

C. Thrust Distribution

The rated thrust on an engine may be calculated by subtracting the sum of all of the rearward forces within the engine from the sum of all of the forward forces within the engine. The compressor, diffuser, combustor and exhaust cone exit areas all exert forward forces while both the turbine and the tail pipe exit areas exert a rearward force.

If the outlet of a particular section exerts more force than is present at its inlet, there is a forward pushing force, but if the inlet of a section (the outlet of the preceding section) exerts more force than is present at its exit, there is a rearward pushing force.

We can use a hypothetical engine to compute the thrust at the various stations and then we will

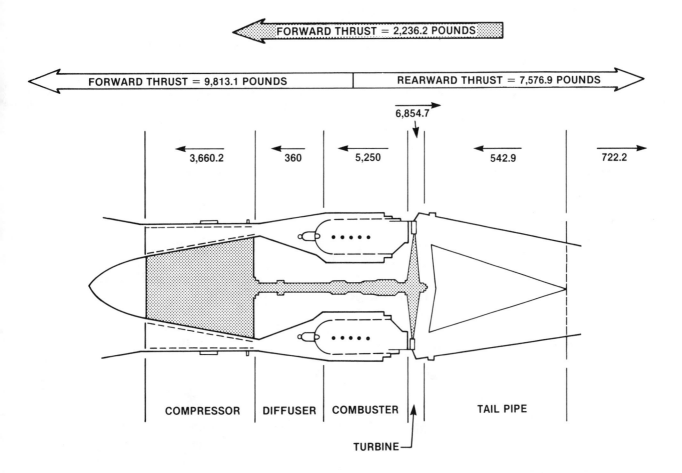

FORWARD THRUST = 2,236.2 POUNDS

FORWARD THRUST = 9,813.1 POUNDS

REARWARD THRUST = 7,576.9 POUNDS

6,854.7

3,660.2   360   5,250   542.9   722.2

COMPRESSOR   DIFFUSER   COMBUSTER   TAIL PIPE

TURBINE

Fig. 4A-4   The sum of all of the forward forces produced in an engine, less the sum of all of the rearward forces, is the resultant force.

17

see that the sum of the forces produced by all of the stations is equal to the total thrust of the engine.

These are the pertinent specifications of our example engine:

Initial velocity = zero

Mass flow of air through the engine = 29 pounds per second

Pressure at the compressor outlet = 55 psig

Area of the compressor outlet = 60 square inches

Velocity at the compressor outlet = 400 feet per second

Pressure at the diffuser outlet = 56 psig

Area of the diffuser outlet = 66 square inches

Velocity at the diffuser outlet = 360 feet per second

Pressure at the combustor outlet = 52 psig

Area of the combustor outlet = 160 square inches

Velocity at the combustor outlet = 1,055 feet per second

Pressure at the turbine outlet = 11 psig

Area of the turbine outlet = 170 square inches

Velocity at the turbine outlet = 605.7 feet per second

Pressure at the exhaust cone outlet = 12 psig

Area of the exhaust cone outlet = 202 square inches

Velocity at the exhaust cone outlet = 593.4 feet per second

Pressure at the tail pipe exit = 5 psig

Area of the tail pipe exit = 105 square inches

Velocity at the tail pipe exit = 1,900 feet per second

1. *Thrust at the compressor outlet*

The compressor section exerts a forward force because there is a greater pressure at its discharge than there is at its inlet where the pressure is zero. A forward force is exerted on the blades, the vanes, and on the outer case by the internal gas pressure buildup at the compressor discharge area.

We can see this by the formula:

$$F_g = (A \times P + \frac{M_s \times V}{g}) - I$$

F_g = Gross thrust of the section, in pounds

A = Area of the section in square inches

P = Pressure in pounds per square inch, gage

M_s = Mass airflow in pounds per second

V = Velocity change in feet per second $(V_2 - V_1)$

g = Gravitational constant, 32.2 feet per second^2

I = Initial pressure force in pounds

$$F_g = (60 \times 55 + \frac{29 \times 400}{32.2}) - 0$$

$$= (3,300 + 360.2) - 0$$

$$= 3,660.2 \text{ pounds, with the force acting forward}$$

A net forward thrust of 3,660.2 pounds occurs here because the compressor outlet is creating 3,660.2 pounds of force and there is no thrust at the inlet of the compressor.

2. *Thrust at the diffuser outlet*

The force present at the diffuser inlet is the same as the compressor outlet (3,660.2 pounds). In our example, the Area of the diffuser outlet is 66 square inches, the Pressure at the outlet is 56 psi gage, the Mass of the airflow is 29 pounds per second, the Velocity is 360 feet per second, and the Initial force is 3,660.2 pounds.

$$F_g = (A \times P + \frac{M_s \times V}{g}) - I$$

$$= (66 \times 56 + \frac{29 \times 360}{32.2}) - 3,660.2$$

$$= (3,696 + 324.2) - 3,660.2$$

$$= 4,020.2 - 3,660.2$$

$$= 360.2 \text{ pounds forward thrust}$$

The net forward thrust of 360.2 occurs because the diffuser outlet is creating 4,020.2 pounds of thrust or force and the diffuser inlet (same as the compressor outlet) is only creating 3,660.2 pounds of force or thrust.

3. Thrust at the combustor outlet

The force present at the inlet of the combustor is the same as that leaving the diffuser, or 4,020.2 pounds.

$$F_g = (A \times P + \frac{M_s \times V}{g}) - I$$

$$= (160 \times 52 + \frac{29 \times 1,055}{32.2}) - 4,020.2$$

$$= (8,320 + 950.2) - 4,020.2$$

$$= 9,270.2 - 4,020.2$$

$$= 5,250 \text{ pounds forward thrust}$$

The forward thrust produced by the combustor is 5,250 pounds. We find this by subtracting the 4,020.2 pounds of thrust that already existed at the inlet of the combustor from the total of 9,270.2 pounds at the combustor outlet.

4. Thrust at the turbine outlet

The procedure here is exactly the same as we have used previously.

$$F_g = (A \times P + \frac{M_s \times V}{g}) - I$$

$$= (170 \times 11 + \frac{29 \times 605.7}{32.2}) - 9,270.2$$

$$= (1,870 + 545.5) - 9,270.2$$

$$= 2,415.5 - 9,270.2$$

$$= -6,854.7 \text{ pounds acting rearward}$$

The turbine *extracts* energy from the gases, or it produces a rearward thrust of 6,854.7 pounds. The thrust at the turbine outlet is 2,415.5 pounds because 9,270.2 pounds of forward thrust was at the turbine inlet.

5. Thrust at the exhaust cone outlet

The force present at the inlet of the exhaust cone is that which was at the outlet of the turbine, and is 2,415.5 pounds. Now, we can find the thrust at the exhaust cone outlet.

$$F_g = (A \times P + \frac{M_s \times V}{g}) - I$$

$$= (202 \times 12 + \frac{29 \times 593.4}{32.2}) - 2,415.5$$

$$= (2,424 + 534.4) - 2,415.5$$

$$= 2,958.4 - 2,415.5$$

$$= 542.9 \text{ pounds, acting forward}$$

The exhaust cone outlet area forms a divergent duct which has a thrust, or force, of 2,958.4 pounds. This is 542.9 pounds more than the thrust at its inlet.

6. Thrust at the tail pipe

This is the rear of the engine and the thrust that exists here is a rearward thrust.

$$F_g = (A \times P + \frac{M_s \times V}{g}) - I$$

$$= (5 \times 105 + \frac{29 \times 1,900}{32.2}) - 2,958.4$$

$$= (525 + 1,711.2) - 2,958.4$$

$$= 2,236.2 - 2,958.4$$

$$= -722.2$$

The tail pipe produces a rearward thrust of −722.2 pounds because it increases the velocity of the exiting gases at the expense of pressure.

7. Resultant thrust

If we add up all of the forward and all of the rearward thrust values, we see that this engine has a resultant forward thrust of 2,236.2 pounds.

Compressor	3,660.2 pounds	
Diffuser	360.0 pounds	
Combustor	5,250.0 pounds	
Exhaust cone	542.9 pounds	
Turbine		−6,854.7 pounds
Tailpipe		− 722.2 pounds
Total	9,813.1 pounds	−7,576.9 pounds

There is a total forward thrust of 9,813.1 pounds and a total rearward thrust 7,576.9 pounds. This gives us a resultant forward thrust of 2,236.2 pounds.

The total thrust we have found by this procedure is the same as we get for the total engine by using the formula:

$$F_g = (A \times P) + \frac{M_s \times V}{g}$$

$$= (105 \times 5) + \frac{29 \times 1,900}{32.2}$$

$$= 525 + 1,711.2$$

$$= 2,236.2 \text{ pounds, forward}$$

D. Fan Engine Thrust

Thrust calculations of a turbofan engine can be computed in the same way as for a turbojet except that the hot and cold stream nozzle thrust values are figured separately and then added together.

We can find the thrust produced by a typical business jet airplane fitted with a turbofan engine by using this typical information: Both the fan and the exhaust nozzle are unchoked. The fan discharge velocity is 800 feet per second, and the discharge from the exhaust nozzle is 1,000 feet per second. The mass airflow through both the fan and the engine is 50 pounds per second. We can find the total thrust produced under static conditions by using this formula:

$$F_g \text{ (total)} = F_g \text{ (engine)} + F_g \text{ (fan)}$$

$$F_g = M_s \frac{(V_2 - V_1)}{g} + M_s \frac{(V_2 - V_1)}{g}$$

$$= 50 \frac{(1,000 - 0)}{32.2} + 50 \frac{(800 - 0)}{32.2}$$

$$= 1,552.8 + 1,242.2$$

$$= 2,795 \text{ pounds of forward thrust}$$

You will notice that in none of these computations have we mentioned humidity as a factor influencing the production of the thrust. The standard day conditions used for all engineering computations such as these specify a humidity of zero percent, but engines are almost never operated under zero humidity conditions. We have not considered the fact that there is humidity in the air because it has so little practical effect on the power or on the thrust produced by a turbine engine.

Seventy-five to eighty percent of the mass airflow through a gas turbine engine is used for cooling, and the moisture in the air has a negligible effect on the cooling process and very little effect on the remaining 20 to 25 percent of the air that is used for combustion.

E. Thrust Horsepower

For a comparison of thrust and horsepower it is possible to convert gas turbine engine thrust to horsepower by using the formula:

$$\text{THP} = \frac{F_n \times \text{aircraft speed (miles per hour)}}{375 \text{ mile-pound per hour}}$$

The constant 375 is the mile-pound-per-hour equivalent of one horsepower. This is found by considering that one horsepower is equal to 33,000 foot pounds of work done in one minute. This is 1,980,000 foot pounds of work per hour. When we divide this by 5,280, the number of feet in one statute mile, we get 375 mile-pounds-per-hour. At 375 miles per hour, one pound of thrust is equal to one horsepower.

This formula shows us that the horsepower produced by a turbine engine increases as the airspeed increases.

Thrust horsepower may be used for turbojet and turbofan engines but it does not apply to the power produced by turboshaft engines. These engines are rated in shaft horsepower which is measured on a dynamometer.

If aircraft speed is expressed in ft/sec the more conventional formula would apply:

$$THP = \frac{F_n \times V_1 \text{ (fps)}}{550 \text{ mile-pound per hour}}$$

As the formula indicates THP can only be calculated in *flight*, thus the use of symbol F_n for thrust applies. The reasoning is that while the aircraft is stationary no energy is expended for propulsion and in the formula V_1 (mph) is zero.

An observation can be made here that at very high aircraft speeds the thrust horsepower of a gas turbine engine can be considerable.

EXAMPLE 1:

The Concorde SST, for instance, if flying at 1,550 mph will have a THP factor of 1,550 ÷ 375, or (4.1) times it net thrust output. On one model thrust per engine at cruise is 10,000 lb, so THP = 4.1 × 10,000). The Concorde then has 41,000 thrust horsepower per engine in cruise. The graph in Fig. 4A-5 illustrates the relationship between THP and aircraft speed (V_1) for the Concorde engine.

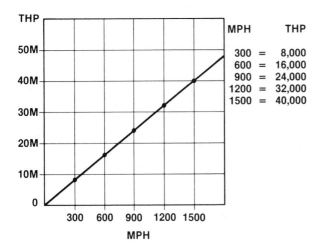

Fig. 4A-5 The relationship between THP and aircraft speeds (V_1) for the Concorde SST engines.

This aircraft would not actually fly at all cruise speeds shown plotted on the graph at rated cruise thrust, but rather the graph indicates the idea that the faster the design speed for a given

thrust the more thrust horsepower equivalent is realized. That is, at 375 mph, one pound of thrust is equal to one horsepower and at 1,550 mph one pound of thrust is equal to 4.1 horsepower.

From this it can be seen that turbine engine development had to wait until faster aircraft could be designed to fill its role as an effective powerplant for high speed flight. If the Concorde for instance was speed limited to 1,200 mph its per engine THP would only be:

THP = 10,000 × 1,200 ÷ 375   THP = 32,000

EXAMPLE 2:

A gas turbine is producing 2,230 pounds of net thrust while flying at 500 miles per hour. What thrust horsepower is being created?

$$THP = \frac{2,230 \times 500}{375 \text{ mile-pound per hour}}$$

$$THP = 2,973.3$$

where: $F_n = 2,230$
mph = 500

It should be noted that THP is not a means of measuring power of a turboshaft engine. Turboshaft engines are rated in shaft horsepower (SHP) as are reciprocating engines and calculated by use of a dynamometer device.

F. Equivalent Shaft Horsepower

Turboprop engines are rated in equivalent shaft horsepower or ESHP. This is the combination of the exhaust thrust converted into thrust horsepower and the shaft horsepower determined by a dynamometer test.

Under static conditions, thrust horsepower may be found by dividing the pounds of jet thrust by 2.6. When we add this to the shaft horsepower measured by a dynamometer we have the equivalent shaft horsepower of the engine.

$$ESHP = SHP + \frac{F_g}{2.6}$$

Let's find the ESHP of a turboprop engine that produces 187.5 pounds of thrust and develops 580 shaft horsepower.

$$ESHP = SHP + \frac{F_g}{2.6}$$

$$= 580 + \frac{187.5}{2.6}$$

$$= 580 + 72.1$$

$$= 652.1 \ ESHP$$

V. GAS TURBINE PERFORMANCE

A. Thermal Efficiency

Thermal efficiency is one of the important considerations in turbine engine performance. It is the ratio of the net amount of work produced by an engine to the amount of energy in the fuel burned, and it can be computed by the formula:

$$T.E. = \frac{SHP}{\frac{F \times 18{,}730 \times 778}{33{,}000}}$$

$$= \frac{\text{Power developed by the engine}}{\text{Power in the fuel}}$$

SHP = Shaft horsepower of the engine

F = Pounds of fuel consumed in one minute

18,730 = Nominal heat energy content of one pound of turbine fuel in Btu

778 = Number of foot pounds of work that can be produced by one Btu of heat energy

33,000 = Number of foot pounds of work per minute in one horsepower

Let's look at an example of finding the thermal efficiency of a turboshaft engine using this formula. The engine in our example produces 700 shaft horsepower while burning 300 pounds of fuel per hour.

The power produced by the engine is 700 shaft horsepower, and the horsepower in the fuel may be found by multiplying the pounds of fuel burned in one minute by the heat energy content of one pound of fuel, in British thermal units, and

the number of foot-pounds of work one Btu will perform. This gives us the number of foot-pounds of work per minute the fuel is capable of performing. Dividing this by 33,000 converts it to horsepower.

$$T.E. = \frac{HP}{\frac{F \times 18{,}730 \times 778}{33{,}000}}$$

$$= \frac{700}{\frac{\frac{300}{60} \times 18{,}730 \times 778}{33{,}000}}$$

$$= \frac{700}{2207.9}$$

$$= 0.317 = 31.7\%$$

This engine converts only 31.7% of the fuel it burns into useful work.

Two of the more important factors that affect the thermal efficiency of a turbine engine are the turbine inlet temperature (TIT) and the compression ratio. In Fig. 5A-1 we see the way the relationship between compression ratio and TIT affects the thermal efficiency of the engine. The highest thermal efficiency is produced with a high turbine inlet temperature, and as technological

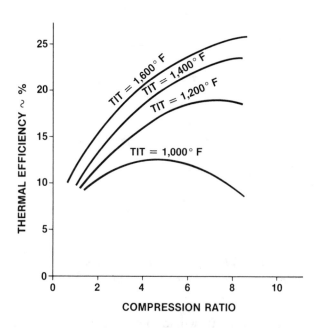

Fig. 5A-1 Effect of compression ratio and turbine inlet temperature on the thermal efficiency of a gas turbine engine.

22

developments bring out new metals that have high strength at high temperatures, we can allow TIT to reach higher values. This will let us allow more of the energy in the fuel to reach the tail pipe and produce thrust.

The walls of the combustors in which the fuel burned are insulated from the burning gases by a cooling film of air. This cooling is effective until the hot gases reach the turbine, but here they must pass through the turbine blades and some cooling air must be mixed with the hot air to reduce its temperature to a value that the turbine blades can tolerate. Heating the gases and then having to cool them lowers the thermal efficiency of the engine.

B. Factors Affecting Engine Thrust

1. Temperature

The thrust produced by a turbine engine is determined as we have just seen, by the mass of air flowing through it, and the temperature of the air entering the engine affects its density. The hotter the air, the less dense it is, and the less mass the air has for its volume. Less thrust is produced when the inlet air is heated. All calculations of thrust made by engine manufacturers are based on standard temperature conditions of 15 degrees Celsius or 59 degrees Fahrenheit, and all performance calculations must be adjusted for non-standard temperatures.

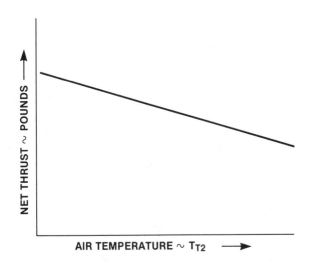

Fig. 5A-2 Effect of air temperature on the thrust produced by a gas turbine engine

2. Altitude

The atmosphere that surrounds the earth is a compressible fluid whose density is varied by both the pressure of the air and its temperature. Air under standard conditions at sea level has a pressure of 14.69 pounds per square inch, and this pressure decreases as the altitude increases. At 20,000 feet it is down to 6.75 psi, and at 30,000 feet it has dropped to 4.36 psi, and it continues to drop at a relatively uniform rate. As the pressure of the air decreases, so does its density.

The temperature of the air also drops with an increase in altitude from its standard sea level condition, and as the temperature drops, the air becomes more dense. The decrease in density caused by the dropping pressure more than overcomes the increase caused by the lowering temperature, and the air density decreases as altitude increases.

At around 36,000 feet, an interesting thing happens. The temperature of the air stabilizes at −69.7° F (−56.5° C) and above this altitude the air no longer has the help of the dropping temperature to increase its density, and at this altitude the air density begins to drop more rapidly with altitude. Because of this, long-range jet aircraft find 36,000 feet an optimum altitude to fly. Below this altitude, the dense air creates more aerodynamic drag, and above this altitude the rapidly dropping density decreases engine thrust output.

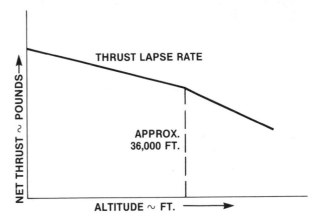

Fig. 5A-3 Effect of altitude on the thrust produced by a gas turbine engine

3. Airspeed

If we will remember, the formula for thrust uses the acceleration of the mass of air through the engine, $(V_2 - V_1)$. As the speed of the airplane through the air (V_1) increases, the amount of thrust produced decreases. We see this in the downward slope of the thrust curve in Fig. 5A-4 that shows the effect of velocity. But there is a compensating effect caused by the air being rammed into the inlet duct as the aircraft moves through the air. This effect is shown as the ram curve. You will notice that this curve is not linear (straight), but it steepens as the airspeed increases. The greater the airspeed, the greater the rate of thrust increases. When we combine the loss from acceleration with the gain from ram, we have the net effect of airspeed on thrust. We see this in the resultant curve.

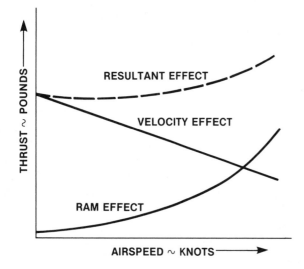

Fig. 5A-4 Effect of airspeed on the thrust produced by a gas turbine engine

4. RPM limits in turbine engines

Engine RPM has a non-linear effect on the thrust produced by an engine. At low RPM the thrust is low, but as the RPM increases, the aerodynamic effect of the compressor moves a greater mass of air through the engine and as the mass airflow increases, so does the thrust.

There is a limit to the engine RPM, however, and this limit is imposed by the aerodynamics of the compressor. As the tips of the compressor blades reach the speed of sound, Mach one or slightly higher, their efficiency drops off. The design of the compressor is such that the blades are operated at such a speed that the airflow over them is not allowed to get into a severe shock stall condition. This accounts for the relatively slow rotational speed of some of the large dia-

meter compressors such as the fan section of the Pratt and Whitney JT9D which turns at around 3,000 RPM, and the very high rotational speed of some of the small diameter turbine engines whose compressor speed may be upward of 50,000 RPM.

It is important that the compressor be operated at an RPM that will keep all of the compressor blade tips operating in their proper Mach number range. We can compute the compressor Mach number by using this formula:

$$\text{Tip speed} = \frac{\pi \times D \times RPM}{60}$$

π = Pi, a constant 3.1416

D = Compressor diameter in feet

RPM = Compressor revolutions per minute

60 = The number of seconds in one minute

This gives the compressor blade tip speed in feet per second.

$$\text{Mach Number} = \frac{\text{Tip Speed}}{\text{Local speed of sound}}$$

Now, by dividing the tip speed we have just found by the local speed of sound we can find the Mach number for the compressor.

Let's use an example of a compressor with a diameter of 1.8 feet turning at 15,500 RPM.

$$\text{T.S.} = \frac{\pi \times D \times RPM}{60}$$

$$= \frac{3.1416 \times 1.8 \times 15,500}{60}$$

$$= 1,460.8 \text{ feet per second}$$

We can find the speed of sound at the compressor blade by multiplying the square root of the absolute temperature by 49.022. The absolute temperature is given in degrees Rankine and is found by adding 460 degrees to the local temperature in degrees Farenheit.

Assuming the local temperature of the compressor stage is 59° F, the local speed of sound is:

$$\text{Speed of sound} = 49.022 \sqrt{T_{abs} (°R)}$$

$$= 49.022 \sqrt{59 + 460}$$

24

$$= 1{,}116.7 \text{ feet per second}$$

$$\text{Mach Number} = \frac{\text{Tip Speed}}{\text{Local speed of sound}}$$

$$= \frac{1{,}460.8}{1{,}116.7}$$

$$= 1.3$$

The tips of the compressor blades are traveling at 1.3 times the local speed of sound.

In order to see the effect temperature has on the Mach number of the air flowing through the engine, we can compute the local speed of sound at the rear stage of the compressor where the air has been heated by compression to 300° F. The compressor diameter remains at 1.8 feet, and the rotational speed is still 15,500 RPM. The tip speed will still be 1,460.8 feet per second.

The local speed of sound will be higher because the air is hotter and the speed of sound depends on the temperature of the air.

$$\text{Speed of sound} = 49.022 \sqrt{T_{abs}(°R)}$$

$$= 49.022 \sqrt{300 + 460}$$

$$= 49.022 \times 27.57$$

$$= 1{,}351.53 \text{ feet per second}$$

$$\text{Mach Number} = \frac{\text{Tip Speed}}{\text{Local speed of sound}}$$

$$= \frac{1{,}460.8}{1{,}351.53}$$

$$= 1.08$$

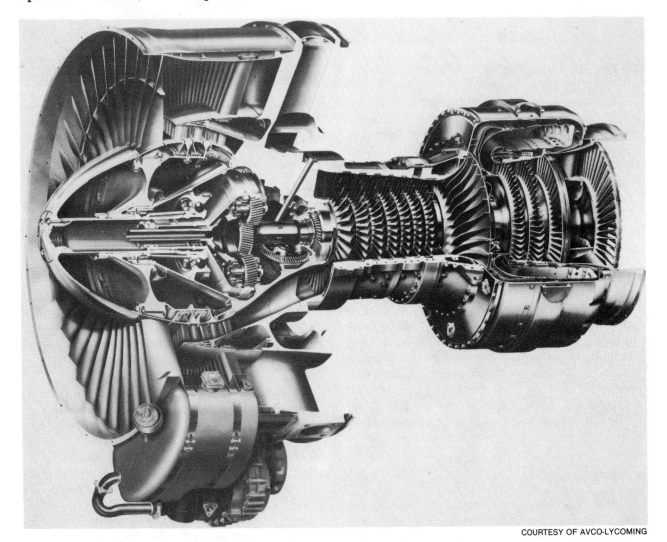

Fig. 5A-5   The Avco-Lycoming ALF502L engine is an example of a high by-pass turbofan engine using both axial flow and centrifugal compressors and reverse flow burners.

5. Efficiency of the turbofan engine

The turbofan engine has replaced the turbojet in most airliners and is now doing the same in many business jet airplanes. This is happening because the turbofan engine is more fuel efficient. If a turbojet and a turbofan engine have the same rated thrust, the turbofan will burn less fuel because of the greater propulsive efficiency of the turbofan. (See Fig. 5A-5 on page 25.)

The maximum thrust for the least fuel flow can be obtained by adding the smallest acceleration to the largest possible mass airflow. The high by-pass turbofan does just that.

VI. TURBINE ENGINE DESIGN AND CONSTRUCTION

A. Turbine Engine Inlet Ducts

1. Principles of operation

The air entrance, or flight inlet duct, is normally considered to be a part of the airframe rather than part of the engine. Nevertheless it is usually identified as engine station one. Understanding the function of the inlet duct and its importance to engine performance makes it a necessary part of any discussion on gas turbine design and construction.

The air inlet to a turbine engine must furnish a uniform supply of air to the compressor so the compressor can operate stall-free, and it must cause as little drag as possible. It takes only a small obstruction to the airflow inside the duct to cause a significant loss of efficiency. If the inlet duct is to deliver its full volume of air with a minimum of turbulence, it must be maintained as close to its original condition as possible, and any repairs to the inlet duct must retain its smooth aerodynamic shape. An inlet cover should be installed any time the engine is not operating to prevent damage or corrosion in this vital area.

a. Subsonic inlets

The inlet duct used on multi-engine subsonic aircraft such as we find in the business and commercial jet aircraft fleet is a fixed geometry duct whose diameter progressively increases from the front to back, as we see in Fig. 6A-2. A diverging

Fig. 6A-1 The inlet duct of a gas turbine engine is critical in providing the proper airflow into the fan and compressor.

duct is sometimes called an inlet diffuser because of the effect it has on the pressure of the air entering the engine. As air enters the inlet at ambient pressure it begins to diffuse, or spread out, and by the time it arrives at the inlet to the compressor its pressure is slightly higher than the ambient pressure. Usually the air diffuses in the front portion of the duct and then it progresses along at a fairly constant pressure past the engine inlet fairings and into the compressor. This allows the engine to receive the air with less turbulence and at a more uniform pressure.

This added pressure adds significantly to the mass airflow when the aircraft reaches its designed cruising speed. At this speed, the compressor reaches its optimum aerodynamic design point and produces the most compression for the best fuel economy. It is at this designed cruise speed that the flight inlet, the compressor, the combustor, the turbine, and the tail pipe are designed to match each other as a unit. If any section mismatches any other because of damage, contamination, or ambient conditions, the engine performance will be affected.

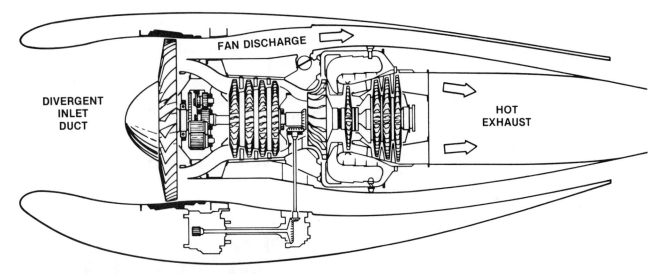

Fig. 6A-2 The subsonic air inlet forms a divergent duct to increase the pressure of the air as it enters the fan or compressor.

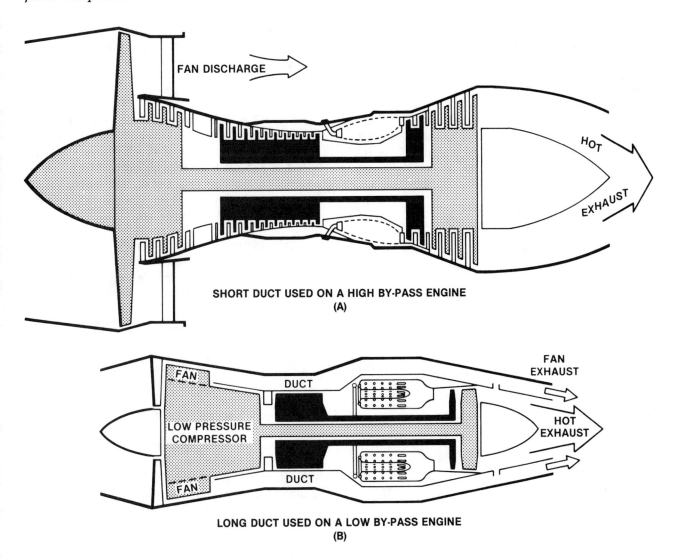

Fig. 6A-3 Airflow patterns for turbofan engines

The inlet for a turbofan is similar in design to that for a turbojet except that it discharges only a portion of its air into the engine, the remainder passes only through the fan.

Fig. 6A-3 shows two common airflow patterns: one is the short duct design used for high by-pass turbofans, and the other is the full duct used on low or medium by-pass turbofan engines. The long duct configuration reduces surface drag of the fan discharge air and enhances the thrust. A high by-pass engine cannot take advantage of this drag reduction concept, however, because the weight penalty caused by such a large diameter duct would cancel any gain.

Ram recovery

When a turbine engine is operated on the ground there is a negative pressure in the inlet because of the high velocity of the airflow. But, as the aircraft moves forward in flight, air rams into the inlet duct and ram recovery takes place. This ram pressure rise cancels the pressure drop inside the duct and the inlet pressure returns to ambient. Ram recovery is said to occur above about 160 miles per hour on most aircraft. From this point pressure continues to increase with aircraft speed and the engine takes advantage of the increasing pressure in the inlet to create thrust with less expenditure of fuel.

b. Supersonic inlets

A variable geometry inlet duct is required on supersonic aircraft to slow the airflow at the face of the compressor to a subsonic speed regardless of the aircraft speed. Most compressors require subsonic inlet airflow if shock waves are to be kept from the rotating airfoils of the compressor.

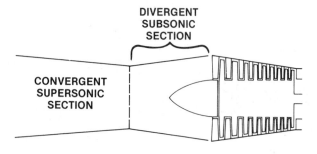

Fig. 6A-4 Convergent-divergent inlet duct used on supersonic airplanes to slow the air entering the compressor to a speed below Mach one

In order to vary the geometry, or the shape, of the inlet duct, a movable restrictor is sometimes used to give this duct a convergent-divergent, or C-D shape. A C-D-shaped duct is necessary to reduce the supersonic airflow to a subsonic speed. This speed reduction is important because at subsonic flow rates, air acts as though it were an incompressible fluid, but at supersonic flow rates it can be compressed, and when it is, it creates shock waves.

Fig. 6A-5 Spikes are used in the inlet ducts of some supersonic airplanes to produce shock waves that slow the air to a subsonic speed before it enters the compressor.

Fig. 6A-6A shows a fixed C-D duct in which the supersonic airflow is slowed by the formation of both oblique and normal shock waves. Once the speed of the inlet air is reduced to Mach one, it enters the subsonic diffuser section where velocity is further reduced and its pressure is increased before it enters the compressor.

The supersonic diffuser-type inlet shown creates a shock wave across the inlet and then directs the air into a variable convergent-divergent portion of the duct whose shape can be changed to accommodate the various flight conditions from takeoff to cruise. Fig. 6A-6B and C shows a variable geometry inlet in both its supersonic and its subsonic position. (See Fig. 6A-6 on page 29.)

c. Bellmouth compressor inlets

Bellmouth inlets are converging in shape and are found on helicopters and other slower moving aircraft which generally fly below ram-recovery speed. This type of inlet produces a great deal of drag, but this is outweighed by their high degree of aerodynamic efficiency (Fig. 6A-7).

When turbine engines are calibrated in test stands they use a bellmouth fitted with an anti-

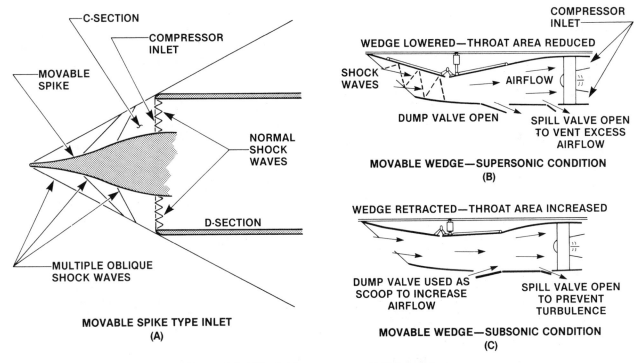

MOVABLE SPIKE TYPE INLET
(A)

MOVABLE WEDGE—SUPERSONIC CONDITION
(B)

MOVABLE WEDGE—SUBSONIC CONDITION
(C)

Fig. 6A-6   Variable geometry C-D inlet duct

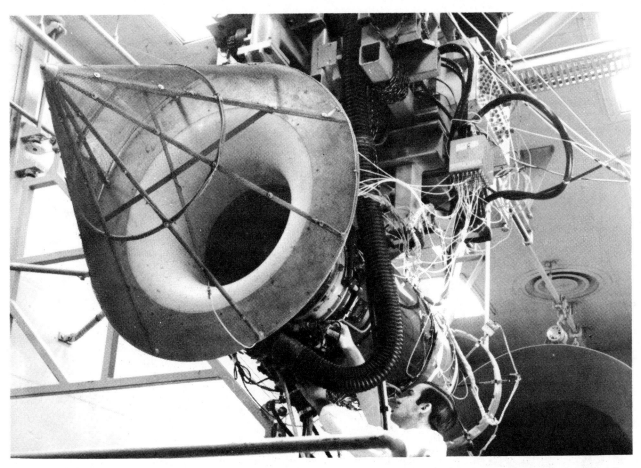

Fig. 6A-7   Bellmouth inlet ducts with anti-ingestion screens are used in engine test cells for calibrating gas turbine engines.

29

ingestion screen. Duct loss is so slight with this design that it is considered to be zero. Engine performance data are collected when the engine is fitted with a bellmouth compressor inlet.

The effect of the inlet duct shape on the aerodynamic efficiency and duct loss is illustrated in Fig. 6A-8. Here we see that a rounded leading edge allows the airstream to use the total inlet

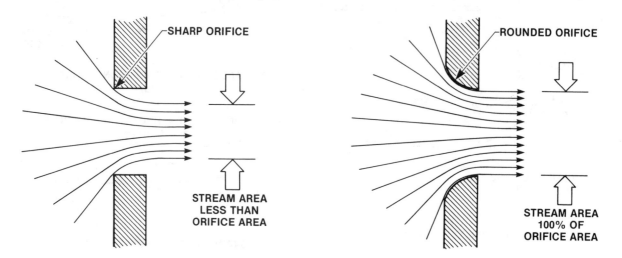

A DUCT WITH A SHARP LEADING EDGE RESTRICTS THE FLOW OF HIGH-VELOCITY AIR.
(A)

ROUNDED LEADING EDGES OF THE INLET DUCT ALLOW FOR A FULL FLOW OF AIR.
(B)

Fig. 6A-8   Effect of the shape of the leading edge of an inlet duct

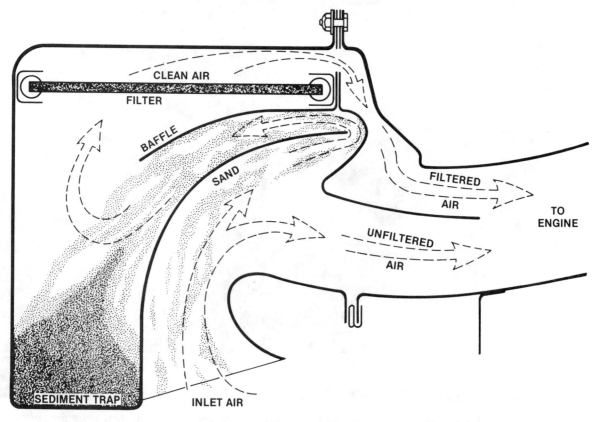

Fig. 6A-9   Sand and dust separator used on helicopter turbine engines

area while the effective area of an inlet having sharp edges is greatly reduced.

d. Compressor inlet screens

The use of compressor inlet screens is usually limited to rotorcraft, turboprop, and ground turbine installations. Inlet screens are seldom used on high mass flow engines because icing and screen fatigue failures have caused so many maintenance problems. Many helicopter engines, however, are fitted with inlet screens to protect them against foreign object ingestion.

Some aircraft are fitted with sand or ice separators, many of which are removable and only used when operating conditions require them. In the sand separator in Fig. 6A-9 inlet suction causes sand particles and other small debris to be deflected by a centrifugal effect into the sediment trap.

B. Accessory Section

The engine-drive gearbox is the main unit in the accessory section. Components and accessories such as the fuel pump, oil pump, fuel control, hydraulic pump, starter and generator all mount on the main gearbox.

The gearbox is usually driven by a radial shaft which meshes with a bevel gear driven by the main rotor shaft but some installations use an intermediate gearbox to drive the main gearbox.

Another common gearbox location is on the front or rear of the engine if the inlet and exhaust

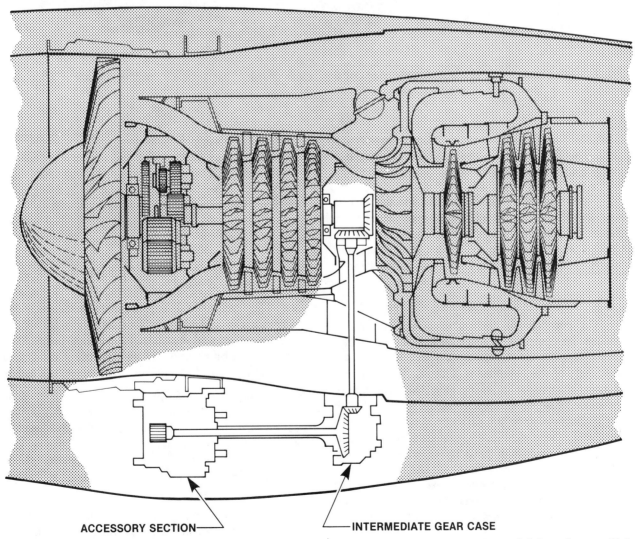

ACCESSORY SECTION —————⟍   ⟋——— INTERMEDIATE GEAR CASE

Fig. 6A-10 Some turbine engines have the accessory section mounted on its waist and driven by a radial shaft through an intermediate gear case.

Fig. 6A-11   Accessories mount on gear cases at both ends of the Pratt and Whitney of Canada PT6 Twinpack used in some twin-engine helicopters.

locations permit. This location is particularly desirable because it allows the narrowest engine diameter and thus the lowest drag configuration.

As a secondary function, the main gearbox acts as a collection point for oil scavenged from the engine before it is pumped back to the oil tank. This allows many of the internal gears and bearings to be lubricated by splash lubrication.

C.   Compressor Section

The compressor section of a turbine engine houses the compressor rotor and the stator vanes and it supplies air in sufficient quantity to satisfy the needs of the combustor. The primary purpose of the compressor is to increase the pressure of the mass of air entering the engine inlet and then to discharge it into the diffuser and the combustors at the correct velocity, pressure, and temperature. The problems associated with these requirements are great because the compressor must move air at a velocity of around 400 to 500 feet per second and increase its pressure by perhaps 20 to 30 times in a space of only a few feet.

The secondary purpose of the compressor is to supply engine bleed air to cool the internal hot

section, and supply heated air for inlet anti-icing. Air is also extracted for such aircraft uses as cabin pressurization, air conditioning, fuel system deicing heat, pneumatic engine starting, and various other functions that require compressed air.

1.   Centrifugal compressors

The centrifugal compressor, sometimes referred to as a radial outflow compressor is the oldest design and it still in use today. Many of the smaller flight engines as well as the majority of gas turbine auxiliary power units use this design.

A centrifugal compressor performs its duties by receiving the air at its center and accelerating it outward by centrifugal force. The air then expands into a divergent duct called a diffuser, and as it spreads out, it slows down and its static pressure increases (Fig. 6A-12).

Centrifugal compressors consist basically of an impeller rotor, a diffuser, and a manifold. The impeller is usually forged from aluminum alloy, and can be either single- or double-sided. The dif-

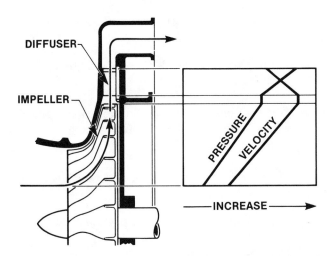

Fig. 6A-12 Centrifugal compressors take the air in near their center and increase both its velocity and pressure. The air then flows into the diffuser where its velocity is decreased and its pressure is further increased.

fuser acts as a divergent duct in which the air spreads out, slows down and increases in static pressure. The compressor manifold distributes the air in a smooth flow to the compressor section.

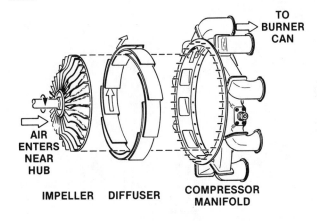

Fig. 6A-13 Centrifugal compressor components

A single-stage, dual-side impeller allows a high mass airflow from a small diameter engine, and it has been used in a number of flight engines in the past for those reasons. This design does not, however, receive the full benefit from ram effect because of the corners the air must turn as it enters and leaves the compressor.

A single-sided impeller does benefit from ram intake and its less turbulent air entry makes it well suited for aircraft installation.

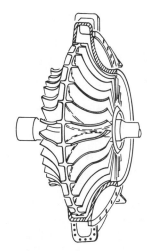

Fig. 6A-14 Single-stage, dual-sided centrifugal compressor

Compression ratios attainable are about the same for both of the single-stage types of centrifugal impellers. More than one stage of compression can be used but the use of more than two stages of single-entry compressors is considered impractical. The energy lost in the airflow as it slows down to make the turns from one impeller to the next, the added weight, and the amount of power needed to drive the compressor all seem to offset the benefits of additional compression by using more than two stages.

The most generally used centrifugal compressor is the single-sided type with either one or two stages.

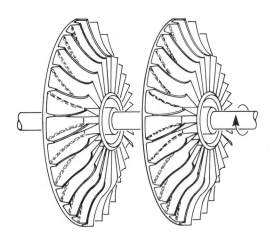

Fig. 6A-15 Two-stage, single-sided centrifugal compressor

Recent developments in centrifugal compressors have produced compression ratios as high as 15:1. In the past, pressures this high could be obtained only with axial flow compressors. Centrifugal compressors are shorter than axial flow compressors and because of their spoke-like design, they can accelerate air faster and immediately diffuse it in the direction of flow.

Tip speed of a centrifugal impeller may reach speeds as high as Mach 1.3, but the pressure within the compressor casing prevents airflow separation and provides a high transfer of energy into the airflow.

A centrifugal compressor may be used in combination with an axial flow compressor, and this is done in some of the smaller flight engines, but all of the larger engines today use axial flow compressors.

Advantages of centrifugal compressors are:

a. High pressure rise per stage - up to 10:1 to 15:1 in a dual stage

b. Good efficiency over a wide rotational speed range, idle to approximately Mach 1.3 tip speed

c. Simplicity of manufacture and relatively low cost.

d. Low weight

e. Low starting power requirements

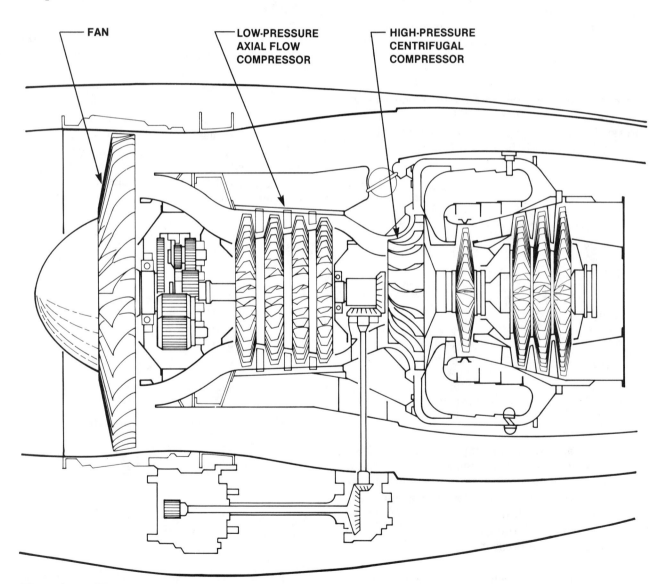

FAN

LOW-PRESSURE AXIAL FLOW COMPRESSOR

HIGH-PRESSURE CENTRIFUGAL COMPRESSOR

Fig. 6A-16   The Garrett TFE731 engine uses an axial flow compressor for the low-pressure stage and a single-stage centrifugal compressor for the high-pressure compressor.

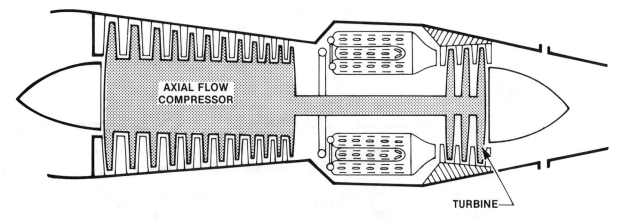

Fig. 6A-17 A single-spool axial flow compressor is used in this engine.

Disadvantages of a centrifugal compressor are:

a. Large frontal area for a given airflow

b. More than two stages are not practical because of the energy losses between the stages.

2. *Axial flow compressors*

There are three types of axial flow compressors: single-spool, dual-spool, and triple-spool. Single- and dual-spool compressors are used in turbojet and turboshaft engines while dual- and triple-spool compressors are commonly used in turbofan engines.

Fig. 6A-17 illustrates a single-spool compressor. This compressor has only one rotating mass.

The compressor, shaft, and turbine all rotate together as a single unit.

Fig. 6A-18 shows how in a multi-spool engine, the turbine shafts attach to their respective compressors by fitting coaxially, one within the other.

The front compressor is referred to as the low pressure, low speed, or N_1 compressor. Its turbine is referred to in the same manner. The rear compressor is called the high pressure, high speed, or N_2 compressor.

The rotor arrangement we see in Fig. 6A-19 is such that the fan is referred to as the N_1, or low speed compressor, the compressor next in line is called the N_2, or intermediate compressor, and the innermost compressor is the high pressure, or N_3 compressor.

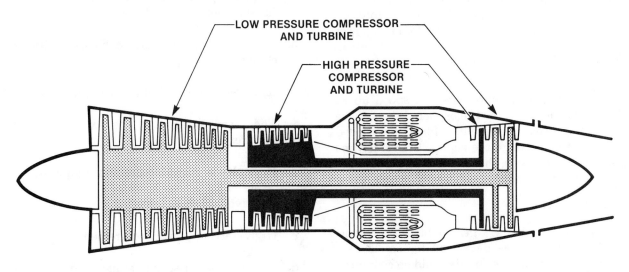

Fig. 6A-18 A dual-spool axial flow compressor is used in this engine.

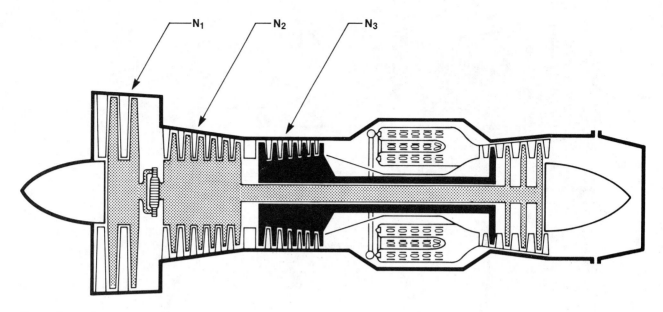

Fig. 6A-19   The geared fan is referred to as N_1, the low pressure compressor as N_2, and the high pressure compressor as N_3.

Dual- and triple-spool compressors were developed for the operational flexibility they afford the engine in the form of high compression ratios, quick acceleration, and better control of the stall characteristics.

For any given power lever setting, the high pressure compressor speed is held relatively constant by the fuel control governor. And, assuming that there is a fairly constant energy level available at the turbine, the low pressure compressors will speed up or slow down with changes in the aircraft inlet conditions resulting from atmospheric changes or flight maneuvers. The N_1 compressor tries to supply the N_2 compressor with a fairly constant air pressure for each power setting by speeding up or slowing down to maintain a constant mass airflow at the inlet of N_2.

Low pressure compressors will speed up as altitude is gained, as the atmosphere is less dense and more speed is needed to force the needed amount of air through the engine. Conversely, as the aircraft descends, the air becomes more dense and easier to compress so the N_1 compressor slows down.

Fig. 6A-20 shows a geared fan-type engine. This compressor was developed for smaller engines so higher turbine speeds could be converted to torque to drive the fan. Also since the fan is geared to the compressor, the compressor is not restricted to fan speed.

Fan tip speed may be allowed to exceed Mach one so the compressor can deliver the correct amount of air. The pressure within the fan duct helps to retard airflow separation from the blades at speeds over Mach one so there is an effective transfer of energy to the air at the required compression ratio.

There are several advantages of the axial flow compressor. They are:

a. High peak efficiencies from ram, created by its straight-through design.

b. High peak pressures attainable by addition of compression stages.

c. Small frontal area and resulting low drag.

The disadvantages of the axial flow compressor are:

a. Difficulty of manufacture and high cost.

b. Relatively high weight.

c. High starting power requirements.

d. Low pressure rise per stage, approximately 1.27:1.

Fig. 6A-20 This geared fan engine uses small diameter compressors that allow high compressor speeds to provide torque to drive the fan.

a. Compression ratio

Small gas turbine engines used in business jet aircraft may have a compression ratio in the order of 6:1 in older models, up to as high as 18:1 in the newer designs. By comparison, the engine used in some of the jumbo jets will compress the air as much as 30 atmospheres.

Compression ratio is found by comparing the discharge pressure of the last stage of compression with ambient air pressure. For example, if the ambient pressure is 14.7 psi and the compressor static discharge pressure is 97 psi, the compression ratio is expressed as 6.6:1.

A typical example of the compression ratio of a dual-spool compressor may be computed as:

N_1 compression ratio = 3:1

N_2 compression ratio = 4:1

Total compression ratio = 12:1

Note that the compression ratio of one compressor is *multiplied* by the other to get the total compression ratio. Normally N_2 will turn at a higher speed than N_1, and because of its small diameter, it will have a higher compression ratio. Therefore, a dual compressor having 10 stages would have a higher compression ratio than 10 stages in a single-spool compressor. This is one of the principle advantages of dual-spool compressors.

Compression ratio of the fan section may be found by dividing the fan discharge pressure by

37

the fan inlet pressure. An example of compression ratio of a fan might be:

Inlet pressure = 14.7 psi

Fan discharge pressure = 26.5 psi

Fan compression ratio = 1.8:1

The rated compression ratio of a gas turbine engine is calculated at standard conditions and full takeoff power and any deviation from these conditions will affect the discharge pressure.

If the mass airflow rating for a given engine is 50 pounds per second under standard-day conditions at full power, the mass airflow will increase as the compressor discharge pressure increases. In fact, the mass airflow change which occurs after takeoff will require the pilot to reduce power to keep from overboosting the engine. The climb to altitude will result in a drop in ambient pressure, but an airplane flying at an altitude where the ambient pressure is only about one third of sea level pressure has ram compression in the engine inlet of approximately 1.5:1, and this helps compensate for the decreased air density. Actually, the engine is designed to operate most efficiently with this lower mass airflow at cruise altitude by a corresponding reduction in fuel flow.

b. *Blades and vanes*

An axial flow compressor has two main elements, the rotor and the stator. The rotor blades force air rearward through each stage which consists of one set of rotor blades and the

following set of stator vanes. The speed of the rotor determines the air velocity in each stage. As the velocity increases, kinetic energy is added to the air. The stator vanes are placed to the rear of the rotor blades to receive the high velocity air and act as diffusers, changing the kinetic energy of velocity into potential energy of pressure. The stators also serve a secondary function of directing the airflow into the next stage of compression.

Compressor blades are constructed with a varying angle of incidence, or twist. This twist compensates for the blade velocity change caused by its radius. The further from the axis of rotation, the faster the blade section travels. The blades also decrease in size from the first stage to the last to accommodate the converging, or tapering, shape of the space in which they rotate.

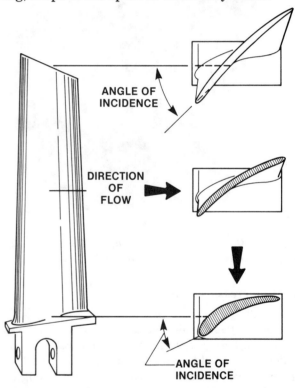

Fig. 6A-22 *Typical compressor blade showing the twist*

The length, chord, thickness, and aspect ratio (ratio of the length to the width) of the compressor blade are designed to suit the performance factors required for a particular engine and aircraft combination.

Axial flow compressors normally have from 10 to 18 stages of compression with the fan con-

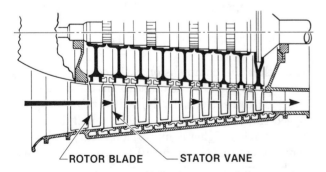

Fig. 6A-21 *The rotor blades impart energy to the airflow through the engine, and the stator vanes act as diffusers, decreasing the velocity and increasing the pressure. The stator vanes also direct the air into the next stage of the rotor blades.*

118

sidered to be the first stage of compression. Some long fan blades have a mid-span shroud fitted to each blade to form a circular ring which helps support the blades against the bending forces from the airstream. The shrouds, however, block some of the airflow, and the aerodynamic drag they produce reduces the efficiency of the fan. The section of the fan blade from the mid-span shrouds to the root, is the compressor blade section for the core engine.

The roots of the compressor blades are often loosely fitted into the compressor disk for ease of blade assembly and for the vibration damping it provides. As the compressor rotates, centrifugal force keeps the blade in its correct position, and the airstream over the airfoil provides a shock absorbing or cushioning effect. These blades are attached to the disk with a dovetail and are secured with a pin and a lock tab or lockwire.

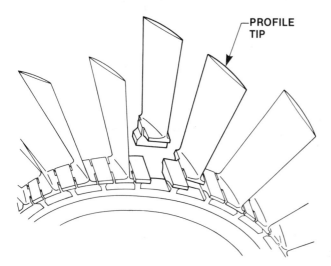

Fig. 6A-23 Dovetail method of securing compressor blades to the disk.

Some blades are cut off square at the tip and these are referred to as flat machine tips. Other blades have a reduced thickness at the tips and these are called profile, or squealer, tips. All rotating machinery has a tendency to vibrate, and profiling a compressor blade increases the natural frequency of the blade. By raising the natural frequency of the blade beyond the frequency of rotation, the vibration tendency is reduced.

Profiling changes the aerodynamics at the tip to produce a smooth axial airflow even if the tip is rotating at speeds beyond the speed of sound and flow separation has started to occur.

On some of the newer engines, the compressor rotor tips are designed to have a tight running clearance and rotate within a shroud tip strip of abradable material. This strip will wear away rather than cause blade damage if contact takes place and the strip is replaced when the engine is overhauled. A high pitch noise can be heard on coastdown if the compressor blade touches the shroud tip strip, and this is the reason profile tips are called squealer tips.

Variable vanes

Stator vanes may either be stationary or may have their angle variable. The inlet guide vanes which are the vanes immediately in front of the first stage rotor blades may also be stationary or variable. The function of the inlet guide vanes is to direct the airflow into the compressor at the most desirable angle. Exit guide vanes are placed at the compressor discharge to remove the rotational moment imparted to the air by the compressor.

3. Combination compressors

To take advantage of the good points of both the centrifugal and the axial flow compressor and eliminate some of their disadvantages, the combination axial-centrifugal compressor was designed. This application is currently being used in many small turbine engines installed in business jet airplanes and helicopters. The combination compressor is specially well suited to engines using reverse-flow annular combustors. The engine diameter is wider to accommodate this type of combustor so there is no disadvantage in using the centrifugal compressor which, by the nature of its design, is much wider than a comparable axial flow compressor. (See Fig. 6A-24 on page 40.)

4. Interstage airflow

Compressor airfoils experience an infinite variety of angles of attack and air densities, and controlling the angle of attack is a design function of the inlet duct, the compressor and the fuel control sensors.

The rotating compressor blades speed up the air in the inlet duct and the air passes through the inlet guide vanes, which changes its angle of flow, but does not change either its velocity or its pressure. The amount the inlet guide vanes

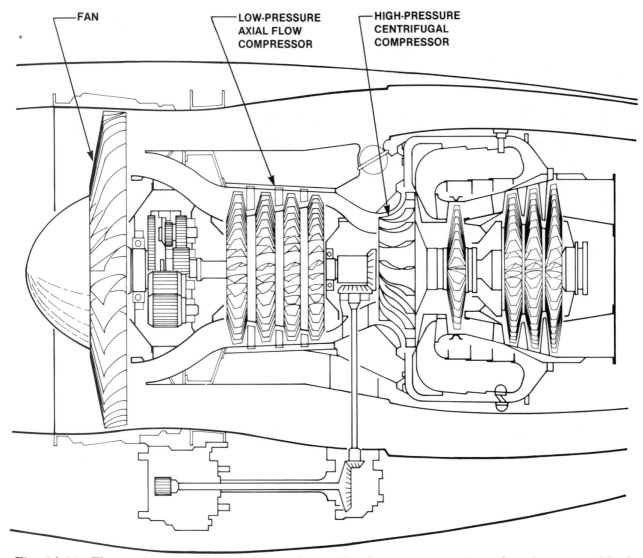

Fig. 6A-24 The combination of axial flow and centrifugal compressors gives the advantage of both types of compressors and is used on some of the engines installed on business aircraft.

change the angle of the air entering the compressor is determined by their position and by the curvature of the vanes. Note that the entering and exiting arrows are the same length, showing that there has been no change in velocity, but only a change in direction.

There are two vector forces acting on airflow. One vector is the ram effect giving velocity to the air entering the compressor. This is shown by the arrow labeled "inlet guide vane effect". The other force is created by the aerodynamic shape of the airfoil (rotor blade). Air is pulled into the compressor and flows over the airfoil in the direction opposite to the blade rotation. This vector is labeled "rotor speed effect".

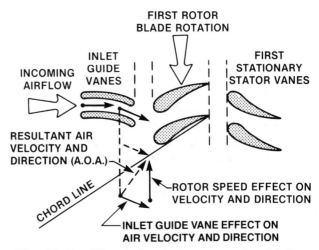

Fig. 6A-25 The air leaving a stage of axial flow compression is at essentially the same velocity as when it entered, but its pressure is increased.

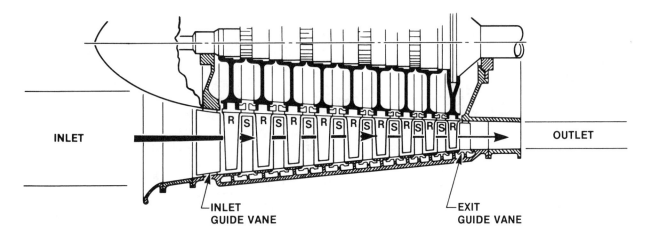

Fig. 6A-26 *To prevent the air velocity decreasing as the pressure increases, the area of the outlet of an axial flow compressor is smaller than its inlet.*

The resultant of the two force vectors gives us the angle of attack, or the angle between the resultant vector and the chord line of the blade. Airflow through the compressor, in spite of the spinning rotor, is relatively straight, with no more than about 180 degrees of rotation as the air passes through the engine. Air leaving the last stage of compression passes through a stationary set of exit guide vanes which straightens the airflow before it enters the combustor (Fig. 6A-26).

The duct, formed by the top, or cambered, side of one blade and the bottom side of the adjacent blade is diverging in shape, and air passing through a diverging duct has its static pressure increased, and work done on the air by the rotating blades increases its velocity.

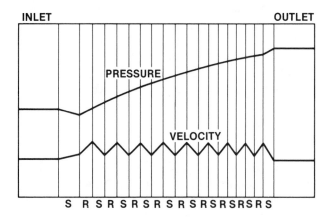

Fig. 6A-27 *Pressure and velocity changes as air passes through each stage of an axial flow compressor.*

As the air leaves the trailing edge of the compressor rotor blades, it flows through a row of stator vanes which also form diverging ducts. But since no energy is added by the stator, the velocity of the air decreases, and its static pressure increases. The action of the compressor rotor blades and the stator vanes continues through all of the stages of compression, and when the air leaves the compressor it has approximately the same velocity it had when it started, but it has a much a higher static pressure (Fig. 6A-27).

The air can flow rearward against the ever increasing pressure only because energy is transferred from the turbine as it drives the compressor. When a compressor blade or a stator vane has a positive angle of attack, the pressure on the bottom of its airfoil shape is higher than the pressure on the top. The high and low pressure areas formed on the airfoils allow air passing through one stage to be influenced by the next stage. This is called the cascade effect.

Fig. 6A-28 shows the high-pressure area of the first stage blade being pulled into the low-pressure area of the stator. Notice that the leading edge of the stator vane faces in the opposite direction as the leading edge of the rotor blade. This arrangement produces a pumping action. The high-pressure area of the first-stage stator vane pumps air into the low-pressure area of the second-stage rotor blade and so on through the compressor.

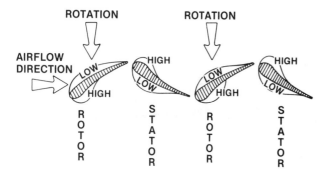

Fig. 6A-28 The cascade effect of the compressor stages increases the pressure as the air flows through the compressor

According to Bernoulli's principle, as pressure builds up in the rear stages of the compressor, its velocity should decrease. To keep the velocity constant, the shape of the path of the gas through the compressor converges.

a. Angle of attack and compressor stall

As we see in Fig. 6A-29, the angle of attack of the compressor blade is determined by the inlet air velocity and the compressor RPM. These two forces combine to form a vector force which gives us the angle of attack of the airfoil. A compressor stall is a condition of airflow when the angle of attack becomes excessive.

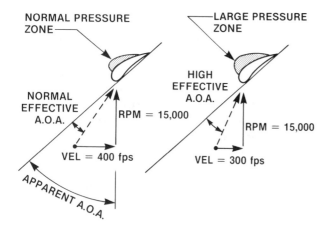

Fig. 6A-29 The angle of attack of an axial flow compressor blade is determined by the rotational speed of the compressor and the velocity of the air flowing through the engine.

A compressor stall causes the airflow to slow down, stop, or even reverse its direction, depending upon the severity of the stall. Stalls can range

from a slight air vibration, or fluttering sound, to a louder pulsating sound, or even to a violent backfire or explosion. Quite often the gages in the cockpit do not show a mild or transient stall condition, and these stalls are usually not harmful to an engine. They often correct themselves after one or two pulsations. But severe stalls, called hung stalls, can significantly impair engine performance, cause loss of power and can even damage the engine. The pilot can identify a stall condition by its audible noise, by fluctuations of the RPM, by an increase in the exhaust gas temperature, or by a combination of these clues.

Compressor stalls may be caused by:

a. Turbulent or disrupted airflow to the engine inlet, which reduces the velocity vector.

b. Excessive fuel flow caused by abrupt engine acceleration. This increases the back pressure of the combustor and reduces the velocity vector.

c. Contaminated or damaged compressor blades or stator vanes.

d. Damaged turbine components which cause a loss of shaft horsepower delivered to the compressor. This decreases the compressor speed and reduces the velocity vector.

The remedy for an acceleration stall is to reduce the power and allow the inlet air velocity and the engine RPM to get back into their proper relationship.

In case of a severe compressor stall or surge caused by a fuel control malfunction or from foreign object ingestion, a reversal of airflow can occur with such force that the compressor blades may be bent enough to cause them to contact the stator vanes. This extreme condition will result in disintegration of the compressor and complete engine failure.

b. Stall, or surge, margin curve

Another way to describe the compressor stall is by the use of a stall, or surge, margin curve. A stall is a *localized* condition, while a surge occurs *across the entire compressor*. Every compressor has a best operating condition for a given compression ratio, speed, and mass airflow. This is commonly called the design point.

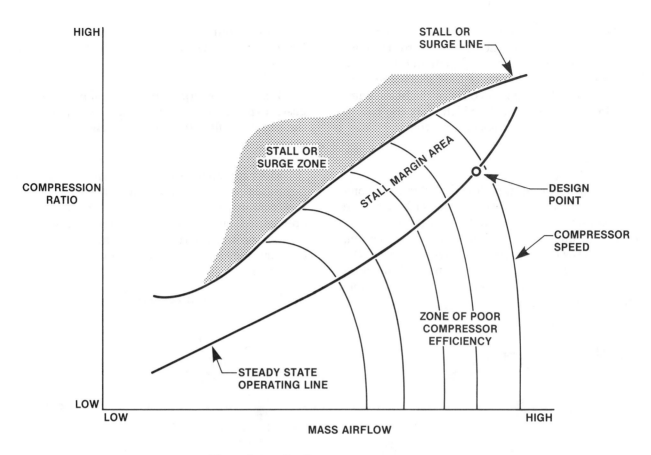

Fig. 6A-30   Stall, or surge, margin curve

In Fig. 6A-30 the steady-state operating line indicates that the engine will perform without stall at the various compression pressure ratios, engine speeds, and mass airflows along the line. This line falls well below the stall zone. The design point is located on this line and it represents the conditions under which the engine will operate for most of its life—that is, at high altitude cruise.

From this curve, it may be seen that for any given compressor speed, only a narrow band of compressor pressure ratios are acceptable. Only those between the steady-state line and the stall zone will provide satisfactory engine operation. This band of compression ratios is called the stall margin.

Also, for any given mass airflow, there exists only a narrow band of compressor pressure ratios which will allow the engine to operate stall free.

If compressor contamination, cold- or hot-section damage, incorrect fuel scheduling or other engine malfunctions cause a significant change in any of the three parameters on the chart, a stall or surge may occur, or if not a stall, at least poor efficiency will result.

The interstage airflow we have discussed has all been at subsonic speed. But some of the newer and more advanced compressors have supersonic airflow with speeds up to Mach 1.3 over some of the airfoils. The airfoil sections of both blades and vanes in compressors of this design have thin leading edges and a twist that forms a slightly convergent-divergent airflow passage, similar to the C-D flight inlet duct we have discussed.

5. *Comparison of axial flow with centrifugal flow compressors*

Axial flow and centrifugal flow compressors both raise the pressure of the air inside the engine. The centrifugal compressor raises the pressure by accelerating the air outward into a single divergent-duct diffuser where according to Bernoulli's principle the air spreads out and slows down, and its pressure increases.

An axial flow compressor raises the air pressure by accelerating the air rearward through many small diffusers or divergent ducts formed by the shape and position of the rotor blades and stator vanes. The trailing edges of the blade pairs also form divergent ducts which start the rise in pressure prior to entry into the stators.

D. Diffuser Section

The diffuser, located directly behind the compressor, provides the space for the air leaving the compressor to spread out. It is in the form of a divergent duct, and is usually a separate section, bolted to the compressor case. The pressure in the diffuser is the highest in the engine, and this high pressure gives the combustion products something to push against.

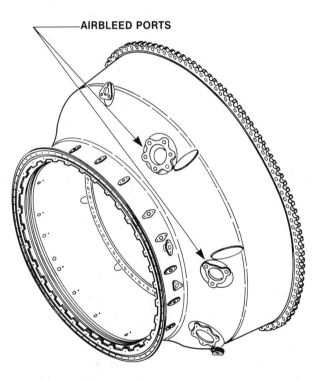

Fig. 6A-31 Typical diffuser section located between the compressor and the combustor. The air is at its highest pressure in the diffuser, and it is from here that bleed air is taken.

E. Combustion Section

The combustors, or burners, in a gas turbine engine have an outer casing, an inner perforated liner, usually made of stamped sheet metal, a fuel injection system, and an ignition system for starting. Heat energy is added to the flowing gases in the burners, and this energy expands the gases and accelerates them as they leave the engine.

When heat energy from the fuel is added, the gases expand, but since the area through which the gas must flow remains the same, the flowing gases speed up.

Most combustors are of the through-flow configuration which is sometimes called a through-flow combustor. Gases entering from the diffuser are ignited and then pass directly through the combustor into the turbine section. The multiple-can, annular, and can-annular combustors are of this type.

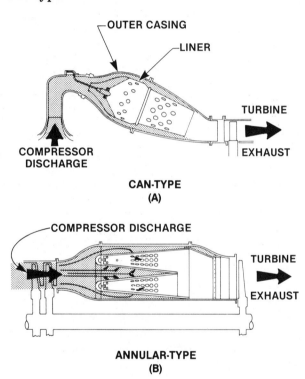

Fig. 6A-32 Through-flow combustors

Another configuration is the reverse-flow annular type where gases leaving the diffuser flow to the rear of the combustor where fuel is sprayed in and ignited. Then the burning gases follow a reverse-S path into the turbine section.

To function properly, the combustors must mix the air and the fuel for efficient combustion. Then it must lower the temperature of the hot combustion products enough that they will not overheat the turbine components. To do this, the airflow through the combustor is divided into

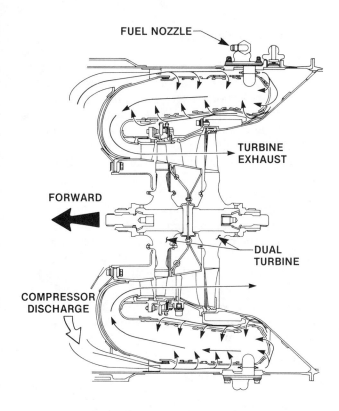

Fig. 6A-33   Reverse-flow combustor

primary and secondary air paths. Approximately 25 to 35 percent of the air is routed to the area around the fuel nozzle for combustion. This is the primary air. The secondary air, or the remaining 65 to 75 percent, forms a cooling air blanket on either side of the liner and centers the flames so they do not contact the metal. The secondary air

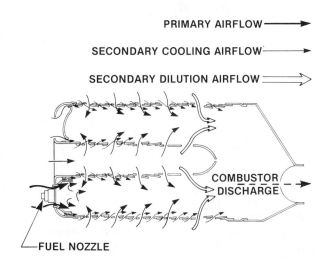

PRIMARY AIRFLOW ⟶

SECONDARY COOLING AIRFLOW ⟶

SECONDARY DILUTION AIRFLOW ⟹

COMBUSTOR
DISCHARGE

FUEL NOZZLE

Fig. 6A-34   Secondary airflow in the combustor cools the burning gases enough that they will not damage the combustor or turbine components.

also dilutes and cools the hot primary air to a temperature that will not shorten the service life of the turbine components.

Developments in recent years have brought out the smokeless, or reduced smoke combustor. Incomplete combustion in the early engines left unburned fuel in the tail pipe where it entered the atmosphere as smoke. By shortening the flame pattern and using new materials that can withstand higher operating termperatures, manufacturers have been able to almost completely eliminate the smoke emissions from turbine engines as more complete combustion occurs.

The secondary air in the combustors may flow at a velocity of up to several hundered feet per second, but the primary airflow is slowed down by swirl vanes, which gives the air a radial motion and retards its axial velocity to about five or six feet per second before it is mixed with the fuel and burned. The vortex created in the flame area provides the required turbulence to properly mix the fuel and the air. This reduction in the airflow velocity is important because of the slow flame propagation rate of kerosene-type fuels. If the primary airflow velocity was too high, it would literally blow the flame out of the engine. As it is, the combustion process is complete in the first third of the combustor length, and the burned and unburned gases then mix to provide an even distribution of heat at the turbine nozzle.

Although flameout is uncommon in modern engines, combustion instability still occurs and, occasionally, a complete flameout. Turbulent weather, high altitude, slow acceleration during maneuvers, and high-speed maneuvers are some of the typical conditions which induce combustor instability which could lead to flameout.

There are two types of flameouts: a lean flameout usually occurs at low engine speed and low fuel pressure, at high altitude where the flame from a weak mixture can be blown out by the normal airflow. A rich flameout occurs during rapid engine acceleration where an overly-rich mixture causes the combustion pressure to increase so much that the compressor airflow stagnates and slows down, or even stops. The interruption of the airflow then causes the flame to go out. Turbulent inlet conditions and violent flight maneuvers can also cause compressor stalls which could result in airflow stagnation and flameout.

Combustor instability sometimes causes small gas pressure fluctuations. These low-pressure cycles cause high fuel flow pulsations, which increase the combustor instability until the pilot makes the necessary adjustments to the flight conditions or to the engine controls.

Combustor efficiency ranges between 95 and 99 percent, which means that 95 to 99 percent of the heat energy in the fuel is released. The combustor efficiency is high, but only about one third of the mass airflow is used for combustion, with the remainder of it used for cooling, to keep the temperatures within acceptable limits for the combustor and the turbine.

1. Multiple-can combustor

This older type of combustion chamber (not commonly used today) consists of a series of outer housings, each with its own perforated inner liner. Each of the multiple combustor cans is actually a separate burner unit, with all of them discharging into the open area at the turbine nozzle inlet. The individual combustors are interconnected with small flame propagation tubes so that when combustion starts in the two cans having igniter plugs, the flame will travel through the tubes and ignite the fuel-air mixture in the other cans.

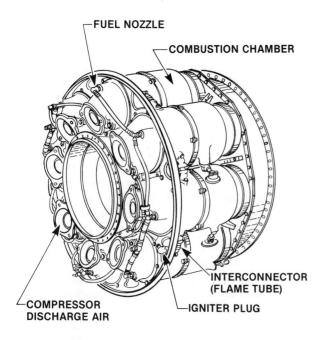

Fig. 6A-35 Multiple-can combustor

2. Annular combustor

The annular combustor consists of an outer housing and a perforated inner liner called a basket, with both parts encircling the engine. Multiple fuel spray nozzles stick out into the basket, and both primary and secondary air for combustion and cooling flow through it in the same way as in the other combustor designs.

Annular combustors are in common use today in both small and large engines. They are the most efficient type from the standpoint of both thermal efficiency and weight, and they are also shorter than the other types. The small amount of surface area requires less cooling air, and makes the best use of the available space, especially for large engines where other types of combustors would be much heavier for the large mass airflow these engines use.

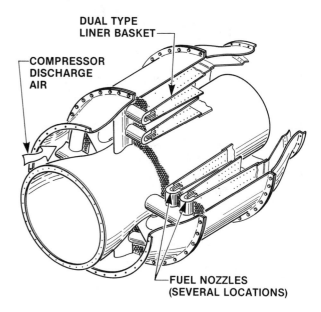

Fig. 6A-36 Annular combustor

3. Can-annular combustor

The can-annular combustor is used for commercial aircraft powered by Pratt and Whitney engines. This type of combustor consists of an outer case with multiple inner liners located radially around the axis of the engine. Flame propagation tubes connect the individual liners and two igniter plugs are used for starting.

The combustor in Fig. 6A-37 uses eight cans, and each can has its own fuel nozzle supporting

46

its forward end. The outlet duct with its eight openings supports the cans at their aft end. An advantage of this type of combustor is its ease of on-the-wing maintenance because the forward half of the outer combustor casing may be unfastened to slide rearward exposing the cans for inspection.

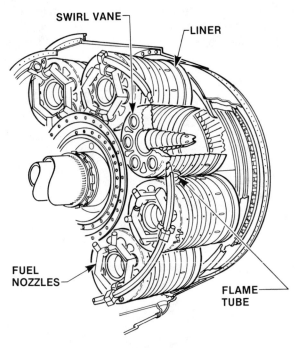

CANS INSTALLED IN THE ENGINE
(A)

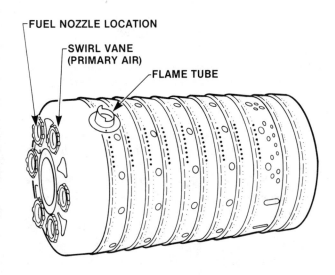

INDIVIDUAL BURNER CANS
(B)

Fig. 6A-37 Can-annular combustor

4. Reverse-flow annular combustor

This design is used by the Pratt and Whitney PT6 and the Garrett TFE 731 and several of the other engines installed in business aircraft. The reverse-flow combustor serves the same function as the through-flow combustor, but it differs by the air flowing around the chamber and entering from the *rear*, causing the combustion gas flow to be in the opposite direction as the normal airflow through the engine.

Notice in Fig. 6A-38 that the turbine wheels are inside the combustor area rather than in tandem, as they are with through-flow combustors. This allows for a shorter and lighter engine, and it also uses the hot gases to preheat the compressor discharge air. These factors help make up for the loss of efficiency caused by the gases having to reverse their direction as they pass through the combustor.

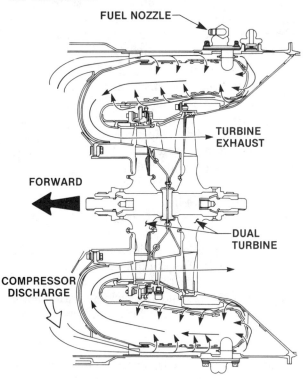

Fig. 6A-38 Reverse-flow combustor

F. Turbine Section

The turbine section is bolted to the combustor, and it contains the turbine wheels and the turbine stators. The turbine transforms a portion of the kinetic energy and heat energy in the exhaust gases into mechanical work, so it can drive the compressor and the accessories.

The compressor *adds* energy to the air by increasing its pressure, and the turbine *extracts* energy by reducing the pressure of the flowing gases. This is done by converting pressure into velocity at the nozzles formed at the trailing edge of the stator vanes and rotor blades. The airflow is directed in a tangential rather than axial direction, and this slows the gas flow and reduces its reactive power; but it adds torque power to the rotor system. The mass of the airflow is not changed by the transfer of energy to the rotor system, but the velocity of the air flowing through the engine is decreased as power is taken to drive the compressor. Turboprop and turboshaft engines have very little reactive thrust in the tail pipe after their multi-stage turbines extract power.

An efficient turbine performs a maximum amount of work with the least fuel consumption, and efficient turbines operate at their design point of temperature and RPM when the compressor is operating at its design point of compression ratio and mass airflow.

The turbine absorbs most of the energy released by the combustion, and it is the most highly stressed component in the engine. The disk, usually a heavy forging of a nickle alloy, must withstand extremely high temperature loading and high rotational speed.

The blades are held in the disk by a method similar to the compressor blades. The stator vanes for the turbine are located ahead of the rotor rather than behind the rotor, as they are in the compressor. The compressor stators act as diffusers to decrease the velocity and increase the pressure of the gas, but the turbine stators act as nozzles, to increase the velocity and decrease the pressure.

Most turbine nozzles operate in a choked condition from cruise to takeoff power. This provides a fairly constant flow of energy to the turbine wheel over its normal operating range. Since the nozzle is choked, the velocity of the gases depends upon their temperature, which affects the local speed of sound. The downstream pressure has little effect on a choked turbine nozzle, as it creates its own back pressure. The turbine stator also directs the gases at the optimum angle to the turbine blades so the wheel will turn with maximum efficiency.

The gas flowing through the turbine stator is at the highest velocity of any point in the engine, and this velocity is controlled by the total area of the opening between the vanes.

Turbines are classified as impulse, reaction, or a combination impulse-reaction type. An impulse turbine causes no net change in the pressure across the turbine wheel. The nozzle guide vanes are so shaped that they change the direction of the gases so they will strike the turbine wheel at the correct angle and at an increased velocity.

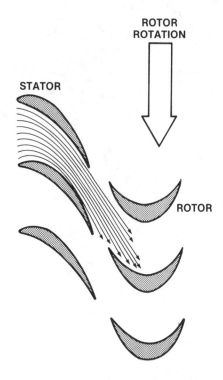

Fig. 6A-39 An impulse turbine is driven by the impulse of hot gases on the turbine blades.

Reaction turbines produce their turning force by an aerodynamic action. The nozzle guide vanes are shaped in such a way that they only direct the gas in the correct direction, they do not increase the velocity of the gas. This gas passes between the blades of the turbine, which form converging nozzles that further increase its velocity. As the gases flow over the airfoil-shaped blades, a force component in the direction of the plane of rotation causes the turbine to spin.

Most turbines are neither totally of the reaction nor of the impulse type, but are of the impulse-reaction type that causes the turbine to spin because of a combination of the impulse pressure and the reaction force of the flowing gases.

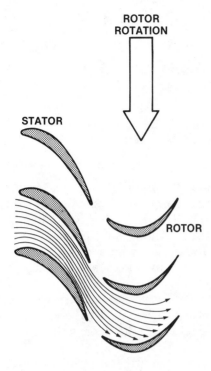

Fig. 6A-40 *When the hot, high-velocity gases flow through an impulse-reaction turbine, an aerodynamic force as well as the impulse force moves the blades in the direction needed to spin the wheel.*

The turbine blades fit into the disk with some form of fastening that allows them to be loose when the engine is cold, but to be firmly attached at operating temperature. The most commonly used method is the fir tree design we see in Fig. 6A-41.

The turbine blades may be either open or shrouded at their ends, and both types of blades may be used in an engine. Open end blades are used in the high-speed wheels, and shrouded blades are found in the wheels having slower rotational speeds. Shrouded blades form a band around the perimeter of the wheel, which helps reduce blade vibration, and the weight of the shrouded tip is offset by the blades being thinner and more efficient.

A knife-edge seal around the outside of the shroud reduces air losses at the blade tip, keeps the airflow in an axial direction, and minimizes

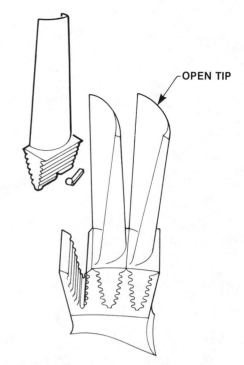

Fig. 6A-41 *The fir-tree method of attaching a turbine blade to the disk allows the blade to be loose when it is cold, but it becomes rigid at operating temperature.*

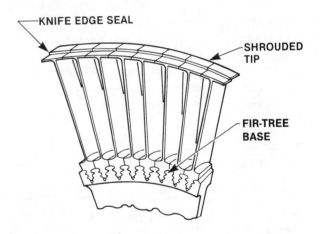

Fig. 6A-42 *Shrouded turbine rotor blades*

radial losses. The knife-edge seal fits with a close tolerance into a shroud ring mounted in the outer turbine case.

Many modern engines have air-cooled turbine stator vanes and rotor blades as we see in Fig. 6A-43. This cooling allows the turbines to operate at much higher temperatures than they could if they were not cooled. Often, only the first stage stator and rotor need to be cooled, as they extract

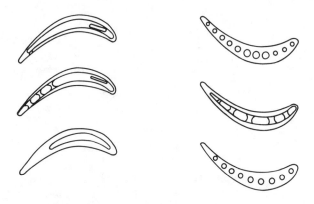

Fig. 6A-43 *Typical hollow inlet guide vanes and first-stage turbine blades. Compressor bleed air flows through the vanes and blades to remove heat. This allows the turbine to operate at a higher temperature.*

enough energy from the flowing gases to lower their temperature enough for the remaining stages to operate uncooled.

The cooling air is taken from the compressor, and after it passes over or through the turbine blades and absorbs heat, it cools the shroud seals and shroud ring. Cooling the turbine components allows the engine to operate at a higher temperature, which measurably increases the thermal efficiency. It does this by reducing the amount of secondary airflow needed to cool the combustion gases to a temperature that would not damage the turbine components.

G. Exhaust Section

1. Exhaust duct and tail cone

The exhaust section is usually located directly behind the turbine section, and it consists of a convergent exhaust duct and an inner tail cone. The exhaust duct attaches to the rear of the turbine housing and converts the gases leaving the turbine into a straight-flowing, high-velocity stream.

The exhaust duct itself is usually a stainless steel, conical tube, supported at its forward end by the turbine exhaust case, and is open at its aft end. The gradually decreasing area of the duct causes the exhaust gases to accelerate. The area of the end of the duct, which is called the exhaust nozzle, determines the velocity of the gases as they leave the engine.

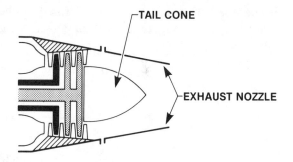

Fig. 6A-44 *The convergent exhaust duct accelerates the exhaust gases as they leave the engine.*

Some of the earlier engines had their RPM or exhaust gas temperature trimmed by varying the exhaust nozzle area with small adjustable tabs called "mice." And some of the larger engines are equipped with thrust reverser nozzles whose exit may be altered by an actuator system.

The velocity at the exit of a convergent exhaust duct is either held to a velocity slightly below the speed of sound, or is choked.

Engines that produce pressure ratios across the exhaust nozzle that are high enough to cause the gas velocity to exceed Mach 1.0, use a convergent-divergent, or C-D, duct. The convergent portion of the duct is of such a design that the gasses reach Mach 1.0 at its throat and then as the supersonic flow passes into the divergent portion of the duct, it further accelerates. The C-D duct converts the pressure energy that is present at the choked convergent nozzle into velocity energy which is effeciently converted into thrust. This exhaust duct is designed as an afterburner and used exclusively on supersonic aircraft.

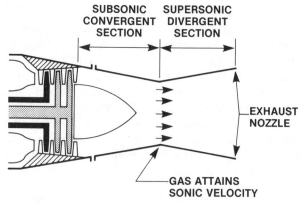

Fig. 6A-45 *A convergent-divergent exhaust duct converts the pressure energy at its choked convergent nozzle into velocity energy.*

50

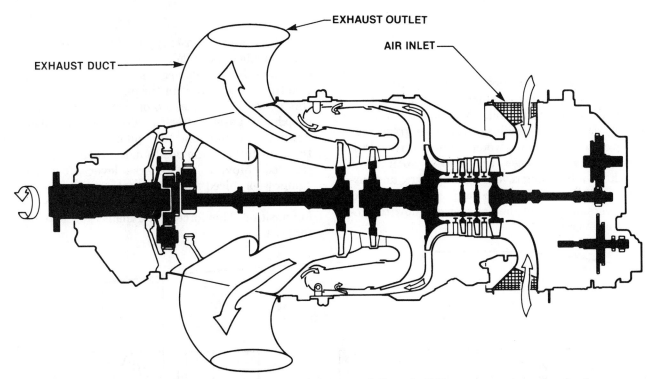

Fig. 6A-46 *The exhaust on this Pratt and Whitney of Canada PT6 engine exits near the front.*

Some of the modern engines have placed the exhaust in front of the turbine rather than in the more conventional tail-end location. In Fig. 6A-46, we see the Pratt and Whitney of Canada PT6 turboprop engine. The air is taken into the engine near the rear, and it flows forward and exits through the exhaust outlets near the front of the engine.

The air enters the Garrett AiResearch ATF 3 turbofan engine through the fan at the front, and

after passing through the axial flow, low-pressure compressor and centrifugal high-pressure compressor, it turns forward and passes through the burner and the high-pressure turbine, the fan turbine, and the low-pressure turbine. Then it again reverses its direction as it flows through two ninety-degree cascades. It exits the engine through the cascades where it mixes with the airflow from the fan. This mixing effectively decreases the noise produced by the exhaust.

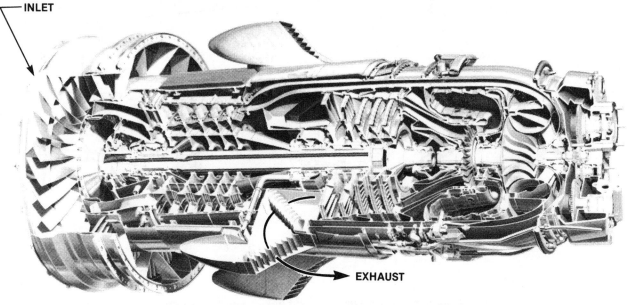

COURTESY OF GARRETT TURBINE ENGINE COMPANY

Fig. 6A-47 *The exhaust on this Garrett ATF3 engine leaves through eight turning vane modules. The exhaust mixing with the duct airflow effectively reduces the noise.*

2. Thrust reversers

Wheel brakes are not effective in providing all of the force needed to slow a heavy and fast jet airplane in its landing roll. The amount of kinetic energy that must be dissipated is so great that the wear on the brakes would be prohibitive. Propeller-driven airplanes are slowed soon after touchdown by using reversing propellers, and turbojet and turbofan-powered airplanes are slowed by reversing part of the exhaust gas flow to produce a reverse thrust.

There are two basic types of thrust reversers: the clamshell type and the cascade type. The clamshell thrust reverser is stowed in such a position that it forms part of the exhaust duct, and in this position it creates no opposition to the flow of exhaust gases. After the airplane has touched down on landing and the pilot has pulled the throttles to their idle position, the reverse lever is released. Moving the reverse lever aft causes compressor bleed air to actuate pneumatic actuators that move the clamshells into the blocking position. This forces the exhaust gases to flow forward.

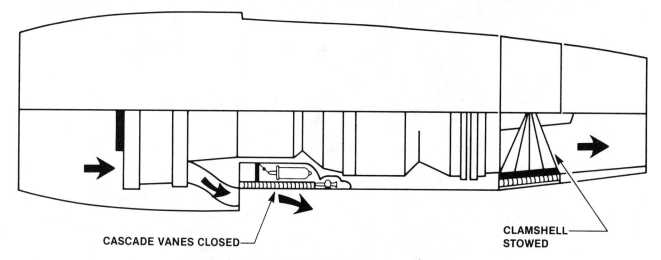

CASCADE VANES CLOSED

CLAMSHELL STOWED

FORWARD-THRUST POSITION—REVERSERS ARE STOWED
(A)

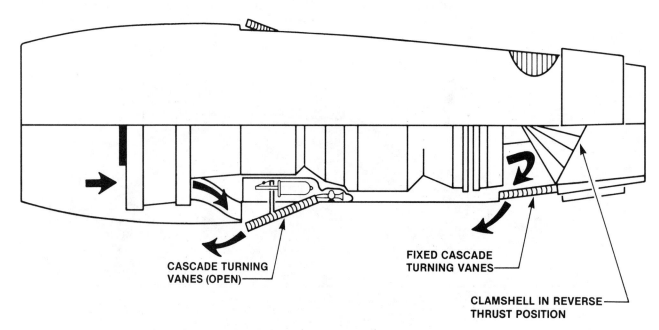

CASCADE TURNING VANES (OPEN)

FIXED CASCADE TURNING VANES

CLAMSHELL IN REVERSE THRUST POSITION

REVERSE-THRUST POSITION—REVERSERS ARE DEPLOYED
(B)

Fig. 6A-48   Fanjet engine thrust reverser

The thrust reversers on a short duct turbofan engine directs the reverse airflow up and down.

In Fig. 6A-48, we see the way the thrust reversers in a turbofan engine operate. In view A, the reversers are in their stowed postion. The fan exhausts overboard through the short ducts, and the exhaust from the core engine flows unrestricted out of the tail pipe. When the reversers are deployed, view B, the fan cascades open and divert the fan discharge forward. And the clamshells in the exhaust of the core engine pivot together so that they block the exhaust and force it through the cascade turning vanes which diverts the gas stream forward.

Thrust reversers are most effective at high speed, and they are generally used only during the first part of the landing roll, while the airplane is moving at a speed of greater than 60 knots.

3. Noise suppressors

Much of the energy released by the burning fuel in a jet engine ultimately causes vibrations of the air which we hear as noise. As airplanes have become larger and flights from densely populated areas more frequent, control of this noise is of real concern.

The majority of the noise produced by a turbojet engine is created as the hot, high-velocity gases mix with the cold, low-velocity air surrounding the engine. This noise has a high intensity and it includes both low- and high-frequency vibrations, with low frequencies being predominant.

If you have ever noticed the approach of a marching band, the sound you hear first is the beat of the bass drum; you hear this for quite some time before you hear the flutes, piccolos, and other high-pitched instruments. The reason is

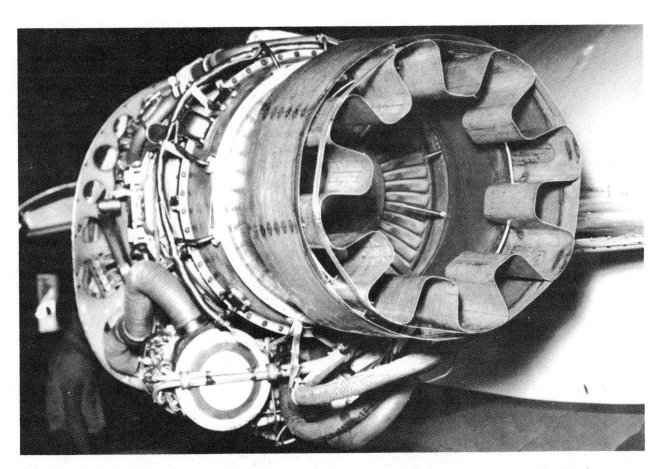

Fig. 6A-49 Noise suppressors fit behind the tail cone of a turbojet engine to convert the low-frequency vibrations of the air into high-frequency vibrations that are more readily attenuated by the atmosphere.

that the higher frequencies are attenuated, or absorbed by the atmosphere, much more than the lower frequencies.

Turbofan engines seldom require noise suppressors because the exhaust from the fan and the hot exhaust from the core engine are lower in velocity and noise. But turbojet engines usually require some form of device behind the tail pipe to break up the noise.

We see in Fig. 6A-49 a corrugated-perimeter type of noise suppressor. This does not decrease the total amount of noise, but it increases the frequency of the noise by breaking up one eddy flow into small high frequency airstreams, so that it will fall off in intensity more than the untreated exhaust. (See Fig. 6A-49 on page 53.)

Turbine Engine Inspection, Maintenance, and Operation

I. INSPECTION AND MAINTENANCE

A. Types of Maintenance

Two classifications of inspections and repairs are performed on turbine engine aircraft by A&P technicians. They are line (flight line) and shop inspections, also line and shop repairs. Although the tendency today is to perform more extensive repairs on the line, the really heavy repairs are performed on the engines removed from the aircraft and in a repair shop.

A good point to remember when working on turbine engines is that they are all lightweight, high-speed machines which are manufactured to very close tolerances; therefore extreme care is required in their maintenance. This includes rigid adherence to established technical procedures, correct use of tools, and especially cleanliness of parts and shop environment.

1. Line maintenance

Line maintenance and heavy maintenance are the basic levels of repairs accomplished on gas turbine engines. Line maintenance encompasses all the repairs that can be made to the powerplant while it is installed in the airplane.

a. Foreign object damage

Much of the damage to the compressor section encountered on the flight line arises from foreign matter being drawn into the engine inlet. Damage to compressor blades results in a compressor geometry change which can cause malfunctions such as performance deterioration, compressor stalls, or even engine failure.

Foreign object damage (F.O.D.) prevention is a concern of all flight line personnel. Sometimes the most harmless looking piece of debris can cause thousands of dollars in maintenance costs to the aircraft owner. The following is a list of suggested F.O.D. prevention methods:

1. The maintenance technicians should keep the ramp and hanger areas clean and impress upon other flight line personnel the importance of cleanliness.

2. Technicians should ensure that articles of clothing and materials in pockets are secure when working on operating aircraft.

3. Technicians and pilots should always check inlets for foreign objects before engine operation and avoid run-ups and taxiing into exhaust blasts of other operating aircraft.

4. Everyone who maintains jet aircraft should ensure that the inlet and exhaust covers are kept in place when the engine is static (not operating), to prevent contamination and windmilling.

Note: Damage incurred in the gas path from material failure of aircraft or engine parts is called Domestic Object Damage (D.O.D.).

b. Compressor field cleaning

Accumulation of contaminants in the compressor reduces aerodynamic efficiency of the blades and reduces engine performance. Two common methods for removing dirt, salt and corrosion deposits are a fluid wash and an abrasive grit blast.

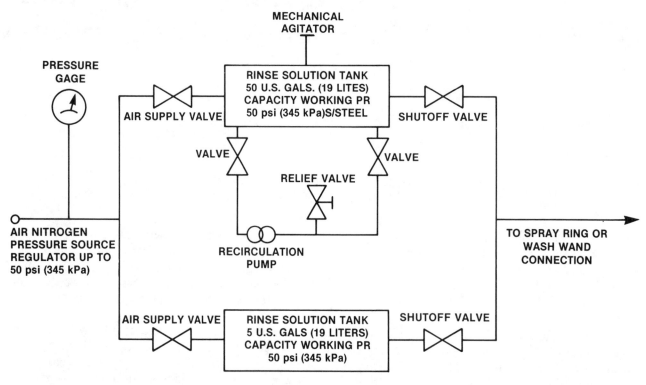

Fig. 1B-1   Schematic of the compressor wash facilities to use when cleaning the compressor on a Pratt and Whitney PT6 turboprop engine

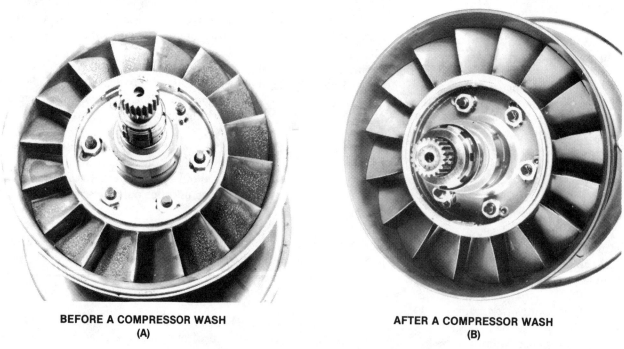

Fig. 1B-2   PT6 turboprop engine compressor

The fluid cleaning procedure is easily accomplished by first spraying an emulsion-type surface cleaner and then applying a rinse solution into the compressor. This is done while the engine is being motored by the starter or during low-speed operation. Fig. 1B-1 shows a Pratt and Whitney PT6 turboshaft performance recovery wash apparatus. It cannot be overstressed that the wash procedure must be performed in strict accordance with the instructions set forth in the manufacturer's maintenance manual.

A second, more vigorous method of field cleaning is to inject an abrasive grit of ground up walnut shells or apricot pits (tradename Carboblast) into an engine operating at high power settings. The amount of material and the operation procedure is prescribed by the manufacturer for each particular engine.

The timely use of fresh water rinsing, where prescribed, to remove salt deposits and also use of inlet and exhaust plugs will greatly reduce the need for these heavy cleaning procedures.

Fig. 1B-3 Foreign object damage and compressor contamination can be minimized by installed inlet plugs in the engine when the aircraft is not ready for flight.

c. Scheduled line maintenance

Inspections such as the 100-hour, annual, and progressive, are frequent tasks performed by the maintenance technician. FAR Parts 43, 65, and 91 describe the scope of these inspections, while the manufacturer publishes the specific inspection procedures for the particular engine being used.

d. Unscheduled line maintenance

The maintenance technician corrects discrepancies found during flight, walk-around inspections, scheduled inspections, while performing ADs (Airworthiness Directives), etc.

2. Heavy, or shop, maintenance

Whenever the engine cannot be repaired in the airplane, it is removed for shop maintenance or for test cell operation and troubleshooting. The FAA requires that this level of maintenance be accomplished only at a manufacturer's facility, or at a certified repair station which has the trained personnel, technical data, and the necessary tooling.

a. Powerplant removal

Powerplants are removed from the aircraft by one of two methods. One method involves lowering the engine from its mounting location using a hydraulically operated installation stand which looks like a large scissor jack. This method

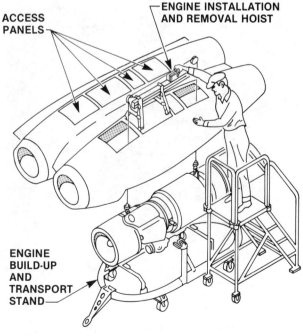

Fig. 1B-4 Engine removal using a two-cable hoist system

is generally used when working with large engines. The other method, more commonly used in general aviation, requires a sling and hoist arrangement to lower the engine into its transportation dolly.

b. Shop maintenance

Once in the shop, the engine to be repaired is usually installed in a maintenance stand. Some stands are on casters while others look like the one in Fig. 1B-5. Many stands are designed to keep the engine horizontal.

Fig. 1B-5 Turbine engine installed on a horizontal work stand

At some point during this procedure, disassembly is usually completed vertically, such as the removal of the compressor. On some engines,

the entire disassembly is accomplished vertically so the weight will assist in component alignment. One standard maintenance practice during disassembly is to cover all openings as they are exposed, using plugs, caps and other suitable material to prevent contamination; also to maintain the utmost in shop cleanliness and safety procedures.

Fig. 1B-6 Turbine engine installed on a vertical work stand

Another general rule during any maintenance applies here: that is never to reuse lockwire, lockwashers, tablocks, cotter pins, gaskets, packing or rubber O-rings; and to reuse locknuts and other fasteners only within the prescribed manufacturer's instructions.

In Fig. 1B-7, we see an expanded view of an entire General Electric CJ-610 engine as it would appear disassembeled during heavy maintenance.

3. Cold section and hot section maintenance

For the purpose of inspection and maintenance, the basic engine is divided into two main sections; the cold section and the hot section. The cold section includes the engine inlet, compressor

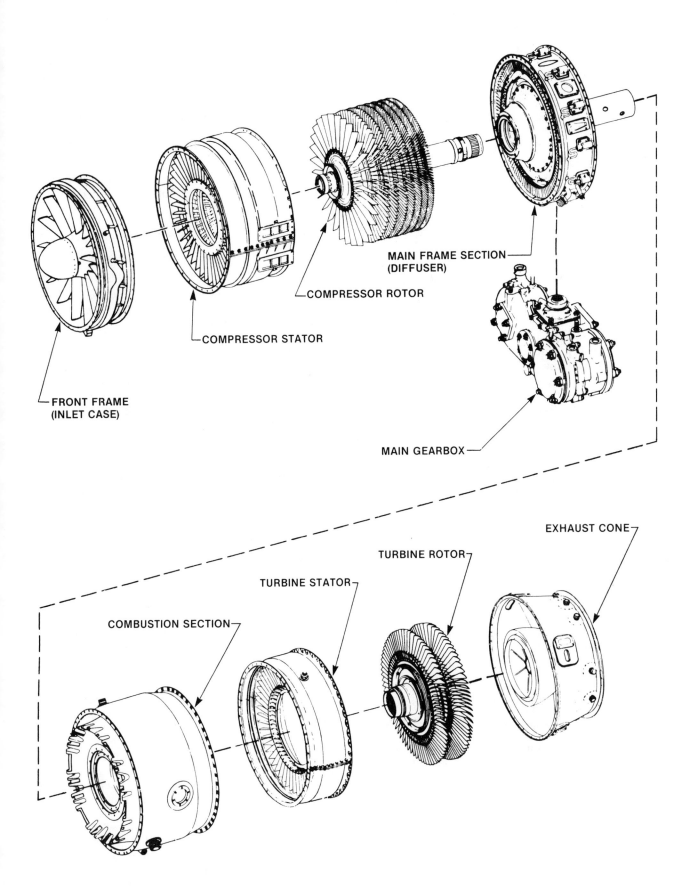

FRONT FRAME
(INLET CASE)

COMPRESSOR STATOR

COMPRESSOR ROTOR

MAIN FRAME SECTION
(DIFFUSER)

MAIN GEARBOX

COMBUSTION SECTION

TURBINE STATOR

TURBINE ROTOR

EXHAUST CONE

Fig. 1B-7 Major subassemblies of a General Electric CJ610 gas turbine engine

59

and diffuser sections. The hot section includes the combustor, turbine and exhaust sections.

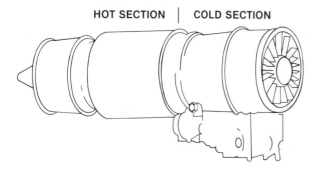

HOT SECTION | COLD SECTION

Fig. 1B-8 Turbine engines are divided into two basic sections, the hot section and the cold section.

The following is a list of terms commonly used to describe inspection findings during gas turbine engine maintenance:

TERM	DEFINITION
Abrasion	Wearing away of small amounts of metal as a result of friction between parts
Blister	The raised portion of a surface, caused by separation of layers of material
Bowing	A bend or curve in a normally straight or nearly straight line
Buckling	Large scale deformation from the original shape of a part, usually caused by pressure or impact of a foreign object, unusual structural stresses, excessive localized heating, or any combination of these
Burn	Discoloration from excessive heat
Burr	A sharp projection or rough edge
Converging	Two or more lines (cracks) which approach one another and which, if allowed to continue, will meet at a single point
Corrosion	A surface chemical action result-in surface discoloration, a layer of oxide, or, in advaced stages, the removal of surface metal
Crack	A fissure, or break in a material
Crazing	Minute cracks which tend to run in all directions. It is often noticed on glazed or ceramic-coated surfaces and is occasionally referred to as "china cracking"
Deformation	Alteration of form or shape
Dent	A smooth, round-bottomed depression
Distress	An all-inclusive term to mean any of the problems included in this listing
Distortion	A twisted or misshaped condition
Erosion	Wearing away of metal and/or surface coating
Flaking	Loose particles of metal on a surface or evidence of removal of surface covering
Frosting	An initial stage of scoring caused by irregularities or high points of metal welding, together with minute particles of metal transferring to the mating surface, giving a frosted appearance
Galling	Chafing caused by friction
Gouging	A removal of surface metal, typified by rough and deep depressions
Grooving	Smooth, rounded indentation caused by concentrated wear
Inclusion	Foreign matter enclosed in metal
Metalization	Coating by molten metal particles sprayed through the engine
Nick	A sharp-bottomed depression with rough outer edges
Peening	Flattening or displacing of metal by repeated blows. A surface may be peened by continuous impact of foreign objects or loose parts
Pitting	A surface condition recognized by minute holes or cavities which occur on overstressed areas. The pits may occur in such profusion as to resemble spalling
Scoring	A form of wear characterized by a scratched, scuffed, or dragged appearance with markings in the direction of sliding. It generally occurs at or near the top of a gear tooth

Scratches	Narrow, shallow marks, or lines, resulting from the movement of a metallic particle or sharp-pointed object across a surface
Scuffing	A dulling or moderate wear of a surface resulting from a slight amount of rubbing
Seizure	A welding or binding of two adjacent surfaces, preventing further movement
Spalling	A separation of metal by flaking or chipping
Stress	Metal failure due to compression, tension, shear, torsion, or shock
Tear	Parent metal torn by excessive vibration or other stresses
Unbalanced	A condition created in a rotating body by an unequal distribution of weight about its axis. Usually results in vibration.
Wear	A condition resulting from a relatively slow removal of parent material. Frequently not visible to the naked eye.

a. Cold section inspection and repair

(1) Compressor blade repair

Minor impact damage to compressor blades can be repaired if the damage can be removed without exceeding allowable limits. When repairs are completed within the prescribed limits, there will be no compressor imbalance and balancing checks are not usually required.

The illustrations in Fig. 1B-9 are of typical gas turbine cold section repairs. This information is general in nature and information contained in the manufacturer's manuals is always followed when making these repairs.

Repair to coded areas is accomplished by a procedure called blending. Blending is a hand method of recontouring damaged blades and vanes using small files, emery cloth and stones. The use of power tools is seldom permitted because of the possibilty of inadvertantly damaging adjacent parts in what is usually a restricted working area. Blending is performed parallel to the length of the blade to minimize stress points and to restore as smooth a shape as possible to the surface. Quite often this procedure can be completed on the flightline if damage is limited to the first one or two stages.

After repair, some manufacturers recommend covering the blend with a felt tip dye marking material or similar solution to identify it as a reworked area. This benefits maintenance personnel who will later see it while performing inlet inspections.

Welding and straightening of rotating airfoils usually requires special equipment and is

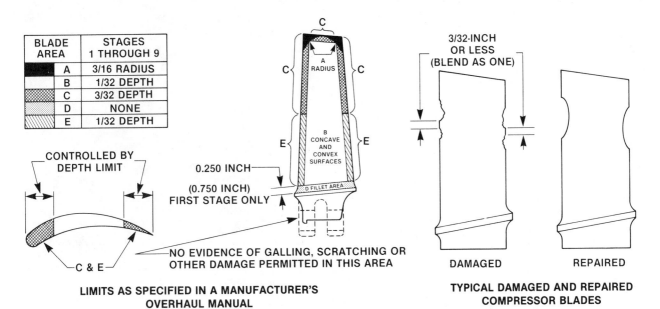

BLADE AREA	STAGES 1 THROUGH 9
A	3/16 RADIUS
B	1/32 DEPTH
C	3/32 DEPTH
D	NONE
E	1/32 DEPTH

CONTROLLED BY DEPTH LIMIT

C & E

0.250 INCH

(0.750 INCH) FIRST STAGE ONLY

NO EVIDENCE OF GALLING, SCRATCHING OR OTHER DAMAGE PERMITTED IN THIS AREA

LIMITS AS SPECIFIED IN A MANUFACTURER'S OVERHAUL MANUAL
(A)

C

A RADIUS

B CONCAVE AND CONVEX SURFACES

E

D FILLET AREA

3/32-INCH OR LESS (BLEND AS ONE)

DAMAGED   REPAIRED

TYPICAL DAMAGED AND REPAIRED COMPRESSOR BLADES
(B)

Fig. 1B-9 Compressor blade repair limits

quite often authorized only at overhaul facilities or by the manufacturer. One new technique called electron beam welding, now permits the reworking of many formerly unserviceable compressor blades. This beam welding procedure is especially useful on titanium from which many blades are made and results in a strength factor equal to a new blade.

Blade replacement is generally allowed but there is usually a restriction placed on the number of blades that may be replaced per stage and in the entire rotor. The blades are moment-weighted and coded to provide a means of exact replacement for maintaining compressor balance. Moment-weight accounts for both the mass weight and the center of balance.

On compressor stages with an even number of blades, two blades can be replaced if one having the appropriate moment-weight is not available. The damaged blade and its opposite blade are replaced together as a pair. If the compressor stage has an odd number of blades, three blades, 120° apart can be replaced as a set under the same circumstances. Under severe restrictions some manufacturers allow replacement by mass-weight and certainly other replacement criteria also exists for specific installations. Some fan blades, for instance, are replaceable on the wing and some only during heavy maintenance.

(2) Stator vane and compressor case repairs

Repairs of slight impact damage to stator vanes and compressor cases are generally made by blending. Cracks are usually weld-repairable with inert gas welding apparatus. Typical observable damage is shown in Fig. 1B-10. If weld beads interfere with fits or airflow, they are ground back to the original contour, as near as possible.

b. Hot section inspection and repair

(1) Combustion section

Cracking is one of the most frequent discrepancies that will be detected while inspecting the combustion section of a turbine engine. The combustion liner is constructed of thin stainless steel material that is subject to high concentrations of heat.

Visual inspection

A common method of checking for misalignment, cracks and other hot and cold section distress on installed engines, is by using an internal viewing device called a borescope, Fig. 1B-11. There are both rigid and flexible types available with different illumination and magnification capabilities. To use the borescope, simply insert the viewing end into special inspection ports or through existing openings such as igniter plug bosses. With this unit, the maintenance technician can easily view internal engine components and determine airworthiness from his visual inspection.

COURTESY OF LENNOX

Fig. 1B-11 Inspection with a borescope of the inside of a turbojet engine

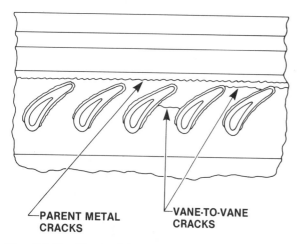

PARENT METAL CRACKS

VANE-TO-VANE CRACKS

Fig. 1B-10 Typical damage to a compressor case that is repairable by welding

During this type of inspection, he looks for distress which is out of the manufacturer's limits, such as obvious cracking, warpage, burning, erosion, and sometimes obscure but tell-tale hot spots. These hot spots are possible indicators of a serious condition, such as malfunctioning fuel nozzles or other fuel system malfunctions, requiring special interpretation.

Another important aspect of borescoping is to check for misalignment of combustion liners. It has been determined that burner can shift as it is called, can seriously affect combustor efficiency and engine performance.

Whether the engine is fit to remain in service or needs to be disassembled, must be determined by the technician. His expertise and training make this important determination possible.

(2) Turbine and exhaust sections

Inspection of the inner turbine section is done with a borescope, or visually, through the tailpipe with a strong light, mirror and a magnifying glass, or by disassembling of the engine. On the flightline, nondestructive inspection techniques such as dye penetrant, Zyglo, etc., are also useful in external inspection.

As in other hot section inspections the technician is most likely to see small cracks caused by compression and tension loading from heating and cooling. Much of this type of distress is acceptable and requires no maintenance action, because after initial cracks relieve the stress no elongation of cracks occurs.

Erosion is another common defect found during inspection. Erosion is the wearing away of metal either from the gas flow across its surfaces or impingement on the surfaces of impurities in the gas flow.

If hot spots are seen, they are generally the first indication of a malfunction in fuel distribution. Too rich a mixture can cause lengthening of the flame until it reaches the turbine nozzles. Also a condition called hot streaking can cause a flame to penetrate through the entire turbine system to the tailpipe. This condition is caused by a clogged fuel nozzle which does not atomize the fuel into a cone-shaped pattern, but rather allows a small fuel stream to flow with sufficient force to cut through the cooling air blanket and impinge directly on the turbine surfaces.

(a) Turbine wheel

Fig. 1B-12 shows turbine blade blend repair limits which are typical of either shop or flight-line maintenance. According to most manufacturers, cracks are never acceptable.

Of particular concern during visual inspections are stress rupture cracks on turbine blade leading or trailing edges. Stress rupture cracks are perceptible as minute hairline cracks at right angles to the blade length. This condition, and rippling of the trailing edge, are indications of a serious over temperature and a special in-shop manufacturer's inspection will probably be required. (See Fig. 1B-12 on page 64.)

To maintain turbine wheel balance, a single turbine blade replacement is generally accomplished by installing a new blade of equal moment-weight. If the blade's moment-weight cannot be matched, the damaged blade and one blade 180° away, are replaced with blades of equal weight; or, the damaged blade and the blades 120° from it are replaced with three blades of equal moment-weight in the same manner as was mentioned for compressor blades.

Turbine blades are rarely replaced on the wing today. However, shop procedures usually allow for entire turbine re-blading after which the rotor is checked on a special balancing device.

Code letters indicating the moment-weight in ounces are marked on the fir-tree section of the blade as shown in Fig. 1B-13 on page 65.

(b) Turbine case, turbine vanes, exhaust section

Certain types of small visible cracks result from thermal stresses and are commonly found after periods of normal operation. The growth of many such cracks with further operation is often negligible, since the cracks in effect relieve the original stress condition. The manufacturer's limits, based on this consideration, allow certain distress conditions to be acceptable. The only consideration for the technician performing the inspection is whether the cracks converge and are likely to permit a portion of the material to break away. This condition can cause impact damage to downstream components as well as result in serious burn-through damage by misdirection of the hot gases. (See Fig. 1B-14 on page 66.)

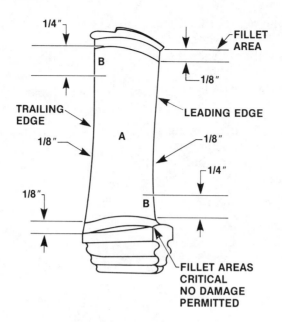

NOTE: RIPPLING OF TRAILING EDGE IS NOT ACCEPTABLE

IDENTIFICATION OF BLADE LOCATIONS
(A)

INSPECTION	BLADE SHIFT	AREA A			AREA B	LEADING & TRAILING EDGES	
		NICKS	DENTS AND PITS	CRACKS	NICKS, DENTS, AND PITS	NICKS, DENTS, AND PITS	CRACKS
MAXIMUM SERVICEABLE	PROTRUSION OF ANY BLADE ROOT MUST BE EQUAL WITHIN 0.015 INCH EITHER SIDE OF DISK	THREE 0.015 INCH LONG BY 0.005 INCH DEEP	THREE 0.010 INCH DEEP	NOT ACCEPTABLE	ONE 0.020 INCH DEEP	ONE 0.020 INCH DEEP	NOT ACCEPTABLE
MAXIMUM REPAIRABLE	NOT REPAIRABLE	THREE 0.015 INCH LONG BY 0.010 INCH DEEP	THREE 0.015 INCH LONG BY 0.010 DEEP	NOT REPAIRABLE	NOT REPAIRABLE	TWO 1/8 INCH DEEP	NOT REPAIRABLE
CORRECTION ACTION	RETURN BLADED DISK ASSEMBLY TO AN OVERHAUL FACILITY	BLEND OUT DAMAGED AREA	BLEND OUT DAMAGED AREA	RETURN POWER SECTION TO AN OVERHAUL STATION	RETURN POWER SECTION TO AN OVERHAUL STATION	BLEND OUT DAMAGED AREA	RETURN POWER SECTION TO AN OVERHAUL STATION

TYPICAL REPAIR LIMITS
(B)

Fig. 1B-12 Typical turbine blade repairs

(c) Creep and untwist

Creep is a term used to describe the permanent elongation which occurs to rotating parts. Creep is most pronounced in turbine blades because of the heat loads and centrifugal loads imposed during operation. Each time a turbine blade is heated, rotated, and then stopped, (this is called one cycle) it remains slightly longer than it was before. The additional length may be only millionths of an inch under normal circumstances or very much longer after an engine overtemperature or overspeed condition. Nevertheless, if the blade remains in service long enough, chances are

144

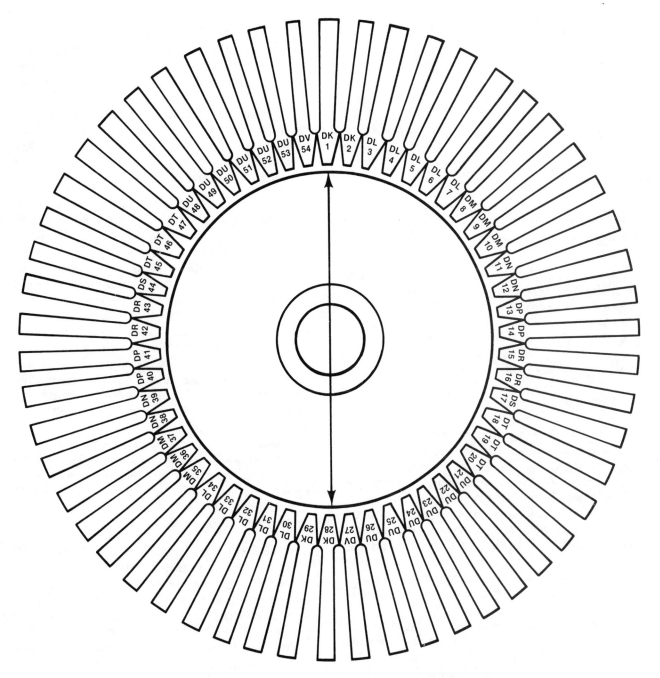

Fig. 1B-13   Typical turbine blade moment-weight coding

that it will eventually make contact with its shroud ring and begin to wear away. When this occurs, an audible rubbing can be heard on engine coastdown. Clearance checks are then taken and appropriate maintenance action is determined.

Untwist occurs in both turbine blades and turbine vanes from gas loads upon their surfaces. Loss of correct pitch effects efficiency of the turbine system and engine performance deterioration results. The check for untwist is generally

possible only after engine teardown when parts can be measured in special shop fixtures.

(d)   Marking of parts

The temporary marking procedure for both hot and cold section parts is to apply layout dye, or to mark with a commercial felt-tip applicator or special marking pencil. A note of caution to observe when marking any hot section parts for identifying distressed areas is *not* to use a

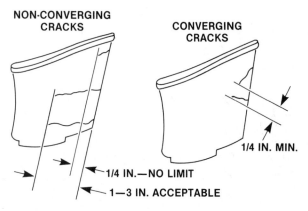

Fig. 1B-14 Typical converging and non-converging crack conditions

substance which leaves a carbon, copper, zinc, or lead deposit. When the metal is heated, these deposits are drawn into the metal and can cause intergranular stress. *Marking with a common graphite-lead pencil is strictly prohibited.*

4. Bearings and seals

a. Bearings

The main bearings of a gas turbine engine are either ball or roller antifriction types. Ball bearings are caged and support the main engine rotor for both axial (thrust) and radial (centrifugal) loads. The roller bearings generally are not caged. And, because they have a greater surface contact area than the ball bearings, they are so positioned to absorb the bulk of the radial loading and also to allow for axial growth of the engine during operation. For this reason tapered roller bearings are seldom used as they do not allow for growth.

Plain bearings are not used as main bearings in turbine engines, as they are in reciprocating engines, because turbines operate at much higher speeds and friction heat buildup would be prohibitive. Plain bearings (bushings), however, are used in some minor locations and in accessories.

The primary loads acting on main bearings come from the following sources: weight of the rotating mass (compressor and turbine) magnified many thousands of times by radial G-forces; axial forces from power changes; gyroscopic effect of heavy rotating masses trying to remain in place as the aircraft changes direction; compression and tension loads between the stationary casings and the rotor system caused by thermal expansion; and vibrations induced by the air-

stream, the aircraft, and the engine itself. The main bearings, then, support the rotor assemblies and transfer the various loads through the bearing housings and support struts to the outer cases and ultimately into the aircraft mountings.

The number of main bearings varies from one engine model to another. One manufacturer may prefer to install three heavy bearings and another manufacturer may use five or six lighter bearings to accommodate the same load factors.

Construction features of ball and roller bearings are shown in Fig. 1B-15. A design feature to note is that the roller bearing inner race is not grooved, allowing the roller freedom to move axially when the engine expands and contracts during operation. The split inner race is a design feature of the ball bearing which allows for ease of disassembly, maintenance, and inspection. Also shown in Fig. 1B-16 is the oil damped bearing which is provided with an oil film between the outer race and the bearing housing to reduce vibration tendencies in the rotor system. The component pieces of turbine bearings are generally identified by serial number to prevent intermixing. Gas turbine quality anti-friction bearings are matched sets of races, cages, etc., to conform to extremely close production tolerances. Remember that the gas turbine is referred to as a high-speed lightweight device, and this requires that vibrations be kept to an absolute minimum even under great G-forces and gyroscopic loads.

The air balance chamber shown in Fig. 1B-16 aids the compressor thrust bearing (No. 2) in combating high gas path pressures which try to push the compressor forward. The balance chamber and the thrust bearing restrain the compressor against this pushing force. Some engines do not need as large a thrust bearing as others or do not use an air balance chamber because the opposite thrust load at the turbine partially cancels out the forward pushing loads on the compressor.

b. Bearing seals

The bearing housing usually contains seals to prevent loss of oil into the gas path. Oil seals are usually of the labyrinth or carbon rubbing types and it is common to see one or the other or a combination of the two in the same engine. The need for this arrangement arises because of temperature gradient differences in the hot and cold sections of the engine. That is, if a labyrinth seal is

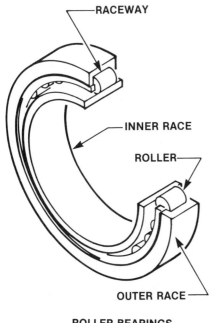

RACEWAY

INNER RACE

ROLLER

OUTER RACE

ROLLER BEARINGS
(A)

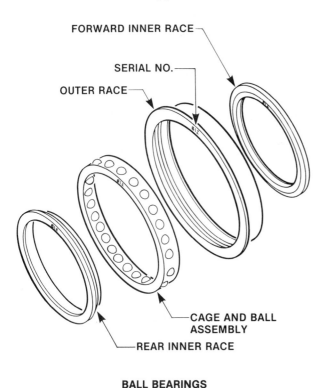

FORWARD INNER RACE

SERIAL NO.

OUTER RACE

CAGE AND BALL
ASSEMBLY

REAR INNER RACE

BALL BEARINGS
(B)

Fig. 1B-15 Turbine engine main bearings

used in some hot locations it might expand and contact its seal land, causing wear. The seal land is the name generally given to the rotating portion of the labyrinth seal.

The two labyrinth seals, as shown in Fig. 1B-16, form a compartment in which the bearing is housed. Air from the gas path that is present outside of the bearing compartment bleeds inward across grooves cut in the labyrinth seal. These grooves form sealing rings in either a concentric path similar to a screw thread or are non-concentric with each ring in its own plane. In any case the *seal dams* formed by the rings allow for a metered amount of air from the engine gas path to flow inward. Pressure in this cavity is often maintained near atmospheric level. This is discussed later in detail in the Lubrication Systems Section of this Intergrated Training Program.

The oil mist created by the oil jet spraying on the rotating bearing is prevented from exiting the bearing compartment by the air entering across the labyrinth seal. The seal pressurizing air then leaves the bearing area, in this particular engine, by way of the scavenge oil system. Most higher compression engines are designed with a separate vent subsystem.

Carbon seals, a blend of carbon and graphite, are similar in function and location to labyrinth seals but not in design. The carbon seal rides on a highly polished surface, while the labyrinth seal maintains an air gap clearance.

The carbon seal is usually spring-loaded and sometimes pressurized with air to create a uniform pressure of the carbon ring against its mating surface, providing the oil sealing capability.

The carbon seal shown in Fig. 1B-17A is classified as a carbon-ring type of seal which rides on a rotating shaft. Another common design is the carbon-face type of seal Fig. 1B-17B. It is similar to those used as drive shaft seals in many fluid carrying accessories. The carbon seals themselves are generally stationary, with their highly polished mating surface called a seal-race attached to and turning with the main rotor system. (See Fig. 1B-17 on page 69.)

The carbon seal will be found where a more positive control over airflow into the bearing sumps is required or where this full-contact-type seal will hold back oil which might at times puddle before being scavenged. Conversely, labyrinth sealing will usually be associated with oil system locations designed with higher vent subsystem pressures. This system is discussed in the Lubrication Systems Section of this Integrated Training Program.

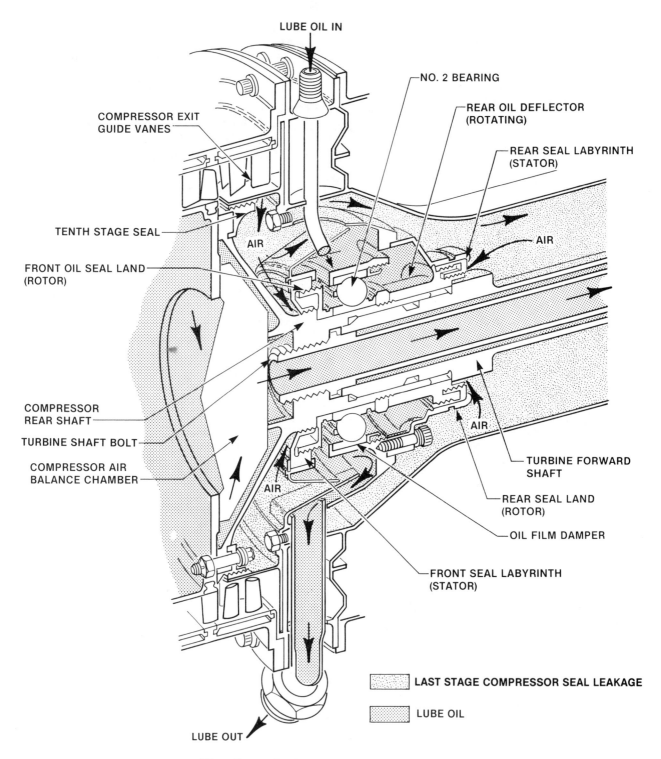

LUBE OIL IN

NO. 2 BEARING

REAR OIL DEFLECTOR
(ROTATING)

REAR SEAL LABYRINTH
(STATOR)

COMPRESSOR EXIT
GUIDE VANES

TENTH STAGE SEAL

AIR

AIR

FRONT OIL SEAL LAND
(ROTOR)

COMPRESSOR
REAR SHAFT

TURBINE SHAFT BOLT

COMPRESSOR AIR
BALANCE CHAMBER

AIR

AIR

TURBINE FORWARD
SHAFT

REAR SEAL LAND
(ROTOR)

OIL FILM DAMPER

FRONT SEAL LABYRINTH
(STATOR)

LAST STAGE COMPRESSOR SEAL LEAKAGE

LUBE OIL

LUBE OUT

Fig. 1B-16 Typical compressor rear bearing

c. Bearing handling and maintenance

Most manufacturers require that bearings be cleaned and inspected in a special environmentally-controlled bearing handling room. This prevents surface corrosion while bearings are out of the engine. Moreover, bearings are usually never left unprotected, to prevent even the slightest damage from occurring. Proper bearing handling during inspection is accomplished with lint-free cotton or synthetic rubber gloves to prevent acid or moisture on the hands contacting the

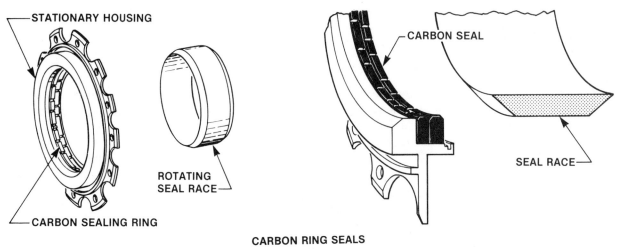

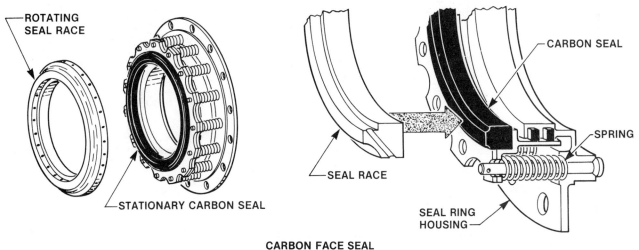

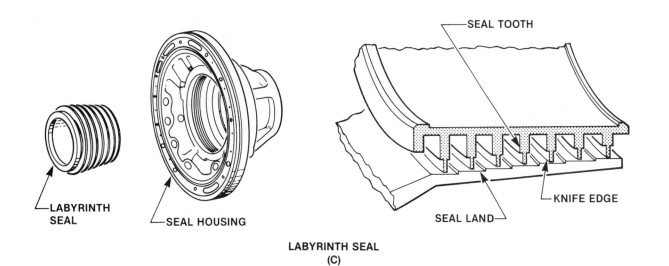

Fig. 1B-17   Typical bearing seals

bearing surfaces. Bearing inspection is usually conducted under strong lights and magnification to determine their serviceability. Only very minor defects are allowable.

During inspection, bearings receive a feel test. This is accomplished by an experienced technician who compares the rotational feel of one bearing against the feel of a new one.

After inspection many measurement checks are performed on the bearings with special measurement devices provided by the manufacturer.

The technician should be made aware of several precautionary measures during bearing inspection. These are as follows:

1. Do not spin dry bearings because dust particles scratch them.

2. Do not blow bearings dry with shop air because the moisture in the air can cause corrosion.

3. Do not vapor-degrease bearings, as vapors support contaminants.

4. Do not use petroleum oils on bearings intended for use in engines which use synthetic oil. A chemical reaction can result.

5. Do not use the shop cleaning vat. Instead immerse bearings in *clean* fluid or wipe them clean, using a lint-free cloth or suitable paper wiper and an approved cleaning solvent.

d. *Magnetism check*

After inspection, bearings are checked for the presence of magnetism with a device called a field detector. If magnetism is present, it must be removed with a suitable degausser to prevent the attraction of foreign ferrous particles into the bearing during engine operation.

e. *Bearing installation*

Bearings are stored in vapor proof paper until ready to install. Also, where appropriate, the inner and outer races are either heated in a clean bath of engine oil or chilled in a refrigerator before being fitted into their installation positions.

f. *Bearing distress terms*

TERM	DESCRIPTION
abrasion	A roughened area caused by the presence of fine foreign material between moving surfaces
brinelling	An indentation sometimes found on the surface of ball or roller bearings, caused by shock.
burning	An injury to the surface caused by excessive heat. This is evidenced by discoloration or, in severe cases, by loss of material.
burnishing	A mechanical smoothing of a metal surface by rubbing. It is not accompanied by removal of material, but sometimes by discoloration around the outer edges of the area.
burr	A sharp projection or rough edge.
chafing	A rubbing action between two parts which have limited relative motion.
chipping	Breaking out of small pieces of material.
corrosion	Breakdown of the surface by chemical action.
fretting	Discoloration which may occur on surfaces which are pressed or bolted together under high pressure. On steel parts the color is reddish brown. On aluminum, the oxide is white.
galling	The transfer of metal from one surface to another, caused by chafing.
gouging	The displacement of materials from a surface by a cutting, tearing, or displacement effect.
grooving	Smooth rounded furrows, such as score marks, where the sharp edges have been polished off.
guttering	A deep concentrated erosion, caused by overheat or burning.

inclusion	Foreign material being enclosed in the metal. Surface inclusions are indicated by dark spots or lines; an inherent discontinuity in the material.
nick	A sharp indentation caused by striking one part against another metal object.
peening	Deformation of the surface cause by impact.
pitting	Small, irregularly shaped cavities in a surface from which material has been removed by corrosion or chipping. Corrosive pitting is usually accompanied by a deposit formed by the action of a corrosive agent on the base material.
scoring	Deep scratches made by sharp edges or foreign particles during engine operation; elongated gouges.
spalling	Sharply roughened area characteristic of the progressive chipping or peeling of surface material, caused by overloading.

5. Torque wrench use

The torque wrench is a tool required for both line and shop maintenance. Before using a torque wrench, the technician should ensure that it is in calibration by means of a weight and lever arm tester. All gas turbine engine manufacturers require careful torquing (application of twisting force) and most recommend a calibration schedule for torque wrenches utilized in maintenance of their engines. A typical schedule is as follows:

1. Micrometer-type torque wrenches — Check once a week

2. Non-set-type torque wrenches — Check once a month

It is not uncommon in larger shops to have a torque wrench calibration tester available in a particular work station where frequent daily use is made of torque wrenches. In this case the wrench can be calibrated daily or even before each use.

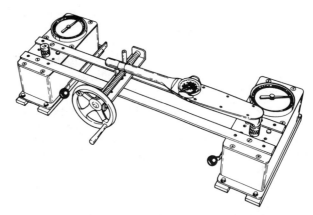

Fig. 1B-18   Torque wrench calibrator

To prevent loss of calibration of micrometer type torque wrenches, always return them to their lowest setting after use. The micrometer torque wrench is the most widely used in the industry because almost every fastener on a gas turbine engine requires a specific torque. And this type wrench can be used quickly and accurately in positions and places difficult to work in with other types. Torque with this wrench is applied by feel. When the wrench breaks away the correct torque is applied and the technician does not have to see the scale.

The beam and dial torque wrenches are said to apply a more accurate initial torque if the user holds the torque for a few seconds after reaching the desired amount to allow a set to occur between the mating parts. The user must, however, be positioned so he will always have an unob-

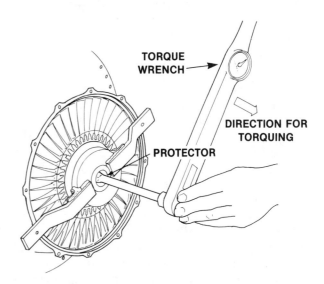

TORQUE WRENCH

DIRECTION FOR TORQUING

PROTECTOR

Fig. 1B-19   Proper use of a torque wrench to install a turbine wheel in a gas turbine engine

structed, straight-on view of the scale, or an inaccurate torque will result.

With the micrometer type wrench, if the manufacturer feels a set must occur between parts for accurate torque application, he will stipulate a procedure of torquing, loosening and retorquing; torquing to minimum value and a second torque to the required value or some such arrangement which will bring about the desired results for his particular needs.

a. Use of torque wrench with extension

Occasionally an extension to a torque wrench is needed. The following information describes the procedure for calculating the indicated torque value as compared to the true torque being applied. (See Fig. 1B-20.)

EXAMPLE:

A torque of 1,440 inch-pounds is required on a part. A special extension having a length of 3 inches from the center of its wrench slot to its square drive is used. The torque wrench measures 15 inches from the center of its handle to the center of its square adapter.

$$R = \frac{L \times T}{L + E} = \frac{15 \times 1,440}{15 + 3} = 1,200$$

With the axis of the extension and torque wrench in a straight line, tightening to a wrench reading of 1,200 inch-pounds will provide the desired torque of 1,440 inch-pounds.

b. Torque procedure

The manufacturer usually specifies the torque to apply and the torquing procedure to use. This procedure must be followed consistently so as not to set up stresses between the mating flanges.

The turbine engine, being cylindrical in shape, has many bolt-ring sets. Often the manufacturer will stipulate torquing the entire ring to the low limit, and then loosening and retorquing to the median value. For example, if the manual torque chart indicates a torque of 90 to 100 inch-pounds, the technician would torque all bolts initially to 90 inch-pounds and then loosen and retorque them all to 95 inch-pounds.

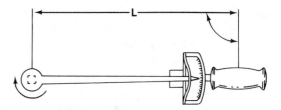

FLEXIBLE-BEAM-TYPE TORQUE WRENCH
(A)

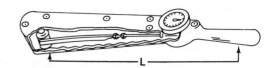

RIGID-FRAME, DIAL-TYPE TORQUE WRENCH
(B)

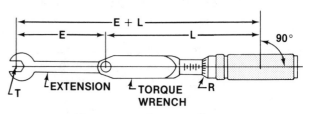

TYPICAL EXTENSION ATTACHED
(CENTER LINES CONNECTING)

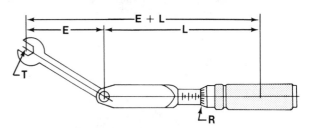

TYPICAL EXTENSION ATTACHED
(CENTER LINES BISECTING)

FORMULA: WHERE:

$$R = \frac{L \times T}{L + E}$$

T = TORQUE DESIRED
E = EFFECTIVE LENGTH OF EXTENSION
L = EFFECTIVE LENGTH OF TORQUE WRENCH
R = INDICATOR TORQUE READING

FORMULAS FOR USE OF AN EXTENSION
(C)

Fig. 1B-20 Use of a torque wrench with an extension

If a special procedure is not prescribed, the standard maintenance practice is to torque the bolts at staggered locations until the required torque value is reached. Also when torquing castellated nuts or nuts and bolts with tab-washers, the *low* torque should be applied and then additonal torque applied up to the *maximum* value if necessary to align the cotter key slot or the tablock slot. If alignment cannot be made, a new nut or bolt should be installed until you can get the right combination of torque and alignment.

Another generally used procedure is to torque one bolt at each 90° location initially, and then stagger torque, at 180° locations, the remaining bolts in the ring.

If a special torque value is not indicated by the manufacturer, the technician should use a standard torque chart.

6. Lockwiring

Lockwiring, or safety wiring as it is sometimes called, is a common line and shop maintenance procedure. It is a method designed to prevent loosening of threaded parts after they have been torqued. This is a significant airworthiness consideration on turbine engines because dangerous air, oil, and fuel leaks can occur at loose flanges and present serious flight hazards.

Care must be taken to select only the correct type and diameter lockwire as recommended by the manufacturer. The wire should not be overtightened, as it work-hardens and could break during service. The twisting for 0.032 inch lockwire is usually recommended at 8 to 10 twists per inch and for 0.041 inch diameter lockwire, 6 to 8 twists per inch.

The commonly accepted lockwiring techniques are indicated in Fig. 1B-21 on page 74.

7. Modular maintenance

A newer concept of line and heavy maintenance is referred to as modular maintenance. This is an engine construction concept, in that the total engine is a set of separate modules. These modules are preassembled, balanced, and designed to be more easily removed and replaced with a minimum of man-hour expenditure. Some modules can also be replaced while the engine remains installed in the airplane.

In Fig. 1B-22, the fan module is shown being removed. The modular concept in this case is a repair that is a combination of line maintenance

and heavy maintenance. This procedure will save an engine replacement and significantly reduce the cost to the customer. The factory-supplied module reportedly provides a better product with an increased service life. However, when rebuilt modules are installed in engines with high service time achieving a good aerodynamic and thermodynamic match is extremely difficult and is becoming a science in itself for repair station and heavy repair shop managers. Because of the nature of modular maintenance, FAR Part 43 is now being interpreted to state that in most cases replacement of modules is considered minor repair; requiring no FAA Form 337. But overhaul of modules is considered a major repair which requires FAA Form 337.

Fig. 1B-22 Modular construction of a turboprop engine makes on-wing maintenance practical.

8. Test cell maintenance

After manufacture or heavy repair, the engine generally requires an integrity check in a specific test facility. This facility is instrumented to provide operational data that is beyond the capability of the instruments in the aircraft cockpit. The test bed, for instance, is equipped to electronically measure the thrust which can be compared to the conventional thrust parameters.

Another primary engine inspection deals with vibration analysis during test cell operation. This is a very important inspection, as it is used to check for component imbalance in the low weight, high speed turbine engine, where rapid wear can occur quickly from even very small induced vibration. While reciprocating engines have a visible vibration characteristic, gas tur-

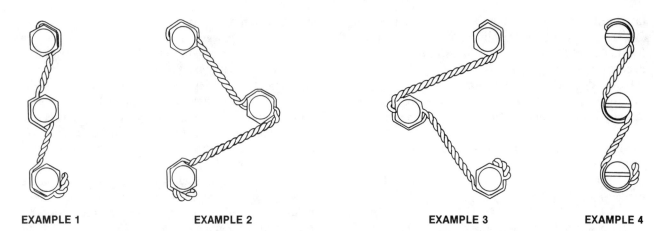

EXAMPLE 1   EXAMPLE 2   EXAMPLE 3   EXAMPLE 4

EXAMPLES 1, 2, 3, AND 4 APPLY TO ALL TYPES OF BOLTS, FILLISTER HEAD SCREWS, SQUARE HEAD PLUGS, AND OTHER SIMILAR PARTS WHICH ARE WIRED SO THAT THE LOSENING TENDENCY OF EITHER PART IS COUNTERACTED BY TIGHTENING OF THE OTHER PART. THE DIRECTION OF TWIST—FROM THE SECOND TO THE THIRD UNIT IS COUNTERCLOCKWISE TO KEEP THE LOOP IN POSITION AGAINST THE HEAD OF THE BOLT. THE WIRE ENTERING THE HOLE IN THE THIRD UNIT WILL BE THE LOWER WIRE AND BY MAKING A COUNTERCLOCKWISE TWIST AFTER IT LEAVES THE HOLE, THE LOOP WILL BE SECURED IN PLACE AROUND THE HEAD OF THAT BOLT.

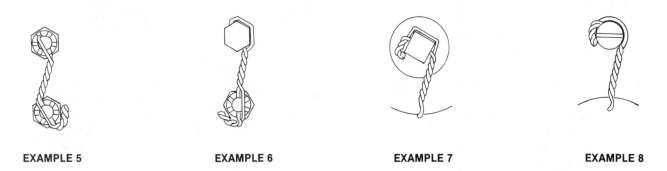

EXAMPLE 5   EXAMPLE 6   EXAMPLE 7   EXAMPLE 8

EXAMPLES 5, 6, 7, AND 8 SHOW METHODS FOR WIRING VARIOUS STANDARD ITEMS. NOTE: WIRE MAY BE WRAPPED OVER THE UNIT RATHER THAN AROUND IT WHEN WIRING CASTELLATED NUTS OR ON OTHER ITEMS WHEN THERE IS A CLEARANCE PROBLEM.

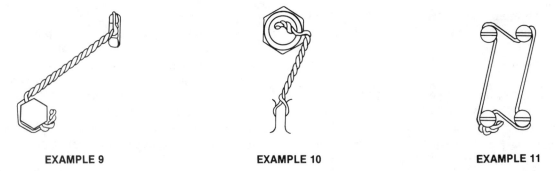

EXAMPLE 9   EXAMPLE 10   EXAMPLE 11

EXAMPLE 9 SHOWS THE METHOD FOR WIRING BOLTS IN DIFFERENT PLANES. NOTE THAT WIRE SHOULD ALWAYS BE APPLIED SO THAT TENSION IS IN THE TIGHTENING DIRECTION.

HOLLOW HEAD PLUGS SHALL BE WIRED AS SHOWN WITH THE TAB BENT INSIDE THE HOLE TO AVOID SNAGS AND POSSIBLE INJURY TO PERSONNEL WORKING ON THE ENGINE.

CORRECT APPLICATION OF SINGLE WIRE TO CLOSELY SPACED MULTIPLE GROUP.

Fig. 1B-21   Examples of proper lockwiring

bine engine vibration limits are along the order of only 3 to 5 thousandths of an inch. Vibration, then, can only be measured by using a meter and a special engine-mounted detector called a vibration transducer. This transducer is simply a small electrical generator which sends a signal to the vibration meter.

If an engine produces a vibration level beyond the limit during test cell operation, the rotating component causing the imbalance is identified and the engine is sent back to the overhaul facility for correction.

Test cells are also used by airlines and repair stations to troubleshoot engine malfunctions, test engines after minor repairs, perform research and development, test after modifications and for various other maintenance support reasons.

When all parameters are recorded, test cell personnel plot the observed readings against standard data to give a permanent record of the test run.

Run data

In order to test run an engine and check its performance on any given day and have meaningful data, all instrument readings are corrected to Standard Day conditions. A comparison can then be made between observed data and present conditions of ambient temperature and pressure and what the engine instruments would read if the engine were being run on a Standard Day of 29.92 in. Hg and 519° R (59° F). Also, to use the same type data in flight recording and monitoring of engine performance, correction factors are programmed into the system which bias ambient conditions at altitude to Standard Day atmospheric conditions. The data then becomes a tool for both maintenance and management to use in determining such factors as economy of continued operation and safety with regard to engine maintenance.

Consider the following run data taken when ambient temperature is 535° R (75° F) and ambient pressure is 30.0 in. Hg.

RPM   15,000

EGT   1,560°R (1,100°F)

Fuel Flow   1,500 lb./hr.

Mass Airflow   65 lb./sec.

To find what these values would be on a standard day, two correction factors are needed, referred to as delta (δ), the pressure correction factor, and theta (θ), the temperature correction factor.

$$\delta t2 = \frac{\text{observed ambient pressure}}{29.92 \text{ in. hg.}}$$

$$= \frac{30.0}{29.92} = 1.003$$

$$\theta t2 = \frac{\text{observed ambient temperature}}{519°R}$$

$$= \frac{535}{519} = 1.031$$

For ease of computations, delta and theta tables are used by Test Cell personnel.

The following formulas can now be used in conjunction with the two correction factors:

$$\text{Corrected RPM} = \frac{\text{observed RPM}}{\sqrt{\theta t2}}$$

$$= \frac{15,000}{\sqrt{1.031}} = 14,778.3$$

$$\text{Corrected EGT} = \frac{\text{observed EGT (°R)}}{\theta t2}$$

$$= \frac{1,560}{1.031} = 1,513.1°R$$

$$\text{Corrected W}_f = \frac{\text{observed W}_f \text{ (PPH)}}{\delta t2 \times \sqrt{\theta t2}}$$

$$= \frac{1,500}{1.003 \times \sqrt{1.031}} = 1,473.4 \text{ PPH}$$

$$\text{Corrected M}_s = \frac{\text{observed M}_s \times \sqrt{\theta t2}}{\delta t2}$$

$$= \frac{65 \times \sqrt{1.031}}{1.003} = 65.8 \text{ lb./sec.}$$

It is a fact that gas turbine engine performance is sensitive to ambient conditions. When conditions are different from standard day the observed readings are not what they would be on a Standard Day. The values that were solved for above, are what the engine would have if it were being run at Standard Day conditions. These values can now be compared with the manufacturer's specifications to determine engine condition. A further explanation of the corrections is as follows:

a. Corrected RPM — At higher than ambient temperature, RPM (compressor speed) is higher because less dense air requires more work to provide the necessary compression ratio.

b. Corrected EGT — In order to increase the RPM, more fuel is required. This of course is what elevates the exhaust gas temperature.

c. Corrected Fuel Flow (W_f) — The increased fuel flow occurs as the fuel control senses a drop in compressor energy output (compression) and substitutes fuel energy to keep the thrust level up.

d. Corrected Mass Airflow (M_s) — When ambient temperature is above STD, mass airflow relative to STD is lower. When ambient pressure is higher, as in this case mass airflow tends to be higher than STD. What can be seen here is more loss of M_s due to temperature effect than M_s gain due to pressure effect. This results in a net reduction of M_s from 65.49 to 65 lb./sec., a value which can be used to more accurately calculate thrust.

9. Time between overhaul (TBO)

Time between overhaul on gas turbine engines varies greatly between engines. General aviation engines usually have a manufacturer's recommended TBO of approximately 3,000 hours, with a half-life hot section inspection in between. Commercial operators do not generally establish rigid TBO limits. In conjunction with the Federal Aviation Administration (FAA), they maintain an inspection schedule and trend analysis of engine performance upon which engine removal for overhaul is based. Using this method, some large engines remain on the wing for many thousands of hours in what is called an on-condition maintenance program. Many operators use cycles rather than hours upon which to base inspection intervals. A cycle is generally considered to be one engine start, run and stop. So from this viewpoint, the traditional hourly TBO is not applicable to some gas turbine engines.

Certain engine parts are however life-dated or life-limited, meaning they have a scheduled time interval between the time they are new or overhauled and the time at which they must be replaced. The time interval can be in operating hours, cycles or calendar time such as months.

For all other non-life-limited parts the concept of on-condition maintenance prevails, meaning that the part will remain in place through successive inspections as long as it appears to be in an airworthy condition.

Cycle counters are installed in many aircraft. They usually function on the principle of electronically recording engine starts.

If counters are not available, flight crew and ground crew must ensure accurate recording of hours, dates or cycles so that all inspections, part changes, etc., are accomplished on time.

Turbine engines are expensive. The smallest flight engine new price is approximately $75,000 and a jumbo aircraft fan engine upwards to $2,000,000. From this standpoint alone, it is evident that good record keeping and proper recording of operating time becomes essential.

10. Troubleshooting

Troubleshooting is a major part of both line and heavy maintenance. Most of the engine systems troubleshooting comes as part of line maintenance and internal engine malfunctions are likely to be a troubleshooting task for either line or shop technicians. The technician in any case should adopt an intelligent sequence of procedures to ensure efficient correction of the problem. Too often a remove and replace philosophy prevails because the engine problem at hand looks similar to one that has happened in the past and the technician makes a snap judgment that is incomplete or incorrect.

A few guidelines to follow include:

1. Find all the facts surrounding the problem by interviewing the flight crew, reading the logbook, reviewing the work order file, etc.

2. Review the same documents for previous related discrepancies.

Troubleshooting Information and Priority Listing.

1. Problem: _No engine oil pressure indication on cockpit gauge on start_

2. Cockpit Indications:

   a. _EPR   OK_

   b. _EGT   OK_

   c. _RPM N1   OK_

   _N2   OK_

   d. _WF   OK_

   e. _Oil Temp.   Slightly high 21°C_

   f. _Fuel Temp.   High 215°C_

   g. _Oil Pressure   0_

   h. _Oil Quantity   OK_

   i. _Warning Lights   None_

3. Other Factors to Consider:

   From flight log: Same problem 4 flights previous. Changed indicator. Normal oil comsumption. Last service – 1 pint. Flight crew changed indicators – #1 to #2 – still no indication.

4. Suspect Causes: (in priority order)

   a. _Low oil quantity (cockpit quantity indicator may be inaccurate)_

   b. _Circuit breaker tripped_

   c. _Defective indicator input signal_

   d. _Defective transmitter or input signal_

   e. _Obstruction in line to oil pump_

   f. _Defective oil pump_

   g. _____

3. Research the maintenance manual thoroughly for system operation and troubleshooting clues.

4. Make a written list in order of priority of possible causes.

5. Troubleshoot by inspecting, testing and if possible, duplicating the malfunction until the cause of the problem is determined.

6. Make the necessary repairs to correct the problem or remove the engine for shop repair.

7. Where possible give recommendations to shop maintenance for possible repairs needed.

II. TURBINE ENGINE OPERATION

There are many instances when the maintenance technician is required to operate a gas turbine engine. Some of these instances are as follows:

1. To duplicate a flight crew reported discrepancy and for troubleshooting.

2. To perform a basic engine or engine system checkout after maintenance.

3. To move an aircraft from one maintenance location to another.

4. To taxi-check an aircraft system.

A. Safety Precautions

The technician must be throughly familiar with the flight line safety precautions previously mentioned for engine trimming; i.e., use of ear defenders for hearing protection, awareness of inlet and exhaust area hazards for protection of both personnel and equipment, and knowledge of adverse weather restrictions which if neglected could result in poor engine performance or possible engine damage. Also, complete familiarity with the manufacturer's checklists and maintenance manuals for turbine aircraft operation becomes a must for safe and accurate performance testing.

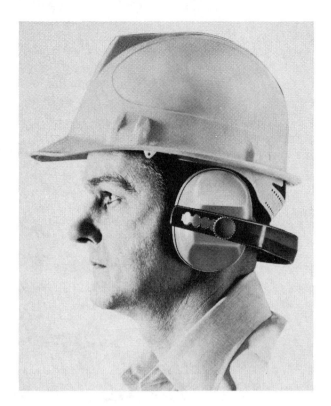

Fig. 2B-1   The high noise level on a jet flight line makes it imperative that effective hearing protectors be used.

B. Engine Runup of Turbojet and Turbofan Engines

Each particular aircraft will have a very specific checklist provided by the manufacturer. The general procedures for operating a turbine engine include, but are not necessarily limited to the following items:

1. Normal operating procedures

a. Prior to engine operation:

(1) Clear the inlet and exhaust areas of personnel and equipment, and clear ramp of debris which could cause foreign object damage to the engine.

(2) Perform a walk-around inspection of the aircraft to ensure complete security of necessary aircraft and engine systems.

(3) Be sure that there is adequate fuel and oil for the runup intended.

(4) Connect ground power unit to aircraft if one is required.

b. *When first entering the aircraft, ensure that:*

(1) Engine master switch is — OFF.

(2) Landing gear handle is in position — WHEELS DOWN.

(3) Seat and brake pedals — ADJUSTED.

(4) Generator switches — OFF.

(5) Power lever — OFF, or fuel control shutoff — OFF (engines with thrust reversers require use of a separate shutoff control).

(6) Starter and ignition — OFF.

(7) Aircraft systems — SAFE FOR ENGINE OPERATION.

c. *To start engine:*

(1) Master switch — ON.

(2) Select battery or external power — ON.

(3) Fuel valves — ON (aircraft system).

(4) Fuel boost — ON (aircraft system).

(5) Starter — ON (starter, ignition and fuel are often time sequenced).

(6) Ignition — ON (usually between 5 and 10% RPM).

(7) Power lever — OPEN (to approximate idle position of fuel shutoff — OPEN). Allow 20 to 30 seconds to lightoff, then abort start.

(8) Ignition and starter — OFF (below idle).

(9) Generator — normally on at this time.

d. *Instrument checks on start cycle:*

(1) Exhaust temperature — WITHIN LIMITS (starting peak and stabilized).

(2) Engine oil pressure — WITHIN LIMITS.

e. *Instrument checks stabilized at idle:*

(1) Percent RPM (generally between 40% to 60%).

(2) Exhaust temperature (EGT, TIT, or ITT)

(3) Fuel flow

(4) Fuel manifold pressure

(5) Oil temperature

(6) Fuel temperature

(7) Oil pressure

(8) Vibration amplitude (large aircraft).

f. *Typical high power checks:*

(1) Engine trim check (EPR or fan speed).

(2) Acceleration and Deceleration time check.

(3) Compressor Bleed Band or Bleed Valve check

g. *Taxi procedure:*

Release brakes and move power lever forward as required for RPM, thrust, and ground speed. Communication with the airport control tower is often required before taxiing.

h. *Normal shutdown procedure:*

(1) Operate engine at prescribed speed for recommended time interval, (usually idle or slightly above for 20-30 seconds).

(2) Power lever and/or fuel lever — OFF (with a quick motion to the stop).

(3) Fuel boost — OFF

(4) Fuel valves — OFF

(5) Generator, battery, external power — OFF

(6) Master switch — OFF

2. *Emergency operating procedures*

a. *Procedure for engine fire during ground start:*

(1) Power lever and/or fuel lever — OFF

(2) Starter — continue to crank to attempt to blow fire out.

(3) Fire extinguisher — ON (if needed)

(4) Master switch — OFF, all other switches — OFF.

(5) Troubleshoot cause

b. *Procedure for hot start:*

(1) Power lever and/or fuel lever — OFF (if temperature attempts to exceed the red line limit).

(2) Master switch — OFF.

(3) Troubleshoot cause

c. *Procedure for failure to start:*

(1) Power lever and/or fuel lever — OFF (if engine does not ignite within the required time period, usually 20-30 seconds after fuel is introduced to the engine).

(2) Troubleshoot cause

Note: If the engine fails to start because fuel flow is terminated inadvertently, DO NOT reopen fuel lever because a hot start will more than likely result. Also allow 30-60 seconds for fuel to drain from combustor and perform an engine purging procedure to clear engine of trapped fuel vapors if necessary.

d. *Engine purging procedure:*

(1) Power — ON

(2) Power lever and/or fuel lever — OFF

(3) Ignition — OFF (pull circuit breaker if necessary)

(4) Starter — ON (usually 15-20 seconds)

e. *Emergency shutdown procedure:*

If engine continues to operate when the power lever of fuel lever is moved to OFF, turn off fuel boost and aircraft fuel valves. The engine will shut down from fuel starvation within 30-60 seconds. This is an emergency procedure *only*

because lubrication of the fuel wetted components will cease and repeated shutdowns in this fashion will reduce system service life.

f. *Flight air starting procedure:*

If an in-flight flameout occurs the starter switch is placed in the AIRSTART position. This by-passes the engine starter and allows ignition only to occur. The engine motors over sufficiently from ram air entering the engine inlet, and electrical ignition relights the mixture when fuel is reintroduced into the combustor.

C. *Engine Runup of Turboshaft Engines*

Turboshaft engine powered aircraft are generally configured with two levers in the cockpit. One to control the engine and one to control the propeller. The function of these control levers is often integrated, with one or the other controlling both the propeller and the engine fuel scheduling at different power settings. On the Garrett TPE-331 engine with a Hartzell propeller, for instance, the power lever controls both the fuel control and the propeller pitch setting in the low speed, beta range. The second lever referred to as the Condition Lever controls the propeller at higher speeds in the alpha range.

The following sequences of operation are typical of the TPE-331 turboshaft engine:

1. *Ground starting procedure*

a. Prestart checks — COMPLETED (similar to turbojet checks).

b. Aircraft electrical power — ON

c. Aircraft fuel valve boost pump — ON

d. Power lever — START GROUND IDLE POSITION (beta range).

e. Condition lever — LOW RPM

f. Battery or external power selector — ON

g. Starter — ON

h. Ignition and fuel — AUTOMATICALLY ON AT 10% RPM.

i. Ignition and starter — AUTOMATICALLY OFF AT 50% RPM.

j. Engine instruments — MONITOR CLOSELY (as per manual).

k. Idle — AUTOMATIC ACCELERATION TO 65% RPM.

l. Condition lever — HIGH RPM (to reset an underspeed governor).

2. *Taxi procedure*

a. Condition lever — HIGH RPM (low blade angle position).

b. Power lever — FORWARD (toward or past flight idle position, alpha range, as necessary for required thrust and ground speed).

NOTE: For takeoff and flight the engine power is set at a predetermined horsepower or torque value by movement of the power lever. The condition lever sets engine speed by changing the propeller blade angle. During flight this lever usually remains at its set position with the engine running at a constant speed. When power changes are required the Power Lever position is adjusted.

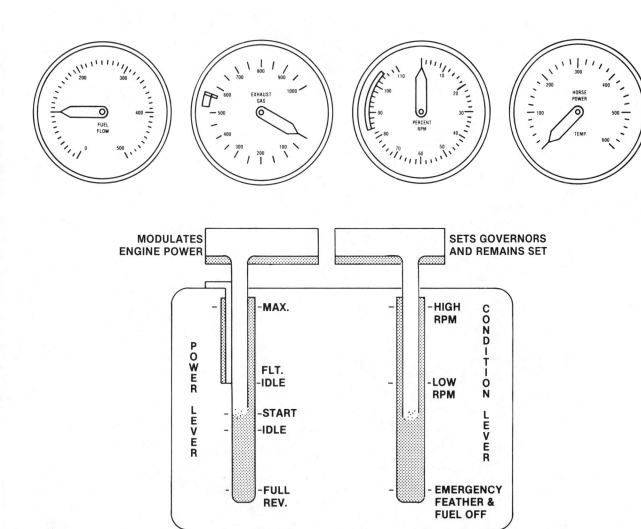

Fig. 2B-2   Cockpit controls and instruments for a Garrett TPE-331 turboprop engine

Book Three

INTRODUCTION

The progress of the internal combustion reciprocating engine has been pretty much paralleled by the developments in the ignition systems for these engines. Automobile engines have the problem of their high rotational speed requiring a great many sparks each second, but aircraft reciprocating engines turn much more slowly than automobile engines and they do not need high speed ignition. However, since aircraft engines operate with such high cylinder pressures, an extremely high degree of reliability is essential.

Dual ignition systems are required to start the combustion in two places in the cylinders so that complete combustion can take place before the cylinder pressure becomes high enough to cause detonation. If either spark plug fails to ignite the fuel-air mixture when the engine is developing full power, catastrophic engine failure can result.

Automobile engines almost universally use some form of battery ignition in which the current flow to produce the high voltage spark comes from the automobile battery. The more modern systems use some form of high-energy capacitor discharge ignition system and many use a type of breakerless system to initiate the spark. Aircraft ignition systems are required by Federal Aviation Regulations, Part 33, to have "at least two spark plugs for each cylinder and two separate electrical circuits *with separate sources of electric energy,* or have an ignition system of equivalent in-flight reliability."

The requirements for two separate sources of electric energy have traditionally been met by

using two separate magnetos. The relatively slow-turning aircraft engines do not require the complexity of the breakerless ignition systems nor the high energy produced by the capacitor discharge system.

A magneto ignition system produces its initial voltage by two self-contained alternating current generators, which have permanent magnets for their rotating fields. As with any permanent magnet generator, the faster the generator turns, the more voltage it produces. The main problem with the magneto system is that of producing a sufficiently intense spark for starting the engine when the magneto is rotating very slowly. We will see that there are several ways we can assist the magneto in producing this spark for starting.

I. TYPES OF IGNITION SYSTEMS

A. *Battery Ignition System*

To overcome the limitation of a magneto producing a weak spark for starting, some of the ignition systems used with early aircraft engines had a magneto to fire one of the spark plugs in each cylinder and a battery ignition system to fire the other spark plug. To start the engine, only the battery system was used, and a good hot spark was provided with this system that allowed the engine to start and build up sufficient speed for the magneto to produce a hot spark. For all normal operation, both magneto and the battery system worked together.

In Fig. 1-1, we have the schematic of a typical battery ignition system.

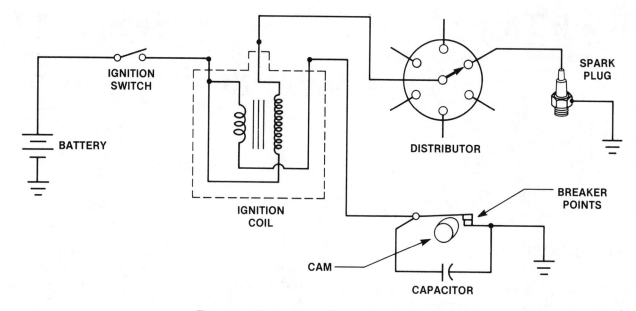

Fig. 1-1   Typical battery ignition system

When the ignition switch is turned on (closed), and the breaker points are closed, current flows from the battery through the coil and the breaker points to ground. As this current flows, a magnetic field builds up around the turns of wire in the primary winding. The lines of flux in this field expand to the maximum as current flows through the coil, and when the breaker points are opened by the cam, the current stops flowing and the magnetic field collapses. As it collapses, the lines of flux cut across the turns in the secondary winding and induce a high voltage in it.

It might be well to review alternating current electricity in Book Three of the General Section of this Integrated Training Program. In this book we see that the amount of voltage induced in a coil by a change of current in an adjacent coil is determined by the ratio of the number of turns of wire in the two coils and by the speed at which the magnetic field collapses. In a typical battery ignition coil there are about 250 turns of wire in the primary winding and some 20,000 turns in the secondary winding.

When the breaker points open, they stop the flow of primary current, and the collapsing magnetic field cuts across the turns of the secondary coil and induces a high voltage in it. One end of the secondary coil is connected to the primary coil, and the other end is carried outside the coil housing through the high-voltage terminal.

The high secondary voltage is carried into the distributor which acts as a selector switch to direct it to the proper spark plug. In the spark plug, the high voltage causes current to flow across the gap and produce the spark we need to ignite the fuel-air mixture.

Inductance is a characteristic of a coil that opposes any change in the rate of flow of current. When the breaker points close, inductance prevents the current reaching its maximum flow rate immediately, and when the points open, inductance tries to keep the current flowing. The amount of voltage induced into the secondary winding is determined by the rate of collapse of the magnetic field around the primary winding.

If the current could stop flowing immediately, there would be a higher voltage induced into the secondary winding. To increase the speed with which the current stops flowing, a capacitor is installed in parallel with the breaker points. When the points open, the current from the battery stops flowing, but the magnetic field around the primary winding begins to collapse and as it collapses, it cuts across all of the turns of the windings and induces a voltage in them that cause an induced current to flow. This induced current flows in the same direction as the battery current.

If there were no capacitor in the circuit, as the points began to open, and the resistance

2

across the points increased, the current flow through the points would cause a spark to jump between the points. This spark ionizes the air between the points and makes it conductive so that current can continue to flow even though the points are open. This flow, in the form of an arc, not only slows down the collapsing of the magnetic field, but it deposits metal from one breaker point to the other and shortens the life of the points.

To speed up the collapse of the field and prevent arcing, a capacitor is installed across the points. As the points begin to open and the resistance across the points begins to increase, the electrons see the capacitor as an easy path to ground and they flow into it. By the time the capacitor is charged enough to prevent the flow of electrons into it, the points are opened sufficiently wide that there will be no arcing across them, and the current stops flowing.

B. Magneto Ignition Systems

All of the modern certificated reciprocating engines for aircraft use magnetos as the source of the sparks to ignite the fuel-air mixture inside the cylinders. In the history of the magneto, there have been several types, some highly efficient and successful, while others met with less than complete success. We will mention the different types here and discuss in detail only the types that are in use today.

1. Low-tension magneto ignition systems

One of the easiest ways to classify aircraft magneto systems is to divide them into high-tension and low-tension systems.

All ignition systems provide a high voltage at the spark plug to jump the gap and ignite the fuel-air mixture. But during World War II, when high-powered reciprocating engine airplanes were flying in the rarified air at high altitudes, the high voltage inside the distributor of the magneto could jump between the distributor electrodes and cause failure of the spark plugs to fire. This caused rough engine operation and loss of power.

To prevent arcing inside the distributor, some magneto manufacturers made their distributors physically larger to increase the distance the spark would have to jump, and others pressurized the distributor with compressed air to make it more difficult for the spark to jump. These measures were only partially effective, but the most success from the arcing, or flashover, problem came with the low-tension magneto ignition system.

In this system, the magneto coil has only a primary winding and there is another coil with a primary and a secondary winding mounted on the cylinder head, near the spark plugs. These units are called the high-tension transformers. Current from the magneto is directed by a carbon-brush distributor to the primary winding of this transformer where it induces a high voltage into the secondary winding. This high voltage is carried to the spark plug through a very short lead. There is no flashover in this system, because the only place the high voltage exists is in the transformer, the short spark plug lead, and the spark plug itself.

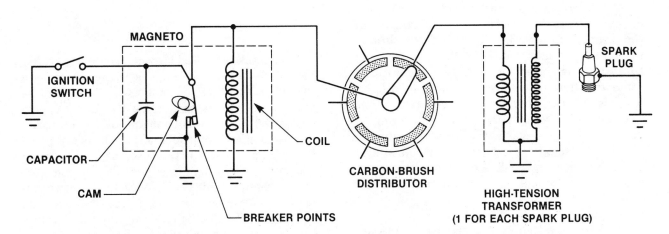

Fig. 1-2 Typical low-tension ignition system for an aircraft engine

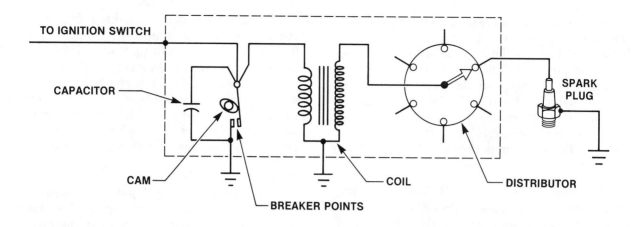

Fig. 1-3 Electrical circuit of a typical high-tension aircraft magneto

Low-tension ignition systems were originally built for the large 18- and 28-cylinder radial engines, but in later years this type of ignition system has been used on some of the horizontally opposed engines. Today, turbojet and turboprop powered aircraft do most of the high altitude flying, and this type of magneto has fallen off in popularity.

2. High-tension magneto ignition systems

All of the ignition systems for the currently manufactured certificated reciprocating engines are of the high-tension type. These systems consist of a magneto with a built-in distributor, an ignition harness connecting the distributor to each of the spark plugs, and the spark plugs themselves.

A high-tension magneto uses a permanent-magnet alternating current generator to produce the primary current. This current flows through the primary winding of the high-voltage step-up coil. A cam-operated set of breaker points interrupts the flow of current through the primary winding of the coil, and the collapse of the magnetic field around the primary winding induces a high voltage in the seconary winding. This high voltage is carried to the distributor rotor by a carbon brush and then to the proper spark plug through the appropriate ignition cable.

The ignition switch grounds the primary circuit to stop the magneto from producing sparks.

C. Types Of Magnetos

1. Shuttle-type magnetos

There have been several types of magnetos used in aircraft ignition systems. The earliest aircraft engines used a shuttle-type magneto, very similar to the magnetos used with the early-day telephones to produce the bell-ringing current. These magnetos used several horseshoe-type permanent magnets to provide the magnetic field. And rotated inside this magnetic field by the engine was the rotor, which included both the primary and secondary winding of the coil, the capacitor, and the cam which opened the breaker points. This type of magneto was replaced with two other types, neither of which had any of the windings in the rotating portion of the magneto.

2. Polar-inductor magnetos

During World War II, when there were many thousands of aircraft engines flying, there were two popular types of magnetos. The Scintilla magneto, which later became the Bendix magneto, used a permanent magnet as its rotor, and the American Bosch Company built a very popular series of magnetos that used as their rotating element a bar of laminated soft iron to complete the magnetic circuit between the poles of two permanent magnets mounted inside the magneto housing. In Fig. 1-4, we see the way this rotating iron bar reversed the direction of magnetic flux through the core of the magneto coil.

4

COIL CORE

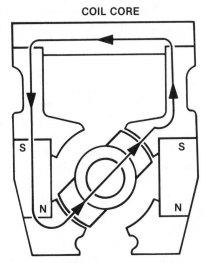

A SOFT IRON ROTOR COMPLETES THE MAGNETIC CIR-
CUIT SO LINES OF FLUX PASS THROUGH THE COIL CORE
FROM RIGHT TO LEFT.

(A)

COIL CORE

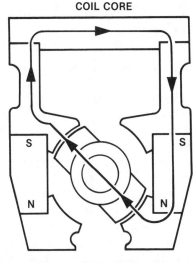

NINETY DEGREES OF ROTOR ROTATION LATER, THE
SOFT IRON CORE COMPLETES THE MAGNETIC CIRCUIT
SO THAT THE FLUX LINES WILL PASS THROUGH THE
COIL CORE FROM LEFT TO RIGHT.

(B)

Fig. 1-4 Magnetic circuit of a polar-inductor magneto.

3. Rotating-magnet magnetos

This is the most popular type of magneto today and is the only one we will discuss in detail. It
has been with us in several forms for almost sixty
years. A permanent magnet having two, four, or
eight poles is rotated inside a laminated iron
frame. The flux from the magnet flows through
the frame and cuts across the primary winding of
the magneto coil that is mounted on the frame.
The cam which opens the breaker points is attached to the rotating magnet shaft, and the gear
that drives the distributor rotor is driven from
the magnet shaft.

4. Double magnetos

The Federal Aviation Administration requires that all certificated aircraft engines have
an ignition system with two separate sources of
electrical energy. A concession has been made in
the interpretation of this regulation that allows
the use of double magnetos. These magnetos have
one housing, one rotating magnet, and one cam,
but there are two separate and independent systems that contain all of the other parts of the ignition system. There are two coils, each having its
own primary and secondary winding, two capacitors, two sets of breaker points, two distributors,

two sets of ignition leads, and of course, two
spark plugs in each cylinder.

COURTESY OF BENDIX

Fig. 1-5 The popular Bendix D-3000 dual magneto

5

II. THE ROTATING-MAGNET MAGNETO

The rotating-magnet magneto has four basic systems we will consider. The mechanical system is that portion of the magneto that includes the housing, the bearings, the oil seals, and all of the non-electrical portions of the magneto. The magnetic system consists of the rotating magnet, the pole shoes, and the core of the magneto coil. The primary electrical system consists of the primary winding of the coil, the breaker points, the capacitor, and the ignition switch. And the secondary electrical system consists of the secondary winding of the coil, the distributor, the ignition harness, and the spark plugs. There is another very important system that must be used with an aircraft magneto, and that is an auxiliary system needed to produce a high-voltage spark that can

be delayed until the piston reaches the top of its compression stroke. This is needed only for starting the engine.

A. The Mechanical System

Most of the magnetos mounted on horizontally opposed engines have the oil seal in the magneto rather than in the engine mounting pad, as is customary in most radial engines. This seal is normally made of a tough, resilient plastic material held snugly against the magneto shaft with a spring. If there is any indication of oil inside the magneto, the magneto must be removed from the engine and this seal replaced.

Most magnetos have the rotating magnet supported in ball or needle bearings. Many of the earlier magnetos required that the bearings be preloaded to increase their life, and preloading

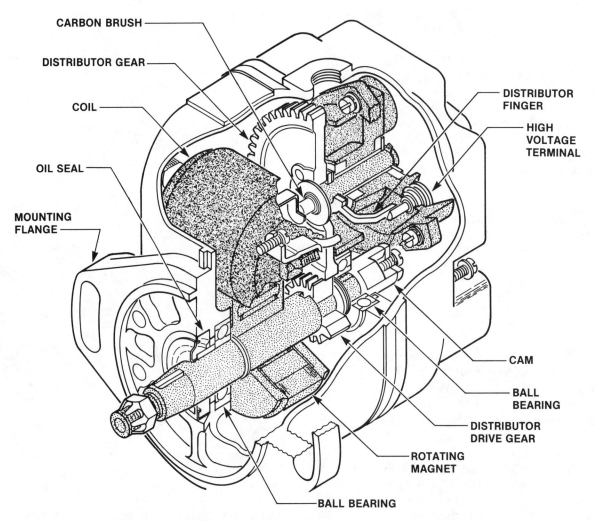

Fig. 2-1 A typical flange-mounted, high-tension, rotating-magnet-type aircraft magneto

6

was done by adjusting the number of shims behind the bearings. But the more modern magnetos use sealed ball bearings pressed onto the shaft. Whatever type of bearing is used, be sure to follow the procedure specified in the manufacturer's overhaul manual in detail when overhauling the magneto or replacing the bearings.

Magnetos are mounted on the engine by methods known as base-mounting or flange-mounting. Base-mounted magnetos are bolted rigidly to the engine accessory case, and the rotating magnet is coupled to the engine drive with a vernier coupling. A vernier coupling is a toothed coupling with more teeth on one side than on the other. By disconnecting the coupling and rotating it, the timing between the magneto and the engine can be changed in very small increments. Flange-mounted magnetos are bolted to the engine with the attachment bolts passing through slots in the magneto case, or it may be held tight to the engine with clamps. Either method allows the magneto to be rotated slightly for fine adjustment of the magneto-to-engine timing.

B. The Magnetic Circuit

Rotating-magnet magnetos may use two-, four-, or eight-pole magnets, but the most popular magnetos for modern four- or six-cylinder aircraft engines use a two-pole magnet made of a very high-grade, high-permeability alloy steel.

The pole shoes for the magnetic circuit are made of laminated steel and are usually cast right into the magneto housing. The core of the magneto coil is also part of the magnetic circuit, and it is wedged tightly against the pole shoe extensions so that it will make a good magnetic path for the flux to flow with a minimum of reluctance. Reluctance, you will remember, is the opposition a material has to the passage of lines of magnetic flux.

In Fig. 2-2, we see the paths the lines of magnetic flux take as they flow through the magnetic circuit of the magneto. In view A, the magnet is in what is called its full-register position, with all of the lines of flux from the magnet passing through the coil core from left to right. In view B, the magnet has rotated 90 degrees and there is no flux in the coil core. This is called the neutral position of the magnet. In view C, the magnet is again in its full-register position, but

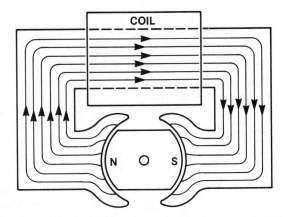

THE MAGNET IS IN ITS FULL-REGISTER POSITION, WITH ALL OF THE LINES OF FLUX PASSING THROUGH THE COIL CORE FROM LEFT TO RIGHT.

(A)

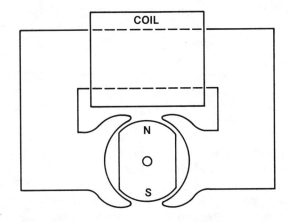

THE MAGNET IS IN ITS NEUTRAL POSITION WITH NO LINES OF FLUX PASSING THROUGH THE COIL CORE.

(B)

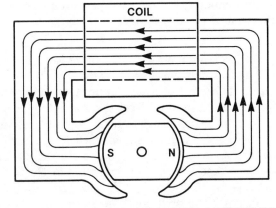

THE MAGNET IS IN ITS FULL-REGISTER POSITION, WITH THE MAXIMUM NUMBER OF LINES OF FLUX PASSING THROUGH THE COIL CORE FROM RIGHT TO LEFT.

(C)

Fig. 2-2 The magnetic circuit of a rotating-magnet magneto

this time the lines of flux pass through the coil core from right to left. The magnetic field has expanded to a maximum, collapsed and expanded again to a maximum, only in the opposite direction. Each time the lines of flux cut across the coil, they induce a voltage in it.

In Fig. 2-3A, we have a lot of the static flux in the magnetic circuit. When the magnet is in the full-register position, the flux is maximum. Then as it rotates 90 degrees, the flux drops to zero, and by the time the magnet has rotated another 90 degrees the flux is again maximum, only this time it is passing through the circuit in the opposite direction. (See Fig. 2-3 on page 9.)

The primary winding of the magneto coil is cut by the flux each time the magnetic field expands and collapses and current is induced in the primary winding.

When the magnet is in its full-register position, there is no change in the intensity of the flux, and no lines of flux cut across the windings of the coil. There is no voltage induced in the primary winding. We see this at the 0-degree, 180-degree, and 360-degree positions of the magnet in Fig. 2-3B. As the magnet continues to rotate, the amount of flux drops off until at the 90-degree point there is no flux; but then it starts to build up immediately after the magnet passes the 90-degree position. The greatest *change* in the number of lines of flux in the core of the coil occurs as the magnet passes through the 90-degree, or neutral, position. At the position of the greatest change in flux, the open-circuit current induced in the coil will be the greatest.

We will remember from our study of electrical generators that when current flows in a conductor, it produces a magnetic field, and the magnetic field produced by the current opposes the magnetic field that caused it. The current generated as the field collapses will produce a magnetic field that will try to sustain the collapsing field. And the current generated as the magnetic field builds up will produce a magnetic field that opposes the buildup.

If we complete the primary circuit so current can flow in it, the flux from the magnetic field produced by current flowing in the primary winding combines with the flux caused by the rotating magnet to produce the resultant flux curve

we see in Fig. 2-3C. When these fields combine, the greatest change in flux does not occur when the magnet passes through its neutral position as it does when there is no primary current, but the maximum rate of change occurs a few degrees of magnet rotation beyond the neutral position. This is very important in the production of an intense spark by an aircraft magneto.

C. The Primary Electrical Circuit

The primary winding on the magneto coil consists of around 180 to 200 turns of relatively heavy copper wire coated with an enamel insulation. It is wound directly over the laminated iron core that forms a part of the magnetic circuit. The inside end of the primary winding grounds to the iron core and also to a ground lead. The outside end of the primary winding connects to the inside end of the secondary winding and is brought out of the coil through a lead that is connected to the insulated breaker point.

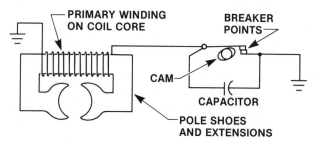

Fig. 2-4 *The primary electrical circuit of a rotating-magnet magneto*

A capacitor of somewhere around 0.3 microfarad is connected across the breaker points.

The breaker points are normally held closed by a leaf-type spring and are opened by the cam that is mounted on the end of the rotating magnet shaft. The cam allows the points to close when the magnet is in its full-register position. As the magnet moves away from the full-register position, the flux through the coil core begins to decrease and a voltage is induced in the primary winding that causes current to flow through the breaker points to ground. This current causes a magnetic field which prevents the flux in the coil core following the static flux curve. The two magnetic fields, the one from the rotating magnet and the one from the primary current, cause the

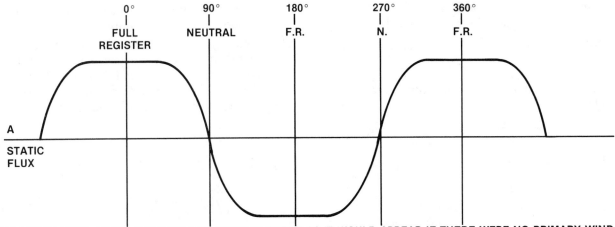

THE STATIC FLUX CURVE, SHOWING THE SMOOTH CURVE AS IT WOULD APPEAR IF THERE WERE NO PRIMARY WIND-ING AROUND THE COIL CORE

(A)

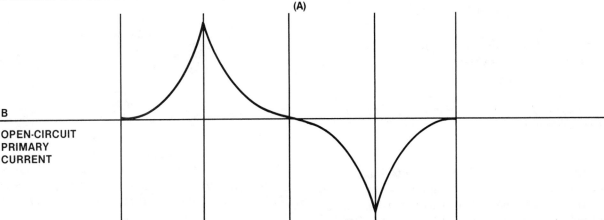

THE OPEN-CIRCUIT PRIMARY CURRENT IS MAXIMUM WHEN THE CHANGE IN STATIC FLUX IS GREATEST, AND ZERO AT THE POINTS WHERE THE STATIC FLUX IS MAXIMUM, BUT IS NOT CHANGING.

(B)

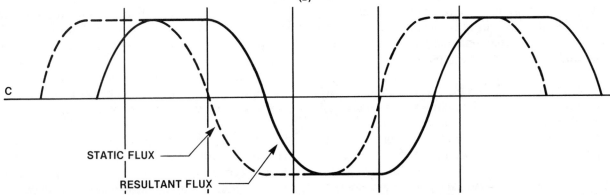

THE MAGNETIC FIELD PRODUCED BY THE PRIMARY CURRENT FLOW OPPOSES THE MAGNETIC FLUX THAT CAUSED IT AND THE RESULTANT FLUX LAGS BEHIND THE STATIC FLUX.

(C)

Fig. 2-3   Production and effect of primary current

flux to follow the resultant flux curve we saw in Fig. 2-3C and in Fig. 2-5.

When the magnet rotates a few degrees beyond its neutral position, the cam opens the breaker points and the primary current stops flowing. This sudden stoppage of the primary current causes the flux in the coil core to drop as we see in Fig. 2-5. Remember that it is the *rate* of flux change that determines the amount of volt-

9

age induced by a magnetic field, and since this flux changes so fast, a high voltage is induced into the primary winding.

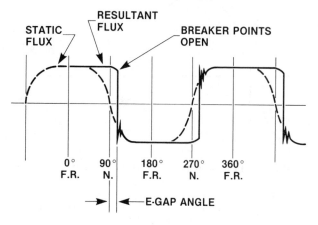

Fig. 2-5 The breaker points in a magneto open when they will cause the greatest change in the resultant flux. This normally occurs a few degrees of magnet rotation beyond its neutral position. The point at which the breaker points open is called the E-gap angle.

Any time moving contact points interrupt a flow of current, an arc is produced between the points. As the points begin to separate, the resistance increases and the current flowing through the resistance produces heat. This heat becomes so intense that it ionizes the air and current flows through it, causing an arc that not only delays the decrease in current flow, but it also transfers metal from one breaker point to the other and can even weld the points together.

To prevent this arcing, a capacitor is installed in parallel with the points. As the contacts begin to open and the resistance across them begins to increase, the electrons see the capacitor as a low-resistance path to ground and flow into it. By the time the capacitor is charged up enough to stop the flow of electrons, the points have opened far enough that no spark can jump across them.

In some modern magnetos, the capacitor serves a dual function. It is in parallel with the breaker points to prevent arcing across them as they open, and it is also in series with the ignition switch to prevent electromagnetic radiation from the primary lead causing radio interference. This type of capacitor serves as a feed-through for the ignition lead. The pigtail inside the magneto connects to the insulated breaker point, and the lead

from the ignition switch in the cockpit connects to the capacitor terminal outside of the magneto. The metal case of the capacitor is connected to the outside of the plates in the capacitor, and any radio frequency energy that is induced into the primary lead by the points opening is carried to ground before it can leave the magneto.

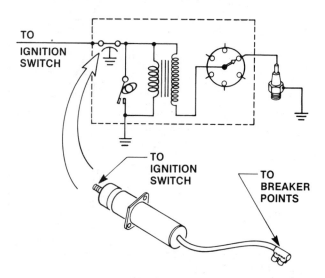

Fig. 2-6 Feed-through capacitor used in the primary circuit of some magnetos. This capacitor is in parallel with the breaker points and is in series between the points and the ignition switch.

D. The Secondary Electrical Circuit

A coil made up of thousands of turns of very fine wire is wound on top of the primary coil in the magneto coil assembly. The inside end of this secondary coil is connected to the outside end of the primary winding, and the outside end is brought out of the coil assembly at a high-voltage contact usually on the side of the coil.

A spring-loaded carbon brush in the center of the distributor gear presses against the high-voltage contact and conducts the high secondary voltage into the distributor rotor. The distributor finger, which is a conductive arm in the rotor, carries the high voltage from the coil to the contact for the proper spark plug lead in the distributor block.

Heavily insulated shielded spark plug leads carry the high voltage from the distributor to the spark plug, where it causes current to flow across the gap in the spark plug and produce the spark that ignites the fuel-air mixture inside the cylinder.

Fig. 2-7 *The high-tension terminal in a coil of an aircraft magneto. The high voltage is picked up from this terminal by a carbon brush that is mounted in the center of the distributor gear.*

E. Auxiliary Systems for Starting

One of the main problems with using magnetos as the source of electrical energy for aircraft engine ignition is the fact that they do not produce a hot spark when they turn at a very slow speed. This requires some form of auxiliary system to provide the spark for starting the engine. This spark must not only have a high intensity, but it should occur later than the normal spark. It should occur about the time the piston reaches top center so the maximum pressure will be built up inside the cylinder just after the piston passes over top center and starts on the way down. This will prevent the engine kicking back when it is being started. Kickbacks can damage the starter.

There are a number of systems that have been used for supplying this hot and late spark for starting. We will consider some of them, and we will discuss in some detail the two systems that are most popular today.

1. Booster magnetos

A system that was quite popular up through World War II was the booster magneto. A separate magneto, usually mounted inside the cockpit and turned with a hand crank, was used to supply the spark for starting. Its output went into the distributor where it was directed to the proper spark plug through a trailing finger. This

finger conducted the high voltage to the cylinder next in firing order to the one that was being supplied by the normal magneto. In a nine-cylinder radial engine whose firing order is 1-3-5-7-9-2-4-6-8, for example, when the magneto was in position to fire cylinder number three, the booster magneto was sending a hot spark to cylinder number one. The piston in cylinder number one had already started down, and the expanding gases caused by this late ignition gave it the needed push to allow the engine to build up enough speed for the regular magneto to take over. The booster magneto was used only until the engine started running properly.

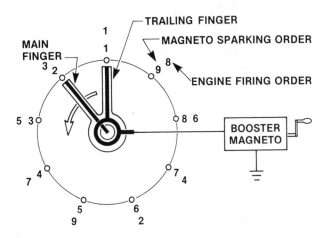

Fig. 2-8 *The high voltage from a booster magneto is distributed to the spark plugs through a trailing finger on the distributor rotor.*

2. Induction vibrators

The booster magneto worked quite well for small single-engine airplanes, but almost all large multi-engine airplanes use a form of induction vibrator. The starter rotates the engine fast enough that kickback is not a problem, and the auxiliary spark is delivered to the spark plugs through the regular distributor finger.

A battery operated vibrator, similar to a buzzer, produces a pulsating direct current any time the engine starter switch is held in the Engage position, and the ignition switch is in the right-magneto position. This pulsating DC goes into the primary coil of the magneto. As long as the breaker points are close, this pulsating DC goes to ground and does nothing inside the magneto. But as soon as the points open, the current must pass through the primary winding of the coil to get to ground. This pulsating current in

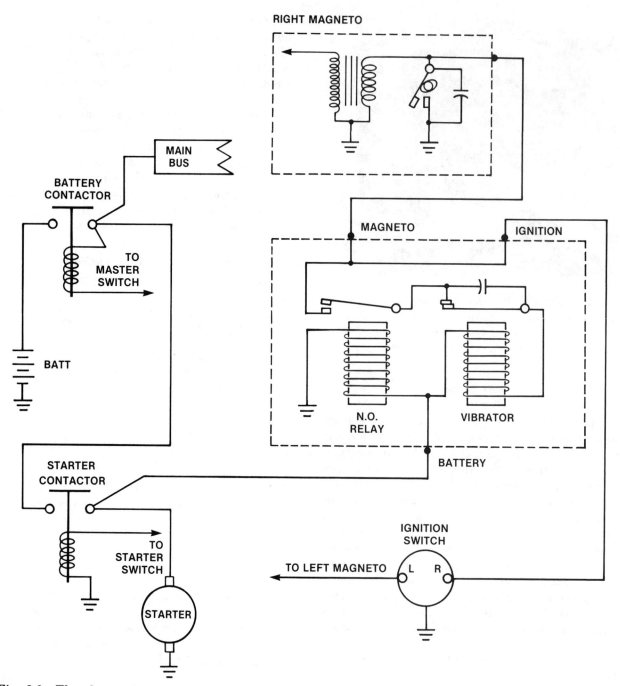

Fig. 2-9 The electrical circuit of an induction vibrator such as is used with some of the large radial engines to provide a hot spark for starting

the primary coil induces a high voltage into the secondary winding and produces a hot spark at the spark plug.

3. *"Shower of Sparks" ignition system*

This is a modern version of the induction vibrator that is used with many of the four-, six-,

and eight-cylinder engines installed in our modern fleet of general aviation aircraft.

This system works essentially the same as the induction vibrator system we have just described, except that the magnetos used with this system have an extra set of breaker points in parallel with the normal points. These points are called the retard points and they are timed so that they open later than the normal points.

12

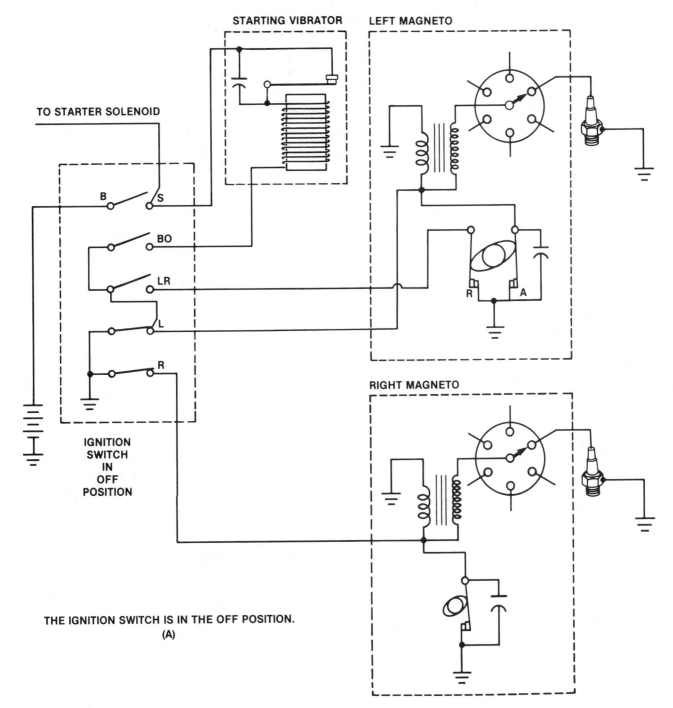

STARTING VIBRATOR   LEFT MAGNETO

TO STARTER SOLENOID

B   S

BO

LR

L

R

IGNITION
SWITCH
IN
OFF
POSITION

RIGHT MAGNETO

R   A

THE IGNITION SWITCH IS IN THE OFF POSITION.
(A)

Fig. 2-10 The electrical circuit of a Bendix "Shower of Sparks" system for providing a hot and late spark for starting an aircraft engine. (1 of 3)

In Fig. 2-10, we have a typical electrical circuit for the Bendix "Shower of Sparks" ignition system. The right magneto used in this system is of standard construction, but the left magneto has two sets of breaker points operated by the same cam. The advance set of points operates in exactly the same way as the points in the right magneto. They are in parallel with the capacitor and are connected to the Left terminal of the ignition switch, and they are in the ignition circuit at all times. The retard points are connectd to the LR (left-retard) terminal of the ignition switch. This portion of the switch is normally open and is closed only when the ignition switch is held in the Start position.

13

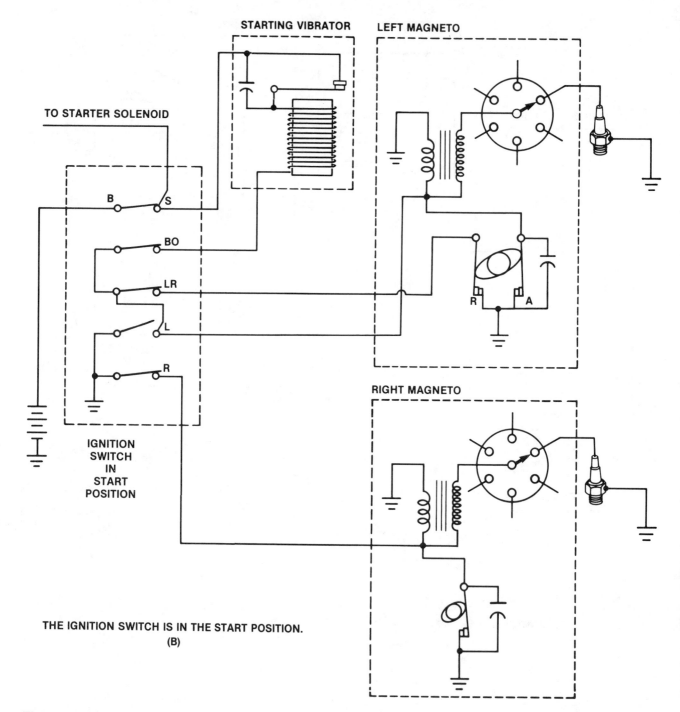

Fig. 2-10 The electrical circuit of a Bendix "Shower of Sparks" system for providing a hot and late spark for starting an aircraft engine. (2 of 3)

A starting vibrator is used with this system to produce pulsating direct current from the DC supplied by the aircraft battery. You will notice in Fig. 2-10B that when the ignition switch is held in the Start position, current from the battery flows through both the contacts and the coil of the starting vibrator. It then flows through the BO and the LR contacts in the ignition switch,

and to ground through both the retard and the advance set of breaker points in the left magneto. As soon as the current flows in the vibrator coil, a magnetic field is produced that pulls the contacts open and interrupts the flow of current so the spring can close the contacts and current will again flow throught the coil.

14

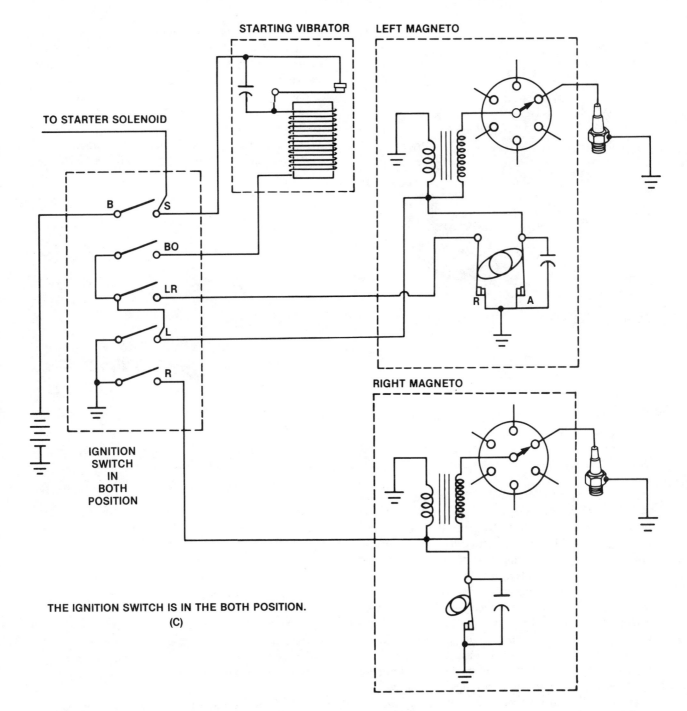

Fig. 2-10 The electrical circuit of a Bendix "Shower of Sparks" system for providing a hot and late spark for starting an aircraft engine. (3 of 3)

This continual making and breaking of the current causes pulsating direct current to flow through the magneto breaker points. The capacitor in parallel with the vibrator contacts prevents the points arcing as they interrupt the flow of current. Electrons flow into the capacitor as the contacts begin to open rather than flowing across the contacts and causing a spark. By the time the capacitor is charged to the battery voltage, the contacts are open far enough that no spark can jump the gap.

When the ignition switch is placed in the Start position, the right magneto is grounded so its normally advanced timing cannot cause a spark that could cause the engine to kick back.

15

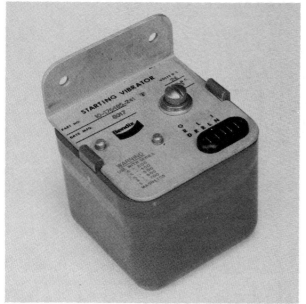

COURTESY OF BENDIX

Fig. 2-11 Starting vibrator, such as is used with the Bendix Shower of Sparks system

The starter is energized so the engine will begin turning and the pulsating DC from the starting vibrator flows to ground through both sets of breaker points in the left magneto.

As the crankshaft turns, the advance points in the left magneto open, and now all of the pulsating DC must flow to ground through the retard points. The crankshaft continues to rotate, and about 20 or so degrees after the advance points open, the piston reaches the top of its stroke and the retard points open. Now, with both sets of points open, the pulsating DC can get to ground by flowing through the primary coil of the left magneto. When this pulsating current flows through the primary coil, it induces a high voltage into the secondary winding, and as long as both sets of breaker points are open, a "shower of sparks" will jump across the spark plug that is connected to the distributor terminal being supplied by the distributor rotor. This procedure continues, distributing to each cylinder hot and late ignition until the engine is running and the ignition switch is returned to the Both position.

When the ignition switch is placed in the Both position, all of the contacts are open. The starter solenoid is de-energized, the starting vibrator gets no more current, the retard breaker points are out of the left magneto circuit, and the primary of neither magneto is grounded.

When the ignition switch is not held in the spring-loaded Start position, it functions as any other ignition switch. In the Off position, both magnetos are grounded. In the Right position, the left magneto is grounded and the primary circuit of the right magneto is open. In the Left position, the right magneto is grounded and the primary circuit for the left magneto is open. In the Both position, the primary circuit for both magnetos are open.

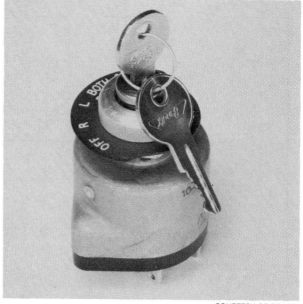

COURTESY OF BENDIX

Fig. 2-12 Ignition switch such as is used with the Bendix "Shower of Sparks" system

4. Impulse couplings

The vast majority of four- and six-cylinder horizontally opposed aircraft engines use an impulse coupling on one or both magnetos to produce a hot and late spark for starting the engine.

An impulse coupling is a small spring-loaded coupling between the magneto shaft and the engine drive gear. When the engine is not running but is turned over by hand or by the starter, flyweights in the impulse coupling contact a stop pin in the magneto housing. This pin holds the magnet shaft still as the engine continues to rotate. By the time the engine has rotated far enough for the piston to reach top center, a projection on the body of the impulse coupling wedges the flyweight off of the stop pin and the spring spins the magneto shaft fast enough to produce a hot and late spark.

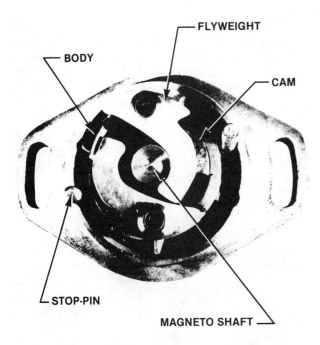

Fig. 2-13 An impulse coupling that mechanically provides a hot and late spark for starting an aircraft engine

This action takes place for every spark the magneto produces until the engine starts running and turning fast enough for centrifugal force on the flyweights to hold them away from the stop pin. Then the impulse coupling acts as a straight coupling for the magneto.

F. Magnetos with Compensated Cams

The magnetos used on some of the large, high-powered radial engines have compensated cams. These are cams with one lobe for each cyl-

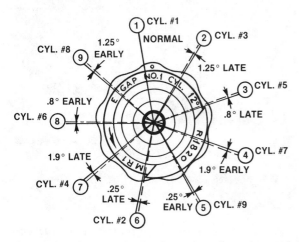

Fig. 2-14 A compensated cam, such as is used in the magnetos installed on some of the high-powered radial engines

inder, ground in such a way that they compensate for the fact that only the master rod rotates around the crankshaft. All pistons except the one on the master rod are attached to link rods that rock back and forth as the engine rotates. The cam is ground in such a way that the breaker points open, not according to the angular position of the crankshaft, but the opening of the points is timed so the pistons will all be the same linear distance from the top of the cylinder when ignition occurs. The lobe for cylinder number one is marked with a dot and a step or slot is cut across the cam so a straightedge can be aligned with a mark in the magneto housing for timing.

Fig. 2-15 The compensated cam and breaker point assembly in a magneto mounted on a nine-cylinder radial engine.

III. MAGNETO INSTALLATION AND SERVICING

A. Preparation for Installation

1. Checking internal timing

When we install a magneto on an engine, there, are *two* types of timing we must consider: the internal timing of the magneto and the timing of the magneto to the engine. Internal timing is the adjustment of the breaker points so they will stop the flow of primary current at exactly the moment the collapse of the magnetic field around the coil will produce the greatest flux change. This position at which the breaker points open is called the "E-gap," or the E-gap angle, and it is the number of degrees beyond its neutral positon the magnet rotates before the points open. This angle is specified by the magneto manufacturer, and it is extremely important that it be accurately set because if it is not, the spark produced by the magneto will be weak.

Many of the Slick Electro magnetos have a timing hole in the housing and in the rotor shaft through which you can slip a timing pin or a sixpenny nail. This pin holds the magnet the correct number of degrees beyond its neutral position.

We can position the magnet in other magnetos by allowing the magnet to find its own neutral position. It has a natural tendency to

"pull in" to its neutral position, and you can easily feel when it is in the neutral position, Attach a timing scale to the magneto housing and clamp a pointer over the cam end of the rotating magnet shaft and set the pointer to the zero mark on the scale.

Fig. 3-2 *Bendix magnetos are put in the correct firing position by using a pointer on the rotating magnet shaft and a calibrated scale attached to the magneto housing.*

Rotate the magnet shaft in the normal direction of its rotation until the pointer aligns with the number of degrees for the E-gap angle specified by the manufacturer. Clamp the magnet in this position with a rotor holding tool.

TIMING PIN OR 6-PENNY NAIL

WHEN TIMING PIN IS IN MAGNETO

POINTS SHOULD BE STARTING TO OPEN

Fig. 3-1 *The magnet in a Slick Electro magneto is held in the correct position for timing by inserting the timing pin or a six-penny nail through a hole in the housing and into a matching hole in the shaft of the rotating magnet.*

Fig. 3-3 *The rotating magnet in a Bendix magneto is held in the correct firing position with a friction lock between the magnet shaft and the magneto housing.*

Connect the magneto timing light across the breaker points and adjust the points until the timing light shows that the points have just opened.

The position of the magnet when the points open is the most important part of the internal timing procedure, but we must be sure that the points close when the magnet is near its full-register position. We do this by being sure that the maximum point opening is within the tolerance allowed by the manufacturer.

Remove the timing pin or loosen the rotor holding tool, and rotate the magnet until the cam follower is on the highest point of the cam lobe; then measure the clearance between the breaker points. The magneto manufacturer has specified the maximum and minimum clearance, and if the clearance does not fall within this tolerance when the E-gap is correctly adjusted, the breaker point assembly will have to be replaced.

In most magnetos the cam either is keyed to the shaft or fits in a slot in the shaft so that its position relative to the magnet is fixed. But in some of the Bendix magnetos, the cam end of the magnet shaft is tapered and the tapered inside of the cam is held tight with a washer and a screw.

The distributor gear is meshed with the drive gear by aligning the timing mark on the distributor gear with the dot or the chamfered tooth on the drive gear. The same distributor gear may be used on magnetos that rotate either clockwise or counterclockwise, and so there are two marks on the gear. Align the "CW" or "RH" mark with the marked tooth on the drive gear if the magneto shaft turns in a clockwise direction as viewed from the drive end. When the distributor gear is properly meshed with the drive gear, a painted tooth or a mark on the face of the gear is visible in the timing hole of many of the Bendix magnetos to indicate that the distributor is in the proper position to fire cylinder number one.

Fig. 3-5 The distributor may be placed in the proper position for installing the magneto on the engine by aligning marks on the distributor gear with an indicator on the magneto housing. These marks are visible through the inspection hole in the magneto housing.

B. Timing the Magneto to the Engine

1. Magnetos without impulse coupling

After the magneto has been carefully checked for its internal timing and physical condition, you are ready to install it on the engine. The oil seal around the magneto drive shaft is usually not installed until after the magneto has been run on the test bench, and at this time you should be sure that it is installed and is in good condition.

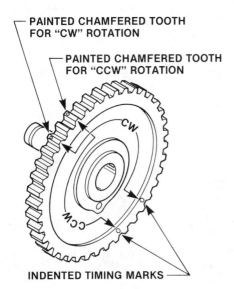

PAINTED CHAMFERED TOOTH
FOR "CW" ROTATION

PAINTED CHAMFERED TOOTH
FOR "CCW" ROTATION

INDENTED TIMING MARKS

Fig. 3-4 The distributor gear of a magneto has marked teeth which are aligned with marked teeth on the distributor drive gear to properly time the distributor to the magnetor.

Put the engine in the position to fire cylinder number one. Locate the compression stroke by removing one of the spark plugs in cylinder number one, and with your finger over the spark plug hole, rotate the engine until you feel air blowing out of the hole, showing that the piston is moving up with both valves closed. Install a Time-Rite indicator in the spark plug hole or use a top-dead-center indicator and a timing protractor. Accurately position the piston the correct number of degrees before top dead center, as is specified by the engine manufacturer.

Fig 3-6B If a Time-Rite indicator is not available, the piston position may be found by using a top-dead-center indicator and a timing protractor.

Fig. 3-6A A Time-Rite indicator screwed into one of the spark plug holes in the engine is a good way to find the position of the piston in the cylinder for the proper installation of the magneto.

Now, with the engine in the position for cylinder number one to fire, hold the magneto with its rotor in the E-gap position and the distributor ready to fire cylinder number one, and slip the magneto drive into the mounting cavity on the accessory section of the engine. Some engines drive the magnetos through rubber drive cushions that are held in pressed steel cushion retainers set into the magneto drive gear. In other engines the magneto drive gear meshes directly with an accessory gear. Whatever system the en-

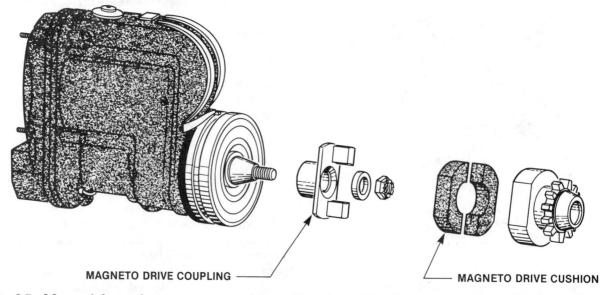

MAGNETO DRIVE COUPLING   MAGNETO DRIVE CUSHION

Fig. 3-7 Many of the modern magnetos are driven through a rubber drive cushion that is held in a slot in the accessory drive gear.

gine uses, be sure that you follow the magneto installation procedures in detail as they are specified by the engine manufacturer.

Secure the magneto to the engine with the appropriate nuts or clamps, but do not tighten them yet. Connect a timing light across the breaker points and gently tap the magneto with a soft hammer to rotate it until the timing light indicates that the breaker points are just opening. Now, you can tighten the nuts or clamps.

Install the second magneto and time it so its breaker points open at the correct time. Most of the more modern engines have the ignition from both magnetos occur at the same time, but some engines have staggered timing. The spark plug that ignites the fuel-air mixture nearest the exhaust valve fires a few degrees of crankshaft rotation before the other one. This is because the mixture in this area is usually diluted and burns slower than the undiluted mixture.

With both magnetos installed and timing lights connected across both sets of breaker points, turn the crankshaft backwards, enough for both sets of breaker points to close, and then slowly rotate the crankshaft in the direction of normal rotation until the timing lights indicate that both sets of breaker points have opened. If the engine uses synchronized timing, the lights should indicate that both sets of points open at the same time. If they do not, loosen the hold-down nuts or clamps of the magneto that is not properly timed and gently tap the magneto until it has rotated in its mount enough for its points to open at the correct time.

2. *Magnetos with impulse coupling*

If the magneto is equipped with an impulse coupling, it must be disengaged before the magneto can be timed to the engine. When installing the magneto, trip the flyweights and hold the impulse coupling in the running position. After the magneto is installed but before you connect your timing light, back the crankshaft up until the impulse coupling engages, and then rotate it ahead until the impulse coupling snaps. Then carefully rotate the crankshaft backward until the breaker points close, but not enough to re-engage the impulse coupling.

Now, connect your timing light across the breaker points and slowly rotate the crankshaft in the direction of normal rotation until the light

indicates that the points are opening. Correct for any improper opening position of the points in the same way as is done with a magneto without an impulse coupling.

An impulse coupling spins the magneto fast enough to produce a large amount of primary current, and it is possible for some timing lights to be damaged if they are connected across the breaker points when the impulse coupling snaps. So do not connect your timing light until the impulse coupling has been snapped.

Some of the Slick Electro magnetos must be "sparked out" to be sure the magneto is in the proper position for installation. Hold the number one spark plug lead about 1/16 inch from the magneto frame and rotate the drive until the impulse coupling snaps and a good healthy spark jumps from the number one lead to the case. This indicates that the magneto is in the position to have just fired the spark plug in cylinder number one. Rotate the magneto back until you can insert the timing pin or six-penny nail in the hole in the housing and the rotating magnet shaft. Now, the magneto is in the correct postion for installation.

3. *Dual magneto timing and installation*

Dual magnetos such as those in the Bendix D-2000 and D-3000 series are timed and installed in much the same way as single magnetos, since they are actually two separate magnetos that share only the same case, rotating magnet and cams. There are several special features, however, that we should consider.

The top of the dual magneto is the side of the housing on which the name plate is located, and the lower set of breaker points are for the left magneto and are used as the reference breaker points when timing the magneto.

We must remove the rear cover so the cams, breaker assemblies, and distributor blocks are accessible. In addition to the two main breaker assemblies, the retard breakers for the "Shower of Sparks" starting system may be stacked on an extended bracket above the left main breaker points, and a second cam is stacked on the shaft to operate them. In some installations an additional set of breaker points may be stacked on an extension bracket over the right main breaker points to operate an electronic tachometer.

COURTESY OF BENDIX

Fig. 3-8   The cam and breaker point assemblies in a Bendix D-3000 dual magneto

The capacitors in these magnetos serve a dual purpose: They are in series with the primary switch lead to filter out radio interference, and they are also in parallel with the breaker points to prevent arcing at the points.

These dual magnetos use a four-pole rotating magnet and it is important that the correct neutral position be selected for the internal timing. The magnet is marked as we see in Fig. 3-9. The magneto is timed with the keyway in the magnet shaft up, and this is indicated by the letter "K" on the magnet being visible in the timing window on the top of the magneto.

Adjust the main breaker points to a clearance of 0.016 inch plus or minus 0.002 inch, when the cam follower is on the highest point of the cam lobe. Then turn the magneto so the proper E-gap angle aligns with the timing mark in the window, and lock the shaft in this position with a friction lock between the shaft and the housing. The cams on these magnetos are not keyed to the shaft and they can be turned for accurate adjustment of the point opening. Loosen the main cam and turn it in the direction of magneto rotation until the approaching lobe just opens the left main breaker points, as is indicated by a timing light across the points. The tips of a pair of Tru-arc pliers inserted

into the two holes in the cam make a good tool for turning it.

Adjust the right main breaker points to open at the same time by slightly loosening their hold-down screws and, using a small drift, gently tap the breaker assembly frame toward the cam until the timing light indicates these points have just opened. Tighten the cam and breaker screws to the recommended torque and loosen the friction

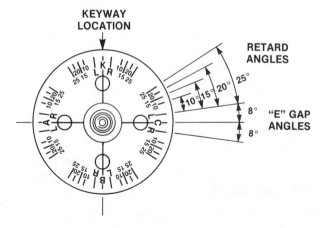

Fig. 3-9   Markings on the rotating magnet in a Bendix D-3000 magneto show the proper retard angles and the proper E-gap angles.

lock holding the magnet. Now, recheck the E-gap and breaker synchronization. Readjust them if necessary by repeating the breaker opening procedure.

Adjust the retard points so they will open a maximum of 0.016-inch plus or minus 0.004-inch, when their follower is on the highest point of the cam lobe. Now, turn the magneto shaft to the proper position of the retard points to open, and lock it in place with the friction lock. Loosen the cam screw just enough that you can move the top cam, and turn it in the direction of magneto rotation until the timing light indicates that the retard points are just opening. Tighten the cam screw to the torque indicated in the service manual; then loosen the friction lock and check the angle at which the points open. It must be within the tolerance allowed by the manufacturer.

Be sure that the capacitor wire is positioned and formed so it will not interfere with the high-tension wells or with the cover screws, and be sure that the high-tension lead grommets are all seated. The cover can now be installed and the cover screws all tightened to the torque recommended in the manufacturer's service manual.

The magneto is installed on the engine by placing the piston in cylinder number one in its proper firing position and the magneto in its proper E-gap position. Marked teeth will be visible in the windows at each end of the magneto housing. These marked teeth will be in the windows every time the distributor is in position to fire cylinder number one, but to be sure that the correct neutral position has been selected, the letter K must be visible in the inspection window in the top of the magneto. If the top of the magneto is inaccessible, however, the inspection plug on the bottom of the magneto may be removed, and the letter B should be visible opposite the timing mark in this window.

When the magneto has been mounted and the cover is in place, use the proper jumper to connect the timing light into the primary circuit.

If it is impossible to get the two sets of points to open at exactly the same instant, the first set of points to open should open at the timing position specified by the engine manufacturer and the later points must open within three degrees of the first. If you cannot get both sets of points to open

within this tolerance, remove the magneto from the engine and recheck the internal timing.

It is possible for magneto timing to drift either early or late, due to cam follower for breaker point wear, and the magneto may be loosened and very slightly bumped to cause the points to open at the correct time relative to the piston position. The manufacturer has specified the maximum amount the magneto can be bumped, and this is indicated by the relationship between the E-gap angle marks and the timing pointer. If the marks are not out of alignment by more than edge-to-edge distance, the magneto has not been bumped too much.

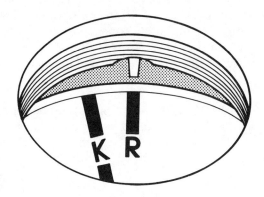

Fig. 3-10 The magneto must not be bumped for timing enough that the pointer and the mark on the rotating magnet are out of alignment more than edge-to-edge distance.

C. Magneto Inspection

The fact that aircraft engines are equipped with dual ignition systems is not primarily for safety in the event one of the systems fails—it is because the high cylinder pressures in modern aircraft engines require the ignition of the fuel-air mixture in the cylinders to occur at two places so the entire mixture can be burned before the pressure inside the cylinder reaches the critical pressure of the fuel.

If the mixture is ignited at only one place, the flame front will move across the cylinder heating and compressing the burning fuel-air mixture until it reaches its critical pressure and temperature, and then it will explode rather than continue to burn evenly. This explosion, known as detonation, can be so severe that pistons can be burned through, heads blown off of the cylinder barrels, and other serious internal damage to the engine can result.

Each preflight inspection requires the magnetos to be checked to be sure that both are operating properly. This is indicated by a uniform drop in RPM when operating on either magneto independently. To perform a proper magneto check, the engine must be warmed up to the proper oil and cylinder head temperature, and the propeller must be in the low-pitch (high RPM) position. Advance the engine to the speed recommended by the engine manufacturer, usually around 1700 RPM, and then turn the ignition switch from Both to Right, and watch the tachometer for the amount of RPM drop. Then place the switch back in the Both position and allow the engine to run on both magnetos to clear the spark plugs that have not been firing; then place the switch in the Left position and note the RPM drop on the tachometer.

The RPM drop when operating on either magneto separately should be somewhere around 125 to 150 RPM, and the engine should run smoothly and evenly when operating on just one magneto.

If the RPM drop is smooth, but is more than the manufacturer recommends, there is a good probability that the engine is operating with either a too rich or a too lean mixture. A sharp drop in RPM beyond the recommended amount is an indication of one or more fouled spark plugs, a defective ignition lead, or the magneto being out of time.

If there is no drop when making the magneto check, there is a possibility that either the magneto switch is malfunctioning or there is a P-lead loose from the magneto. In either case, the magneto is "hot" when the switch is off, and this should be corrected immediately before someone is injured. To check for this condition, idle the engine and turn the magneto switch Off. If one or both of the magnetos are "hot," the engine will not die, but if the switch and the P-leads are all working as they should, the engine will stop firing. If the engine does stop firing, don't turn the switch back on, as this could cause a backfire and damage the engine; rather, allow the engine to stop, then restart it.

If the engine does die when the switch is turned off and there is either no magneto drop or a drop of less than the manufacturer specifies as normal, there is a possibility that the magneto is timed too early.

When a magneto is operating, there are two places of normal wear that can shift the magneto timing. The plastic cam follower that rides on the cam and pushes the breaker points open will wear, and when it wears, the cam must rotate farther before it can open the points. Cam follower wear causes the timing to drift late, or retard the spark. The breaker points also wear, or erode, and as they wear, the timing will drift early, or become more advanced. One popular series of magnetos has been designed so the wear of the cam and the breaker point erosion should essentially balance each other and the timing should drift a minimum amount during the life of the magneto.

The most wear, and thus the greatest amount of timing drift, usually occurs within the first few hours of operation after a magneto is new or after new breaker points have been installed. For this reason, most aircraft manufacturers require you to check the magneto-to-engine timing on the first 25-hour inspection and then on each 100-hour inspection.

If the RPM drop is found to be excessive on a magneto check and a check of the timing shows that it has drifted late, there is a misconception that this can be corrected by loosening the magneto in its mount and bumping it back into time. And, unfortunately, this sometimes works. The magneto manufacturers have, in some cases, given a maximum amount the magneto can be bumped to bring it back into time after it has drifted off. This is usually a very small amount; see Fig. 3-10.

When either the breaker points or the cam follower wear enough to cause the magneto-to-engine timing to be off, the only *proper* way to correct this is to remove the magneto and re-time it internally. Remember that for the maximum amount of primary current to be interrupted, the breaker points must be opened at the correct E-gap angle, and when either the breaker points or the cam follower wears, the points will no longer open at the proper E-gap angle, and the magneto will have to be re-timed.

After correcting the internal timing drift caused by points or cam follower wear, be sure to check the maximum amount of breaker point opening. If both the correct position of point opening and the maximum amount of opening cannot be obtained, the breaker assembly must be replaced.

The breaker points in a magneto are subject to the most difficult operating conditions in the magneto, and any failure on their part can prevent the magneto operating.

The breaker points in some of the older magnetos could be removed from the magneto and dressed with an oil stone to remove pits and surface irregularities. However, on modern magnetos the breaker assembly is a single unit and when there is a problem with the points it is the practice to replace the entire assembly. This assembly consists of the two breaker points, the cam follower with its lubricating felt, and the mounting bracket.

Breaker points are now almost all made of tungsten. In the past, some breaker points were made of platinum, but the high cost of this metal has caused tungsten to be used instead.

Examine the points for their condition. Normal points should have a smooth, flat surface with a dull gray, sandblasted, or frosted, appearance over the area where electrical contact is made. In the process of interrupting the primary current, there will be some transfer of metal from one point to the other and after a short time minor irregularities will develop on the surface of the points. This is not harmful if the pits and mounds are small, but if there is any possibility that any of the pits could go through the tungsten, the assembly should be replaced.

No attempt should ever be made to file or otherwise treat the surface of the breaker points, but it they appear oily they can be cleaned by pulling a piece of clean kraft paper through them.

The felt pad that supplies oil to the cam follower should be checked for the proper amount of oil. It should leave your fingers oily when it is squeezed. If it does not, it should be oiled according to the recommendations in the manufacturer's service manual.

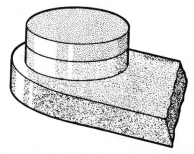

POINTS THAT ARE SMOOTH ON TOP AND HAVE A FROSTED APPEARANCE ARE OPERATING NORMALLY.

(A)

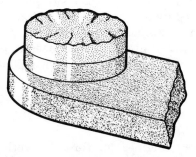

AFTER A TIME OF NORMAL OPERATION, SOME METAL WILL TRANSFER FROM ONE BREAKER POINT TO THE OTHER. THIS IS A NORMAL CONDITION OF BREAKER POINT WEAR.

(B)

THIS BREAKER POINT SHOWS A MOUND EXTENDING NOTICEABLY ABOVE THE SURROUNDING SURFACE. THERE IS A CORRESPONDING DEPRESSION IN THE OTHER POINT, AND THIS INDICATES EXCESSIVE WEAR. BREAKER POINTS SHOWING THIS CONDITION OF WEAR MUST BE REPLACED.

(C)

Fig. 3-11   Conditions of the breaker points in a magneto

IV. SPARK PLUGS

The spark plug is one of the least appreciated components in an engine, and yet it is one whose importance far outweighs its seeming simplicity.

The one function of a spark plug in a reciprocating engine is to provide an insulated electrical terminal, or electrode, inside the combustion chamber. High voltage is supplied to this terminal from the magneto and a spark jumps from the insulated electrode to a ground electrode to ignite the fuel-air mixture.

This seems to be a simple enough job for the spark plug, but the combustion gas temperature may be as high as 3,000 degrees Fahrenheit, and the temperature changes rapidly and radically. The pressure inside of the cylinder can be as high as 2,000 psi. The job of the spark plug is further complicated by the fact that the voltage supplied by the magneto can be as high as 18,000 volts, and during a typical 100-hour operating period, the plug must spark about seven million times. From all of these complications, we can see that aviation spark plugs do indeed have a hard job to perform.

Aircraft engines up through World War I were equipped with spark plugs similar to those used in automobiles and motorcycles, but when radios began to be used in airplanes, the static caused by these open spark plugs interfered with the reception so much that shielded spark plugs were developed. These shielded spark plugs were encased in a metal housing which was connected to the metal-braid shielding around the spark plug lead. The electrical radiation caused by the spark jumping the gap is received by this shielding and it passed harmlessly to the engine so it cannot interfere with the radio.

The insulation used in the early spark plugs was a form of porcelain, but in the early 1930's, the spark plug manufacturers found that thin laminations of natural mica provided better insulation than porcelain, and for the next ten years or so mica was pretty much the standard insulation for aviation spark plugs.

Near the beginning of World War II, the need for spark plugs that could be made in vast quantities and the need for more uniform insulation than the natural mica provided caused the development of aluminum oxide as a spark plug insulator. This has become the standard insulation material used in spark plugs today.

There are very few aircraft today that are not equipped with some form of radio, and because they are, almost all spark plugs are of the shielded type. Shielding prevents radio interference, but it does not do this without causing some problems. Remember in our early study of electricity, we saw that a capacitor is nothing more than two conductors separated by a dielectric, or insulator, and capacitors store electrical energy. The shielded ignition lead between the magneto and the spark plug is a capacitor with the center wire and the shielding being the two conductors and the insulation around the wire the dielectric. When the spark occurs, the electrical energy that is radiated is received and stored in the capacitance of the shielded lead. And then when the spark jumps the gap in the spark plug, it ionizes the air in the gap and provides a low resistance path for the energy stored in the shielded lead to keep the spark going long after the mixture is ignited. This long duration spark wears away the spark plug electrodes, widens the gap, and shortens the life of the spark plug. To prevent this type of damage, most of the modern spark plugs have a built-in resistor that limits the after firing caused by the capacitance of the shielding.

Spark plug development was next challenged by the lead deposits left in the firing-end cavity of the spark plug by the tetraethyl lead that is used in aviation gasoline to allow higher cylinder pressures to be used without the fuel-air mixture detonating. Fine-wire electrodes made of platinum allowed the firing end of the spark plug to be made more open than the massive nickle alloy electrode allowed, so the lead-containing combustion products could be scavenged from the spark plug before they solidified into harmful lead deposits. For severe lead contamination, platinum electrodes have been replaced with iridium electrodes which are less susceptible to lead contamination than platinum.

And finally, the problem of lead deposits is so severe in low-powered airplane engines that have been forced to use fuel with a higher lead content that they were originally designed to use, that a new generation of aviation spark plug with more exposed electrodes for better scavenging has been developed.

Now that we have an appreciation of the developments aviation spark plugs have gone through, let's study them in some detail.

A. *Size and Shielding*

Aviation spark plugs are made in two sizes of shell threads, 14-mm and 18-mm. The 14-mm

26

shell has a 13/16-inch shell hex, and the 18-mm shell has a 7/8-inch hex. With the exception of the Franklin engine, of which there are only a few still in operation, all modern aircraft engines use the 18-mm spark plug, and all of the spark plugs we will discuss in this text are of the 18-mm size.

Only some of the older, low-powered engines use unshielded ignition, and even many of the special-use airplanes that have no radios use shielded spark plugs and shielded harness because they are more readily available than unshielded ignition components.

There are two types of shielded spark plugs used in aircraft engines. The older type harness has a 5/8-24 thread on the shield, and the newer

type, called the high-altitude or All-weather spark plug, uses a 3/4-20 thread on the shield. The ceramic insulator in the All-weather spark plug does not extend to the top of the shell, but there is room for the resilient grommet on the ignition lead to form a water-tight seal that prevents rain entering the terminal end of the spark plug and causing it to misfire.

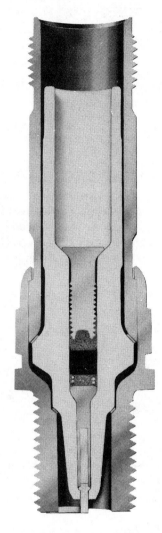

Fig. 4-2 A fine-wire electrode shielded spark plug with All-weather, or 3/4-20, shielding

B. Reach

The length of the threads on the shell classifies a spark plug according to its reach. A short-reach spark plug has 1/2-inch of threads on its shell, and a long-reach spark plug has 13/16-inch of threads.

The reason for the two different reaches is the physical construction of the cylinder heads into

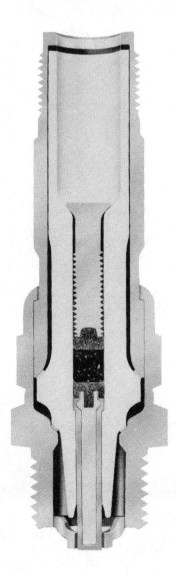

Fig. 4-1 A massive-electrode shielded spark plug with 5/8-24 shielding

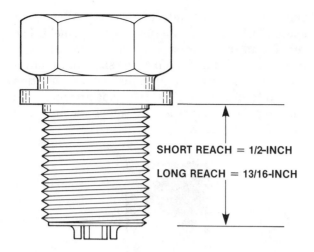

Fig. 4-3 *The reach of a spark plug is the length of the threaded portion of the shell.*

SHORT REACH = 1/2-INCH

LONG REACH = 13/16-INCH

which the spark plug screws. When the recommended spark plug is properly installed in an aircraft engine, the bottom of the threads is flush with the inside of the cylinder head. If a long-reach spark plug is installed in an engine requiring one with a short reach or if the correct spark plug is installed without a gasket, the end of the threads will stick out into the combustion chamber. Heat will damage the bottom threads and carbon will fill the thread grooves so that removal of the spark plug will be difficult, and there is a good probability of either damaging the threads or loosening the bushing or the helicoil insert when the spark plug is removed.

If the reach is too short for the engine or if two gaskets are used, threads in the bushing inside the cylinder will be exposed and they will fill with carbon. When the correct spark plug is screwed into the hole, it will bottom on the carbon and the correct torque will be reached without the spark plug seating against its gasket. This loose installation will allow hot gases to leak past the threads and will damage the cylinder head.

Some cylinder head temperature gages take their reading from a gasket-type thermocouple under one of the spark plugs. If this type of pickup is used, be sure that *no* regular gasket is used with the thermocouple gasket.

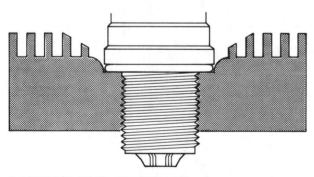

A LONG-REACH SPARK PLUG INSTALLED IN AN ENGINE REQUIRING A SHORT-REACH SPARK PLUG

(A)

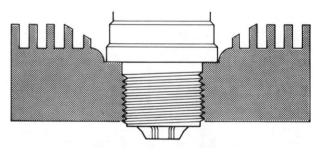

THIS SPARK PLUG DOES NOT HAVE ANY WASHER, AND THE BOTTOM THREAD ON THE SHELL STICKS INTO THE COMBUSTION CHAMBER OF THE ENGINE

(A)

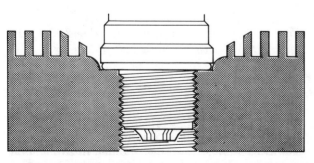

A SHORT-REACH SPARK PLUG INSTALLED IN AN ENGINE REQUIRING A LONG-REACH SPARK PLUG

(B)

Fig. 4-4 *Installation of spark plugs with improper reach*

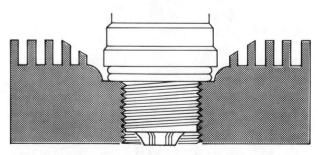

TWO GASKETS ARE INSTALLED ON THIS SPARK PLUG. THE THREADS ON THE SPARK PLUG INSERT WILL BECOME CONTAMINATED WITH CARBON

(B)

Fig. 4-5 *Spark plug installation, with improper use of washers*

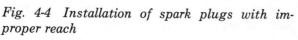

If a short-reach spark plug has been used in a long-reach engine or if two gaskets have been used, before the correct spark plug is installed, you should clean all of the carbon out of the bottom threads by running the correct thread clean-out tool into the hole.

C. Heat Range

The heat range of a spark plug refers to the ability of the insulator and center electrode to conduct heat away from its tip. Hot spark plugs have a long heat path and the tip operates at a high temperature. A cold spark plug has a short path for the heat to flow from the tip into the shell and then into the engine. The loss of this heat allows the spark plug insulator nose core to run cooler. A cold spark plug is used in a hot-running, high-compression engine, and a hot spark plug is used in an engine whose cylinder temperatures are reasonably low.

It is extremely important when selecting the spark plug for an engine to choose the correct

heat range for the operating conditions. The Type Certificate Data Sheet for the engine lists all of the spark plugs that are approved for the engine, and it is important that you do not install any spark plug that is not on the approved list.

The heat range of a spark plug has become of extreme importance in some of the smaller engines that were designed to operate on 80-octane aviation gasoline and are not forced to use 100-octane gasoline. The higher octane fuel contains from four to eight times as much tetraethyl lead as the lower octane fuel, and it is this extra lead that fouls the spark plugs.

When an engine is operating at low power, the hottest part of the spark plug, which is the insulator nose core, has a temperature of about 800° F, and the deposits that form in the firing-end cavity of the spark plug are primarily carbon deposits. The amount of these deposits is determined by two things: by the idling fuel-air mixture ratio and by the presence of any oil that may leak past the valve guides or the piston rings.

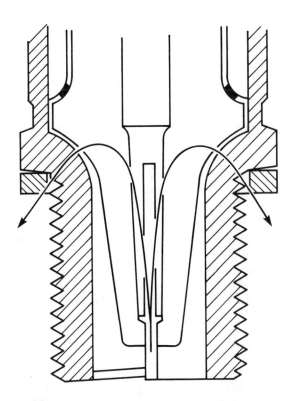

A HOT SPARK PLUG HAS A LONG PATH FOR THE HEAT TO FLOW FROM THE NOSE CORE INSULATOR TO THE SHELL

(A)

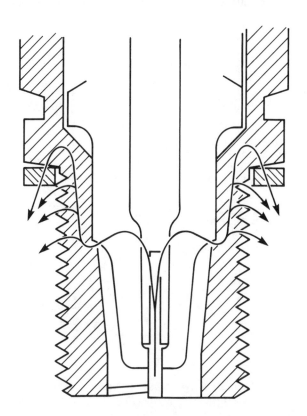

A COLD SPARK PLUG HAS A SHORT PATH FOR THE HEAT TO FLOW FROM THE NOSE CORE INSULATOR TO THE SHELL.

(B)

Fig. 4-6 Heat range of a spark plug

29

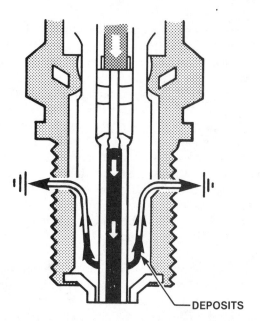

DEPOSITS

CONDUCTIVE PATH FORMED
BY COMBUSTION PRODUCTS

**DEPOSITS IN THE FIRING END CAVITY OF A SPARK PLUG
ALLOW THE VOLTAGE TO LEAK OFF TO GROUND BE-
FORE IT CAN BUILD UP HIGH ENOUGH TO JUMP THE
GAP.**

(A)

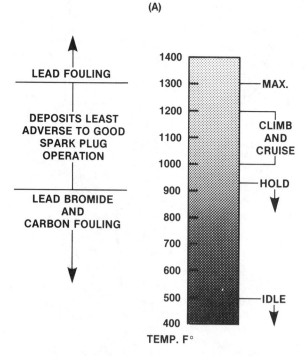

**TEMPERATURES OF THE NOSE CORE INSULATOR IN A
SPARK PLUG FOR THE VARIOUS OPERATING CONDI-
TIONS.**

(B)

*Fig. 4-7 Types of spark plug fouling that occur
in a spark plug depend upon the temperature of
the nose core insulation.*

When the insulator nose core operates at a temperature for more than 900° F, few carbon deposits form in the spark plug, and the ethylene dibromide in the gasoline is able to effectively scavenge the lead from the cylinders. Deposits that form in a spark plug when its insulator nose core is operating with temperatures between 900° and 1,300° F are low in volume and low in electrical conductivity.

Deposits left in the spark plug firing end cavity when the insulator nose core temperature is up around 1,300° F are primarily lead oxide, and they are not normally too objectionable; but lead oxide has a strong affinity for silica, and if the engine ingests any dust or sand, the silica will unite with the lead oxide and act as a flux, lowering its melting point several hundred degrees and converting the lead oxide into lead silicate which is conductive at high temperatures. This lead silicate provides a path for the high voltage to leak off without a spark jumping the gap.

Because of the nature of the deposits that form in the firing-end cavity of a spark plug, it is extremely important that the correct heat range be chosen to allow the insulator nose core to operate at temperatures in the range of between 900° and 1,300° F.

D. Resistor

Shielded ignition, while eliminating the problem of radio interference, causes another problem, that of accelerated electrode erosion in the spark plugs. the shield acts as a capacitor storing electrical energy that is released when the spark jumps the gap in the spark plug. As this spark tries to die away, the energy that is stored in the capacitance of the harness is returned, and it continues to supply energy to sustain the spark. This long-duration spark is not needed since the fuel air mixture is already ignited, and it does cause the spark plug electrodes to erode faster than they should.

To minimize the problem caused by this sustained spark, modern shielded spark plugs have a resistor of about 1,500 ohms installed inside the spark plug insulator between the spark plug lead and the center electrode. This resistor reduces the duration of the spark while in no way affecting the production of the initial spark.

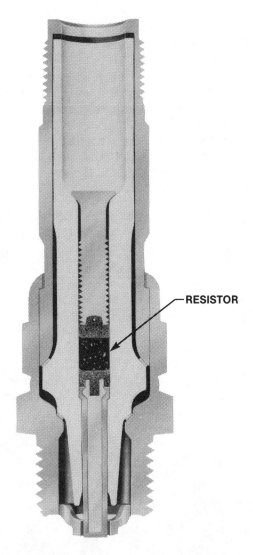

Some of the early spark plugs have a single ground electrode welded on the shell similar to the typical automotive spark plug. To improve reliability, later spark plugs have two, three, and even four ground electrodes made of a nickle alloy swaged into the firing end of the spark plug. They were pressed over near the center electrode to form the gap for the spark to jump.

Fig. 4-9 Typical electrodes in a massive-electrode spark plug

Fig. 4-8 The resistor in this spark plug increases the electrode life by shortening the duration of the spark.

E. Electrodes

We have left the electrodes of the spark plug until last because they are its most important consideration. It is because of the electrodes that the spark plug exists. They must be designed to allow the spark to jump at the lowest voltage, and the spark must be intense enough to provide positive ignition for the fuel-air mixture. The electrodes must have a maximum operating life without changing the spacing of the gap, and they must also be of such a design that they produce the minimum opposition to the flow of gases in and around the center insulator so the lead-forming gases can be scavenged before they solidify and form deposits.

The severe operating conditions under which our military aircraft operated during World War II brought out the fine-wire spark plug in which two very fine platinum or iridium wires were staked into the open end of the firing end cavity and were moved over near a similar wire which acted as the insulated center electrode. This small area electrode allows the maximum opening in the firing end cavity for the most effective scavenging of the lead-rich combustion deposits, and it requires the lowest voltage to produce a spark. This form of electrode minimizes the possibility of ice bridging that has caused so much hard starting of engines operating in sub-zero conditions.

The center electrode of massive-electrode spark plugs is made of a special erosion-resistant nickel alloy tube, or sheath, filled with copper. The copper increases the rate of heat transfer away from the tip end of the electrode, and the nickle sheath protects the copper from cavitation and from copper runout.

The center electrode of platinum fine-wire spark plugs have a small platinum alloy tip locked into an inconel alloy center electrode stem,

31

Fig. 4-10 Typical electrodes in a fine-wire-electrode spark plug

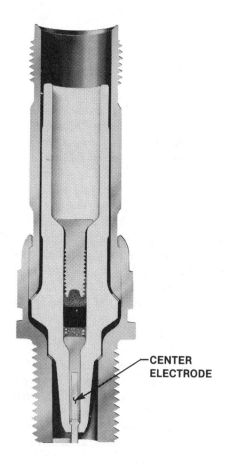

Fig. 4-11 The center electrode in a fine-wire spark plug is made of iridium or platinum alloy, locked into a center electrode stem which is filled with silver.

CENTER ELECTRODE

which is filled with silver for the best conduction of heat and electrical current.

The small size of the platinum electrodes allows the spark to jump at a lower voltage than it can jump from the massive electrodes. The electrons gathering on the small area of the fine-wire electrode have a much greater density than the same number of electrons on the much larger area of the massive electrodes, and they will be forced across the gap at a lower voltage.

Platinum fine-wire electrodes solved many of the problems of lead fouling and ice bridging, but platinum can be contaminated by lead at the high temperatures encountered by the spark plug electrodes, and it was found that iridium, an element very similar to platinum, was not so susceptible to lead-induced erosion, and today fine-wire spark plugs are available with either platinum or iridium electrodes. Iridium spark plugs cost more than platinum plugs, but in conditions of heavy lead fouling, it is quite likely that their longer life will make them less expensive in the long run.

Fig. 4-12 The iridium electrode on the right has not been eroded by the severe lead contamination that has attacked the platinum electrode on the left.

F. Spark Plug Construction

A special high-strength steel shell and shield are machined with a very close tolerance to form a holder for the spark plug insulator. After the center insulator is assembled into the shell and the shield, they are bonded together to form a gas-tight seal. The insulator which holds the contact for the lead wire, the center electrode, and the resistor is made of a special aluminum oxide ceramic which is close to the diamond in hardness, and it has carefully controlled thermal characteristics. The ease with which this material conducts heat makes it possible to have a long insulator tip for a given heat range. And the long, narrow tip provides the maximum size cavity to help minimize lead fouling.

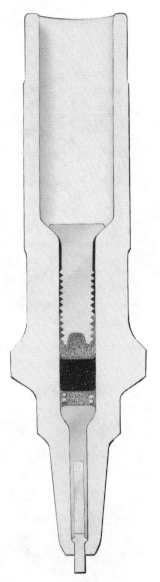

Fig. 4-13 *The insulator in an aircraft spark plug is made of aluminum oxide, and it is near diamond in hardness.*

The center wire seal is made of powdered copper or silver metal and glass or ceramic, and it forms an intimate bond between the resistor and the center electrode.

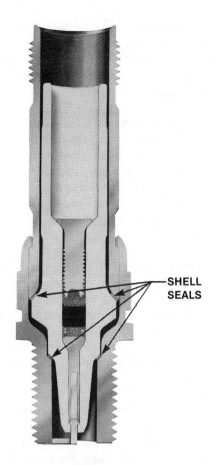

Fig. 4-14 *Spark plugs are sealed with nickel washers between the insulator and the shell and with a conductive seal of powdered metal and glass to bond the center electrode to the insulator.*

The insulator is sealed to the shell with nickel gaskets, and the shield is lock-crimped and bonded to the shell so the spark plug is essentially one piece for maximum transfer of heat.

G. *Spark Plug Servicing*

1. *Removal*

It is extremely important that aircraft spark plugs be serviced regularly because the accumulation of deposits in the firing-end cavity can allow the high voltage to leak off to the ground across the deposits rather than building up high enough to cause a spark to jump across the gap. On the other hand, as the gaps erode wider, the magneto must produce an increasingly higher voltage, and it is possible that the voltage requirements may become higher than the magneto is capable of producing.

Spark plug servicing begins when the spark plug is removed from the engine. This operation should be done carefully, as it is possible to damage the insulator by improper removal of the lead terminal.

Hold the lead with one hand and loosen the terminal nut with the proper crowfoot or open end wrench. When the terminal nut is unscrewed all of the way, remove the terminal from the cavity by pulling it straight out. If it cocks to one side, there is danger of breaking either the insulator inside the spark plug or the terminal insulator itself.

Loosen the spark plug from the cylinder by using a deep socket wrench. Special six-point sockets are made for removing spark plugs, and if one of these or a twelve-point socket is used, it must square on the hex of the spark plug, because if it is cocked to one side there is a good chance that the insulator inside the spark plug will be damaged.

As the spark plugs are removed from the cylinders, put them in a tray with numbered holes for each spark plug. By using this type of tray, you will know the hole from which each spark plug was removed, and since the spark plugs tell so much about the condition of the cylinders, knowing from which cylinder each spark plug was removed will help us locate the troubles the spark plugs indicate.

2. Visual examination

When the spark plugs are all out of the engine, examine them carefully. A spark plug that has been operating normally and that is not wornout will have an appearance similar to that in Fig. 4-16. The insulator will be covered with a dull brown deposit and there will not be an excessive amount of buildup in the firing-end cavity. If the spark plug is in this condition, it can be cleaned, regapped, tested, and reinstalled in the engine.

**MASSIVE-ELECTRODE SPARK PLUG
(A)**

**FINE-WIRE SPARK PLUG
(B)**

Fig. 4-16 Appearance of a spark plug that has been operating normally and is in a condition to be cleaned and replaced

Fig. 4-15 When removing the spark plugs from an engine, be sure to put them into a rack with numbered holes so you will be able to identify the hole from which the spark plug was removed. This knowledge will help you troubleshoot the engine.

If the spark plug has been operating normally but is worn too much for reinstallation, the insulator will be covered with the brown deposit, and there will not be an excessive amount of buildup in the cavity, but the electrodes will be worn to about one-half of their normal diminsions. When spark plugs are worn in this way, they should be replaced with new ones.

Fig. 4-17 *A massive-electrode spark plug that is worn excessively. The spark plug should be replaced when the electrodes wear to one-half of their original dimensions.*

If any of the spark plugs show signs of severe erosion and both the center electrode and the ground electrodes are worn far more that they should be, there is a possibility that the engine has fuel metering problems. In a carbureted engine there is the possibility of an induction air leak, and in a fuel-injection engine, there is a good chance that the injector nozzle in the cylinder from which the spark plug is removed is partially plugged, causing the cylinder to run with an excessively lean mixture, and the cylinder has possibly had some detonation. When this kind of wear is found, the inside of the cylinder should be inspected with a borescope to check it for indications of further damage.

When the inside of the firing-end cavity is filled with a hard clinker-like deposit, it is an indication that the spark plug has been operating too cold for the amount of tetraethyl lead in the fuel. All of the deposits must be removed and the spark plug cleaned, gapped, tested, and reinstalled. If all of the spark plugs show this kind of

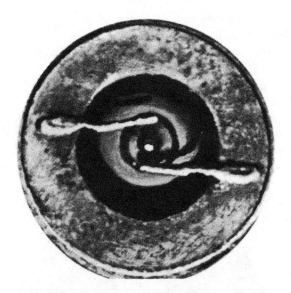

Fig. 4-18 *This fine-wire spark plug has been worn excessively, and its condition indicates possible detonation in the cylinder from which it was removed.*

fouling, you should check into the possibility of installing a set of spark plugs with a hotter heat range. But be very sure that the hotter spark plug is approved for the engine before you install it.

Excessive lead fouling can be caused by uneven distribution of the fuel in the induction system and the proper use of carburetor heat may minimize this type of lead fouling. But, before recommending this action to a pilot, be very sure

Fig. 4-19 *This massive-electrode spark plug has lead deposits in the firing-end cavity. This could indicate improper fuel vaporization or perhaps the use of fuel with too high a lead content.*

that he understands the correct way to use carburetor heat, as its improper use can cause detonation and can severely damage the engine.

Carbon fouling shows up as a soft, black, sooty deposit in the spark plug. It is caused by operating with an excessively rich idling mixture and from excessive ground idling. If this type of problem shows up too much, there is a possibility that the spark plug has too cold a heat range for the operating conditions.

Fig. 4-20 *The carbon deposits in this spark plug indicate that the engine has been operating with too rich an idle mixture.*

Fig. 4-21 *The oil fouling in this spark plug indicates that there is excessive oil leaking into the cylinder from which this spark plug was removed. The oil could have come from a worn valve guide or from a broken piston ring.*

Oil fouling is caused by an excessive amount of oil leaking into the cylinder from either worn or broken piston rings, from a valve guide that has excessive clearance, or from a leaking impeller oil seal. This type of fouling requires a careful inspection of the engine for the source of the excess oil and the damage must be repaired before the aircraft can be approved for return to service.

If there is a hard glaze on the nose insulator of a spark plug, there is a good possibility that there is a leak around the induction system filter and sand has gotten into the engine. This glaze is caused by the silicon fluxing the lead deposits and causing them to flow out over the spark plug insulator. The glaze is non-conductive at low temperatures, but becomes conductive at high temperatures. Silicon glaze is difficult to completely remove, and since it is non-conductive at low temperatures, a spark plug that is not completely clean may test good in the spark plug tester, but fail in the engine when it is hot.

Examine the terminal barrel insulation for cracks and if any are found, discard the spark plug. Carefully check the insulation in the firing-end cavity, and if it is cracked, discard the spark plug. It is better to find all of the obvious damage before spending time cleaning it.

3. Cleaning

Degrease the spark plugs either by spraying them with varsol or Stoddard solvent or by putting them in a vapor cabinet where vapors of trichlorethylene can soften all of the oily deposits. Whatever method you use, be sure that no solvent gets into the terminal cavity. After the spark is clean, dry it with an air blast.

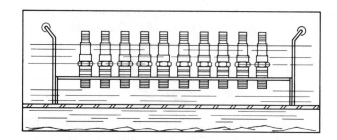

Fig. 4-22 *When cleaning spark plugs, soak the firing-end cavity with trichlorethylene. Do not allow any of the solvent to get into the terminal-end cavity.*

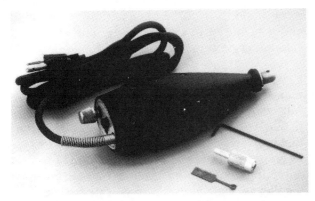

Fig. 4-23 *Hand-held vibrator cleaning tool, used to break lead deposits out of the firing-end cavity of a spark plug.*

Use a vibrator-type cleaning tool with the proper cutter blade to remove all of the lead deposits from the firing-end cavity. Hold the spark plug over the cutter blades, and with the blades vibrating, lightly work the plug back and forth over the vibrating blades so all of the lead deposits will be loosened. Do not force the plug against the cutters; just allow them to chip away at the deposits.

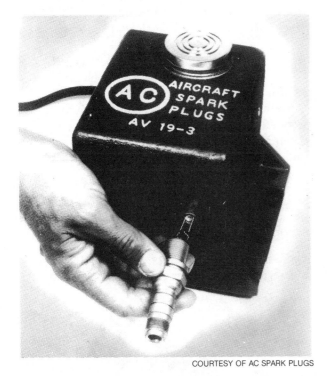

Fig. 4-24 *A vibrator cleaner such as this bench-mounted model makes fast work of removing lead deposits when cleaning large quantities of spark plugs.*

After all of the lead deposits are broken out of the firing-end cavity, very lightly blast the cavity with an abrasive blast. Put the spark plug in the proper size adapter in the spark plug cleaning machine and wobble it in a complete circle for only about three to five seconds. As much as five seconds of abrasive blasting can erode the electrodes of a fine-wire spark plug as much as two to three hundred hours of engine operation.

Fig. 4-25 *This combination spark plug cleaner and tester is typical of those used in aircraft maintenance shops.*

Use only glass beads or aluminum oxide as an abrasive, and never use silica sand. The abrasive should be fresh and clean, and it should be replaced after cleaning about 75 to 100 spark plugs. A water trap should be installed in the air line just before the cleaner, and it should be periodically drained to prevent contaminating the abrasive with water.

When cleaning the spark plugs on a cleaning machine, be sure to follow the instructions of the manufacturer of the machine for the correct operating air pressure. Some manufacturers recommend 80 pounds per square inch for cleaning fine-wire spark plugs and 100 psi for cleaning massive-electrode spark plugs.

After the firing end of the spark plug has been abrasive-blasted, blow it out with a blast of clean air to remove all of the particles of abrasive grit. And when the spark plug is clean, carefully inspect it with a lighted magnifying glass. Carefully check for the condition of the insulator and the electrodes, and be sure that every bit of the abrasive is out of the cavity.

Fig. 4-26 *A lighted magnifier is a handy tool to use to examine the firing-end and the terminal-end cavities of a spark plug when it is being cleaned.*

Clean the terminal-end cavity with either a swab damp with trichlorethylene or a cleaning tool and one of the approved insulator cleaning compounds. If a cleaning compound is used, be sure that all of its residue is removed by the method specified by the manufacturer of the compound.

Clean the threads at both ends of the spark plug with a steel brush, and when they are perfectly clean, inspect them and remove any damaged portions of the threads with a fine triangular file.

4. Gapping

Setting the proper gap between the center and ground electrodes of a spark plug is one of the most important parts of spark plug servicing, and you must use the proper equipment to do a satisfactory job.

For massive-electrode spark plugs, screw the plug into the proper gapping tool and very carefully move the ground electrode over so it will be parallel to the center electrode and the correct distance away as is measured with a wire-type feeler gage. Be sure to position the ram of the gap adjusting tool so the ground electrode will be parallel to the center electrode rather than angling over to it. If it is not parallel, the gap will widen as soon as a very small portion of the ground electrode wears away.

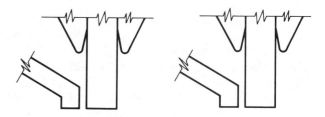

Fig. 4-28 *When gapping a massive-electrode spark plug, move the ground electrode over so the sides of the gap are parallel. If the ground electrode is tilted, a small amount of wear will increase the gap to an unsatisfactory dimension.*

The wire gage must not be between the electrodes when you move the ground electrode over, as this will place a side load on the center electrode and will crack the ceramic. Move it over very slightly and check the gap. If it is not close enough, very carefully move it a slight bit more, and check it with the wire gage.

It is not recommended that a gap be widened as it has been inadvertently closed too much, as damage to the nose ceramic or to the electrode can result.

Fine-wire spark plugs are easier to gap than massive-electrode spark plugs, but extreme care

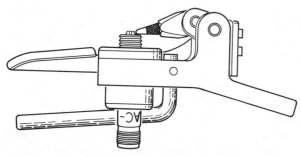

Fig. 4-27 *A gapping tool used to adjust the gap spacing of massive-electrode spark plugs*

Fig. 4-29 *A handy tool to adjust and measure the electrode gap in a fine-wire spark plugs*

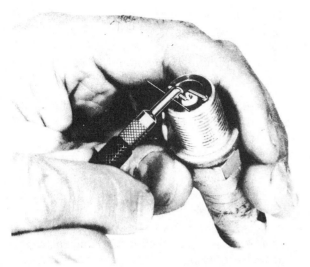

Fig. 4-30 The use of the gapping tool to adjust the gap in a fine-wire spark plug

must be exercised when adjusting the gaps, because both platinum and iridium are extremely brittle and the small electrodes can be broken if they are improperly handled. Use a special gapping tool and very carefully move the ground electrodes over until they are the correct distance away from the center electrode. Do not bend the ground electrodes too far, as they should never be moved back.

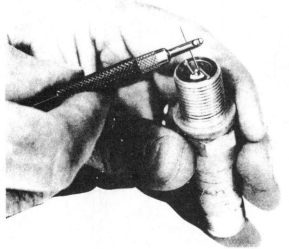

Fig. 4-31 The use of the gapping tool to measure the gap in a fine-wire spark plug

5. Testing

When the spark plugs have been cleaned and gapped, they should be tested in one of the testers sold by the spark plug manufacturers for this purpose. The most accurate tester is the bomb tester in which a 200-psi charge of nitrogen is used in the test chamber and a high voltage is applied to the terminal. If the spark plug fires consistently under pressure, it will operate satisfactorily in the engine.

Fig. 4-32 Bomb testers provide the most accurate means of determining the condition of a spark plug.

Most maintenance shops use the spark plug tester that is a part of the cleaning machine, and these testers operate with a lower air pressure and are actually comparison testers. They compare the reconditioned spark plug with a new spark plug. This is the type of tester shown in Fig. 4-25.

The spark plug is screwed into the tester finger-tight, the high-voltage lead is connected to the shielding barrel contactor, and the air pressure is adjusted to the value specified for the spark plug gap being used. When the test switch button is depressed, a spark should jump the gap consistently. The spark plug is a loose fit in the adapter, so some air can leak out during the test. This leakage removes the ionized air from the test chamber and it also allows better control of the air pressure.

6. Preservation and storage

If the spark plugs are not to be reinstalled in the engine right away, protect the threads on both the shell and the shielding barrel with a light coat of rust-inhibitor oil and then put the spark plug in a carton similar to the ones in which new spark plugs are received.

In humid areas, it is a good idea to store cleaned spark plugs in a closed cabinet with a low-wattage light bulb inside to protect the spark plugs from moisture.

7. Installation

The spark between the electrodes of an aircraft spark plug acts in much the same way as the arc used for welding. Metal is taken from one electrode and deposited on the other. In spark plugs, we are mainly concerned with the metal that is removed. We can extend the service life of a spark plug by minimizing the amount of electrode erosion caused by the metal transfer.

When a spark plug fires positive, the ground electrode will be worn away than the center electrode, but when it fires negative, the center electrode will be worn most severely.

An engine with an odd number of cylinders fires its spark plugs with different polarity each time, and the spark plugs used these engines wear uniformly. But horizontally opposed engines all have an even number of cylinders, and each spark plug will fire with the same polarity every time. To even out the wear on these spark plugs, each one should be replaced in the cylinder next in firing order from the one from which it was removed, and the spark plugs removed from bottom holes should be replaced in a top hole. In a six-cylinder engine with a firing order of 1-6-3-2-5-4, the spark plug removed from cylinder number one, top, should be replaced in cylinder number six, bottom. The one from six, bottom should be replaced in three, top. Follow the sequence shown in Fig. 4-34.

THIS SPARK PLUG HAS BEEN SPARKING NEGATIVE, AND THE CENTER ELECTRODE IS MOST SEVERELY ERODED.

(A)

THIS SPARK PLUG HAS BEEN SPARKING POSITIVE, AND THE GROUND ELECTRODES ARE MOST SEVERELY ERODED.

(B)

Fig. 4-33 Wear of the spark plug electrode caused by failure to rotate the spark plugs to the next cylinder in firing order each time the spark plugs are serviced

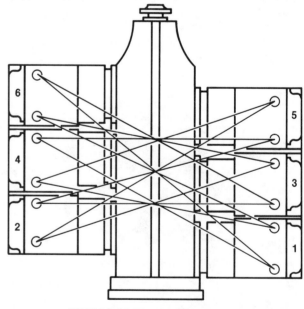

FIRING ORDER 1-6-3-2-5-4

REMOVE PLUG FROM	1T	6B	3T	2B	5T	4B	1B	6T	3B	2T	5B	4T
REPLACE PLUG IN	6B	3T	2B	5T	4B	1T	6T	3B	2T	5B	4T	1B

Fig. 4-34 When replacing the spark plugs in an engine after they have been serviced, rotate them to the next cylinder in firing order, and from top to bottom.

Before replacing the spark plugs, the threads on the shell should be treated with some anti-seize compound. Very sparingly apply a bit of this compound to the *second* thread from the firing end. Never put any on the first thread, as there is a possibility that the compound may run down onto the electrodes and foul them. Some operators feel that the anti-rust oil is sufficient anti-seize protection, but if a regular compound is used, use it sparingly. No anti-seize compound should be used on the barrel terminal threads.

Fig. 4-35 *When replacing spark plugs in the engine, use only a very small amount of thread lubricant on the second thread from the end of the shell. Never put any lubricant on the end thread.*

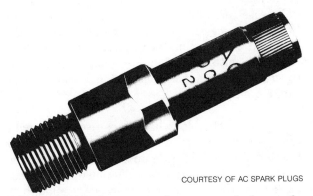

COURTESY OF AC SPARK PLUGS

Fig. 4-36 *Carbon may be removed from the threads in a spark plug hole by using a special 18-mm thread cleaning tool. Pack the threads and grooves in the tool with grease to hold the carbon chips and prevent their falling into the cylinder.*

Give the spark plug one final inspection before installing it. Slip a new solid copper gasket over the firing-end threads and screw the spark plug into the cylinder head. The spark plug should screw in all the way to the gasket with just your fingers. If it does not, there is a possibility that there is carbon in the threads of the adapter and they should be cleaned out with the proper spark plug thread cleanout tool.

When all of the spark plugs have been screwed into the cylinder heads and all are resting on their gaskets, use a properly calibrated torque wrench and, putting a smooth and even pull on the wrench, tighten the spark plugs to the torque recommended in the aircraft service manual. This torque is very important, and no attempt should be made to install spark plugs by feel.

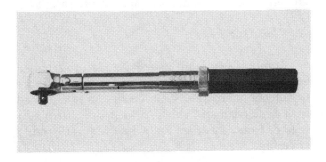

Fig. 4-37 *Always use an accurately calibrated torque wrench when installing spark plugs in an aircraft engine. Never attempt to install the spark plugs by feel.*

If the engine is equipped with a spark plug gasket type of thermocouple for the cylinder head temperature gage, there should not be a regular gasket used with the thermocouple gasket.

When all of the spark plugs have been installed and properly torqued, you should install the spark plug leads. Wipe the lead terminal sleeve, usually called the cigarette, with a clean lint-free cloth moistened *(not wet)* with acetone or MEK. Inspect the cigarette for any indication of cracks of signs of burning. Be sure that the spring on the end of the terminal is secure and the lead is making good contact with the spring.

Slip the terminal sleeve straight into the spark plug barrel and tighten the nut finger-tight, and then about one-eighth of a turn more with a properly fitting wrench.

V. IGNITION HARNESS

There may be enough high voltage generated in a magneto to cause an adequate spark to jump the gap in the spark plug, but if this high voltage is not delivered to the spark plug without losses, engine performance will deteriorate.

Modern aircraft carry a considerable amount of electronic equipment, and the communications and navigation systems must be able to receive signals from ground stations without being interfered with by extraneous sources of radio frequency electrical energy. Since all high-voltage ignition systems act as very effective radio transmitters, the RF energy generated by the spark must be contained within the harness and grounded. Unless this is done, there will be enough energy radiated to impair radio reception.

A. Construction

Ignition leads are usually made of stranded copper of stainless steel wire enclosed in a rubber or silicone insulation. This insulation is enclosed in a braided metal shield, which is in turn protected from abrasion by a tough plastic outer covering.

Fig. 5-1   A typical stranded-wire ignition lead. The stranded ignition wire is surrounded with a high-voltage insulation and is encased in a braided wire shield. This shield is protected with a plastic abrasion shield.

Slick Electro, Inc., produces a very popular ignition harness that, instead of using stranded wire, uses a continuous spiral of wire which is impregnated with a high-voltage silicone rubber insulation.

There have been two sizes of ignition leads used in aircraft engine ignition harnesses, 5-mm and 7-mm, but practically all of the modern harnesses use the smaller size lead. Many of the older radial engines used unshielded wire run inside separate shielding. When it was necessary to replace a lead, the old lead was pulled from the shielding and a new lead was pulled through in its

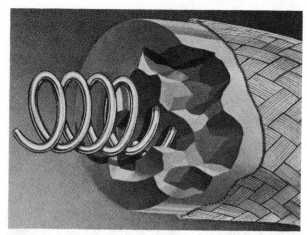

Fig. 5-2   The coiled conductor in a Slick Electro ignition lead. The high-voltage insulation impregnated into the coiled wire is covered with a braided-wire shield.

place a lead, the old lead was pulled from the shielding and a new lead was pulled through in its place. With the more modern installations, the entire lead, shielding and all, is replaced.

Modern ignition harnesses are available with terminal ends to fit either the standard 5/8-24 shielded spark plug or the 3/4-20 All-weather spark plug. Many operators, when having to replace both the spark plugs and the harness at the same time, are converting the ignition system to the All-weather configuration.

The large bend radius required by the older 7-mm ignition leads made angled lead terminals important, and there are elbows with 70°, 90°, 110°, and 135° bends, so the lead can be installed without straining the terminal end. The smaller leads may also use the angled elbows, but some harnesses are fitted with straight terminals. When it is necessary to have an angle between the lead and the terminal, a bracket is used to hold the lead with an adequate bend radius. The bracket prevents the lead being strained. (See Fig. 5-3 on page 43.)

Some of the older harnesses used a phenolic or a ceramic tube with a coil spring at its end for the terminal connection inside the spark plug. These are called cigarettes. The insulation is cut back from the stranded conductor far enough to allow the wires to stick through the small hole in the end of the terminal, and the ends of the wires are fanned out to provide a good electrical contact and to secure the cigarette to the wire. (See Fig. 5-4 on page 43.)

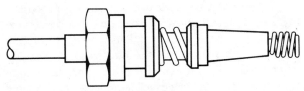

STRAIGHT TERMINAL FOR USE IN AN ALL-WEATHER
SPARK PLUG

(A)

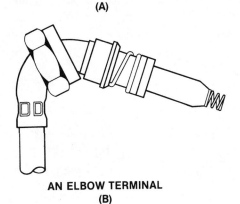

AN ELBOW TERMINAL

(B)

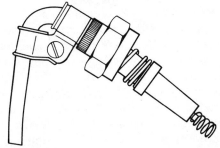

A STRAIGHT TERMINAL HELD IN AN ANGULAR POSITION
WITH A BRACKET

(C)

Fig. 5-3   Typical spark plug lead terminals

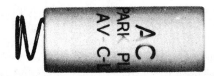

Fig. 5-4   A ceramic terminal insulator for an ignition harness

Some modern harnesses use silicone rubber for the terminal connectors and the terminal is crimped to the wire rather than spreading the strands. The springs screw over the end of the terminal, and they may be replaced when they become damaged. (See Fig. 5-5.)

One of the main sources of radio interference from a shielded ignition harness comes from improper securing of the shielding at the ends of the leads. The shielding is secured by clamping it between an outer and an inner ferrule, and the two ferrules are pressed together so the shielding becomes an integral of the lead, and it is electrically grounded at each end. (See Fig. 5-6.)

If a single spark plug lead becomes damaged, it can be replaced individually without having to replace the entire harness. The various harness manufacturers furnish detailed information describing the approved methods for installing both the spark plug terminal and the magneto terminal. This information must be followed, and the special tools that are required for this operation must be used.

MAGNETO END   SPARK PLUG END

Fig. 5-5   The typical components of an ignition lead

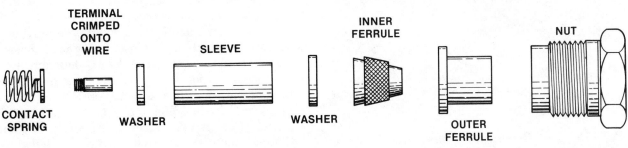

TERMINAL
CRIMPED
ONTO
WIRE

SLEEVE

INNER
FERRULE

NUT

CONTACT
SPRING

WASHER

WASHER

OUTER
FERRULE

Fig. 5-6   The components on the magneto end of an ignition lead. The center conductor is crimped into the terminal, and the shielding is clamped between the inner and the outer ferrules.

There are replacement ignition leads available that have one end already secured to the cable, and all of the parts needed for the other end are furnished in the kit. The cable is cut to the correct length, and the fittings are installed to give you the correct length of replacement lead.

COURTESY OF SLICK ELECTRO

Fig. 5-7 Typical tools used to install the magneto-end terminals on Slick Electro ignition leads.

B. Testing

When troubleshooting an ignition problem, we often blame the magnetos or spark plugs, when the problem is actually somewhere between them, in the harness. If an ignition lead has a high-resistance leak to ground, the voltage may leak off to ground before it can build up high enough to jump the gap in the spark plug.

This kind of electrical leakage often occurs only when the harness is hot, and this makes the source of the trouble hard to find with normal troubleshooting procedures.

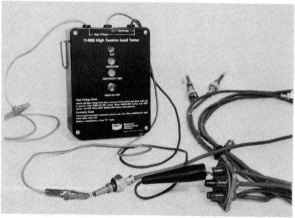

COURTESY OF BENDIX

Fig. 5-8 The Bendix high-tension lead tester. This tester is used to determine if the electrical leakage in an ignition lead is excessive

There are a number of harness testers on the market that will show when a harness has an excessive electrical leakage. These testers usually place a high voltage on the harness and actually cause it to break down if there is a weak spot in the insulation. Excessive leakage is then shown by the illumination of an indicator light or, in the case of one popular tester, by the extinguishing of a spark that is jumping across a visible spark gap.

VI. GAS TURBINE ENGINE IGNITION SYSTEMS

Modern gas turbine engine ignition systems are generally of the high-intensity, intermittent duty, capacitor-discharge type. There are two common classifications of the capacitor ignition systems, the high voltage and low voltage systems. Both draw sufficiently high current to cause heat damage to their units, so they have a restricted duty cycle of a certain time duration, followed by a cooling-off period. After a normal start, ignition is no longer needed and the system is deactivated. At this time the flame within the combustor acts as an ignition source for continuous combustion.

With either the *high* or *low* voltage system, two igniter plugs are usually incorporated in the engine combustor at the 4 o'clock and 8 o'clock positions. The typical system consists of two transformers (exciter) units, two ignition leads and two igniter plugs. The two sets may be housed separately or as one unit with two output leads.

The *main* ignition system is used primarily during *ground starting* and is then turned off. A secondary function of this system is to provide a standby protection against in-flight flameout which might occur at takeoff, landing, bad weather operation, or when operating in anti-ice bleed air mode. Some aircraft provide for alternate use of the left plug or the right plug from the main system at full transformer capacity but for limited time periods which must be scheduled by the pilot. The pilot can select ignition, in this case, whenever he wishes.

On some newer installations a *pressure sensitive* device in an auto-relight system will turn on main ignition automatically whenever internal pressures within the engine drop to a certain value. Another popular configuration is called the *continuous duty* system. This system generally contains a separate, low power discharge to one igniter plug which again can be selected by the pilot. This system can be operated for as long as the need remains for a self-relight capability in flight.

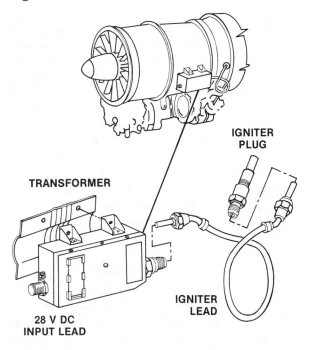

Fig. 6-1 One unit of a dual ignition system for a Pratt and Whitney JT-12 turbine engine

A. Special Handling Requirements

The term, "high intensity" infers that a lethal charge is present and that turbine ignition systems require special maintenance and handling as prescribed by the manufacturer. Typical of these procedures are as follows:

1. Ensure that the ignition switch is turned off before performing any maintenance on the system.

2. To remove an igniter plug, disconnect the transformer input lead, wait the time prescribed by the manufacturer (usually 1-5 minutes), then disconnect the igniter lead and ground the center electrode to the engine. The igniter lead and plug are now safe to remove.

3. Exercise great caution in handling damaged transformer units. Some contain radioactive material, cesium-barium 137, on the air gap points for calibrating the discharge point to a preset voltage.

4. Ensure proper disposal of unserviceable igniter plugs. If they are the semi-conductor type that contains aluminum oxide and beryllium oxide, the usual method is to place the plugs in a closed container and bury them.

5. Before a firing test of igniters is performed, the technician must ensure that the combustor is not fuel-wetted, as a fire or explosion could occur.

6. Do not energize the system for troubleshooting if the igniter plugs are removed. Serious overheating of the transfomers can result.

B. Joule Ratings

Turbine ignition systems carry a joule rating. A joule is defined as watts multiplied by time, whereby one joule per second equals one watt. The time factor for plug firing is very short, in the millionths of seconds. Ignition of atomized fuel occurs very rapidly, in milliseconds, so a long term spark is not necessary. A mathematical explanation of why turbine ignition systems are considered lethal is as follows:

EXAMPLE:

A turbine ignition system with a four-joule rating, ionizing voltage of 2,000 volts, spark time to jump the igniter plug air gap of 20 microseconds: Find the wattage and amperage carried by this circuit.

If: $J = W \times T$

Then: $4 = W \times 0.00002$

$W = 4 \div 0.00002$

$W = 200,000$

If: Watts $=$ volts $\times$ amps

Then: $200,000 = 2,000 \times$ amps

Amps $= 200,000 \div 2,000$

Amps $= 100$

where: J $=$ Joule rating
W $=$ Watts expended
T $=$ Time for spark to jump gap

Some ignition systems of the high voltage type are rated as high as 20 joules with 2,000 amps output. This power is possible from the physically small generating system described because of the very short duration of time required to cause flashover at the igniter plug gap.

C. Types of Ignition Systems

1. Low voltage DC input system

Fig. 6-2 shows the low voltage output, DC input system. Note that only one igniter plug is fired by this circuit. So two identical circuits are usually included in a single transformer unit to supply power to the two igniter plugs most engines require.

Sequence of events:

1. With the cockpit switch open, a permanent magnet will hold the points closed by attraction of the point armature.

2. When the cockpit switch is closed, current flows from the battery negative side, up through the primary coil, across the points, and to the battery positive terminal.

3. As electromagnetic force builds, the points are pulled open, stopping current flow. This action is repeated approximately 200 times per second, producing a pulsating DC voltage. As the points open, there is a tendency for the current to jump the points gap. A capacitor across the points prevents arcing by offering a path of least resistance for this current flow.

4. When the current first flows through the closed points, bottom to top of the primary coil, a pulse is produced in the secondary coil. This pulse attempts to flow in the opposite direction, from the bottom of the secondary coil out of the ground, and up through the capacitor to the top side of the secondary coil. But, the rectifier blocks current flow in this direction. The current path is also blocked by the discharge tube which is an open circuit at this time.

5. A second stronger pulse occurs when the points open and the primary field collapses into the secondary, creating a voltage of greatly increased value. Secondary current flows in the opposite direction, from the top of the coil and through the rectifier, to allow electrons to store on the top plate of the storage capacitor. The current path is completed as free electrons are pushed from the bottom plate of the capacitor, out of the ground, to the bottom side of the secondary coil. The half-wave rectifier or blocking diode is used to change secondary coil induced alternating current into direct current.

6. After repeated cycles, a charge will build on the negative (top) side of the storage capacitor, sufficient to overcome the gap in the discharge tube. The initial current surge ionizes the air gap (makes it conductive) and allows the capacitor to discharge fully to the igniter plug.

7. In a low voltage system, the igniter plug is referred to as a self-ionizing or shunted-gap-

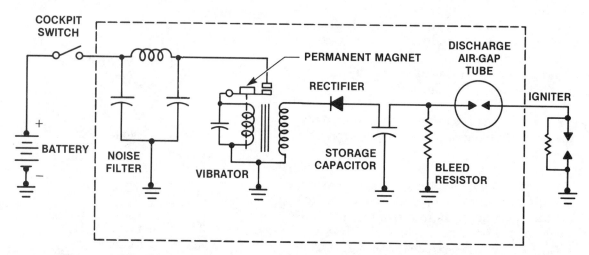

Fig. 6-2 The low voltage DC input ignition system used for a turbine engine

type plug. The firing end of the plug contains a semi-conductor material which bridges the gap between the center electrode and the ground electrode (provided by the igniter plug outer casing). The plug fires when the current flows from the storage capacitor through the center electrode, the semi-conductor, the outer casing, and then back to the lower positive plate of the capacitor. As current flows initially through the semi-conductor, heat builds up creating an increased resistance to current flow. The air gap also becomes heated sufficiently to ionize and the current takes the path of least resistance across the ionized gap, fully discharging the capacitor and creating a high energy capacitive discharge spark.

8. The bleed resistor is present in the circuit to allow the capacitor to bleed off its charge when the system is de-energized. It also protects the circuit from overheating if power is turned on when no igniter plug is installed.

2. *High voltage AC input system*

The following electrical circuit is of a high voltage system with an alternating current input voltage. (See Fig. 6-3 on page 48).

Sequence of events:

1. Alternating current of 115 volts, 400 cycles is supplied to the primary coil of the power transformer. Note the absence of the vibrator point system which is not needed with an AC input system. In some DC systems, the points were a constant source of malfunction and the AC system was developed to counteract this problem.

2. During the first half cycle, the primary coil induces approximately 2,000 volts in the secondary coil and current flows from the minus $(-)$ side of the coil, out of the ground, into the ground at rectifier tube A, through resistor R_1 through the doubler capacitor and back to the plus $(+)$ side of the secondary coil. This leaves the left side of the doubler capacitor charged to 2,000 volts. Rectifier tube B blocks any other path of the first half cycle current flow.

3. During the second half cycle, the primary coil induces another 2,000 volts in the secondary coil and current flows from the minus $(-)$ side of the coil, through the doubler capacitor where it now has a total of 4,000 volts, through resistor R_2, through rectifier tube B, through the storage capacitor, out of the ground, returning to the plus $(+)$ side of the secondary coil. Note that the presence of resistor R_1 and the gas discharge tube causes current to seek the path of least resistance, which is straight back to the secondary coil. Rectifier tube A also blocks current flow from the doubler capacitor and R_1 to ground, ensuring current flow to the storage capacitor.

4. Repeated pulses charge the storage capacitor until it reaches a voltage of approximately twice that produced by the power transformer. When this voltage reaches a preset value, the spark gap in the trigger discharge tube ionizes and breaks down so a part of the charge can flow to ground through the primary coil of the high-voltage transformer and the trigger capacitor. This surge of current through the primary induces a voltage into the secondary coil of the high-voltage transformer of approximately 20,000 volts, sufficient to ionize the igniter plug air gap and complete a path back to the storage capacitor. The trigger spark, which occurs by action of the high-voltage transformer and trigger capacitor, creates a low resistance path and allows both the trigger capacitor and storage capacitor to fully discharge at the igniter plug, creating a second high-intensity spark. The high energy spark is needed to blast carbon deposits from the electrodes, and also to vaporize fuel globules at the firing end sufficiently to ignite the fuel-air mixture in the combustor either on the ground or in flight.

D. *Comparison of AC and DC Systems*

a. The AC input system is said to have a better extreme climate reliability than the DC input system, and the absence of the point system is said to have an effect of increased service life for the transformer unit.

b. The operational cycle of a typical AC main system is 10 minutes on, 20 minutes off (for cooling). The DC main system heats up more rapidly and a typical operational cycle of a system with the same joule rating as the AC

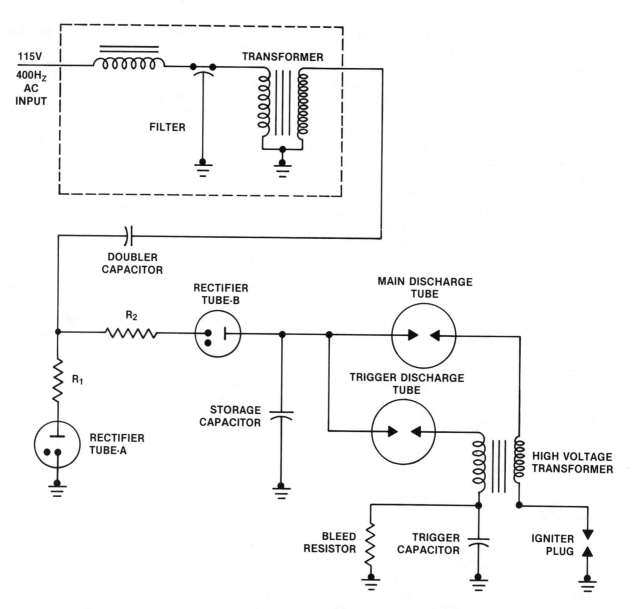

Fig. 6-3   The high-voltage AC input ignition system used for a turbine engine

system might be 2 minutes on, 3 minutes off, or 2 minutes on, 20 minutes off.

c. In flight, both AC and DC main systems can be switched to one igniter plug for one time cycle, then to the other igniter plug. This procedure could be repeated as long as in-flight ignition is required.

d. If an AC continuous ignition system is installed there is usually no time limit. The pilot does switch from right to left plug, however, for equal amounts of time in the interest of plug service life. There are no DC continuous systems.

e. The DC system remains in popular use, especially when no auxiliary power unit is installed and a battery input voltage is all that is available for starting. On larger aircraft, the APU has a battery DC input ignition and the main engines can then use the APU's alternator output for powering an AC input system.

E. *Comparison of High Voltage and Low Voltage Systems*

a. The case for high or low voltage systems appears in many instances to be simply a matter of the manufacturer's preference, or the choice is given to the customer whether a high or low voltage system is installed. It is common to see either system in the same model aircraft.

b. However, as previously mentioned, there is a need for the high voltage system in some engines to counteract a tendency of carbon accumulation on the firing tip of the igniter plug. Also some engines need a high blast effect to better vaporize fuel droplets during starting. This may be especially needed for reliable high altitude relight capability. The low voltage system might be less effective from this standpoint in a particular combustor.

c. The low voltage system has a lower weight factor than the high voltage system and a higher expected exciter box service life with the absence of trigger transformer and capacitor circuit. It is also less hazardous to maintain for this reason.

F. Igniter Plugs

1. Types of igniter plugs

Igniter plugs for gas turbine engines differ considerably from spark plugs for reciprocating engines. The gap at the igniter plug tip is much wider and the electrode is designed to withstand a much higher intensity spark. The igniter plug is also less susceptible to fouling because the high energy spark removes carbon and other deposits each time the plug fires. The construction material is also different, in that the igniter plug is made of a very high quality, nickel-chromium alloy for its corrosion resistance and low coefficient of heat expansion. For this reason, it is many times more expensive than a spark plug.

Many varieties of igniter plugs are available, but usually only one will suit the needs of a particular engine. The igniter plug tip must protrude properly into the combustor in each installation and, on long ducted fan engines especially, must be long enough to mount on the outer case, pass through the fan duct, and penetrate the combustor.

Igniters for high and low voltage systems are not interchangeable and care should be taken to be sure the manufacturer's recommended igniter plug is used.

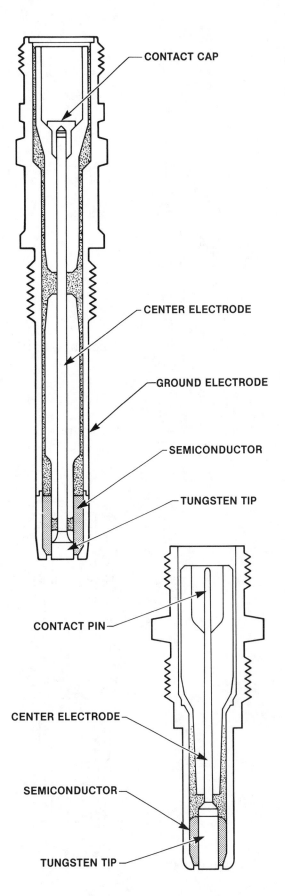

Fig. 6-4 *Typical low voltage igniter plugs*

49

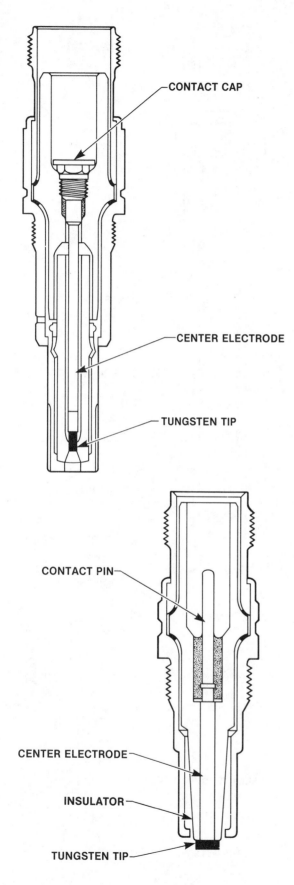

Some smaller engines incorporate a glow plug type of igniter rather than a spark igniter. This glow plug is a resistance coil of very high heat value and is said to be designed for extreme low temperature starting. A typical system is seen in some models of the Pratt & Whitney PT6 turboshaft engine.

The glow plug is supplied with approximately 2,000 volts DC to turn the coil yellow hot. The coil is very similar in appearance to an automobile cigarette lighter. Air directed up through the coil mixes with fuel dripping from the main fuel nozzle. This is designed to occur when the main nozzle is not completely atomizing its discharge at low flow conditions during engine starting. The influence of the airflow on the dripping fuel acts to create a hot streak or torch-type ignition. After fuel is terminated, the air source serves to cool the igniter coil all the time the engine is being operated.

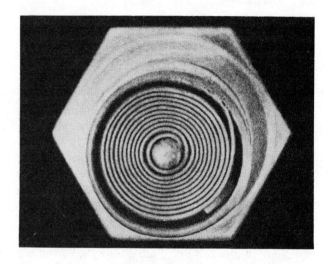

Fig. 6-6   Glow-plug-type igniter plug

2. *Cleaning and inspection of igniter plugs*

a. *High voltage igniter plug checks*

The high voltage igniter plug shown in Fig. 6-7 is cleaned on the outer case with a soft brush and solvent and the ceramic portion is cleaned with a felt swab and solvent. The electrode tip may be cleaned with solvent, but abrasive grit cleaning is never permitted as the ceramic insulator between the electrode and the case would be damaged. After cleaning, dry air is used to blow off the remaining solvent and the plug is ready for inspection.

Fig. 6-5   Typical high voltage igniter plugs

GAP DESCRIPTION	TYPICAL FIRING END CONFIGURATION	CLEAN FIRING END
HIGH VOLTAGE AIR SURFACE GAP	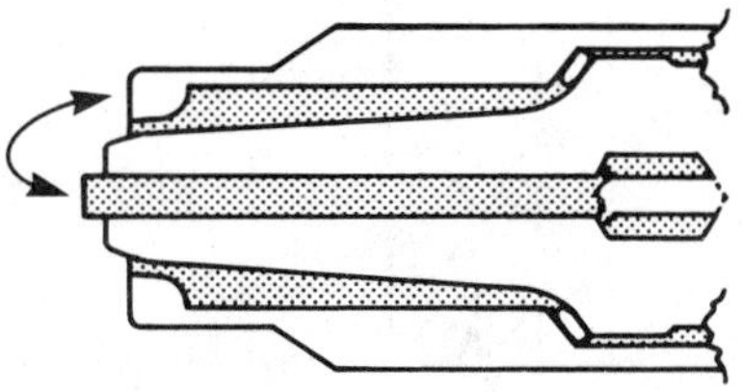	YES
HIGH VOLTAGE SURFACE GAP	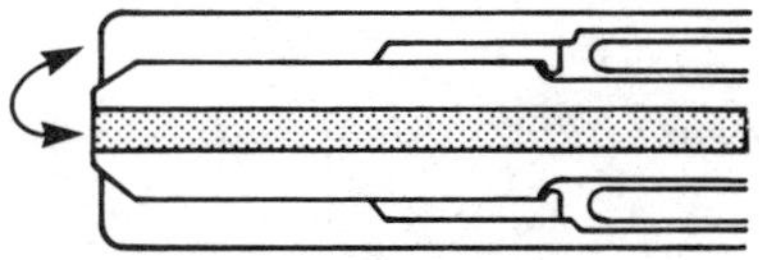	YES
HIGH VOLTAGE RECESSED SURFACE GAP	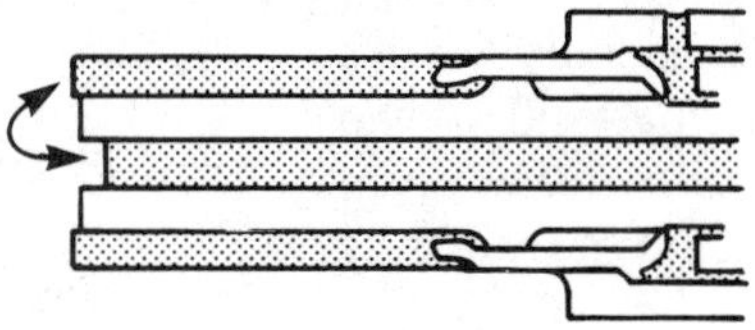	YES
LOW VOLTAGE SHUNTED SURFACE GAP	SEMI-CONDUCTOR 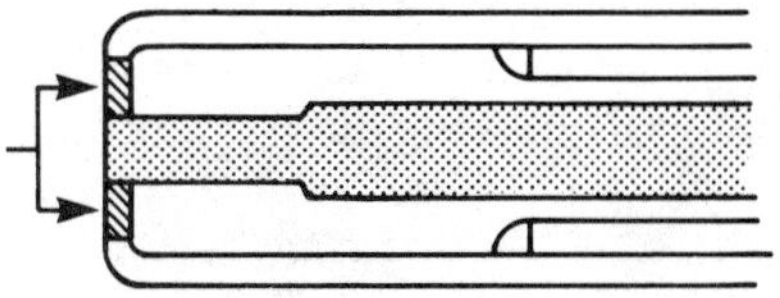	NO
LOW VOLTAGE GLOW COIL ELEMENT		YES

Fig. 6-7   Firing ends of typical turbine igniter plugs

Igniter plug inspection generally consists of a visual inspection and a measurement check with a mechanic's scale or suitable depth micrometer. After visual inspection, an operational check is performed. A typical operational check might include connecting the conditioned plug to its circuit lead, with the plug positioned outside of the engine and firing it to compare its spark intensity with that of a new plug. During this check, extreme care must be taken to completely discharge the capacitor(s) before handling the center electrode.

Another check typical of installed igniters, is to fire the plugs one at a time, using the aircraft circuit breakers and listening near the tailpipe for the sharp snapping noise associated with good ignition spark. The spark interval can also be timed during this check. The usual spark rate is from 0.5 to 2.0 sparks per second. The maintenance manual will specify the exact rate for each particular system. Here again, caution must be exercised to ensure no fuel vapors are present in the combustor during the check or a serious fire could result.

51

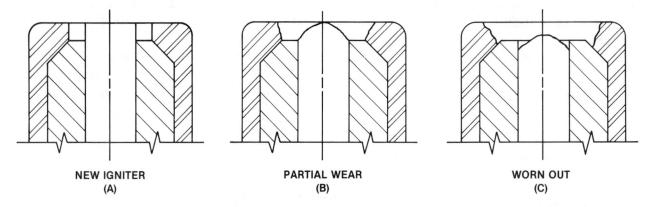

NEW IGNITER
(A)
    PARTIAL WEAR
(B)
    WORN OUT
(C)

Fig. 6-8   Wear conditions of low voltage shunted surface gap igniter

b.   Low voltage igniter plug checks

The self-ionizing, or shunted-gap igniter plug, as it is called, is generally cleaned only on its outer casing. The semiconductor material at the firing end is easily damaged and manufacturers seldom permit any type of cleaning regardless of the carbon buildup.

Measurement checks are not usually accomplished on the firing tip, since quite often the semiconductor consists of only a very thin coating over a ceramic base material, making it impossible to clean or to measure. The operational checks mentioned above are typical of the low voltage plugs as well as the high voltage plugs.

G.   Troubleshooting Turbine Ignition Systems

Logical troubleshooting procedures should be used when hunting for a problem in a turbine ignition system. The chart in Fig. 6-9 outlines the procedures to use when the system produces no spark, when the spark is weak, or when there is too long an interval between sparks.

1. NO IGNITER SPARK WITH THE SYSTEM TURNED ON.		
POSSIBLE CAUSE	CHECK FOR	REMEDY
A. IGNITION RELAY	CORRECT POWER INPUT TO TRANSFORMER UNIT	CORRECT RELAY PROBLEMS, REFER TO STARTER-GENERATOR CIRCUIT.
B. TRANSFORMER UNIT	CORRECT POWER OUTPUT, OBSERVING IGNITION SYSTEM CAUTIONS.	REPLACE TRANSFORMER UNIT
C. HIGH TENSION LEAD	CONTINUITY OR HIGH RESISTANCE SHORTS WITH OHMMETER AND MEGER-CHECK UNIT	REPLACE LEAD
D. IGNITER PLUG	1. CRACKED INSULATOR OR DAMAGED SEMICONDUCTOR	REPLACE PLUG
	2. HOT ELECTRODE EROSION	REPLACE PLUG
2. LONG INTERVAL BETWEEN SPARKS.		
POSSIBLE CAUSE	CHECK FOR	REMEDY
POWER SUPPLY	WEAK BATTERY	RECHARGE BATTERY
3. WEAK SPARK		
POSSIBLE CAUSE	CHECK FOR	REMEDY
IGNITER PLUG	CRACKED CERAMIC INSULATION	REPLACE PLUG

Fig. 6-9   Troubleshooting turbine ignition systems

Powerplant Electrical Systems

INTRODUCTION

Electricity and electrical systems are such an important part of all phases of aircraft operation that we emphasize it in this Integrated Training Program. In the General Section, we cover the fundamentals of both direct current and alternating current electricity. In the Airframe Section, that portion of aircraft electrical systems and components that are normally serviced and maintained by an airframe mechanic are discussed. In this Powerplant portion of the Integrated Training Program we are concerned with that portion of the electrial system that is normally maintained by the powerplant mechanic.

There is no real hard and fast division between the airframe and powerplant electrical systems, and different operators handle this in different ways. Many of the airlines and large operators have electrical specialists who maintain all of the electrical systems and components. Others consider those portions of the electrical system that are on the engine side of the firewall to be within the domain of the powerplant mechanic and everything else as a job for the airframe mechanic. Still others consider the electrical power generation system and the motors that directly affect engine operation to be part of the powerplant electrical system and all else as part of the airframe electrical system.

This training program has been prepared to furnish enough background information to allow a mechanic to interpret manufacturers' service manuals and, when given the proper equipment and information, to be able to properly service systems or components according to the manufacturers' specification. For this reason, there is some overlap between the three sections of this program, but the emphasis follows the Airframe and Powerplant Mechanics examination format used by the Federal Aviation Administration. In Book Three of the Powerplant Section of this Integrated Training Program, we have covered Ignition Systems, and in Book Five, we discuss the installation of electrical wiring and components that are related to the powerplant. In this book we are concerned with electrical generators and alternators with their controls and with electric motors of the types that are used in powerplant systems.

I. DIRECT CURRENT POWER SYSTEMS

A. *Magnetism*

Anytime electrons flow in a conductor, two things happen. Heat is produced by the friction, or resistance, of the conductor as it opposes the flow of electrons and, more important, a magnetic field surrounds the conductor. The more electrons there are flowing, the stronger the magnetic field will be.

This magnetic field is of extreme importance because it can cause electrons to flow in another conductor even when there is no electrical connection between the two. It can produce a mechanical force with another current-carrying conductor or with a permanent magnet.

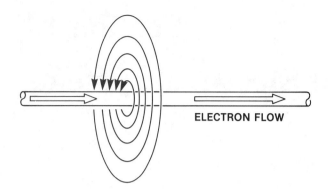

Fig. 1-1 When electrons flow in a conductor, heat is produced in it and a magnetic field surrounds the conductor.

1. Permanent magnets

Before we get too far into our discussion of generators, let's briefly review what we have learned about magnetism.

All matter is composed of atoms, made up of a nucleus that contains positive electrical particles called protons and electrically neutral particles called neutrons. Spinning around the nucleus are negatively charged particles called electrons. These electrons rotate about their axes at an extremely high speed as they circle the nucleus, in much the same way the earth spins about its axis as it rotates about the sun.

Atoms of materials that contain iron behave in a special way: The electrons that spin around the nucleus in atoms of iron and some of the other elements can have their spin axes pulled into alignment if the material is held in a magnetic field.

Rather than the electrons around individual atoms aligning independently, the atoms are affected in clumps, or domains. When the external magnetic field becomes strong enough, the electrons around the atoms in a domain will snap into alignment; then other domains will follow.

In a piece of unmagnetized iron, the domains are oriented in a random fashion and there is no resultant magnetism. But, when the iron is placed in a strong magnetic field, the domains in the iron will all snap into alignment and the iron will become magnetized.

The magnetic domains in soft iron will not remain in alignment after the magnetizing force has been removed, and soft iron is said to have a very low retentivity. This characteristic is important for the iron cores in relays and electromagnets because we want them to lose their magnetism as soon as we shut off the current. But, in other devices we want the part to remain a magnet after the magnetizing force is removed. Hard steel holds its magnetism, and it is said to have a high retentivity. The magnetism that remains in it after the magnetizing force is removed is called residual magnetism. Aluminum, nickel, cobalt, and some other elements are often alloyed with iron to increase its retentivity when it is used for permanent magnets.

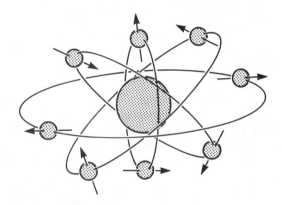

THE SPIN AXES OF THE ELECTRONS IN THE ATOMS OF A PIECE OF UNMAGNETIZED IRON ARE ORIENTED IN A RANDOM DIRECTION.

(A)

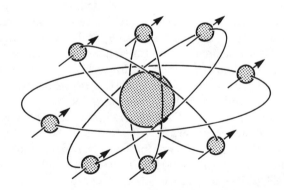

THE SPIN AXES OF THE ELECTRONS IN THE ATOMS OF A PIECE OF MAGNETIZED IRON ARE ALIGNED IN THE SAME DIRECTION. THIS PRODUCES A MAGNETIC FIELD.

(B)

Fig. 1-2 Magnetization by alignment of electron spin axes.

THE MAGNETIC DOMAINS IN A PIECE OF UNMAGNET-IZED IRON ARE ORIENTED IN A RANDOM FASHION.
(A)

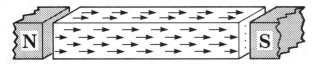

WHEN THE IRON IS PLACED IN A STRONG MAGNETIC FIELD, THE DOMAINS IN THE IRON WILL ALIGN WITH THE EXTERNAL MAGNETIC FIELD.
(B)

Fig. 1-3 Orientation of magnetic domains in iron

A permanent magnet always has two poles, a north pole and a south pole. The names of these poles were chosen when it was seen that a suspended magnet would align with the earth's magnetic field. The end that pointed toward the north star was called the north-seeking pole, or, more simply, the north pole, and the other end was called the south pole.

If a bar-shaped permanent magnet is covered with a flat sheet of paper and iron filings are sprinkled over it, the filings will align themselves much as we see in Fig. 1-4.

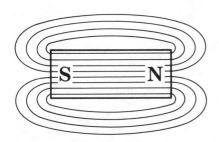

Fig. 1-4 The lines of magnetic flux surrounding a permanent magnet leave the magnet at its north pole and enter at its south pole. All of the flux lines form complete loops.

There have been a number of interesting facts discovered about the force that affects these filings. Briefly they are:

1. The force that acts on the filings is called magnetic flux, or magnetic lines of force.

2. All flux lines form complete loops, leaving the magnet at its north pole and passing by the easiest route to the south pole. They then pass directly through the magnet. The flux lines always leave the poles at right angles to the pole faces.

3. Flux lines cannot cross each other but they can either spread out or group together.

4. Flux lines oppose each other and when possible will spread out to get as much space as possible between them. This force is countered by a force that tries to keep the loops of flux as small as possible.

If we place two bar magnets near each other with the north pole of one near the south pole of the other, the lines of flux will pass from one magnet to the other as we see in Fig. 1-5. The loops of flux, in trying to become as short as possible, create a strong force of attraction between the two magnets that try to pull them together.

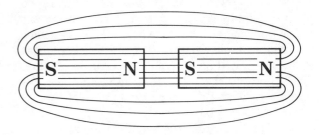

Fig. 1-5 When two bar magnets are placed near each other with their opposite poles next to each other, the lines of flux will loop the two magnets, and as these loops of flux try to shorten, they will pull the magnets together.

The force acting on the magnets is determined by the number of lines of flux between the magnets and the distance between the magnets. If we decrease the separation to one-half of the original distance, the force will increase *four* times. If we decrease the separation to one-third of the original distance, the force of attraction will be *nine* times as great. This relationship between the amount of pull and the separation is known as the inverse square law. The strength of the attraction between two magnets of unlike polarity varies as the inverse square of the distance between them.

If we turn one of the magnets around so the north poles of the two magents are adjacent, the lines of flux from the two magnets will oppose each other and, rather than looping the two magnets, will produce a force that tries to move the magnets apart. This opposing force depends upon the number of lines of flux and by the separation between the magnets. The variation in strength with separation obeys the inverse square law.

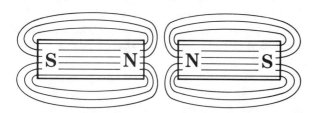

Fig. 1-6 When two bar magnets are placed near each other with their like poles together, the lines of magnetic flux will repel and push the magnets apart.

Most practical applications of magnets do not use simple straight bar magnets, but rather use some form of horseshoe magnet. A horseshoe magnet is simply a bar magnet that has been bent so its north and south poles are near each other. This shortens the separation between the poles and increases the flux density between them.

In Fig. 1-7, we see the way the flux lines leave the north pole of a horseshoe magnet and spread out between the poles and then come back together as they enter the south pole. The lines of flux pass through air between the poles, and air opposes the passage of the lines of flux. The ease

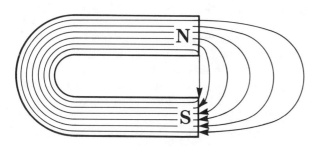

Fig. 1-7 Magnetic lines of flux leave the north pole of a horseshoe magnet and travel by the easiest path to enter the magnet at its south pole.

with which a material allows the passage of flux is called the permeability of the material. Air is used as the reference and it has a permeability of one. Iron has a permeability of several thousand, and it allows the lines of flux to pass through it *far more easily* than they can pass through air.

If we hold a small piece of iron in front of the magnet, all of the lines of flux, in their attempt to find the easiest though not necessarily the *shortest* path between the poles, will pass from the north pole into the iron, through it, and back into the south pole of the magnet. A strong force of attraction is set up between the iron and the magnet as the loops of flux try to shorten. The iron will pull up tight against the poles of the magnet and be held there by the loops of flux.

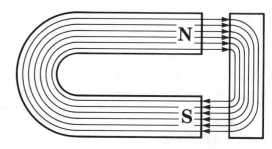

IF A BAR OF SOFT IRON IS PLACED IN THE MAGNETIC FIELD OF A HORSESHOE MAGNET, THE HIGH PERMEABILITY OF THE IRON WILL CAUSE ALL OF THE LINES OF FLUX TO PASS THROUGH THE IRON.

(A)

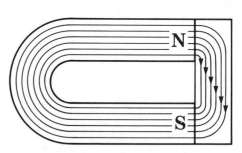

THE LINES OF FLUX, IN THEIR ATTEMPT TO BECOME AS SHORT AS POSSIBLE, WILL PULL THE IRON TIGHT AGAINST THE POLES OF THE MAGNET.

(B)

Fig. 1-8 Attraction of an iron bar to a horseshoe magnet

2. Electromagnets

The magnetic lines of flux from a permanent magnet exist because the electrons spinning around the nucleus of the iron atoms are able to maintain the alignment of their spin axes without an external magnetizing force being maintained on the iron.

The electrons in a copper wire align their spin axes when they are forced to flow through the wire. The alignment of the axes of these electrons causes a magnetic field to surround the wire. This field is maintained only as long as current flows, and the strength of the field is determined by the amount of current that is flowing. The direction the lines of flux encircle the wire is determined by the direction in which the electrons flow in the wire.

If we run a current-carrying wire vertically through a piece of paper and sprinkle iron filings on the paper, the filings will align themselves in concentric circles around the wire. The greater the current, the further the rings will extend out from the wire.

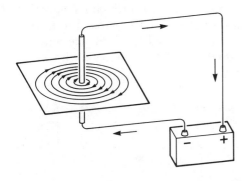

Fig. 1-9 When electrons flow through a conductor, the spin axes of the electrons line up and form a magnetic field around the conductor.

We cannot tell the direction the lines of flux are traveling by looking at the iron filings, but a law has been formulated that describes the direction they travel. If we grasp the wire with our left hand in such a direction that the thumb points in the direction of electron flow, from negative to positive, our fingers will encircle the wire in the same direction as the lines of flux. In Fig. 1-9. the electrons are flowing upward through the wire so the lines of flux encircle the wire in a clockwise direction.

If we have two pieces of wire parallel to each other carrying current in the same direction, the loops of magnetic flux that encircle the wires travel in the opposite direction between the wires. They will cancel each other between the wires,

but add their strength outside of the wires, as we see in Fig. 1-10. The loops will become as short as possible, and in doing this they will exert a force on the wires that will try to pull them together. The strength of the magnetic field surrounding the wire has been increased by our adding the second wire.

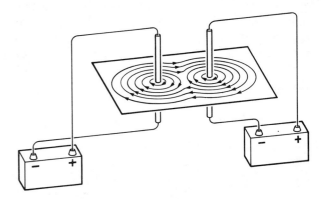

Fig. 1-10 When electrons flow through two parallel conductors in the same direction, the magnetic lines of flux that surround each conductor will join and produce a force that tries to pull the two conductors together.

We can gain the advantage of additional wires carrying current in the same direction by winding our single wire into a coil, as we see in Fig. 1-11. Each turn of the coil has a magnetic field surrounding it, and the fields around each turn add to produce a resulting magnetic field that surrounds the entire coil.

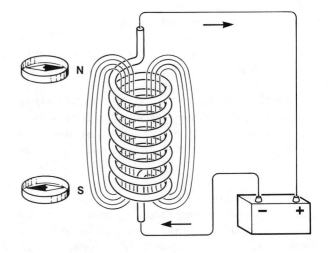

Fig. 1-11 A coil of wire carrying an electrical current forms an electromagnet that acts in the same way as a permanent magnet.

5

If we hold a small magnetic compass near the coil, the north end of the compass magnet will point to one end of the coil and the south end will be attracted to the other end. This proves that our coil of current-carrying wire has become an electromagnet.

We can find the north pole of an electromagnet by holding the coil in our left hand so our fingers encircle the coil in the same direction the current flows, from negative to positive. Our thumb will point to the north pole of the electromagnet.

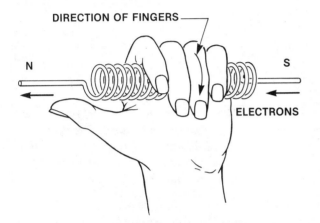

Fig. 1-12 If we grasp an electromagnet with the left hand in such a direction that the fingers surround the coil in the same direction electrons flow, the thumb will point to the north pole of the electromagnet.

The electromagnet formed by winding a piece of wire into a coil is relatively weak because the lines of flux are so spread out, but we can increase its strength by filling the inside of the coil with soft iron. The permeability of the iron is so much greater than that of air that the lines of flux will concentrate in the iron and produce a very strong magnetic field at the ends of the coil. The retentivity of the iron is very low so as soon as we stop the flow of current through the wire, the magnetic field will drop practically to zero and we will no longer have a magnet.

The strength of an electromagnet is determined by the number of ampere turns; that is, by the number of turns in the coil multiplied by the number of amps of current flowing through it. In practical magnetic components, we normally vary the current flow through the coil to control the strength of the magnet.

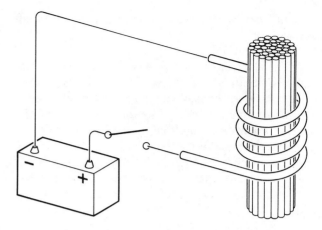

Fig. 1-13 The strength of an electromagnet may be increased by filling the inside of the coil with a bundle of iron wires that have a very high permeability.

B. Generator Action

If we move a wire through a magnetic field in such a way that it cuts across the lines of flux, some of the flux will encircle the wire, and any time a magnetic field encircles a wire, current will flow in it.

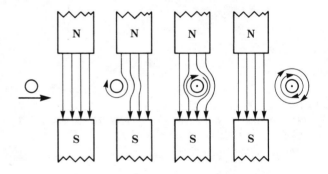

Fig. 1-14 When we move a conductor through a magnetic field in such a way that it cuts across lines of flux, some of the flux will encircle the conductor and cause current to flow in it.

The amount of current that flows in the wire is determined by the rate at which the lines of magnetic flux are cut. This current may be increased by either increasing the number of lines of flux between the magnet poles or by moving the wire through the lines of flux faster. Both methods are used in practical generators.

Even though we can produce current in a wire by simply moving it back and forth in a magnetic field, this is not a very practical generator. To convert some of the mechanical power produced by an aircraft engine into electricity, we use a rotary generator. This is a device in which a coil of wire, actually several coils, is turned inside a magnetic field.

In Fig. 1-15, we see the basic principle of a rotary generator. The coil of wire is turned by the engine and it rotates within a magnetic field. As it turns, both sides of the loop cut the magnetic flux. If the loop turns counterclockwise, the upper half, the side that is shaded, moves to the left in front of the north pole of the magnet, while the lower side moves to the right in front of the south pole.

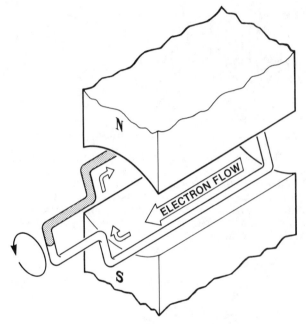

Fig. 1-15 If a loop of wire is rotated in a magnetic field, the voltage induced into the wire will cause electrons to flow.

The left-hand rule for generators helps us visualize the direction the current will flow in the wire. If the left hand is held as we see in Fig. 1-16, with the forefinger pointing in the direction of the lines of flux, from the north pole to the south pole, and the thumb pointing in the direction of the conductor movement, the second finger will point in the direction of the current flow.

Using the left-hand rule on Fig. 1-15, we see that the current moves *away from* the left end of the coil in the upper (shaded) part of the winding and *toward* the left end in the lower part of the winding.

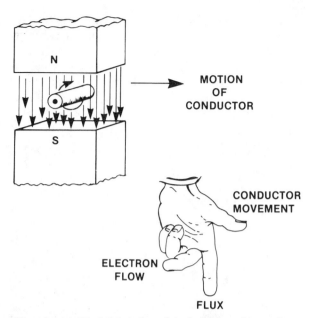

Fig. 1-16 If the left hand is held as shown here, with the forefinger pointing in the direction of the flux, from north to south, and the thumb pointing in the direction of conductor movement, the second finger will point in the direction the electrons will flow in the conductor, from negative to positive.

In the half of a revolution when the upper part of the coil is passing in front of the north pole and the lower part is passing in front of the south pole, the current flows in the direction shown by the arrows. But, as soon as the coil rotates so the light part is moving in front of the north pole and the dark part in front of the south pole, the current reverses and flows in the direction opposite the arrows. A rotating coil produces alternating current, but this is not what we want.

1. The armature

All rotating generators produce alternating current, but in order to use it to charge batteries and to operate direct current machinery, we must convert it into DC. We do this before it leaves the generator with a mechanical switch called a commutator.

Instead of the coil being made into a closed loop, the loop is open and the two ends are connected to a metal ring. This ring is cut so that the half that is connected to the dark side of the coil does not touch the half that is connected to the light side. One conducting brush, usually made of a carbon compound, rides on each of the commutator segments as this ring is called.

As the coil rotates in the magnetic field, electrons move away from the upper brush through the dark part of the coil and toward the lower brush through the light side. Since this is a source of electrical power, the brush through which the electrons leave is the negative brush and the one through which the electrons enter the coil is the positive brush.

The rotating coil in a practical generator does not consist of only one turn of wire, nor does it have only one coil in the magnetic field. The armature, as the rotating part of an aircraft DC generator is called, is made of a core of laminated sheet iron stampings stacked tightly on the shaft. Slots in the periphery of the core are filled with coils of copper wire and the ends of the coils are brought out and soldered into copper commutator bars. The bars are mounted on the shaft and are insulated from both the shaft and from each other. One end of a coil attaches to one bar and the other end is soldered to the adjacent bar. Each coil laps over the preceding one. This method is called lap winding.

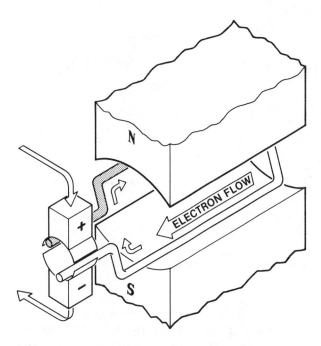

Fig. 1-17 If the loop of wire rotated in the magnetic field is opened and brushes contact the wire, the induced current may be taken out of the conductors.

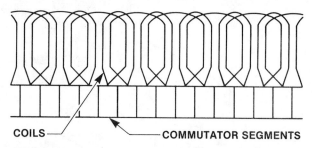

ONE END OF EACH COIL ON THE ARMATURE OF A DC GENERATOR ATTACHES TO A COMMUTATOR SEGMENT. THE OTHER END OF THE COIL ATTACHES TO THE ADJACENT SEGMENT. ONE END OF TWO COILS ATTACH TO EACH SEGMENT AND THE COILS LAP OVER ONE ANOTHER, WITH THE SIDES OF TWO COILS IN EACH SLOT IN THE CORE.

(A)

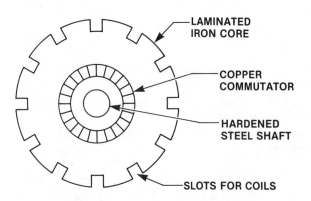

Fig. 1-18 The core of an aircraft DC generator is made of a stack of thin laminated iron disks mounted on a steel shaft. Slots cut in the periphery of the core hold the coils of the armature winding.

THE COMMUTATOR ON THE ARMATURE OF A DC GENERATOR IS MADE OF COPPER SEGMENTS INSULATED FROM ONE ANOTHER AND FROM THE SHAFT.

(B)

Fig. 1-19 Coils and commutator of a DC generator

After the coils are installed in the slots, wedges are slipped in place to hold them, and on some of the high-speed armatures steel bands are wound around the armature over the top of all of the coils to hold them in place against the high centrifugal force that is imposed on them.

Carbon brushes mounted in spring-loaded brush holders ride on the commutator as it rotates to pick up the curent as it is produced.

2. The field

The magnetic field which is cut by the rotating coils is produced in aircraft DC generators by an electromagnet. In this explanation, we will discuss a two-pole field, but high-current and high-voltage generators may use as many as eight poles.

The body of the generator is a soft steel tube into which are fastened the steel pole shoes. The field coils fit tightly around the pole shoes which are held in place with screws from the outside of the case. The body, which is also called the field frame, or the yoke, not only performs the very important function of completing the magnetic circuit between the pole shoes, but it also supports the two end frames which hold the bearings in which the armature rides.

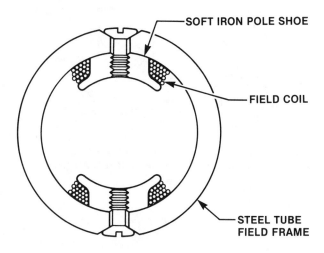

Fig. 1-20 The field coils of a DC generator are held into the field frame by soft iron pole shoes.

The strength of the electromagnetic field can be varied to maintain the voltage output of the generator at the desired level. This field strength is controlled by the voltage regulator that varies the current through the field as a function of the generator output voltage.

3. Armature and field connections

An aircraft DC generator produces its electricity by rotating coils of wire inside an electromagnetic field, and the electromagnetism is produced by some of the current that is generated in the armature. This causes the generator to be self-excited.

There are three ways the armature and the field coils can be connected, and each method has advantages and disadvantages.

In a series-connected generator, the armature and the field coils which are wound with a relatively few turns of heavy wire are connected in series so all of the current that flows in one must flow in the other. This arrangement has some rather unique advantages, but it is difficult to control the output voltage. Since all of the load current flows through the field winding, any increase in load current increases the strength of the field which increases the voltage. When the load current drops, the output voltage also drops. The voltage may be controlled although not too effectively, by using a variable resistor in parallel with the field winding. The output voltage may be decreased by shunting some of the current through the resistor. Because of the difficulty in controlling the voltage, series generators are not used in aircraft electrical systems.

In the parallel, or shunt, connection, the load current is taken from the armature and a small portion of the armature current is passed through the field coils. The voltage regulator is an adjustable resistor between the armature and the field and by varying the amount of current allowed to flow in the field, we can adjust the strength of the field magnets to control the voltage produced in the armature. This is the arrangement that is most generally used.

A compound generator has both a series and a shunt field. All of the current that flows in the armature coils also flows through the series field winding and only a small current, that which is used to control the voltage, flows through the shunt field winding. The strength of the series magnetic field varies directly with the amount of output current, and this field is used to reduce brush arcing by a method we will discuss when we get to large and high-output generators.

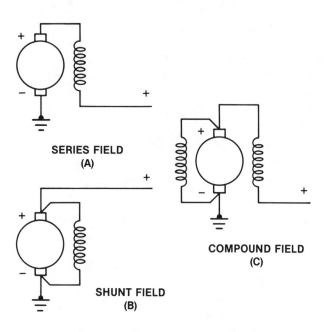

Fig. 1-21 *Armature to field connections for DC generators:*

SERIES FIELD
(A)

SHUNT FIELD
(B)

COMPOUND FIELD
(C)

4. Voltage control

Shunt-connected generators used on light aircraft have two ways the field may be connected to the armature. One end of the field winding may be connected to the insulated brush, and the voltage regulator is connected between the external field connection and ground. Generators with this connection are often called A-circuit generators.

B-circuit generators have an internal connection between the field winding and the grounded brush and the voltage regulator is installed between the field and the armature terminals.

Large and high-powered generators have both ends of the field winding brought out of the generator either at threaded terminals or in a quick-disconnect plug. Large terminals or sockets in the plug bring out the armature lead and the end of the series field. These generators nearly always use a carbon pile voltage regulator connected between the shunt field and the armature output, and the other end of the shunt field is grounded.

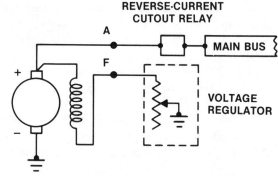

THE VOLTAGE REGULATOR IS BETWEEN THE FIELD AND GROUND.
(A)

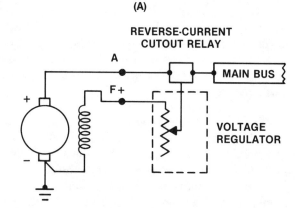

THE VOLTAGE REGULATOR IS BETWEEN THE FIELD AND THE GENERATOR ARMATURE.
(B)

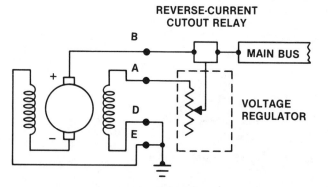

THE VOLTAGE REGULATOR IS BETWEEN THE POSITIVE END OF THE FIELD AND THE GENERATOR ARMATURE.
(C)

Fig. 1-22 *Voltage regulator placement*

Residual magnetism in the generator field frame provides the magnetic field to start the generator producing current when the engine begins to turn. The field is in parallel with the armature and current from the armature flows through the field winding. This increases the strength of the magnetic field which continues to increase the amount of field current and output voltage. The voltage will rise until the voltage regulator inserts a resistance in the field circuit that reduces the field current enough to hold the voltage at the required regulated value.

5. High-output generators

The two-pole generators we have just described are used on many small aircraft, and they are made with outputs of up to 50 amps at 12 volts; however most of them are of a smaller size, usually having a current output of 25 to 35 amps. Larger aircraft have 24-volt electrical systems and use generators with four, six, or eight poles that can produce as much as 300 amps. Many of the new small aircraft have 24-volt electrical systems, but they use alternators, which we will discuss in detail later in this text.

In Fig. 1-23, we see the field arrangements of several types of DC generators. The two-pole generator has one north pole and one south pole and the lines of flux pass from the north pole shoe through the laminated iron core on which the armature coils are mounted to the south pole shoe and back through the soft steel generator frame. Two brushes carry current to and from the armature. A four-pole generator has two sets of pole shoes, two north and two south poles. The north and south poles are next to each other so the lines of flux join them as we see in Fig. 1-23B. The armature delivers its current through four brushes.

Six- and eight-pole generators have their poles arranged as we see in Fig. 1-23C and D, and these large generators usually have additional poles, which we will discuss later. The reason for using more than two sets of poles is to increase

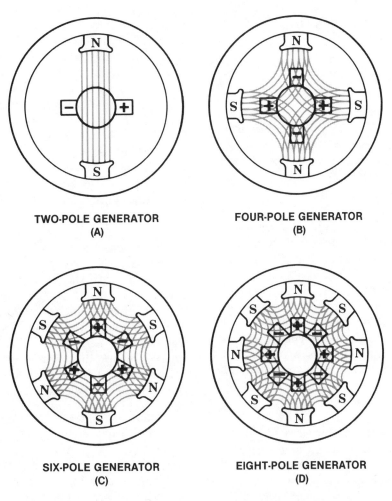

TWO-POLE GENERATOR
(A)

FOUR-POLE GENERATOR
(B)

SIX-POLE GENERATOR
(C)

EIGHT-POLE GENERATOR
(D)

Fig. 1-23   Field flux patterns in DC generators

11

the generator output. The coils on the armature are all in parallel and in one revolution of the armature they all cut the flux of each set of magnetic poles.

a. Brush placement

The placement of the brushes with relation to the field poles is critical, since the brushes must short between two of the commutator segments when there is a minimum of voltage between the two.

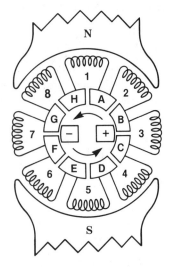

Fig. 1-24 Brush location in a two-pole DC generator. The two brushes short the commutator segments that are connected to coils passing parallel to the lines of field flux. There is no voltage difference between the shorted segments.

In Fig. 1-24, we have a simplified diagram of the armature of a two-pole generator showing the relationship between the coils and the magnetic field. Coils 1 and 5 are cutting across a maximum number of lines of flux, and the voltage between the ends of the coils is maximum. The ends of coil 1 are connected to commutator segments A and H, and the ends of coil 5 are connected to segments D and E. The polarities we assign to these coils with this rotation makes segment H negative with respect to A and segment E negative with respect to D.

Coils 2, 4, 6, and 8 are cutting across lines of flux, but they are cutting it at an angle, so the voltage they produce is less than that in coils 1 and 5. There is a voltage difference between segments A and B, with B more positive than A. Coil 8 is making segment G negative with respect

to H. Coil 6 is making segment F negative with respect to E, and coil 4 is making D negative with respect to C. Coil 2 makes B positive with respect to A and coil 4 makes C more positive than D.

Coils 3 and 7 are moving parallel to the lines of flux, and since they are not cutting across any flux there is no voltage generated in them, and there is no voltage difference between B and C or between F and G. But the voltage produced in coils 2, 1, and 8 is in series, making B positive with respect to G. At the same time, coils 4, 5, and 6 are producing three voltages in series, and they make segment C positive with respect to F. The positive brush contacts segments B and C, and the negative brush contacts segments F and G. Electrons leave the armature through the negative brush and return through the positive brush.

b. Armature reaction

We have just seen that the brushes should be installed so they will be perpendicular to the lines of flux. This will cause them to always short the commutator segments that are connected to the coils that are not cutting across any lines of flux. If there were no current flowing in the armature coils, the lines of flux would pass straight through the pole shoes, as we see in Fig. 1-25A. The line drawn perpendicular to these lines of flux that passes through the brushes is called the neutral plane, and when the brushes are on the neutral plane, the segments they short will have the same voltage and there will be no arcing as the brushes short them.

When current flows in the armature coils, a magnetic field encircles each winding, and this field reacts with the field between the pole shoes and distorts it, as we see in Fig. 1-25C. The neutral plane, which by definition is perpendicular to these lines, is no longer lined up with the brushes and there will be arcing as the brushes short the commutator segments. This shifting of the neutral plane by the current flow in the armature is called armature reaction.

There have been several remedies used to minimize brush arcing. In the smaller, low-power generators, the brushes are located by their design so they will be in a position where there is the minimum arcing when the rated load current flows in the armature. Some of the medium output generators have movable brush rigging and

12

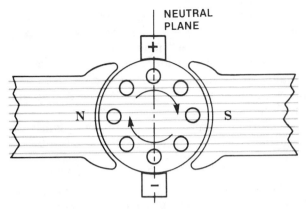

THE NEUTRAL PLANE IS PERPENDICULAR TO THE LINES OF MAGNETIC FLUX. WITH NO CURRENT FLOW IN THE ARMATURE COILS, THE NEUTRAL PLANE IS MIDWAY BETWEEN THE TWO POLE SHOES.

(A)

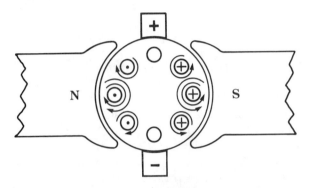

WHEN CURRENT FLOWS IN THE ARMATURE, A MAGNETIC FIELD SURROUNDS EACH OF THE CURRENT-CARRYING CONDUCTORS.

(B)

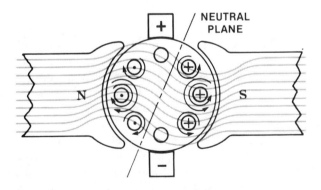

THE REACTION BETWEEN THE MAGNETIC FIELDS AROUND THE CURRENT-CARRYING CONDUCTORS IN THE ARMATURE AND THE FIELD BETWEEN THE POLE SHOES SHIFTS THE NEUTRAL PLANE IN THE DIRECTION OF ARMATURE ROTATION.

(C)

Fig. 1-25 Shifting of the neutral plane by armature reaction

the brushes are adjusted on the test bench so they produce a minimum of arcing when rated load current flows.

Some large high-output generators have an extra set of poles between the main poles. These interpoles, as they are called, carry a series winding. The magnetic field from the interpoles opposes the shift of the neutral plane caused by the armature reaction, and it keeps the neutral plane in line with the brushes.

Armature reaction causes the neutral plane to shift in the direction of armature rotation, and the interpoles installed between the main poles are wound in such a direction that armature current flowing through them gives the interpole a polarity that is the same as the shunt-field pole *ahead* of it. In Fig. 1-26, the armature is rotating clockwise and the interpole at the 10:30 position is a south pole and the one at the 12:00 position is a north pole. (See Fig. 1-26 on page 14.)

When the generator is operating and current is flowing, armature reaction shifts the neutral plane in the direction of armature rotation, but the interpole exerts a magnetic effect that opposes the shift and the neutral plane remains lined up with the brushes. Armature reaction is caused by the current flow in the armature coils and it varies as the load changes. But, since the interpole windings are in series with the armature, their strength also varies with the amount of load current and they effectively cancel armature reaction under all load conditions.

In addition to the series winding around the interpoles, some high-output generators have compensating windings. These are part of the same winding that encircle the interpoles, and they consist of a couple of turns of heavy wire embedded in the faces of the main field poles and are wound so they align with the interpoles. They increase the effectiveness of the interpoles in reducing brush arcing.

C. DC Generator Controls

All DC generators, whether of the small low-output, two-pole type or the largest eight-pole generator with series-wound interpoles, must have three types of controls. They must all have some method to disconnect them from the aircraft electrical system when they are not operating, and to connect them when the gener-

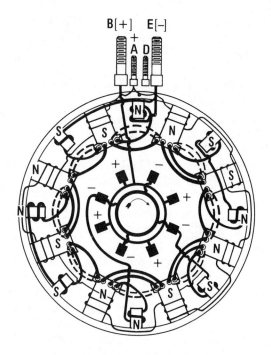

B[+]   E[−]

INTERPOLES ARE LOCATED BETWEEN EACH OF THE
MAIN POLES AND CARRY SERIES WINDINGS. THE COM-
PENSATING WINDINGS ARE LAID IN GROOVES CUT IN
THE FACE OF THE MAIN POLES AND AID THE INTER-
POLES IN KEEPING THE NEUTRAL PLANE LINED UP
WITH THE BRUSHES.

(A)

B

INTERPOLES AND
COMPENSATING WINDINGS

E

A

D

SHUNT FIELD WINDINGS

THE INTERPOLES AND COMPENSATING WINDINGS ARE
IN SERIES WITH THE ARMATURE.

(B)

Fig. 1-26   Interpoles and compensating windings

ator output is higher than the battery voltage. This prevents current from the battery flowing through the armature and discharging the
battery. Since these generators are self-excited,
there must be some method of reducing the field
current as the generator speed increases. The

voltage regulator controls the amount of current
that flows through the field windings to maintain
the output voltage at the level needed to supply
the aircraft load and keep the battery charged.
Finally, there must be some method or device to
prevent the generator producing more than its
rated current. This may be as simple as a fuse-
type device or it may be an automatic circuit that
decreases the field excitation if the output current exceeds its rating.

Generators used in multi-engine installations
have another control. They have a paralleling circuit that forces the generators to share the electrical load equally. It decreases the voltage of any
generator that is carrying more than its share of
the load, and at the same time it increases the
voltage of a generator that is not carrying its
share.

1.   Controls for low-output generators

a.   Vibrator-type generator controls

The vibrator-type voltage regulator is commonly used in a three-unit generator control
panel, which is quite similar to the one in
automobile electrical systems that use a generator rather than an alternator. This three-unit
control consists of a reverse-current cut-out relay,
a voltage regulator, and a current limiter, all
housed in the same metal housing, and it is
generally mounted on the engine side of the
firewall.

b.   Reverse-current cutout relay

When the engine is not running or anytime
the generator is producing a voltage lower than
the battery voltage, the reverse-current cutout
relay keeps the generator disconnected from the
system bus.

In Fig. 1-27, we see a typical aircraft electrical power system for a light single-engine
airplane using a generator. The double-pole,
single-throw master switch energizes the battery
contactor and connects the generator field to the
generator control when it is closed. The ammeter,
you will notice, is between the bus and the battery so it will show whether current is flowing into the battery to charge it or flowing out of the
battery to operate some of the electrical load that
is connected to the bus. The armature lead from
the generator is connected to the G terminal of

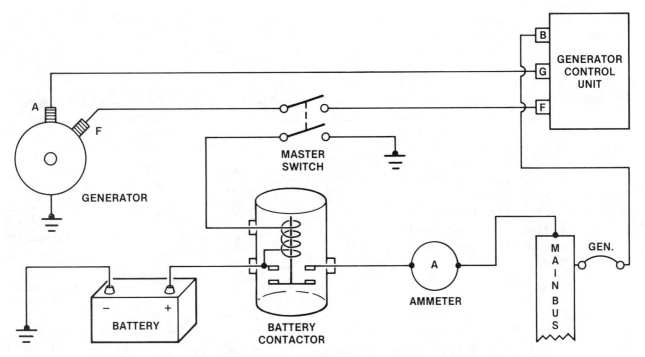

Fig. 1-27   Typical DC generator installation for a light, single-engine airplane

the control unit, and the *B* terminal connects to the system bus through the generator circuit breaker.

We see the reverse-current cutout relay portion of the control unit in Fig. 1-28. The relay itself consists of a soft iron core around which is wound two coils. One coil is wound of many turns of fine wire, and this is the voltage coil, electrically connected between the generator terminal and ground to sense the generator output voltage. The other winding has only a few turns of heavy wire, and it is connected between the generator terminal and the normally-open contacts of the relay. When the generator is not

operating, the spring-loaded contacts are open so no current can flow from the battery to the generator. When the master switch is closed and the engine is started, the generator begins to produce current because of the residual magnetism in the field frame. This current flows through the field coil and increases the output voltage. As soon as the output voltage rises to the value for which the reverse-current cutout relay is set, the current through the voltage coil produces a magnetic field strong enough to close the contacts and connect the armature to the bus and to the battery through the battery contactor. When current flows from the generator to supply the electrical load and charge the battery, the magnetic

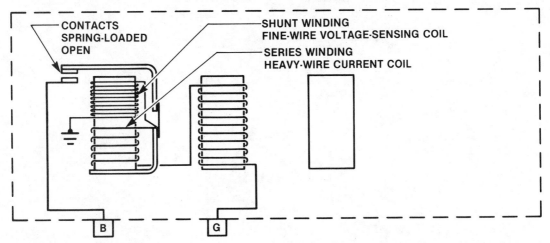

Fig. 1-28   Reverse-current cutout relay used in a vibrator-type generator control unit

field caused by the current flowing in the current coil increases the pull on the contacts and holds them tightly closed.

If the generator voltage drops below that of the battery as it does when the engine is idling or stopped, current flows from the battery, through the series current coil, and into the armature to ground. This reverse flow would soon discharge the battery if it were allowed to continue. But, current flowing through the current coil from the battery to the generator produces a magnetic field that opposes the field produced by the voltage coil. With the magnetic pull canceled, the spring snaps the contacts open and disconnects the generator from the battery.

c. Voltage regulator

In Fig. 1-29, we have the circuit of a vibrator-type voltage regulator that is used with an A-circuit generator, one in which the field winding is connected inside the generator to the "hot" brush and the voltage regulator is between the field winding and ground.

The voltage regulator is wound with a coil consisting of many turns of fine wire, and the coil is connected to the armature output at the frame of the reverse-current cutout relay. The contacts

for the voltage regulator are normally closed and are opened by the magnetic pull when the generator output voltage rises to the value for which the regulator is set.

The field circuit is completed through both the voltage regulator and the current limiter contacts which are in series.

When the generator begins to turn, residual magnetism in the frame furnishes enough flux to start the current building up. This current flows through the field winding, and the output voltage rises. We can visualize this action more easily if we think in terms of conventional current (flow from positive to negative) than if we use the more correct electron flow, and since we will use this direction of flow when we discuss the solid-state voltage regulator, we will also use it here. Conventional current flows from the positive brush, through the generator field, through both sets of contacts (the current limiter and the voltage regulator contacts), and then to ground. This field current causes the voltage to rise to the value for which the voltage regulator is set, and at this point the magnetic pull of the voltage regulator coil becomes strong enough to open the voltage regulator contacts. Current can no longer flow directly to ground, but it must flow to ground

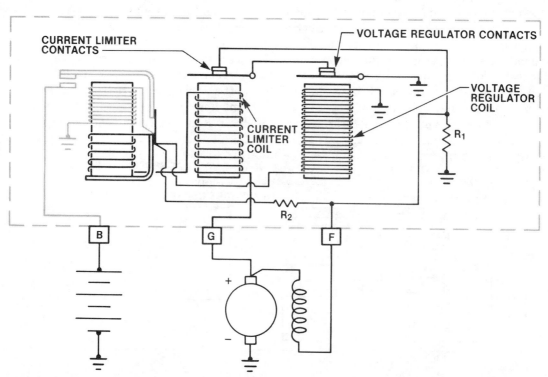

Fig. 1-29   Vibrator-type voltage regulator that is located between the generator field and ground

through resistor R_1, which reduces the field current enough to drop the output voltage.

The contacts vibrate open and closed between 50 and 250 times a second to interrupt the generator field current. Each time the current is interrupted, the magnetic flux around the field coil collapses and induces a voltage in the field winding that tries to keep the current flowing through the contacts as they are opening. This causes an arc which burns the points and can weld them together. To prevent this arcing, resistor R_2 is connected between the field terminal and the armature output. The surge of induced voltage can circulate back through the field coil and be dissipated by the resistor.

The generator control in Fig. 1-30 is identical to the one we have just discussed, except that it is used with a B-circuit generator, one in which the regulator is installed between the armature and the field winding.

Current (we are still talking of conventional current) flows from the connection at the frame of the reverse-current cutout relay, through both the voltage regulator and the current limiter contacts to the field terminal of the control unit, through the generator field winding to ground at the grounded brush. When the output voltage rises high enough, the voltage regulator contacts open so the field current must flow through resistor R_2 which decreases the field current and drops the output voltage. The voltage regulator points vibrate open and closed to maintain the voltage at the desired level, and each time they open, the collapsing field tries to maintain the flow across the points. Resistor R_1 is in parallel with the field coil and it carries the surges of induced current to ground.

There are several variations to the basic vibrator type voltage regulator. In Fig. 1-31, we have a voltage regulator with an accelerator winding. There are two windings on the voltage regulator core. The shunt winding operates in exactly the same way we have seen in the two previous illustrations, but wound over the top of the shunt winding, there is a series winding through which all of the field curent flows. When the generator output voltage rises to the value for which the regulator is set, the combined pull from the magnetic fields of the two coils pull the voltage regulator contacts open. As soon as the points open, the magnetic field from the series, or accelerator, coil is lost, and the spring is able to quickly close the contacts for the next cycle. The accelerator winding, as its name implies, accelerates the action of the voltage regulator.

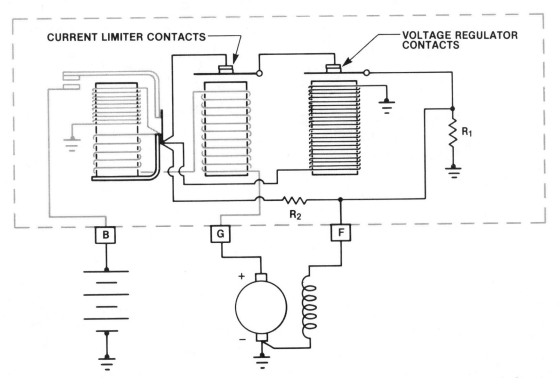

Fig. 1-30 *Vibrator-type voltage regulator that is located between the generator field and the generator armature*

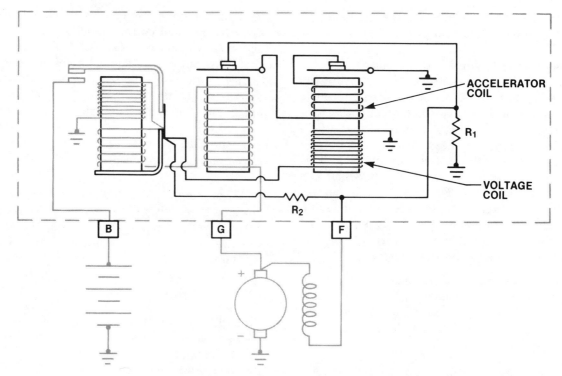

Fig. 1-31 Vibrator-type voltage regulator with an accelerator winding

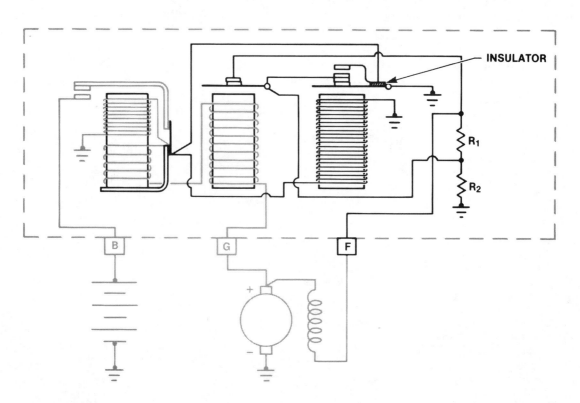

Fig. 1-32 Vibrator-type voltage regulator with double contacts for use with generators producing a large amount of field current.

18

Some generators whose field current requirements vary over a wide range use voltage regulators with a double set of contacts. For operation when the generator speed is low, the bottom set of points controls the voltage in exactly the same way we have seen in the previous regulators. But, when the generator speed is high, there is enough field current produced to keep the generator output above that desired, even when the open points put resistors R_1 and R_2 in the field circuit.

When high field current flows, the armature on which the contacts are mounted pulls down enough to close the upper contacts, and the circuit is completed. This connects the two ends of the generator field and puts resistor R_1 in series with the field. Intermittently shorting the generator field in this way decreases the field strength and allows good voltage regulation.

When vibrator-type voltage regulators are used in the electrical system of twin-engine aircraft, they have a paralleling coil on the voltage regulator.

In Fig. 1-33A, we see the way the paralleling coil operates. It is connected in parallel with the regular voltage coil, but rather than grounding inside the voltage regulator, the end of the coil goes outside of the regulator through terminal P to the contacts inside the paralleling relay. The paralleling coils for both generators tie together by the two sets of contacts in the paralleling relay. (See Fig. 1-33 on page 20.)

The two coils in the paralleling relay attach to the G terminals of the voltage regulators, and when the generators produce enough voltage to close the reverse-current cutout relays, they also close the paralleling relays.

Let's consider a condition in which both generators are connected to the bus and the output of the left generator drops below that of the right. Current flows through the paralleling coil of the right voltage regulator in the direction that aids the shunt coil in opening the voltage regulator contacts. It continues through the contacts of both paralleling relays and through the paralleling coil in the left voltage regulator, then to ground through its shunt coil. The reverse flow through the left engine's paralleling coil cancels some of the magnetic pull produced by the shunt

coil and the contacts remain closed long enough to build up the output of the left generator.

When the output of either generator drops below the voltage needed to hold the paralleling relay closed, it will open and disconnect the paralleling coils so no current can flow from the operating generator through the inoperative one.

d. Current limiter

The current limiter shown in Fig. 1-29 decreases the voltage output of the generator by opening its field circuit and inserting a resistor anytime the current output of the generator exceeds the rating the manufacturer has set. The series winding on the current limiter core carries all of the load current from the armature. When this current produces enough magnetic attraction to open the contacts, the field current is reduced in exactly the same way as is done when the voltage regulator contacts open.

2. Controls for high-output generators

a. Carbon-pile voltage regulators

High-output generators require more field current than can be effectively controlled with vibrator-type voltage regulators and these are usually controlled with a carbon-pile voltage regulator.

A stack of thin disks made of pure carbon are held in a ceramic tube inside an aluminum housing. Pressure is held on the carbon pile, as this stack of disks is called, by a spring. All of the field current flows through the carbon pile, and to control the amount of field current, an electromagnet mounted in the regulator frame pulls against the armature which is a heavy steel disk in the center of the spring. The strength of the electromagnet is determined by the generator output voltage.

Almost all aircraft carbon pile voltage regulators are mounted on a plug-in base and are installed in a circuit such as we see in Fig. 1-34. The positive terminal of the generator armature is connected to a separate switch-type reverse-current cutout relay, and from this point a conductor is taken to the regulator terminal B. One end of the carbon pile connects to this terminal, as well as one end of the voltage coil in the electromagnet. The other end of the voltage coil con-

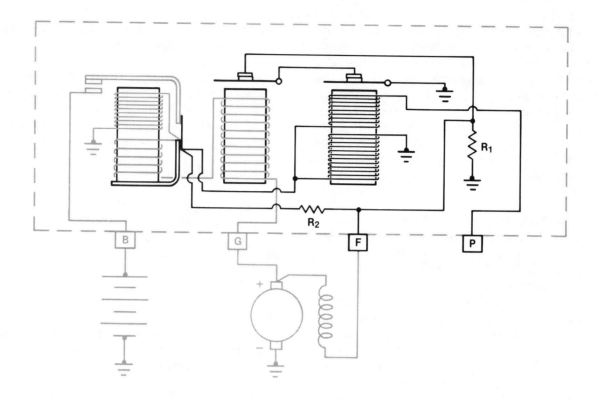

PARALLELING COIL ON THE VOLTAGE REGULATOR
(A)

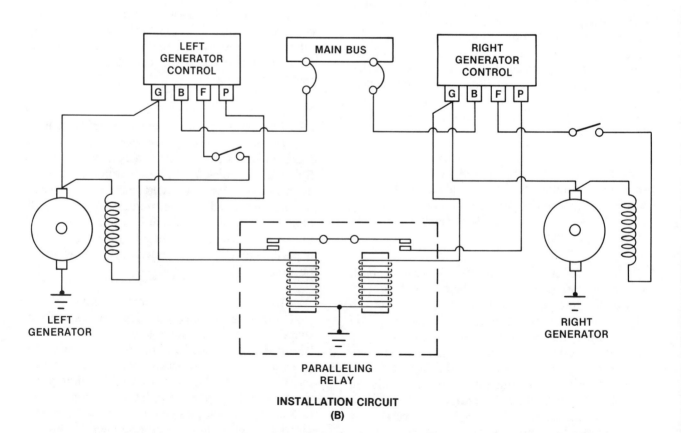

INSTALLATION CIRCUIT
(B)

Fig. 1-33   Paralleling circuit for twin-engine aircraft using vibrator-type voltage regulators

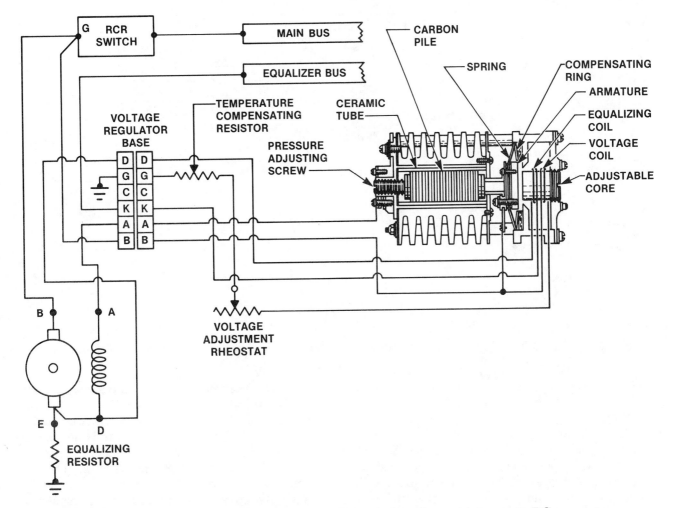

Fig. 1-34 Carbon pile voltage regulator control circuit for a large, high-output DC generator

nects to the wire-wound voltage-adjustment rheostat and then to a temperature compensating resistor and to ground through terminal *G*.

Field current is picked up from the generator armature at the reverse-current relay. It enters the regulator through terminal *B*, flows through the carbon pile and leaves the regulator at terminal *A*. It enters the generator through its terminal A and flows through the field coil to ground. In this illustration, you will notice that the generator is actually grounded through the equalizing resistor. This is a very low resistance device which we will explain shortly, but for now consider terminals *E* and *D* both to be at ground potential.

When the generator output voltage is low, the spring holds the carbon pile compressed and the maximum amount of current flows through the field. But when the voltage rises to the value for which the regulator is set, current through the voltage coil produces a pull on the armature that

loosens the pile and increases its resistance. As the resistance increases, the field current decreases and the generator output voltage drops.

The vibrator-type voltage regulator intermittently places its resistor in the field circuit, but the carbon pile voltage regulator changes the value of the resistor that is always in the circuit.

The voltage adjustment rheostat is used for the very fine adjustment of the generator voltage needed to parallel the generators installed on multi-engine aircraft. There are two main adjustments to the regulator that must be made on a test stand under controlled conditions: They are the pressure the spring holds on the carbon pile, and the space between the core of the electromagnet and the armature.

The current through the carbon pile causes it to operate quite hot, and the housing is usually finned in order to dissipate as much heat as possi-

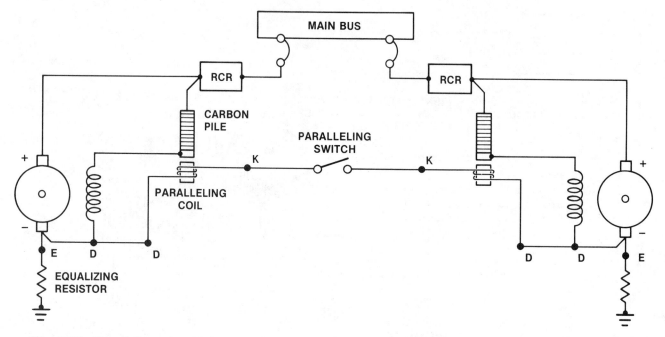

Fig. 1-35 Paralleling circuit used with carbon pile voltage regulators and high-output DC generators

ble. Carbon has a negative temperature coefficient. This means that when carbon gets hot, its resistance decreases. Just below the spring in the voltage regulator housing there is a compensating ring made of a bimetallic material that warps as the regulator gets hot, to decrease the pressure on the stack. This compensates for the change in resistance caused by the change in stack temperature. There is also a compensating resistor in series with the voltage adjustment rheostat. Its temperature coefficient is opposite to that of the rheostat so the external resistance in the voltage coil circuit remains constant as the temperature changes.

When two generators are mounted on a twin-engine aircraft so that they feed the same system bus, they must share the load equally; and to assure that they do, the regulators are fitted with a paralleling, or equalizing, circuit. The generators are grounded through an equalizing resistor that is of such value that it has a 1/2-volt drop across it when the generator is producing its rated current. The top end of this resistor is connected to generator terminal E, and this is in turn connected to terminal D of the voltage regulator. The equalizing coil is wound around the core of the electromagnet along with the voltage coil. The other end of the equalizing coil leaves the regulator through terminal K and goes to the paralleling switch or relay and then to the equalizing circuit of the other generator.

When the generators are producing exactly the same voltage and sharing the current load equally, the voltage drop across the two equalizing resistors is the same and there is no current flow through the equalizing circuit. But, if the voltage of the left generator rises slightly for some reason, it will produce more current and the voltage drop at the top end of its equalizing resistor will increase and cause current to flow through the equalizing circuit, increasing the pull of the electromagnet in the left regulator and decreasing the pull of the one in the right regulator. This will decrease the voltage of the left generator and bring the voltage of the right generator up so they will share the load equally.

b. Reverse-current cutout relay

Rather than using a simple reverse-current cutout relay that closes when the generator voltage rises to a preset voltage and opens at a lower preset voltage, most high-output generators use a differential-voltage reverse-current relay that combines the action of the cockpit-controlled generator switch with the reverse-current relay. It connects the generator to the bus when the generator voltage is a given amount higher than the battery voltage and disconnects it when the current begins to flow from the battery back through the generator.

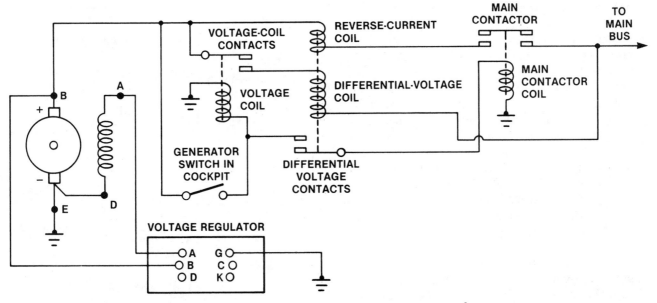

Fig. 1-36 Differential-voltage reverse current relay

In Fig. 1-36, we see the way this type of device operates. When the engine is started, the generator switch in the cockpit is closed, and as soon as the generator output voltage rises to the correct value, the magnetic field produced by the voltage coil closes the voltage coil contacts. This allows current to flow from the generator output through the differential-voltage coil to the main bus to which the battery is connected. As long as the battery voltage is higher than the generator voltage, current flows from the battery to the generator through this coil. But as soon as the generator voltage rises a specified amount above that of the battery, this coil will produce enough magnetic pull to close the differential-voltage contacts. This allows current to flow through the main contactor coil and close the main contactor to connect the generator to the main bus. Current flowing from the generator to the bus through the reverse-current coil strengthens the pull of the differential-voltage coil and holds the differential-voltage contacts closed.

When the generator voltage drops below that of the battery, current flows from the battery through the reverse-current coil to ground through the generator. This reverse flow through the reverse-current coil cancels the pull of the differential-voltage coil and the differential voltage contacts are pulled open by their spring.

If the generator should malfunction, it can be disconnected from the bus by opening the generator switch. When this switch is open, energiz-

ing current to the main contactor coil is shut off. Then these contacts open and disconnect the generator from the main bus.

c. *Current limiter*

The current limiter used with large high-output generators is usually a fuse-type device mounted in the engine compartment near the generator. It is essentially a slow-blow fuse that will carry a momentary current overload, but will open the circuit if the overload is held for a sustained period of time.

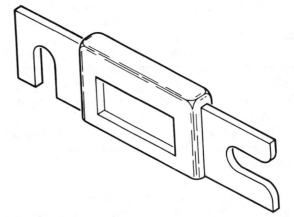

Fig. 1-37 Current limiter of the type used with a large, high-output DC generator

D. *DC Generator Service and Maintenance*

If a DC generator is unable to keep the aircraft battery charged, and if the ammeter does not show the proper rate of charge, you should first check the aircraft electrical system that is associated with the battery and with the generator. Physically check every connection in the generator and battery circuit and electrically check the condition of all of the fuses and circuit breakers. Check the condition of all ground connections for the battery, battery contactor, and the generator control units. When you have determined that there is no obvious external problem and the generator armature does turn when the engine is cranked, you should next determine which is at fault, the generator or the generator control. One of the easiest ways of determining which unit is not operating is to connect a voltmeter between the G terminal of the voltage regulator and ground and check the voltage output of the generator. Even if the field winding is open, or the voltage regulator is malfunctioning, the generator should produce residual voltage—that is, the voltage produced by the armature cutting across the residual magnetic field in the generator frame. This should be about one to two volts.

If there is no residual voltage, it is possible that the generator need only to have its field flashed to restore the residual magnetism to the field frame so the generator can start producing current to energize its field. To flash the field of a generator, you must momentarily run current through the field coils in the same way it normally flows.

If the field winding of the generator is connected to the insulated brush, the voltage regulator is between the field and ground, and to flash the field you should momentarily short between the *Battery* and the *Generator* terminals of the voltage regulator with the engine shut off. This sends a surge of current from the battery through the field coils to ground through the voltage regulator, and this should magnetize the field frame enough to start the generator producing current.

If the field winding connects to the grounded brush, the regulator is between the field and the armature, and to flash the field you should momenetarily short between the *Battery* and the *Field* terminals of the voltage regulator.

Some 24-volt generators can be damaged by this current flow through the armature, and before flashing their field you should insulate the

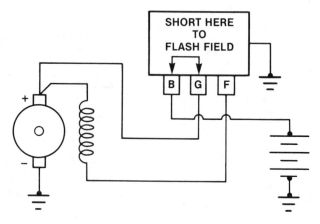

FOR GENERATORS WITH THE VOLTAGE REGULATOR BETWEEN THE FIELD AND GROUND
(A)

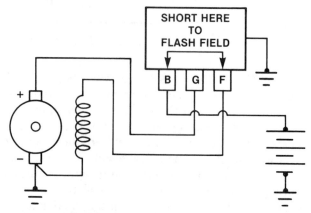

FOR GENERATORS WITH THE VOLTAGE REGULATOR BETWEEN THE FIELD AND THE GENERATOR ARMATURE
(B)

Fig. 1-38 Method of restoring residual magnetism to the field frame of a DC generator:

brushes from the commutator by slipping a piece of paper between the brush and the commutator. Be sure to follow the recommendations of the generator manufacturer when working with these high-output generators.

If the generator produces residual voltage but no output voltage, the trouble could lie with the generator or with the regulator, and to determine which, run the engine at a speed high enough for the generator to produce an output and bypass the voltage regulator with a jumper wire. If the regulator is between the field and ground, short between the *Field* terminal of the regulator and ground.

If the regulator is between the field and the armature, short between the *Field* terminal and the *Battery* terminal of the voltage regulator. Your voltmeter should be connected between the *Generator* terminal of the regulator and ground, and all electronic equipment in the aircraft should be turned off to be sure that any momentary overvoltage condition will not damage any installed equipment.

If the generator produces voltage with the regulator shorted, the problem is with the voltage regulator and it should be carefully examined. Be sure that the regulator is properly grounded, as a faulty ground connection will prevent its functioning properly.

It is possible to service and adjust some vibrator-type generator controls, but because of the expense of the time involved and the test equipment needed to do it properly, most servicing is done by replacing a faulty unit with a new one. If you should choose to service a malfunctioning generator control, be sure to follow the manufacturer's service instructions in detail.

If the generator does not produce an output voltage when the regulator is bypassed, remove the generator from the engine and disassemble it. There are three portions of the generator that could prevent it producing electricity: the field, the armature, and the brushes.

Check the field windings for an open circuit by connecting a 110-volt test lamp in series with the windings. If there is an open circuit, the lamp will not light. Now, connect the test probes between one end of the field winding and the field frame to check for shorts to ground. There should be no continuity: the lamp should not light.

To check the field for shorts, connect a battery of the proper voltage across the field coils and measure the current which must be within the limits specified in the manufacturer's test specifications.

Visually check the armature for indication of a worn or bent shaft, for any separated laminations in the armature body, or for bare wires in the winding. Hold one of the 110-volt test probes on the armature shaft and run the other probe around all of the commutator segments. There should be no connection and the test lamp should

CURRENT DRAW TEST

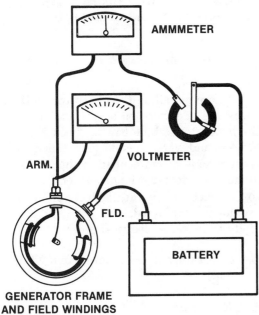

GROUND TEST

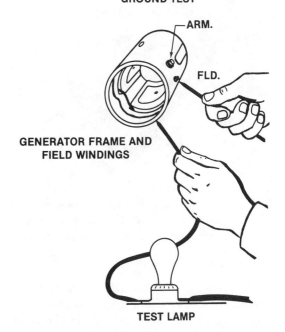

TO TEST FOR SHORTED TURNS IN THE FIELD COIL, MEASURE THE CURRENT DRAWN BY THE FIELD AT A SPECIFIC VOLTAGE.

(A)

TO TEST FOR SHORTS TO GROUND, HOLD ONE TEST PROBE TO ONE END OF THE FIELD WINDING AND PLACE THE OTHER TEST PROBE ON THE GENERATOR FRAME. THE TEST LAMP SHOULD NOT LIGHT.

(B)

Fig. 1-39   Generator field test

not light. Hold one test probe on one commutator segment and make contact with all of the other segments with the other probe. The lamp should remain on, showing continuity through all of the armature coils.

Put the armature in a lathe or in a special turning fixture, and true up the commutator by cutting away only enough metal to remove all traces of burning and all of the brush tracks. Undercut the mica between the segments by about the thickness of the mica and then smooth the edges of all of the segments with very fine sandpaper. Be sure that you do not use emery paper or emery cloth because it is conductive and can short between the commutator bars.

Place the armature on a growler and hold a hacksaw blade or a thin strip of steel slightly above the armature while you rotate it on the growler. If there are any shorts in the armature winding, the blade will vibrate vigorously.

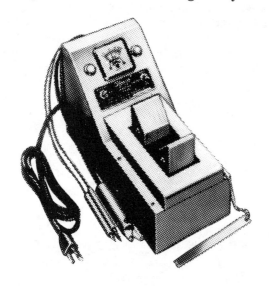

Fig. 1-40 A growler is used to test the armature of a DC generator for shorts. A current is induced into the armature coils by the alternating current in the growler.

Check the brushes by comparing them with a new brush and replace them if they are worn excessively. Reassemble the armature in the field, install the two end frames and put the brushes in the brush holders. Wrap a piece of fine sandpaper around the commutator with the grit side out, and rotate the armature so the abrasive will wear away the brush and form it to fit the contour of the commutator. This will allow the least arcing when the generator is producing its rated current.

Be sure to blow all of the sanding dust out of the housing so it cannot get into the bearings.

When the generator is assembled, connect it as shown in Fig. 1-41 and run it as a motor to check for freedom of armature rotation and for the correct voltage and current as are specified in the manufacturer's test specifications.

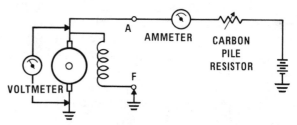

FOR GENERATORS WITH THE FIELD INTERNALLY CONNECTED TO THE INSULATED BRUSH
(A)

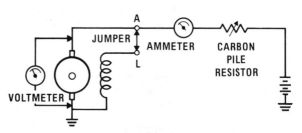

FOR GENERATORS WITH THE FIELD INTERNALLY CONNECTED TO THE GROUNDED BRUSH
(B)

Fig. 1-41   Motoring test for DC generators

Before installing the generator on the engine, check the generator oil seal if one is used, and if the generator is belt-driven, check the mounting bracket and when the belt is installed, adjust the tension as is recommended in the aircraft service manual. As soon as the generator is connected into the electrical circuit, flash the field as we have previously explained, and the generator is now ready to produce electrical power.

E.   Direct Current Alternators and Their Controls

1.   DC alternators

Two of the limitations of generators for aircraft installations are the limited number of pairs of poles that can be used, and the fact that the load current is produced in the rotating member and must be brought out of the generator through brushes.

Alternators solve these two problems, and since they produce three-phase AC and convert it into direct current with built-in solid-state rectifiers, their output at low engine speed allows them to keep the battery charged even when the aircraft is required to operate on the ground with its engines idling—as it must often do when waiting for instrument clearance before takeoff.

Almost all reciprocating engine-powered general aviation aircraft now use alternators with either a 14-volt or 28-volt output. These alternators are available with outputs of up to 95 amperes.

Alternators do exactly the same thing as generators. They produce alternating current and rectify it, or convert it into direct current, before it leaves the device. The field coils are in the stationary part of a generator and the load current is produced in the rotating element, and is taken out

through brushes which ride on the surface of the segmented commutator. In an alternator, field current is taken into the rotor through brushes which ride on smooth slip rings. The AC load current is produced in the fixed windings of the stator, and after it is rectified by six solid-state diodes it is brought out of the alternator through solid connections.

a. The rotor

The rotor of an alternator consists of a coil of wire wound on an iron spool between two heavy iron segments that have interlacing fingers around their periphery. Some rotors have four fingers and others have as many as seven. Each finger forms one pole of the rotating magnetic field.

The two ends of the coil pass through one of the segments and each end of the coil is attached to an insulated slip ring. The slip rings, segments, and coil spool are all pressed onto a hardened steel rotor shaft which is either splined or has a key slot machined in it. This shaft can be driven from an engine accessory pad or fitted with a pulley and belt driven from an engine drive. The slip-ring end of the shaft is supported in the housing with a needle bearing and the drive end with a ball bearing.

Two carbon brushes ride on the smooth slip rings to bring current into the field and carry it back out to the regulator.

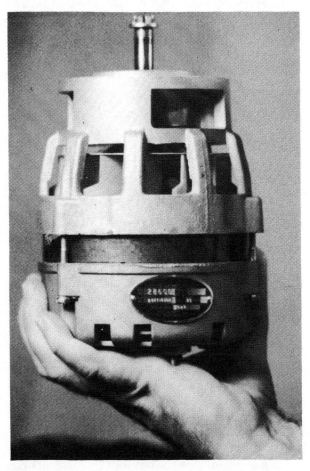

Fig. 1-42 DC alternators are used in almost all of the modern aircraft that require a low or medium amount of electrical power.

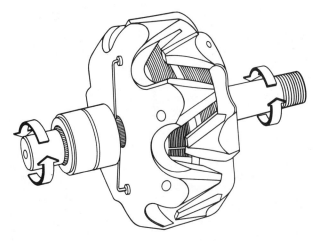

Fig. 1-43 DC alternator rotor. Direct current from the voltage regulator energizes the drum-type coil in the core of this rotor. The magnetic field produces alternate north and south poles of the soft iron interlacing fingers formed by the two end pieces.

b. The stator

The stator coils in which the load current is produced are wound in slots around the inside periphery of the stator frame, which is made of thin laminations of soft iron. There are three sets of coils in the stator and these sets are joined into a Y-connection with one end of each set of windings brought out of the stator and attached to the rectifier.

Fig. 1-44 The stator of a DC alternator consists of a laminated soft iron frame in whose slots the coils for the three-phase output windings are placed.

Fig. 1-45 The coils which are in the stator windings are connected in a Y-connected three-phase arrangement.

With the stator wound in the three-phase configuration, there is a peak of current produced in each set of windings every 120 degrees of rotor rotation. The AC produced in the stator looks much like that we see in Fig. 1-46A, and after it is rectified by the three-phase, full-wave rectifier, the DC output is much like that in Fig. 1-46B.

The large number of field poles and the equally large number of coils in the stator cause the alternator to put out its rated current at a low alternator RPM.

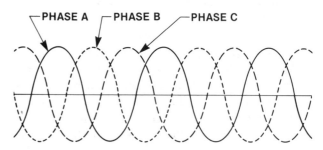

WAVEFORM OF THE THREE-PHASE ALTERNATING CURRENT PRODUCED IN THE STATOR WINDING OF A DC ALTERNATOR.

(A)

RECTIFIED OUTPUT OF THE THREE-PHASE AC IN THE STATOR PRODUCES A RELATIVELY SMOOTH DIRECT CURRENT, WITH A LOW-AMPLITUDE, HIGH-FREQUENCY RIPPLE.

(B)

Fig. 1-46 Output of a DC alternator

c. The rectifier

The three-phase, full-wave rectifier is made up of six heavy-duty silicon diodes. Three of them are pressed into the slip-ring end frame, and the other three are pressed into a heat sink that is electrically insulated from the end frame.

In Fig. 1-47, we see the way this rectifier converts the three-phase AC output of the stator into direct current. Since we are dealing with solid-state devices whose symbols use arrowheads, it is

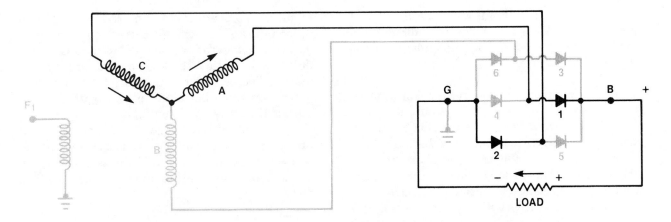

CURRENT FLOW WHEN PHASE A IS POSITIVE WITH RESPECT TO PHASE C

(A)

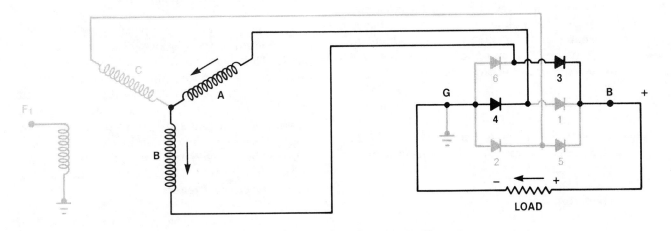

CURRENT FLOW WHEN PHASE B IS POSITIVE WITH RESPECT TO PHASE A

(B)

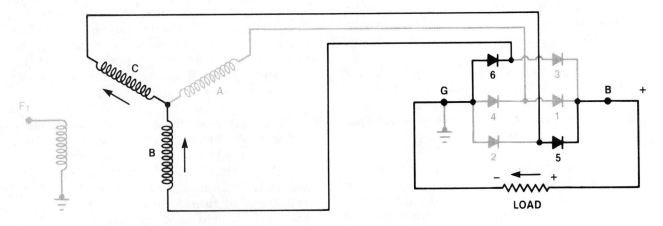

CURRENT FLOW WHEN PHASE C IS POSITIVE WITH RESPECT TO PHASE B

(C)

Fig. 1-47   Three-phase, full-wave rectifier used in a DC alternator:

29

easier to follow the action if we think in terms of conventional current that follows the direction in which the arrows point. This imaginary flow is from positive to negative.

At the instant the output terminal of winding *A* is positive with respect to the output end of winding *C*, current flows through diode *1* in the heat sink, through the load, and back through diode *2* that is pressed into the alternator end frame. From this diode, it flows back through winding *C*. As the rotor continues to turn, winding *B* becomes positive with respect to winding *A* and current flows through diode *3*, the load, and back through diode *4* and winding *A*. Next, *C* becomes positive with respect to *B* and current flows out through diode *5* and back through *6*. In this way the current flows through the load in the same direction all of the time. The terminal connected to the heat sink is the positive terminal of the alternator and the end frame into which the anodes of diodes *2*, *4*, and *6* are pressed becomes the negative terminal of the alternator.

For a review of semiconductor diode action, refer to Section XI, of Book Three of the General Section of this Integrated Training Program, the section entitled "Converting Alternating Current into Direct Current."

2. Alternator controls

The voltage produced by an alternator is controlled in exactly the same way it is controlled in a generator, by varying the field current. When the alternator output voltage drops, the voltage regulator senses it and increases the field current to strengthen the magnetic field and increase the voltage output. It decreases the field current when the voltage rises above the regulated value.

Large generators use carbon pile voltage regulators to do this, and small generators use vibrator-type controls. But solid-state controls are often used with alternators, so we will discuss this type of control here.

Alternator control requirements are different from those of a generator for several reasons. An alternator uses solid-state diodes for its rectifier and since current cannot flow from the battery into the alternator there is no need for a reverse-current cutout relay. The field of an alternator is excited from the system bus whose voltage is limited either by the battery or by the voltage regulator, so there is no possibility of the alternator putting out enough current to burn itself out, as a generator with its self-excited field can do. Because of this, there is no need for a current limiter. There must be one control with an alternator, however, that is not needed with a generator, and that is some means of shutting off the flow of field current when the alternator is not producing power. This is not needed in a generator since its field is excited by its own output. An alternator uses either a field switch or a field relay.

Most modern aircraft alternator circuits employ some form of overvoltage protection to remove the alternator from the bus if it should malfunction in such a way that its output voltage rises to a dangerous value.

In Fig. 1-48 we see a typical alternator circuit in a modern single-engine general aviation airplane. The negative output of the rectifier is grounded through the alternator end frame, *G*, and the positive end of the rectifier is taken out through terminal *B*. A capacitor is often installed between the output and ground to absorb variations of voltage that occur in the system and cause radio interference. The output of the alternator is connected to the main bus through the alternator circuit breaker.

Current flows through the alternator half of the master switch to energize the field relay inside the voltage regulator. This allows current for the field to flow from the main bus through the alternator regulator circuit breaker, through the overvoltage sensor to the *Batt* terminal of the voltage regulator and directly to the alternator field through regulator terminal F_2. Inside the alternator it flows through the field and returns to the regulator through terminal F_1. Inside the regulator it passes to ground through the emitter-collector circuit of the transistor.

The master switch on many airplanes equipped with alternators is a split-type switch that allows both the battery contactor to be energized and the alternator field circuit completed with one switch movement. The battery half of the switch can be turned on without the alternator half, but the alternator cannot be turned on without turning the battery on. The alternator half of the switch, however, can be turned off without affecting the battery portion of the switch. This allows the alternator to be shut off in

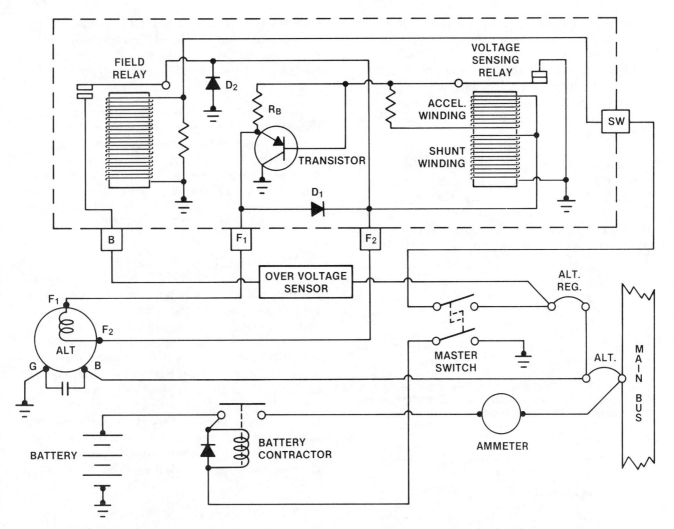

Fig. 1-48   An aircraft alternator electrical system using a transistorized voltage regulator

flight if it malfunctions without turning off all of the electrical power to the aircraft.

a.   Transistorized voltage regulators

Alternator output voltage is controlled by varying the field current. When the voltage rises above the desired value, the field current is decreased and when the voltage drops too low, the field current is increased. This may be done in low-output alternators with vibrator-type controls which interrupt the field current by opening the contacts. But a much more efficient means of voltage control has been devised: that of using a transistor to control the flow of field current. The vibrating points operate in exactly the same way they do in a normal vibrator-type voltage regulator, but instead of the field current flowing through the contacts, only the transistor base current flows through them. This is so small compared with the field current which flows through

the emitter-collector portion of the transistor that there is no arcing of the contacts. This type of regulator is called a transistorized voltage regulator, and we see a typical circuit of one in Fig. 1-48.

Before we discuss the operation of the transistorized voltage regulator, it will be well for us to review the way a transistor functions as a switch. In Section XII of Book Three of the General Section of this Integrated Training Program, we have discussed the operation of transistors, but here, rather than going into the physics of semiconductor devices, we will review the practical way they operate.

In Fig. 1-49A we have the symbols used for the two types of bipolar transistors, the PNP and the NPN. For our purposes here, these two types of devices operate in the same way, with the only difference being the polarity of the voltage used with them. This, of course, affects the direction the current flows through them.

When analyzing semiconductor circuits it is easier to visualize what is happening if we think in terms of conventional current, a flow from positive to negative, as it allows us to follow the arrows in the devices.

In the PNP transistor, the load current flows through the transistor from the emitter to the collector. But this flow can take place only when there is a flow of current from the emitter through the base, and this flow can occur only when the emitter is more positive than the base.

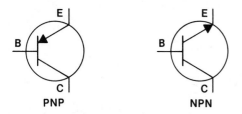

BIPOLAR TRANSISTOR SYMBOLS. THE ARROWHEAD IS ON THE EMITTER, AND IT POINTS IN THE DIRECTION OF CONVENTIONAL CURRENT FLOW.

(A)

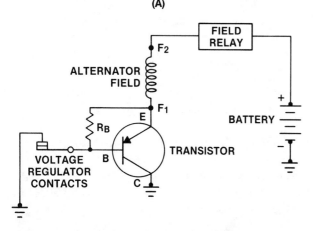

ALTERNATOR FIELD CURRENT FLOWS FROM THE EMITTER TO THE COLLECTOR OF THE TRANSISTOR ONLY WHEN THE VOLTAGE REGULATOR CONTACTS ARE CLOSED, ALLOWING BASE CURRENT TO FLOW FROM THE EMITTER THROUGH THE BASE TO GROUND.

(B)

Fig. 1-49 Operation of a transistorized voltage regulator:

In Fig. 1-49B, we see a simplified operation of the field current flow through the transistor. One side of the field winding is connected to the battery bus through the field relay, and the other side of the field connects to the emitter of the transistor. This makes the emitter positive. The

collector is connected directly to ground and the base is connected to ground through the normally closed voltage regulator points. The base and the emitter are connected through resistor R_B.

When the field relay is closed, current flows through the alternator field and through resistor R_B to ground. The voltage dropped across R_B causes the base of the transistor to be less positive than the emitter and current flows from the emitter to the base. This small flow of current causes a much larger flow from the emitter through the collector. This is the field current.

When the voltage rises to such a value that the magnetic pull caused by the current flowing through the shunt winding and the accelerator windings on the voltage regulator opens the contacts, there will be no more flow through resistor R_B to cause a voltage drop and the voltage on the base will be the same as that on the emitter. This stops the flow of base current which stops the flow of current from the emitter through the collector. The results in the alternator are the same as if the field current flowed through the contacts, but since the field current is not interrupted by moving contacts there is no arcing and the contacts have a much longer life.

The two diodes D_1 and D_2 are used to protect the circuit from spikes of voltage when the current flow in an inductive circuit is interrupted. When the transistor stops the flow of current in the field, the magnetic field collapses and induces a high back voltage into the voltage regulator. This could damage the transistor. But when the field discharge diode, D_1, is subjected to this high back voltage, it breaks down and conducts in its reverse direction long enough to pass the induced current harmlessly back into the field so no damage will be done to the circuit. The suppression diode, D_2, allows any spikes of voltage may appear in the circuit to pass to ground before they build up high enough to damage the transistor.

b. Transistor voltage regulator

The transistorized voltage regulator is a stop in the right direction, but solid-state devices replace all of the moving parts in the transistor voltage regulator we have in Fig. 1-50.

32

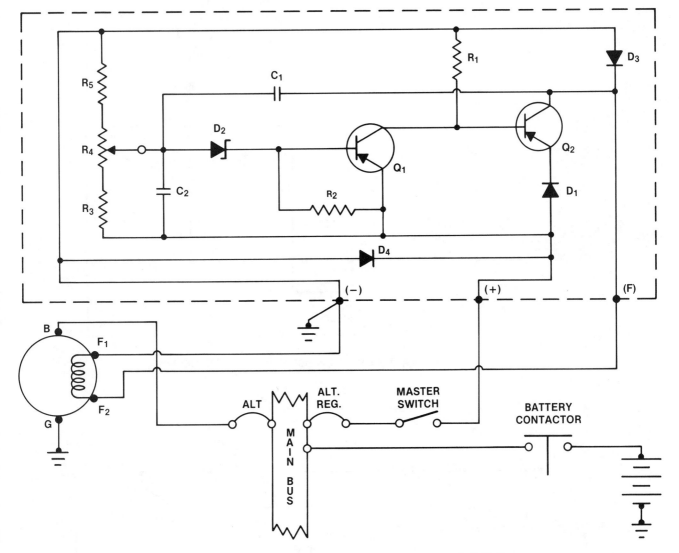

Fig. 1-50   A transistor voltage regulator for an aircraft DC alternator

In order to completely understand this popular voltage regulator, we will break it down by circuits. In Fig. 1-51A we have the portion of the voltage regulator through which the field current flows. Positive voltage is picked up from the main bus at the alternator regulator circuit breaker. It passes through the alternator portion of the master switch and enters the regulator through the (+) terminal. It flows through diode D_1, transistor Q_2 from emitter to collector, and out through terminal F to the alternator field and back to the (−) terminal of the regulator where it passes to ground. You must remember that electrons flow in the opposite direction, but since it is so much easier to visualize, we will use conventional current flow. The field discharge diode, D_3, is across the field winding to pass to the opposite side of the field the spikes of induced voltage that

occur each time the field current is stopped by transistor Q_2. The suppression diode D_4 passes any spikes of voltage that reach the voltage regulator to ground so they cannot build up high enough to damage the transistor.

In Fig. 1-51B we have the sensing portion of the transistor voltage regulator. The voltage sensing device is the zener diode D_2. This special type of diode acts as a normal solid-state diode allowing conventional current flow from the anode to the cathode (in the direction of the arrow), but blocking any flow in the opposite direction until its breakdown voltage is reached. When this voltage is put across the diode in the opposite direction, it will break down and conduct. Any semiconductor diode will do this, but zener diodes are designed to break down at a specific

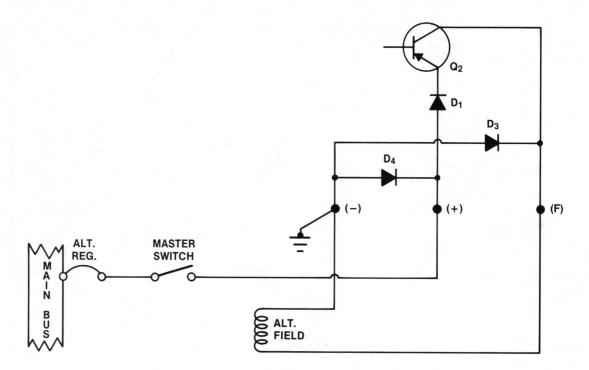

THE FIELD CURRENT PORTION OF A TRANSISTOR VOLTAGE REGULATOR
(A)

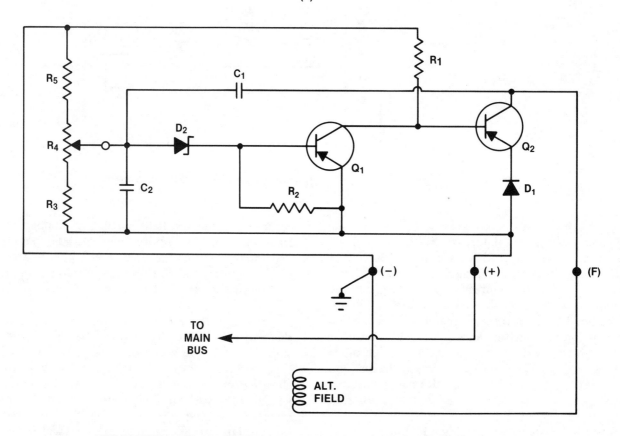

THE CONTROL AND SENSING PORTION OF A TRANSISTOR VOLTAGE REGULATOR
(B)

Fig. 1-51 Circuit breakdown of a transistor voltage regulator

voltage, and they can be operated with reverse current flowing without damage as would occur with a normal diode operating in its breakdown region.

When the alternator output voltage, which is the voltage on the bus, is below the regulated value, current flows from the bus through diode D_1, transistor Q_2, and through the field coils to ground. Transistor Q_2 is turned on because its emitter is more positive than its base.

When field current flows through diode D_1 it produces about a one-volt drop which puts the emitter of Q_2 one volt below the input voltage. The only flow through R_1 is that which flows through the base of Q_2 and this is too small to cause enough voltage drop to bring its base up to its emitter voltage.

When the alternator output voltage rises to the value for which the regulator is set, the voltage drop across the zener diode causes it to break down and conduct. Current flows through resistor R_2, diode D_2, resistors R_4 and R_5 to ground. The voltage drop caused by current flowing through resistor R_2 causes the base of transistor Q_1 to become negative with respect to its emitter and this transistor turns on. Now current flows through Q_1, from its emitter to its collector, to ground through resistor R_1. Current flowing through R_1 produces enough voltage drop to bring the base of transistor Q_2 up to the same voltage as its emitter, and Q_2 shuts off, stopping the flow of current.

There are two capacitors in this voltage regulator. Capacitor C_1 is the feedback capacitor and is connected between the field terminal and the middle of the voltage divider (R_3, R_4, and R_5). It accepts current when the inductance of the field opposes the increase of current as the transistor begins to conduct and returns this current as soon as the field current stops increasing.

Capacitor C_2 is a filter capacitor across part of the voltage divider circuit to provide a steady reference voltage for the voltage sensing zener diode.

c. Alternator control unit

One of the latest developments in aircraft electrical system controls is the alternator control unit that incorporates into one sealed unit the solid-state voltage regulator, and overvoltage sensor and a low-voltage warning circuit. With this unit, if an overvoltage condition occurs, the control unit will open the field circuit and the alternator voltage will drop to essentially zero. The low voltage warning light on the instrument panel will then notify the pilot that the alternator is no longer producing electrical power.

F. DC Alternator Service and Maintenance

If an alternator fails to keep the battery charged, you should first determine that all of the alternator and battery circuits in the aircraft electrical system are properly connected and that there are no open fuses or circuit breakers in the system and that the battery is fully charged. There should be battery voltage at the *B* terminal of the alternator and at the *Batt* or (+) terminal of the voltage regulator.

It is extremely important in an alternator installation that the battery be connected with the proper polarity, and any time an external power source is connected to the aircraft it must have the correct polarity, as improper polarity can burn out the rectifying diodes.

The solid-state diodes in an alternator are quite rugged and have long life when they are properly used, but they can be damaged by excessive voltage or by reverse current flow. For this reason an alternator must never be operated without being connected to an electrical load, as the voltage can rise high enough to destroy the diodes.

Alternators receive their field current from the aircraft bus and do not depend upon residual magnetism to get them started. Since there is no need for residual magnetism, alternators must never have their field flashed, or polarized, as this action is called.

Most modern alternator controls are solid-state devices and many are encapsulated so they are not repairable. Servicing of these systems consists merely of isolating the offending component and replacing it with a new component.

To aid in systematic troubleshooting, some manufacturers have made test equipment available that can be plugged into the aircraft electrical system between the alternator regulator and the aircraft system to indicate by the use of indicator lights whether the trouble is in the voltage regulator, the overvoltage sensing circuit, or in the alternator field or output circuit. By using this type of test equipment, much time can be saved and unnecessary replacement of good components can be avoided. If systematic troubleshooting indicates that the alternator is at fault, it can be disassembled and repaired.

There are basically two problems that prevent an alternator producing electrical power. The most likely is an open or shorted diode in the rectifying circuit and there is a possibility of an open circuit in the field.

To check for a shorted diode, measure the resistance between the B terminal of the alternator and ground. Set the ohmmeter on the R × 1 scale and measure the resistance. Then reverse the ohmmeter leads and measure the resistance again. With one measurement, the batteries in the ohmmeter forward bias the diodes and you should get a relatively low resistance reading. When you reverse the leads, the batteries reverse-bias the diodes and you should get an infinite reading. If you do not get an infinite or a very high reading, one or more of the diodes are shorted.

You cannot detect an open diode with this type of ohmmeter test, because the diodes are in parallel and the ohmmeter cannot detect one diode that is open when it is in parallel with other diodes that are good. But, if the alternator does not produce sufficient output voltage and everything else appears in good condition, you should disconnect the diodes from the circuit and test them individually with an ohmmeter. A good diode will have an infinite or very high resistance when measured with an ohmmeter in one direction and a relatively low resistance when the ohmmeter leads are reversed. An open diode will give a high resistance reading with both positions of the ohmmeter leads, and the diode will have to be replaced.

The field winding may be checked with an ohmmeter and its resistance must be within the range specified in the alternator service manual. If it is not, the field will have to be replaced.

Alternator brushes seldom give problems because they ride on smooth slip rings and carry only a very small amount of current. But they should be checked to be sure that they are not

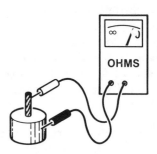

WHEN THE OHMMETER REVERSE-BIASES THE DIODE, ITS RESISTANCE WILL BE HIGH.

(A)

WHEN THE OHMMETER FORWARD-BIASES THE DIODE, ITS RESISTANCE WILL BE LOW.

(B)

Fig. 1-52 *Method of checking an alternator rectifier diode:*

worn excessively and that they ride smoothly in the brush holders.

G. Starter-Generators

Most small turbine engines are equipped with starter-generators rather than separate starters and generators. This effects an appreciable weight saving, as both starters and generators are quite heavy and they are never used at the same time.

The armature of a starter-generator is splined to fit into a drive pad on the engine, rather than being connected through a clutch and drive jaws, as starters are.

Starter-generators are equipped with two or three sets of field windings. In Fig. 1-53, we have a schematic of a typical starter-generator. The generator circuit consists of the armature and a series field around the interpoles and a shunt field for generator control. A series motor field is wound around the pole shoes inside the field frame, and the end of this winding is connected to the C terminal.

For starting, current from the battery or external power unit flows through the series winding and the armature. As soon as the engine starts, the start relay disconnects this winding and connects the generator circuit to the aircraft electrical system.

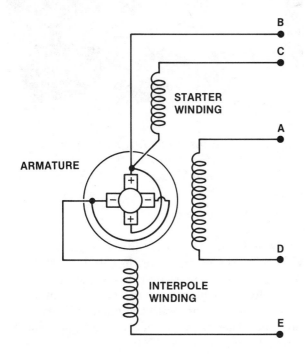

Fig. 1-53 Internal wiring diagram of a typical starter-generator.

II. ALTERNATING CURRENT POWER SYSTEMS

Direct current is normally used as the main electrical power for aircraft, because it can be stored and the aircraft engines can be started from the battery. But since large aircraft require elaborate ground servicing facilities and external power sources, they do not need to be started from their batteries and they can take advantage of the weight saving provided by using alternating current for their main electrical power.

Electrical power is the product of the voltage times the current, and since direct current systems use low voltage, their current must be large to produce sufficient power to operate heavy electrical loads. The wire needed to carry this current is large and heavy, and to reduce the weight needed for an electrical system, alternating current of 120 or 208 volts is used. The lower current needed with this higher voltage allows the use of much smaller wires.

Alternating current has the advantage over direct current in that its voltage and current can easily be stepped up or down. If we need to carry current for a long distance, we can pass the AC through a step-up transformer to increase the voltage and decrease the current. The high voltage AC can be conducted to the point it will be used through a relatively small conductor, and at its destination it is passed through a step-down transformer where its voltage is lowered and its current is stepped back up to the value we need.

It is an easy matter to convert AC into DC when we need direct current to charge batteries or to operate variable speed motors. All we need to do is pass the AC through a series of semiconductor diodes. This changes the AC into DC with realatively little loss.

We should very briefly review some of the important characteristics of alternating current so we can better understand the way we can use it most effectively.

Inductance is a characteristic of a conductor that produces a back voltage when current changes its rate of flow or direction. Since alternating current is constantly changing its rate and periodically reversing its direction of flow, there is always a back voltage being produced in an AC circuit. This back voltage causes an opposition to current flow we call inductive reactance. The higher the frequency of the AC, the greater the rate of change, and the greater this opposition will be. Commercial alternating current has a frequency of 60 cycles per second, or 60-hertz, but weight saving is so important in aircraft electrical systems, that these systems use 400-hertz AC. At this higher frequency the inductive reactance is high and the current is low. Motors can produce their torque when wound with smaller wire and transformers can be made much smaller and lighter for use with this higher frequency.

Inductive reactance varies with the frequency and this makes it important that the frequency of the alternating current produced by an aircraft generator be constant. If the frequency drops off, the inductive reactance will decrease and the current will increase to a point that the generator could burn out. Almost all aircraft installations have some provision to decrease the generator

37

field current if the frequency drops off. If it gets below a minimum frequency, the generator will automatically be disconnected from the system.

A. AC Generating Systems

The frequency of the AC produced by an AC generator is determined by the number of poles and the speed of the rotor. Many of the 400-hertz AC generators used in aircraft electrical systems have eight poles and rotate at 6,000 RPM. The frequency of the AC may be found by the formula:

$$F = \frac{P}{2} \times \frac{N}{60}$$

F = Frequency of the AC in hertz

P = Number of poles in the rotating field

N = Rotational speed of the generator in RPM

To provide a constant frequency as the engine speed varies, many engine-driven aircraft AC generators are connected to the engine through a hydraulically operated constant speed drive unit, a CSD. These drive units normally consist of an axial-piston variable-displacement hydraulic pump driven by the engine, supplying fluid to an axial-piston hydraulic motor which drives the generator. The displacement of the pump is controlled by a governor which senses the rotational speed of the AC generator. When the generator speed is below 6,000 RPM, the speed required to produce 400-hertz AC, the governor directs oil into the pump control portion of the CSD to cause the pistons to pump more fluid to the motor to speed it up. If the speed becomes too high, the governor will cause the pump output to decrease. This governor action holds the output speed of the generator constant and maintains the frequency of the AC at 400-hertz, plus or minus the tolerance which the aircraft electrical system engineers have found they can allow.

1. Integrated drive generator

Some of the most modern jet aircraft produce their alternating current with a generator similar to the one we see in Fig. 2-1. This unit is called an Integrated Drive Generator, an IDG, and it includes a constant speed drive unit in the housing with the generator.

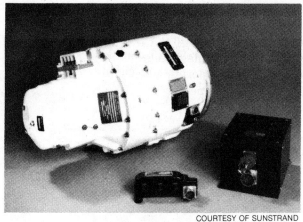

Fig. 2-1 A Sunstrand Integrated Drive Generator with its control unit and a current transformer assembly

a. Constant-speed drive

The spline that fits into the engine drive rotates a carrier shaft inside the axial-gear differential. This carrier shaft drives the axle shafts on which two planet gears spin. One of these planet gears meshes with the fixed-unit ring gear and the other with the output ring gear that is splined to the generator rotor.

The speed ratio between the input shaft and the output is 2:1 when the fixed-unit ring gear is not turning. If the fixed-unit ring gear is rotated

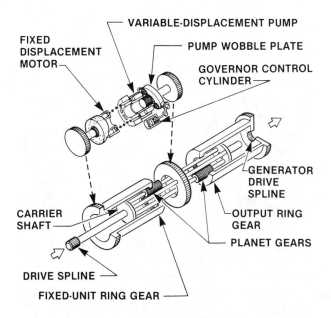

Fig. 2-2 Constant-speed drive axial-gear differential used in the Sunstrand Integrated Drive Generator

in the opposite direction to the carrier shaft, the planet gears will speed up and increase the speed of the output. But if the fixed-unit ring gear is rotated in the same direction as the carrier shaft, the planets will slow down and will drive the output at a slower speed.

An axial-piston, variable-displacement hydraulic pump is driven by a gear on the carrier shaft, and its output is fed directly into a fixed displacement axial-piston hydraulic motor. The output of the pump is determined by a hydraulic signal from the governor

When the speed of the drive is correct, the governor hydraulic pressure in the control cylinder holds the wobble plate in the pump at the correct angle to prevent the pump producing a flow to the motor, and with the motor not turning, its drive gear holds the fixed-unit ring gear stationary.

When the engine speeds up and drives the generator at too high a speed, the governor which is driven by the output of the CSD unit senses the overspeed condition, and its oil pressure in the control cylinder tilts the wobble plate so the pump will send fluid into the cylinders of the motor in such a direction that it will drive the fixed-unit ring gear in the same direction as the variable unit and the planet gears will drive the output at a lower speed.

If the engine slows down, the governor oil pressure will cause the control cylinder to tilt the wobble plate so the pump will drive the motor and rotate the fixed-unit ring gear in the direction *opposite* the variable unit, and the output will speed up.

The Integrated Drive Generator contains its own oil supply, oil filter, and oil pumps. The oil in this unit not only provides the working fluid for the constant speed control, but it also provides lubrication and cooling for both the drive unit and the generator. (See Fig. 2-3 on page 40.)

b. Generator controls

The state-of-the-art generator control unit used with the Integrated Drive Generator includes circuits that regulate the voltage, limit the current, provide over- and under-voltage and over- and under-frequency protection as well as differential current protection.

Current transformer assemblies sense the current in each of the three phases and send their output to the generator control unit.

2. Electronic converters

Another way of maintaining a constant frequency with a changing generator speed is to convert the output of the AC generator into direct current with a solid-state rectifier and then, using a semiconductor oscillator and amplifier, convert the DC into 400-hertz, sine-wave AC.

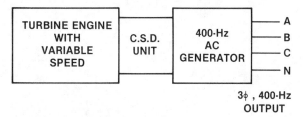

USING A GENERATOR COUPLED TO THE ENGINE THROUGH A CONSTANT-SPEED DRIVE UNIT.
(A)

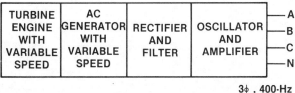

USING AN ELECTRONIC CONVERTER TO PRODUCE AC FROM DIRECT CURRENT WITH AN OSCILLATOR AND AMPLIFIER. THE DC IS THE RECTIFIED OUTPUT OF THE VARIABLE-SPEED AC GENERATOR.
(B)

Fig. 2-4 Methods of obtaining constant frequency alternating current from a variable speed generator:

B. Reactive Power

The true power produced by an AC generator is the product of the voltage and that portion of the current that is in phase with the voltage, and this is expressed in watts or kilowatts. It is this power that determines the amount of useful work the electricity can do.

AC generators are rated, however, not in watts, but in volt-amps, which is a measure of the apparent power being produced by the generator. The reason for using this rating is that it is the

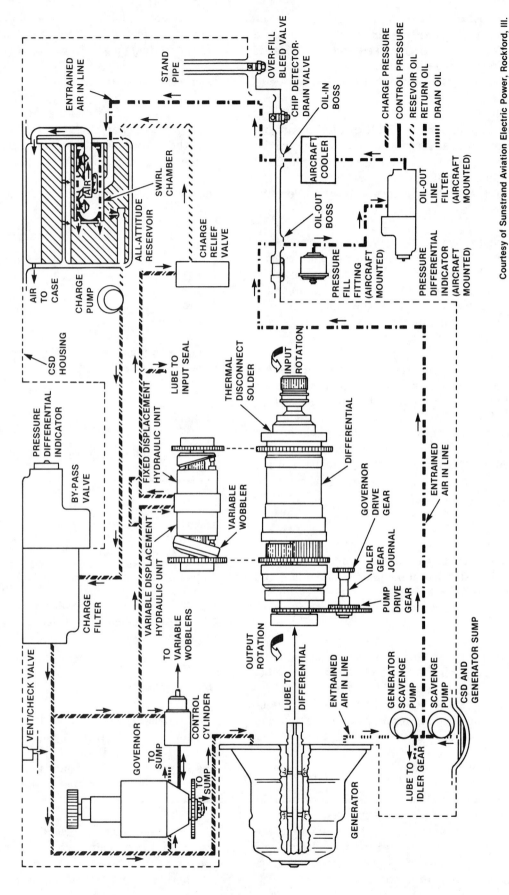

Courtesy of Sunstrand Aviation Electric Power, Rockford, Ill.

Fig. 2-3 Lubricating and control oil flow for the Sunstrand Integrated Drive Generator

heating effect of the current in the generator windings that limits the generator output, and this current flows in the winding whether it is producing power or not.

Reactive power produced by an AC generator is measured in volt-amps-reactive (VAR) or in kilovolt-amps-reactive (KVAR) and is the product of the voltage and the amount of current that is *not* in phase with the voltage. It is important when operating AC generators in parallel that they share equally the reactive load as well as producing the same amount of true power. A circuit in the generator control unit measures the amount of reactive current produced by each generator and reduces the excitation of any generator that is producing too much reactive

current. The flight engineer has instruments on his panel that normally show the kilowatts of power being produced by each generator, but by pressing a switch beside the instruments they can be made to indicate the reactive power in KVAR produced by the generator.

C. Phase Rotation

It is important when installing an AC generator on an aircraft engine that the phase rotation be correct. Most generators are connected so that their phase rotation is A-B-C. This means that the voltage in phase A rises first, then the voltage in phase B, and finally the voltage in phase C. If the phase rotation is backward, the generator

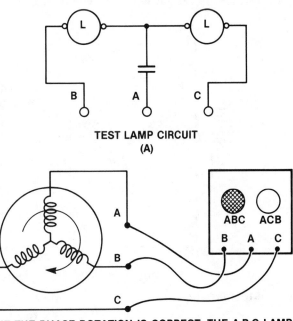

TEST LAMP CIRCUIT
(A)

IF THE PHASE ROTATION IS CORRECT, THE A-B-C LAMP WILL BE BRIGHT AND THE A-C-B LAMP WILL BE DIM.
(B)

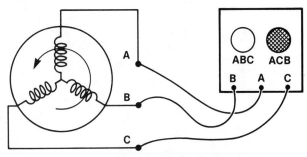

IF THE PHASE ROTATION IS NOT CORRECT, THE A-B-C LAMP WILL BE DIM AND THE A-C-B LAMP WILL BE BRIGHT.
(C)

Fig. 2-5 Generator control panel for a Boeing 727 jet transport aircraft

Fig. 2-6 Determining the phase rotation of a three-phase AC generator:

cannot be synchronized with others, and it it is operating independently, any three-phase AC motor will run backward. To determine the phase rotation of a generator, we can use a phase rotation test lamp which consists of two incandescent bulbs and a capacitor connected as we see in Fig. 2-6A. The current through the capacitor leads the applied voltage, and the lamp connected across the phase *next* in rotation will burn more brilliantly than the lamp across the phase behind the reference phase.

The test lamps are connected to the generator with the test leads A, B, and C connected to the generator terminals A, B, and C. If the phase rotation is A-B-C, the A-B-C light will burn bright and the A-C-B light will be dim. If the generator has been connected so the phase rotation is backward, the A-C-B light will be bright and the A-B-C light will be dim. No AC generator installation is finished until you have checked the phase rotation and know that it is right.

D. AC Generators

Most of the AC generators used in the large jet-powered aircraft are of the brushless type and are usually air cooled. Since brushless generators have no current flow between brushes and slip rings, they are quite efficient at high altitude where brush arcing could be a problem.

In Fig. 2-7, we have a schematic of a brushless generator. The exciter field current is

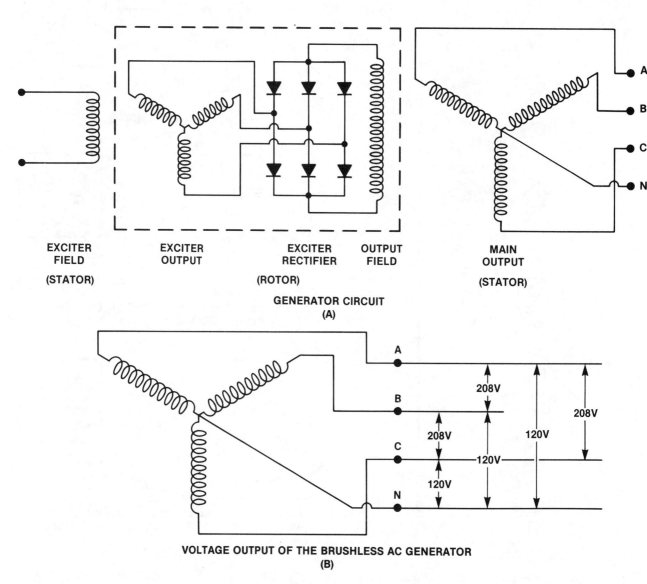

Fig. 2-7 Brushless AC generator

brought into the generator from the voltage regulator. Here it produces the magnetic field for the three-phase exciter output. Permanent magnets furnish the magnetic flux to start the generator producing an output before field current flows. The voltage produced by these magnets is called residual voltage. The output from the exciter is rectified by six silicon diode rectifiers, and the resulting DC flows through the output field winding. The exciter output winding, the six diodes, and the output field winding are all mounted on the generator shaft and rotate as a unit. The three-phase output stator windings are wound in slots in the laminated frame of the generator housing, and their ends are connected in the form of a Y with the neutral and the three-phase windings brought out to terminals on the outside of the housing. These generators are usually designed to produce 120 volts between

any of the phase terminals and the neutral terminal and 208 volts between any of the phase terminals.

E. AC Generator Instrumentation and Controls

The generator control unit used with an AC generator maintains the voltage at a constant value by controlling the amount of current fed into the exciter field winding. It also controls the amount of excitation current to maintain an equal reactive load distribution between generators connected in parallel, and it will decrease the field excitation current if the generator frequency drops enough that the output current becomes excessive because of the decrease in inductive reactance.

In addition to the automatic controls, AC generators have several indicators and controls

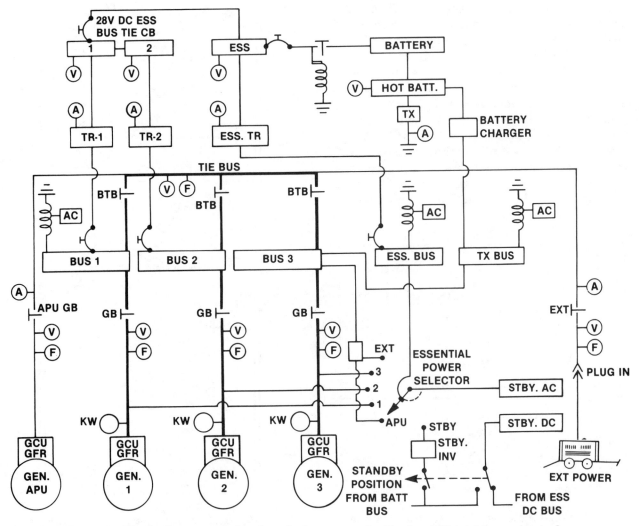

Fig. 2-8 Block diagram of the electrical system of a Boeing 727 jet transport airplane

43

257

that are used by the flight engineer. In Fig. 2-8 we have a block diagram of the electrical system of a Boeing 727. There are three engine-driven 40-KVA, three-phase generators that can be connected to the main tie bus in such a way that they share the electrical load. A fourth generator of the same type is driven by the auxiliary power unit and it can power the system when none of the engine-driven generators are connected to the bus. Each generator is provided with its own control panel, Fig. 2-5, that contains a KW-KVAR meter, a switch, and indicator light for control ling and indicating the condition of the bus tie breaker and the generator field breaker. There is also a knob with which the flight engineer can vary the output of the constant-speed drive unit to bring the frequency of the generator to exactly the same frequency as that of the other generators so they can be synchronized to connect them to the tie bus.

Above the generator control panels are the constant-speed drive controls that consist of switches that allow the flight engineer to disconnect the CSD if there is any indication of generator malfunction. Below these switches is a CSD oil temperature indicator for each engine and a switch that allows the flight engineer to cause the meter to read either the temperature of the oil entering the CSD or the rise in temperature of the oil as it passes through the CSD. A low-pressure warning light beside the disconnect switch warns the flight engineer if the CSD unit ever loses its oil pressure. The constant speed

drive units can be disconnected in flight, but they can be reconnected only on the ground.

An AC control panel beside the generator control panels holds an AC voltmeter and a frequency meter with a selector switch so the flight engineer can monitor the voltage and frequency of any of the three engine-driven generators, the tie bus, the APU, or the external power source. There is a button that allows the flight engineer to change the reading of the voltmeter so it will indicate the residual voltage. If the generator is being driven by the CSD, but there is no field excitation from the voltage regulator, the generator will show residual voltage which is produced by the permanent magnets in the generator frame.

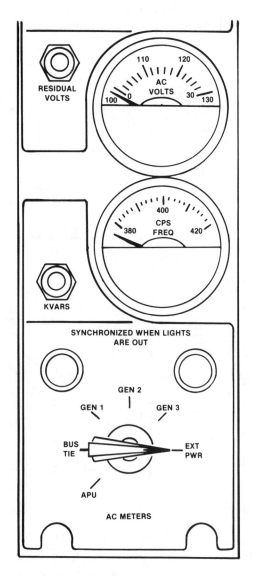

Fig. 2-10 AC control panel at the flight engineer's station of a Boeing 727

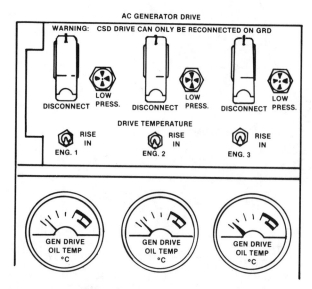

Fig. 2-9 Constant-speed drive control panel in a Boeing 727 aircraft

Before a generator can be connected to the tie bus with another generator, it must be synchronized; that is, its frequency and voltage must be the same as the frequency and voltage of the bus. When one of the generators is ready to be connected to the tie bus, the selector switch on the AC control panel is turned to that generator and the two synchronizing lights are watched. If the two lights flash alternately, the phase rotation between the generator supplying the tie bus and the one being readied to connect are opposite, and the generator must not be connected. If the lights burn steadily, the generators are not synchronized and the synchronizing control can be adjusted to get the lights to turn off, or blink at their slowest rate. When both lights are out, the bus tie breaker can be closed to connect the generator to the tie bus.

III. ELECTRICAL MOTORS

A. DC Motors

Electric motors are such an important part of a powerplant installation that it is wise to review some of their principles before we discuss the motors that are used on the engine side of the firewall.

All electric motors operate on the principle of reaction between two magnetic fields. For a motor to rotate and produce torque, one of the magnetic fields must have its polarity reversed as the motor shaft rotates.

We have seen in other parts of this Integrated Training Program that when electrons flow in a conductor, their spin axes line up and produce a magnetic field around the conductor. The strength of this magnetic field is determined by the amount of electron flow, and the direction the lines of flux encircle the conductor is determined by the direction of electron flow. The left-hand rule we associate with motors states that if a conductor is grasped with the left hand in such a way that the thumb points in the direction of electron flow, from negative to positive, the fingers will encircle the conductor in the same direction as the lines of magnetic flux.

Another convention regarding magnetic flux tells us that lines of flux leave the north pole of

the magnet and enter the magnet at its south pole.

When a current-carrying conductor is placed in the magnetic field of a stationary magnet, the flux from the two fields will concentrate on one side of the conductor and apply a force that will try to move it. The direction of the force can be visualized by applying the right-hand rule for motors. If we hold the fingers of the right hand as we see them in Fig. 3-1, with the forefinger pointing in the direction of the lines of flux, from the north pole to the south pole, and the second finger pointing in the direction of electron flow in the conductor, the thumb will point in the direction the combined magnetic fields will move the conductor. The amount of force is determined by the combined strength of the two magnetic fields.

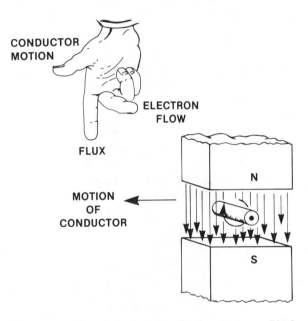

Fig. 3-1 *The right-hand rule for motors: If the right hand is held as shown with the forefinger pointing in the direction of the magnetic flux, and the second finger pointing in the direction of electron flow in the conductor, the thumb will point in the direction the conductor will be forced to move.*

In order to gain the reversing magnetic field we need in a DC motor, we use a commutator and brush arrangement on the armature that is much the same as the commutator we have discussed on the DC generator. When electrons enter the negative brush of the armature coil in Fig. 3-2A, they flow through the dark side of the winding and cause lines of flux to surround the winding in a counterclockwise direction, and the flux lines around the lower side of the coil will be in a clockwise direction. The result is that the upper conductor is forced to the right and the lower conductor to the left. This causes the armature to rotate in a clockwise direction. As the armature rotates, the commutator segment connected to the light side of the coil moves under the negative brush and the one on the dark side moves under the positive brush. This reverses the direction of

electron flow in the armature coil but keeps the electron flow direction the same, as the conductors pass in front of the stationary magnetic field poles. The large number of coils in the armature of a practical electric motor causes the conductors moving in front of the magnetic field poles to always carry the maximum amount of current so that the motor can produce a maximum amount of smooth torque.

The basic way of classifying DC motors is by the type of stationary fields they use. Some of the smaller DC motors use permanent magnets for the field, but most of the motors used in powerplant electrical systems use electromagnets for the field, and these may be connected in parallel or in series with the armature, or the motor may be compound-wound and use both shunt and series fields.

1. Shunt-field motors

Shunt-field motors are used for most low-torque applications where constant speed is desirable. The field windings are made up of many turns of relatively small wire, and they are connected in parallel with the armature. When current flows in the armature, it also flows in the field coils, but since the two are in parallel, the majority of the current will flow through the low-resistance armature. The current through the field coils produces the magnetic flux that causes the armature to rotate. As it rotates, it cuts across the field flux and a voltage is generated in the armature windings. This voltage, according to Lenz's law, has a polarity opposite that which caused the armature to rotate. This back voltage, or counter electromotive force (CEMF), decreases the amount of voltage available to force current through the armature, and the motor will operate at a fixed speed. If the load on the motor causes the armature to slow down, the rate at which the

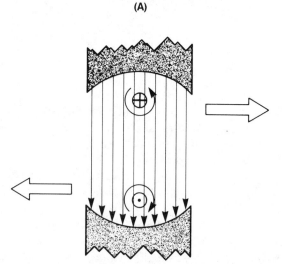

ELECTRONS ENTER THE NEGATIVE BRUSH AND CAUSE A MAGNETIC FIELD TO SURROUND THE CONDUCTORS IN THE COIL.

(A)

THIS MAGNETIC FIELD PRODUCES A TORQUE WHICH CAUSES THE ROTOR TO ROTATE IN A CLOCKWISE DIRECTION.

(B)

Fig. 3-2 Current flowing in the rotating coil of a DC motor produces torque that causes the coil to rotate.

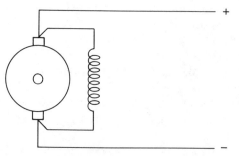

Fig. 3-3 Shunt-connected DC motor

260

armature coils cut across the field flux will decrease, and this will decrease the CEMF. But since the source voltage remains constant, the motor will get more current through the armature and it will regain its speed. If the load decreases, the armature will speed up and its coils will produce more CEMF. This reduces the current in the armature and the motor will slow back to its original speed.

Speed of a shunt-wound DC motor may be controlled by putting a variable resistor in either the armature circuit or in the field circuit.

Increasing the resistance of the armature decreases the armature current, and the motor will slow down. The main disadvantage of using armature resistance to control the speed is that a considerable amount of power must be dissipated in the speed-control resistor. A much smaller speed-control resistor may be used if it is placed in the shunt field circuit. Increasing the resistance of the shunt field circuit will decrease the field strength and thus decrease the CEMF pro-

duced in the armature, and the motor will speed up. Speed control by varying the shunt field strength is usable only for light motor loads because it causes commutation difficulties when the load is high. Resistors can be used to control DC motor speed, but most of the modern speed control devices use a silicon controlled rectifier (SCR) in a form of electronic control.

The direction of rotation of a DC shunt-wound motor can be reversed by reversing the direction of current flow in either the armature or the field winding, but not in both at the same time. This may be done by using a reversing switch or by using two sets of field coils and a selector switch. Both methods are used in practical motor installations.

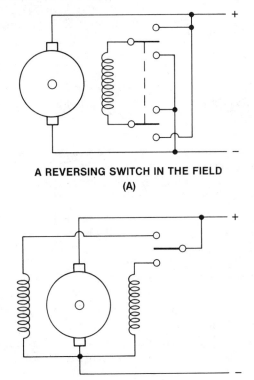

A REVERSING SWITCH IN THE FIELD

(A)

A SELECTOR SWITCH IN A MOTOR HAVING TWO OP-POSITELY WOUND SHUNT FIELDS

(B)

Fig. 3-5 Method of reversing a shunt-connected DC motor.

2. *Series-field motors*

Many of the electric motors used in aircraft powerplant installations require a relatively high starting torque, and for this reason they are series-wound motors. In a series motor, the field coils consist of a relatively few turns of heavy

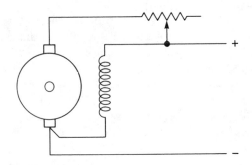

DECREASING THE CURRENT THROUGH THE ARMATURE DECREASES THE SPEED OF THE MOTOR.

(A)

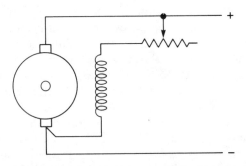

DECREASING THE CURRENT IN THE SHUNT FIELD DECREASES THE CEMF WHICH INCREASES THE SPEED OF THE ARMATURE.

(B)

Fig. 3-4 Speed control in a shunt-connected DC motor.

wire, and they are in series with the armature. All of the current that flows in the armature also flows in the field.

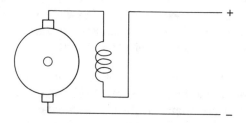

Fig. 3-6 *Series-connected DC motor*

The torque produced by a series motor decreases as the speed increases. When the current first begins to flow in a series motor, there is no CEMF and current pushed by all of the source voltage produces the magnetic fields that cause torque. But as soon as the armature begins to rotate, its turns cut across the field flux and produce a large amount of CEMF, which decreases the current through the motor and thus decreases its torque. When the armature is loaded, the speed decreases and so does the CEMF. Current through the motor increases and more torque is produced to bring the speed back up. Series motors are used for all aircraft engine starters because they need lots of torque at low speed, and they always operate into a mechanical load.

Series motors must always be operated into some kind of mechanical load. If they are operated without a load, they will continue to accelerate until they reach such a speed that centrifugal force will likely cause the armature to throw a winding loose, and this can destroy the motor.

3. Compound-field motors

If a motor must have a relatively high starting torque and at the same time its speed must be controllable, it may be compound-wound. That is, it may have both a shunt and a series winding. The relationship between the effect of the two fields determines the characteristics of the motor. If the series field produces the most magnetic flux, the torque will increase as the speed decreases, but if the series field is weak compared with the shunt field, the torque will remain relatively uniform as the speed changes.

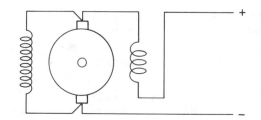

Fig. 3-7 *Compound-connected DC motor*

B. AC Motors

In Book Five of the Airframe Section of this Integrated Training Program we have discussed the principle of AC induction motors. We saw in this portion of the text that a rotating field may be produced in a single-phase motor by using a high-resistance start winding or by placing a capacitor in series with the start winding. For small, low-torque motors, a shading coil may be wound around two of the poles of the stator. We also studied the production of the rotating field in a three-phase induction motor.

Single-phase induction motors are used perhaps more than any other type of electric motor. They are used for tape recorders, record players, fans, refrigerators and air conditioners, drill presses, lathes, air compressors—for more applications than we can name. But AC motors that are used for aircraft powerplant applications normally require a high torque, and for this reason, when AC motors are used they are usually three-phase induction motors or repulsion motors.

1. Repulsion motors

The rotor of a repulsion motor resembles the armature of a DC motor, as it has a large number of coils and a commutator. The stator is similar to that used in an induction motor. The brushes that ride on the commutator do not bring current into the rotor, but they short-circuit the coils in the rotor. When current flows in the stator winding, it induces a voltage in the rotor that causes current to flow through the shorting brushes and produce a magnetic field that *opposes* the field of the stator. This repulsive force causes the rotor to turn with a relatively high torque. As the rotor turns, the brushes ride over the commutator and keep the polarity in the armature at the same location rather than rotating as the magnetic field around the bars in a squirrel cage rotor rotates.

Powerplant Electrical Installation

I. ELECTRICAL INSTALLATIONS

Electrical components and electrical installations account for a large part of the complete powerplant of a modern aircraft. Large aircraft have electrical systems that are capable of supplying as much as a quarter of a million watts of electrical power, and these systems use complex three-phase alternating current generators and their controls. Even small training airplanes have alternators, batteries, motors, and all of the controls and wiring needed for efficient operation.

In the General Section of this Integrated Training Program, we have discussed the fundamentals of both AC and DC electricity. In the Airframe Section we have covered the installation of the electrical system as it is used for airframe functions, and in Book Four of the Powerplant Section, we studied the generators, alternators, and motors used in the powerplant portion of an aircraft. In this book we are concerned with the installation of the wiring and circuit components that are primarily used in the electrical systems that relate directly to the powerplant portion of a modern aircraft.

The electrical installation in an aircraft consists of a source of electrical energy and a load to use the energy. Between the source and the load, there must be conductors of sufficient size to carry all of the current without overheating or without producing an excessive voltage drop. There must also be adequate circuit protection devices to assure that the proper amount of current is available at the load when it is needed and to assure that excessive current cannot flow in the circuit. Excessive current can damage electrical equipment and produce enough heat to create a fire hazard.

A. Wire

When the wire for an electrical system is chosen, there are several factors that must be considered:

(1) The wire must be of sufficient size that the current flowing through it will not produce enough heat to damage the insulation.

(2) The wire must be large enough that its resistance will not cause the current flowing through it to produce a voltage drop which will lower the voltage at the load beyond the allowable limit.

(3) The wire and its terminals must be sufficiently strong and installed in such a way that vibrations will not cause the wire to break.

(4) The insulation on the wire must have sufficient electrical strength to prevent electrical leakage and it must be strong enough mechanically that it is not likely to be damaged by abrasion.

(5) The wire must be as light in weight as is consistent with the other requirements.

1. Wire type

Because of the vibration inherent in all aircraft, the wiring between the components must be of the stranded type, but the wire used inside electrical and electronic components where vibration is no problem is usually solid. Installation and servicing of this type of wiring is done by electronic specialists and is not considered to be part of the electrical system servicing done by A&P technicians.

The majority of the wire used in an aircraft electrical system is stranded copper wire that meets specifications MIL-W-5086. This wire is made up of strands of annealed copper that are covered with a very thin coating of tin to prevent the copper oxidizing. The wire is covered with an insulation of polyvinyl chloride or nylon and, in some instances, with a glass cloth braid. The insulation on most of the MIL-W-5086 wire will withstand 600 volts, but one type of this wire (Type IV) has an insulation that has a dielectric strength of 3,000 volts.

SOLID CONDUCTORS ARE USED WHERE VIBRATION IS NO PROBLEM.

(A)

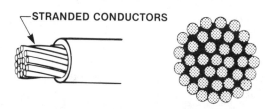

THE MAJORITY OF THE WIRING IN AN AIRCRAFT ELECTRICAL SYSTEM USES STRANDED CONDUCTORS.

(B)

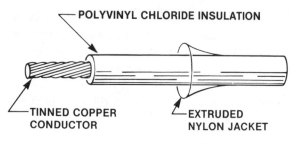

TYPICAL MIL-W-5086 WIRE.

(C)

Fig. 1A-1 Typical wire used for aircraft electrical system installation

MIL-W-5086 wire is used in aircraft installations in sizes from AN-0000, called four aught, to AN-22. Smaller size wire is available and is used in some electronic equipment, but because of its low physical strength it is not used in airframe or powerplant wiring installations.

When large amounts of electrical current must be carried for long distances in an aircraft, a weight saving is accomplished by using MIL-W-7072 aluminum wire instead of copper. Aluminum wire can carry about two thirds as much current as the same size copper, and it weighs only about one third as much.

Aluminum wire has two basic limitations. It requires special techniques for installing the terminals, and it is easily crystallized by vibration; so breakage is a problem. But, when aluminum wire is properly terminated and protected from vibration, it makes an effective weight-saving installation.

The insulation used on MIL-W-7072 aluminum wire is the same as that used on MIL-W-5086 copper wire.

A handy rule of thumb for replacing copper wire with aluminum wire is to always use an aluminum wire that is two wire gage numbers larger than the copper wire it is replacing. Remember when using this rule that the larger the number, the smaller the wire. If we want to replace a piece of four-gage copper wire with aluminum, we must use at least a two-gage wire.

Because of the small amount of weight that can be saved with the smaller size wires, and because of the danger of the wire breaking from vibration, aluminum wire of sizes smaller than six-gage is not approved for use in aircraft electrical systems.

2. Wire size

Wire size is measured according to the American Wire Gage (AWG) system, with the smaller numbers used to identify the larger wire. A wire gage, similar to the one in Fig. 1A-2, can be used to measure the size of solid conductors. The width of the slot is the size of the wire, and the circular cutout at the end of the slot is merely for clearance, not for measuring.

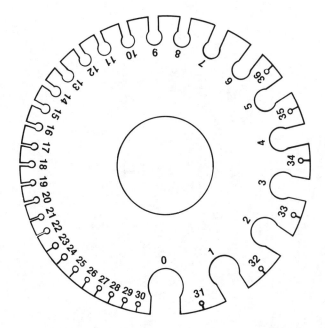

Fig. 1A-2 *The American Wire Gage is used to indicate the size of electrical wire. The width of the parallel slot is the diameter of the wire.*

The amount of current a wire can carry is determined by its cross-sectional area, and rather than square units being used for this measurement, the circular mil is used to express the conductor area: One circular mil is the area of a conductor whose diameter is one mil, or one thousandth of an inch. For example, a 20-gage wire has a diameter of 0.032 inch or 32 mils. Its area is 1,024 circular mils.

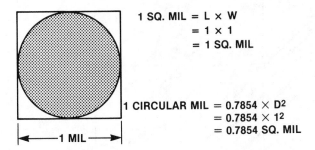

Fig. 1A-3 *The circular mil is used to measure the area of a round conductor.*

The area of a circle is 0.7854 times the square of its diameter so, as we see in Fig. 1A-3, the area of a round conductor in circular mils is 0.7854 times its area in square mils. A 20-gage wire whose diameter is 32 mils has a area of 1,024 circular mils, but this area is only 804.2 square mils.

Diameter = 0.032 inch, or 32 mils

Area = 1,024 circular mils

= 804.2 square mils

When we replace a wire in an aircraft electrical system, we must use the same size and type of wire as was originally used; but if we are making an original installation, we must choose a wire of a size that will carry the current without overheating and will not produce more than the allowable voltage drop for the length of the wire used and for the current it carries.

NOMINAL SYSTEM VOLTAGE	ALLOWABLE VOLTAGE DROP VOLTS	
	CONTINUOUS OPERATION	INTERMITTENT OPERATION
14	0.5	1.0
28	1.0	2.0
115	4.0	8.0
200	7.0	14.0

Fig. 1A-4 *The allowable voltage drops for typical aircraft electrical systems*

In Fig. 1A-4, we have the allowable voltage drop for both intermitttent and continuous operation in 14-, 28-, 115-, and 200-volt systems. If we install a cowl flap motor in a 28-volt system, , we must use a wire large enough that it will not cause more than a two-volt drop between the bus and the motor. We are allowed two volts because a cowl flap motor is an intermittent load. If we were installing a fuel boost pump that operates continually, we would be allowed only a one-volt drop.

In the wire chart in Fig. 1A-5, we can find the wire size we must use when we know whether the load is continuous or intermittent, when we know the voltage of the system and the length of the wire, and when we know if the wire is routed by itself or if it is in a bundle with other wires. Let's take the example of a 28-volt boost pump motor that draws six amps continously: The wire must be 40 feet long, and we will run it in a bundle with other wires.

To find the size wire we need, first locate the diagonal line that represents six amps and follow it down until it intercepts the horizontal line representing 40 feet in the 28-volt column. This in-

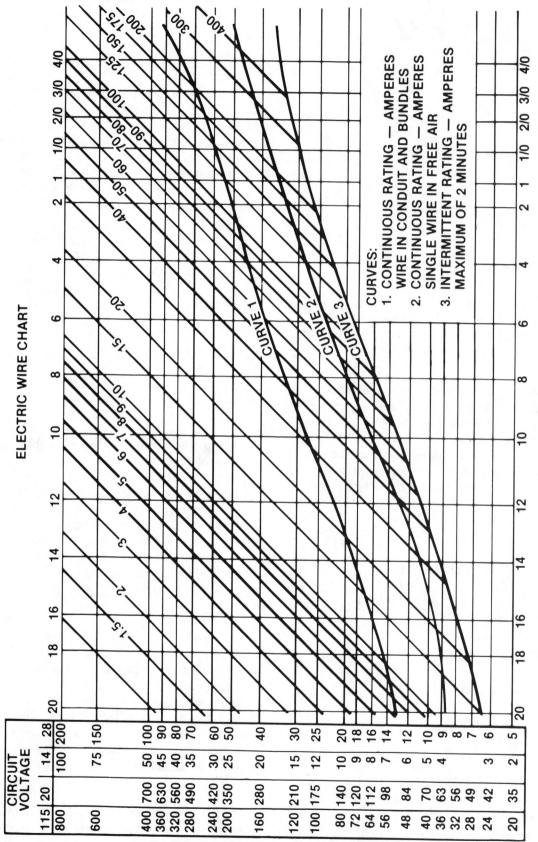

Fig. 1A-5   Electric wire chart

tersection takes place between the 16-and 14-gage wire size vertical lines. We should use the larger of the two wires, and so we will choose a 14-gage wire. Notice that this intersection is above all three of the condition curves, which means that we can safely run six amps of current in this size wire without it overheating, and there will be only a one-volt drop in the 40 feet of wire.

You will notice that there are four columns of figures for the circuit voltages. This is in keeping with the information we have in Fig. 1A-4. A 28-volt system is allowed a one-volt drop, a 14-gage system can only have a half-volt drop, and the length of wire for the allowable voltage drop is only one half of that allowed for a 28-volt system. A 115-volt system is allowed a four-volt drop, so the length of wire is four times that in the 28-volt column. And in a 200-volt system we are allowed seven volts, so a wire 280 feet long would be needed before the voltage drop caused by six amps of current would be excessive for this system.

The three condition curves allow us to determine whether or not we can run the wire in a bun-

dle or if it must be run by itself. If the intersection of the diagonal current line and the vertical line representing the wire gage falls above curve 1, the wire is satisfactory from the current standpoint to be routed in a bundle, even when the current flows through it continuously. For example, a 14-gage wire can carry up to 17 amps continuously in a bundle.

Between 17 and 29 amps can be carried in a 14-gage wire continuously if the wire is routed by itself in free air so that the heat can be carried away from it. Between 29 and 32 amps can be carried in a 14-gage wire if the flow is intermittent, with a maximum duration of flow of two minutes. A 14-gage wire should not carry more than 32 amps of current under any conditions, as this much current will overheat it.

The information we find in this chart is verified in the table of Fig. 1A-6. Fig. 1A-7 gives us the same information on the current carrying capability, resistance, area, and weight of MIL-W-7072 aluminum wire.

WIRE SIZE SPECIFICATION MIL-W-5086	SINGLE WIRE IN FREE AIR MAXIMUM AMPERES	WIRE IN CONDUIT OR BUNDLED MAXIMUM AMPERES	MAXIMUM RESISTANCE OHMS/1000 FT. (20° C)	NOMINAL CONDUCTOR AREA CIRCULAR MILLS	FINISHED WIRE WEIGHT POUNDS PER 1,000 FEET
AN—20.................	11	7.5	10.25	1.119	5.6
AN—18.................	16	10	6.44	1,779	8.4
AN—16.................	22	13	4.76	2,409	10.8
AN—14.................	32	17	2.99	3,830	17.1
AN—12.................	41	23	1.88	6,088	25.0
AN—10.................	55	33	1.10	10,443	42.7
AN—8...................	73	46	.70	16,864	69.2
AN—6...................	101	60	.436	26,813	102.7
AN—4...................	135	80	.274	42,613	162.5
AN—2...................	181	100	.179	66,832	247.6
AN—1...................	211	125	.146	81,807	
AN—0...................	245	150	.114	104,118	382
AN—00.................	283	175	.090	133,665	482
AN—000...............	328	200	.072	167,332	620
AN—0000.............	380	225	.057	211,954	770

Fig. 1A-6 Current-carrying capacity of copper electrical wire

5

WIRE SIZE SPECIFICATION MIL-W-7072	SINGLE WIRE IN FREE AIR MAXIMUM AMPERES	WIRE IN CONDUIT OR BUNDLED MAXIMUM AMPERES	MAXIMUM RESISTANCE OHMS/1000 FT. (20° C)	NOMINAL CONDUCTOR AREA CIRCULAR MILLS	FINISHED WIRE WEIGHT POUNDS PER 1,000 FEET
AL—6....................	83	50	0.641	28,280	....
AL—4....................	108	66	.427	42,420	....
AL—2....................	152	90	.268	67,872	....
AL—0....................	202	123	.169	107,464	166
AL—00..................	235	145	.133	138,168	204
AL—000................	266	162	.109	168,872	250
AL—0000..............	303	190	.085	214,928	303

Fig. 1A-7   Current-carrying capacity of aluminum electrical wire

3.   Wire identification

There are so many wires in a modern airplane that troubleshooting an electrical system malfunction would be greatly hindered if some way were not used to identify each of the wires.

J —SYSTEM IN WHICH THIS WIRE IS USED
14—INDIVIDUAL WIRE NUMBER
C —SECTION OF WIRE FROM POWER SOURCE
20—WIRE SIZE ~AWG SIZE

TYPICAL WIRE MARKING
(A)

C — FLIGHT CONTROL SYSTEMS
D — DEICING AND ANTI-ICING SYSTEMS
E — ENGINE INSTRUMENTS
F — FLIGHT INSTRUMENTS
H — HEATING AND VENTILATION SYSTEMS
J — IGNITION
K — ENGINE CONTROL SYSTEMS
L — LIGHTING
M — MISCELLANEOUS
N — GROUND NETWORK
P — DC POWER SYSTEM
Q — FUEL AND OIL
R — RADIO
V — INVERTERS
W — WARNING SYSTEMS
X — AC POWER SYSTEMS

CIRCUIT IDENTIFIERS
(B)

Fig. 1A-8   Wire identification system

Many of the aircraft manufacturers have their own system for identifying wires, but one standard identification system is shown in the example in Fig. 1A-8. The wire is stamped about every 12 to 15 inches along its length with the number J14C-20. From this identification we know that this wire is in the magneto circuit (the letter J tells us this). It is wire number 14 in this system, and it is the third individual segment of this wire from the magneto. We know this from the letter C. The -20 indicates that it is a 20-gage wire. If the wire went to ground, it would have the letter N at the end of the identification.

A two-letter identification code is used by one popular aircraft manufacturer, and this code is shown in Fig. 1A-9. A wire in the ignition circuit would have the two-letter identification JA. (See Fig. 1A-9 on page 7.)

Wire bundles can be marked by using pressure sensitive tape around the bundle with the number marked on it or with a piece of polyvinyl chloride sleeving with the bundle number stamped on it. This sleeving is tied around the bundle. Bundles are marked near the points at which they enter and leave a compartment. (See Fig. 1A-10 on page 7.)

B.   Wiring Installation

There are two basic types of wiring installation used in aircraft: open wiring and wiring in a conduit. In open wiring, the wires are installed without any type of protective covering. It is fast

A — ARMAMENT
B — PHOTOGRAPHIC
C — CONTROL SURFACE
 CA — AUTOMATIC PILOT
 CC — WING FLAPS
 CD — ELEVATOR TRIM
D — INSTRUMENT (OTHER THAN FLIGHT OR ENGINE
  INSTRUMENT)
 DA — AMMETER
 DB — FLAP POSITION INDICATOR
 DC — CLOCK
 DD — VOLTMETER
 DE — OUTSIDE AIR TEMPERATURE
 DF — FLIGHT HOUR METER
E — ENGINE INSTRUMENT
 EA — CARBURETOR AIR TEMPERATURE
 EB — FUEL QUANTITY GAGE AND TRANSMITTER
 EC — CYLINDER HEAD TEMPERATURE
 ED — OIL PRESSURE
 EE — OIL TEMPERATURE
 EF — FUEL PRESSURE
 EG — TACHOMETER
 EH — TORQUE INDICATOR
 EJ — INSTRUMENT CLUSTER
F — FLIGHT INSTRUMENT
 FA — BANK AND TURN
 FB — PITOT STATIC TUBE HEATER AND STALL WARNING
  HEATER
 FC — STALL WARNING
 FD — SPEED CONTROL SYSTEM
 FE — INDICATOR LIGHTS
G — LANDING GEAR
 GA — ACTUATOR
 GB — RETRACTION
 GC — WARNING DEVICE (HORN)
 GD — LIGHT SWITCHES
 GE — INDICATOR LIGHTS
H — HEATING, VENTILATING AND DEICING
 HA — ANTI-ICING
 HB — CABIN HEATER
 HC — CIGAR LIGHTER
 HD — DEICING
 HE — AIR CONDITIONERS
 HF — CABIN VENTILATION
J — IGNITION
 JA — MAGNETO
K — ENGINE CONTROL
 KA — STARTER CONTROL
 KB — PROPELLER SYNCHRONIZER
L — LIGHTING
 LA — CABIN

LB — INSTRUMENT
LC — LANDING
LD — NAVIGATION
LE — TAXI
LF — ROTATING BEACON
LG — RADIO
LH — DEICE
LJ — FUEL SELECTOR
LK — TAIL FLOODLIGHT
M — MISCELLANEOUS
 MA — COWL FLAPS
 MB — ELECTRICALLY OPERATED SEATS
 MC — SMOKE GENERATOR
 MD — SPRAY EQUIPMENT
 ME — CABIN PRESSURIZATION EQUIPMENT
 MF — CHEM O_2 — INDICATOR
P — DC POWER
 PA — BATTERY CIRCUIT
 PB — GENERATOR CIRCUITS
 PC — EXTERNAL POWER SOURCE
Q — FUEL AND OIL
 QA — AUXILIARY FUEL PUMP
 QB — OIL DILUTION
 QC — ENGINE PRIMER
 QD — MAIN FUEL PUMPS
 QE — FUEL VALVES
R — RADIO (NAVIGATION AND COMMUNICATIONS)
 RA — INSTRUMENT LANDING
 RB — COMMAND
 RC — RADIO DIRECTION FINDING
 RD — VHF
 RE — HOMING
 RF — MARKER BEACON
 RG — NAVIGATION
 RH — HIGH FREQUENCY
 RJ — INTERPHONE
 RK — UHF
 RL — LOW FREQUENCY
 RM — FREQUENCY MODULATION
 RP — AUDIO SYSTEM AND AUDIO AMPLIFIER
 RR — DISTANCE MEASURING EQUIPMENT (DME)
 RS — AIRBORNE PUBLIC ADDRESS SYSTEM
S — RADAR
U — MISCELLANEOUS ELECTRONIC
 UA — IDENTIFICATION — FRIEND OR FOE
W — WARNING AND EMERGENCY
 WA — FLARE RELEASE
 WB — CHIP DETECTOR
 WC — FIRE DETECTION SYSTEM
X — AC POWER

Fig. 1A-9 Circuit function and specific circuit code letter identification for electric wiring.

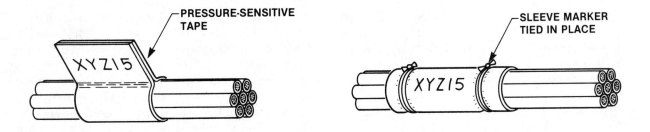

Fig. 1A-10 Methods of identifying wire bundles in an aircraft electrical installation

and easy to install and it makes servicing of the wiring simple, but it has a definite disadvantage in wheel wells and engine nacelles as it provides no protection from abrasion or from heat.

Polyvinyl chloride tubing, which is usually called spaghetti, is often used to enclose a bundle of wires to protect them from abrasion and moisture. This is not considered as conduit and it must be used with caution where the wires are subjected to a great deal of hard usage.

Wire bundles, whether run in the open or inside tubing, should be tied together every three to four inches with a double wrap of waxed linen or nylon lacing cord, secured with a clove hitch and a square knot. The bundles should be securely clamped to the structure, using clamps lined with a non-metallic cushion.

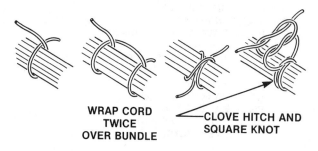

WRAP CORD TWICE OVER BUNDLE

CLOVE HITCH AND SQUARE KNOT

Fig. 1A-11 Method of making a spot tie on an electrical wire bundle

The bundles should be supported from the structure of the aircraft and not from any fluid line. If a wire bundle is to run parallel to any line carrying a liquid fluid such as fuel, oil, or hydraulic fluid, the wire bundle should be routed above the fluid line so that a leak will not drip fluid into the wires. If it is at all possible, the wires should be at least six inches above the fluid lines.

Wire bundles should not be run any closer than three inches to any control cable, and if there is any possibility that the wire could ever touch the cable, some form of mechanical guard must be installed that will keep the wire bundle and the cable separated.

When the wire bundles are to pass through a hole in a bulkhead in the aircraft structure, the bundle should be clamped to a bracket to hold it centered in the hole. If there is less than 1/4-inch

between the wire bundle and the edge of the hole, the hole should have a rubber grommet installed, to protect the wire from being cut if it should contact the metal.

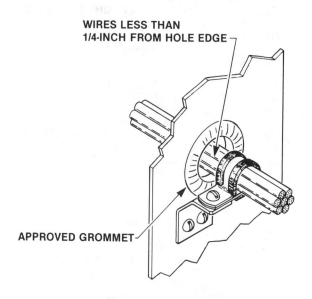

WIRES LESS THAN 1/4-INCH FROM HOLE EDGE

APPROVED GROMMET

Fig. 1A-12 When wire bundles must pass through a bulkhead or frame, they must be supported with a cushion clamp, and a grommet must be installed in the hole to protect the wires from chafing.

When making up a wire bundle, first connect one end of the wires to the terminal strips or into the connector; then arrange the wires into the neatest and most compact bundle you can form. Be generous with the use of spot ties as you form the bundle. When all of the wires are in the bundle, use a spot tie every three to four inches along the run of the bundle. Keep all of the wires parallel by using a plastic comb such as the one in Fig. 1A-13 on page 9.

If the bundle is to be encased in polyvinyl chloride tubing, you must, of course, put the bundle in the tubing before the final end is connected. New tubing resists the wire bundle sliding through it and you can make the job easier if you blow a spoonful of talcum through the tube to help the wire slide through. One easy way of putting a long wire bundle through vinyl tubing is to blow one end of a piece of lacing cord through the tube with compressed air and then tie the end of the bundle to the lacing cord and pull it through the tubing.

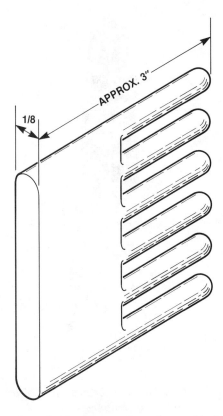

1/8

APPROX. 3"

Fig. 1A-13 A plastic comb such as this is a handy tool for keeping the wires straight when making up a wire bundle.

Some wire bundles are laced, rather than secured with spot ties. But, lacing is not a good practice for wires installed in the aircraft structure and especially in the powerplant area, since a break in the lacing will loosen the bundle. Individual spot ties made with lacing cord or with nylon straps such as the patented TY-RAP are much better than lacing for holding wire bundles together.

Wire bundles may be enclosed in either rigid or flexible metal conduit when they pass through an area where they are likely to be chafed or crushed.

The inside diameter of the conduit must be about 25% larger than the the maximum diameter of the wire bundle it encloses, and when you are figuring the size of the conduit to use for a wire bundle, you must remember that the nominal diameter of a conduit is its outside diameter; so you must subtract twice the wall thickness of the conduit from its outside diameter in order to find its inside diameter.

The conduit must be supported by clamps from the aircraft structure, and there must be a 1/8-inch drain hole at the lowest point in each run of the conduit, so that any moisture that condenses inside the conduit can drain out.

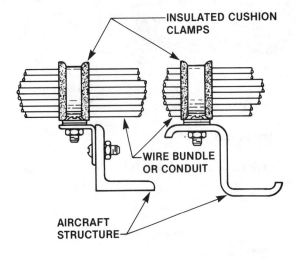

INSULATED CUSHION CLAMPS

WIRE BUNDLE OR CONDUIT

AIRCRAFT STRUCTURE

Fig. 1A-14 Methods of supporting a wire bundle from the aircraft structure

When the conduit is made up, all sharp edges and burrs must be removed and all of the bends must be made using a bend radius that will not cause the tube to kink, to wrinkle or to be excessively flattened.

C. Wiring Termination

1. Terminals

When wires are to be attached to studs on a terminal strip, they are terminated with a crimp-on-type solderless terminal. In the smaller sizes—that is for wires up through ten-gage—these terminals are pre-insulated with the color of the insulator identifying the size of wire they fit. Terminals with red insulation are used on wire sizes 22 through 18, blue insulation is used on terminals for 16- and 14-gage wire, and yellow terminals fit 12- and 10-gage wire. Wires larger than 10-gage use uninsulated terminals, and after the terminal is crimped in place, a piece of vinyl tubing is slid over the barrel of the terminal, and it is tied in place with lacing cord.

To install a terminal, the insulation on the wire is stripped back enough to allow the end of the wire to extend through the barrel of the terminal when the end of the insulation is against

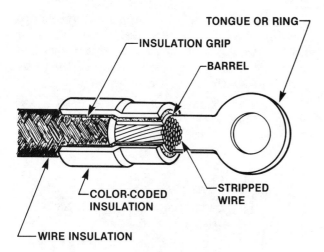

Fig. 1A-15 Pre-insulated solderless terminals for aircraft electrical wire

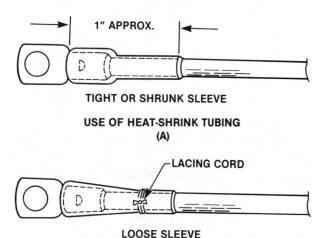

TIGHT OR SHRUNK SLEEVE

USE OF HEAT-SHRINK TUBING
(A)

LOOSE SLEEVE

INSULATION WITH VINYL TUBING HELD IN PLACE WITH A SPOT TIE OF LACING CORD
(B)

Fig. 1A-16 Methods of insulating solderless terminals on an electrical wire

the barrel. When the terminal is crimped with the proper tool, the joint between the wire and the terminal will be as strong as the wire itself.

There are two basic types of hand-operated crimping tools. The small type shown in Fig. 1A-17 is used only for occasional repair work, but the ratchet type is preferred, as it assures that all crimps will have the proper amount of pressure applied.

The handles on the ratchet-type crimping tool will not release until the jaws have moved close enough together to properly compress the ter-

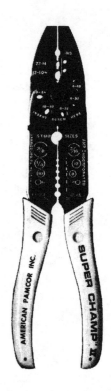

Fig. 1A-17 Combination crimping tool and wire stripper for installing solderless terminals on aircraft electrical wire

minal barrel. Only ratchet-type crimping tools are used in aircraft production shops, and these tools are periodically calibrated to assure that every terminal installed with them will be properly compressed.

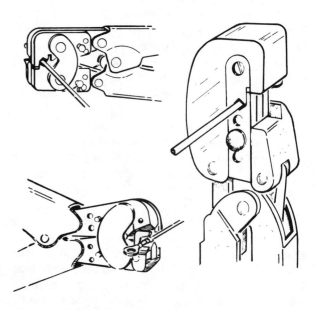

Fig. 1A-18 Heavy-duty crimping tool for installing preinsulated solderless terminals on aircraft wires.

Terminals on wires of 0-gage through 0000-gage are usually installed with pneumatic squeezers that exert enough pressure on the terminal barrel to produce a joint that is as strong as the wire itself.

Aluminum wire presents some special problems when installing the terminals. Aluminum wire oxidizes when it is in contact with the air, and the oxide is an electrical insulator. To assure a good electrical connection between the wire and the terminal, the inside of the aluminum terminal is partially filled with a compound of petrolatum and zinc dust that will grind the oxide film from the wire when the terminal is compressed. The

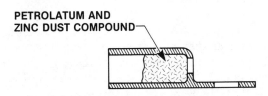

THE TERMINAL IS PARTIALLY FILLED WITH A COMPOUND OF ZINC DUST AND PETROLATUM.

(A)

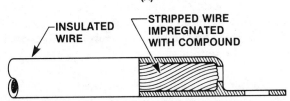

THE INSULATION IS STRIPPED FROM THE END OF THE WIRE, AND THE BARE PORTION OF THE WIRE IS INSERTED INTO THE TERMINAL.

(B)

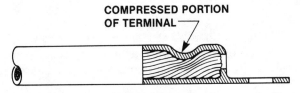

THE TERMINAL IS COMPRESSED WITH A PNEUMATIC SQUEEZER.

(C)

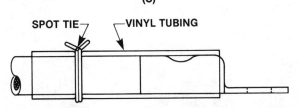

A PIECE OF VINYL TUBING IS TIED IN PLACE OVER THE TERMINAL.

(D)

Fig. 1A-19 Installation of a wire terminal on large-gage aluminum wire

petrolatum will then cover the aluminum and prevent the oxygen in the air from causing new oxides to form.

2. Terminal strips

Most of the terminal strips in an aircraft electrical system are of the barrier type and are made of a strong paper-base phenolic compound. The smallest terminal stud allowed for electrical power systems is a number ten, and many of the circuits in the powerplant portion of the electrical system are electrical power circuits.

When attaching the terminal end of wires to the terminal strips, fan the wires out from the bundles so they will align with the terminal studs and stack the terminals on the studs as shown in Fig. 1A-20. Notice that the ring is not in the center of the terminal, but is to one side of the wire so there is a correct way of stacking them.

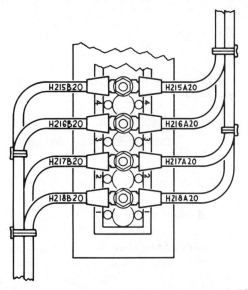

METHOD OF FANNING WIRES FROM A BUNDLE TO ATTACH TO A TERMINAL STRIP

(A)

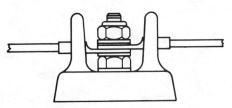

METHOD OF STACKING TWO TERMINALS ON A SINGLE LUG ON A TERMINAL STRIP

(B)

Fig. 1A-20 Electrical wires attached to a terminal strip

There should never be more than four terminals on any one stud. If it is necessary to connect more than four wires to one point, you can join two or more studs with cadmium-plated copper bus straps and use enough studs to connect all of the wires.

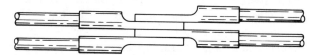

METHOD OF STACKING FOUR WIRES ON A SINGLE LUG ON A TERMINAL STRIP.

(A)

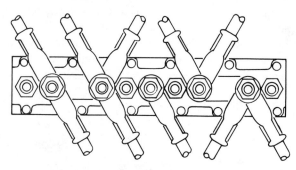

BUS STRAPS MAY BE USED TO JOIN TWO LUGS ON A TERMINAL STRIP.

(B)

Fig. 1A-21   Wires stacked on a terminal strip

3. Junction boxes

Most of the junction boxes installed in the powerplant section of an aircraft are made of aluminum alloy or of stainless steel and are mounted in such a way that they minimize the possibility of water getting into the box and causing elec-trical shorts or corrosion. Most of the junction boxes in the powerplant area are mounted vertically so that any washers or nuts dropped during servicing will not fall under the wires where they could cause a short and maybe even a fire.

The power should always be off of the aircraft when you are working in an electrical junction box, but since this is not always possible, you must use extreme care that you do not short-circuit any of the terminals with your tools. Be sure to remove your rings and wrist watch, as it is possible to short accidentally between two terminals and get a severe burn.

4. MS connectors

Wiring used to connect portions of an aircraft electrical system that are not removed for servicing is normally done with wires connected to the studs on a terminal strip. But if the equipment is likely to be removed for servicing, it is connected with one of the AN or MS connectors. The individual wires are soldered into the pots in the ends of the contacts, or in some of the newer type of connectors, tapered pins are crimped onto the end of the wires, and the pins are wedged into tapered holes in the end of the contact.

To install a soldered connector to a wire bundle, first disassemble the connector and slip the wire clamp, the coupling ring, and assembly nut over the bundle. These components, naturally, cannot be installed after the connector is soldered in place.

Strip the insulation from about one-half inch of the end of the wire with a wire stripper. Do not use a knife to strip the insulation from the wire

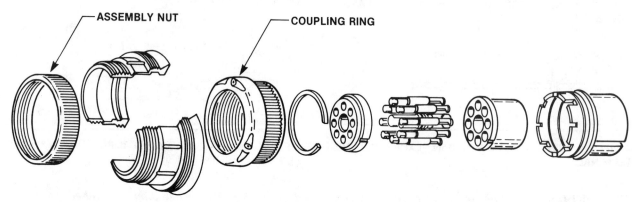

Fig. 1A-22   An exploded view of a typical MS connector

because of the danger of nicking some of the strands; nicked strands are sure to break and increase the resistance of the joint.

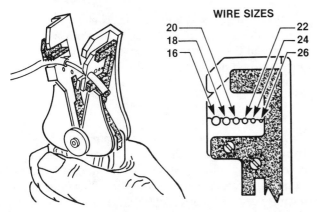

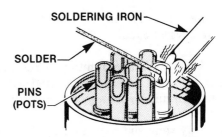

THE SOLDER POT IS HEATED AND FILLED WITH SOLDER.
(A)

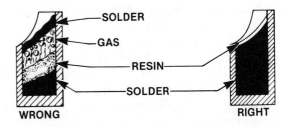

IF THE POT IS NOT HEATED SUFFICIENTLY, RESIN WILL BE ENTRAPPED IN THE SOLDER.
(B)

Fig. 1A-23 This type of wire stripper will not nick the strands of wire when the insulation is removed.

Slip a piece of vinyl tubing about one inch long over the end of the wire and solder the wire into the solder pot. To properly solder the wire in the pot, first fill the pot with solder and then heat it with a soldering iron until the solder melts. Then insert the wire into the molten metal. Remove the soldering iron and hold the wire still until the solder solidifies—the solder is hard when it loses its shiny appearance and looks frosted. If the wire is moved before the solder solidifies, it will have a granular appearance and the joint will have a high resistance.

When all of the wires are soldered into the pots, inspect the connector for the condition of the solder, looking especially for any solder that may bridge any of the contacts or for too much or too little solder. The solder in a properly soldered connection should completely fill the pot and have a slightly rounded top. The solder must not wick up into the strands of the wire, and there must be about 1/32-inch of bare wire between the top of the solder pot and the end of the insulation. Slip the vinyl tubing down over the solder pot and tie the wires together with a spot tie of lacing cord to prevent the insulation tubing slipping up and leaving the contacts uninsulated. Reassemble the connector and clamp the wires in the wire clamp. (See Fig. 1A-24.)

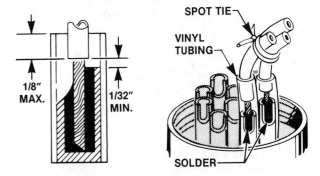

THERE SHOULD BE AT LEAST 1/32-INCH OF BARE WIRE BETWEEN THE TOP OF THE SOLDER AND THE END OF THE INSULATION. NO SOLDER SHOULD WICK UP INTO THE STRANDS OF THIS WIRE.
(C)

Fig. 1A-24 Proper method of attaching wires into the solder pots on an MS connector

D. Wire Bundle Routing and Clamping

The wiring harnesses that are installed in an aircraft at the factory are made on a jig board, and they are preformed with all of the bends they will have when they are installed. This provides the neatest and most secure installation. If a number of wires in a bundle have been damaged and must be repaired, it is usually a good idea to order a complete new harness from the factory and install it in exactly the same way the damaged harness was installed. But, if you are making an original installation, there are some considerations you should follow:

(1) If possible, limit the number of wires in a single bundle. Since it is possible for a single fault to damage all of the wires in a bundle, it is better

13

to use more than one bundle than to risk a fault disabling several circuits.

(2) Do not bundle ignition wires, shielded wires, or wires that are not protected with either a fuse or a circuit breaker with other wires.

(3) Any time you have extra pins or sockets in a connector, put spare wires in them, because if one wire becomes damaged, it is much easier to switch from a damaged wire to one of the spares than it is to run a new wire in the bundle. Use a wire of the maximum size the contact will support, and number it with the number of the contact to which it is connected. Be sure to label it as a spare and include this wire on the wiring diagram for the installation, showing that a spare wire is available.

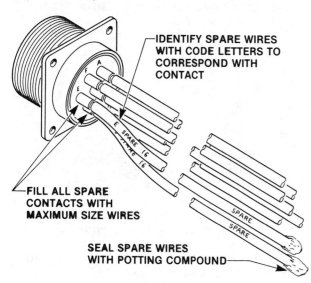

Fig. 1A-25 *If there are spare contacts in the connector, solder spare wires into them and route the spares with the bundle. Be sure to note the presence of these spares in the wiring diagram.*

(4) Clamp the bundle to the aircraft structure using insulated cushion clamps, and when the bundle must make a bend, be sure to use a bend radius that does not cause the wires on the inside of the bend to bunch up. A good rule of thumb is to use a bend radius of about ten times the diameter of the bundle.

(5) The bundle should be run tight enough that there is not more than about one-half inch of slack between the support points when normal hand pressure is put on the wires between the supports. But there must be sufficient slack after the last support that will allow you to replace the terminal if need be, and enough slack to prevent vibration or shifting of the equipment putting strain on the wires.

(6) Wiring installations in the engine compartment are often subjected to high temperatures, and any wires that pass through hot, or potentially hot, areas should have high-temperature insulation.

E. Bonding and Shielding

Because of the large amount of radio and electronic equipment carried in a modern airplane, it is important that all of the electromagnetic fields radiating from any motors or wires carrying alternating current be intercepted and grounded. And, the electrical charges that build up on any components must be carried to the main aircraft ground before they become high enough to cause a spark as they pass from the component to ground.

1. Bonding

There are two basic ways to prevent electromagnetic interference; bonding and shielding. Bonding consists of electrically connecting all components in the aircraft together so there will be no insulated components on which electrical charges can build up. In the airframe section, all of the control surfaces are connected to the main aircraft structure with braided bonding straps so the resistance of the control hinge will not insulate the surface from the main structure. In the powerplant portion of an aircraft, we must bond all of the shock-mounted components to the main structure with bonding straps.

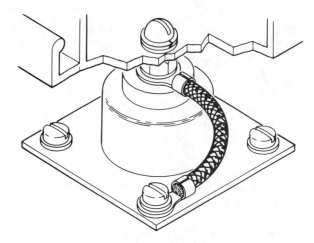

Fig. 1A-26 *A bonding strap of adequate electrical capacity should be used to jump any insulated shock mount.*

Any electrical component that uses the aircraft structure as the return path for its current must be bonded to the structure. Use bonding straps that are large enough to carry all of the return current without producing a voltage drop in excess of that allowed by the aircraft manufacturer. A rule of thumb that will usually keep the bonding straps large enough is to be sure that the resistance between any component and the aircraft structure is no more than three milliohms (0.003 ohms).

The bonding braid must be long enough that it will not interfere with the free movement of the component, and it must be of a material that will not produce an electrode potenetial difference with the structure that could cause corrosion. Aluminum alloy jumpers are usually used to bond aluminum alloy components together, and cadmium-plated copper jumpers are used for steel components.

Aircraft engines are mounted in rubber shock mounts, and a large flow of current from the starter must return through the engine mount to the main aircraft structure. Be sure there is a heavy bonding strap between the engine side of the mount and the airframe side. If this strap is missing or broken, there is a danger of fire, as fluid lines such as the fuel lines and primer lines must carry all of the return current from the starter.

2. Shielding

Shielding is used in an aircraft electrical system to intercept the radiated electrical energy, from either wires carrying alternating current or wires in the ignition system that have radio frequency energy from the spark superimposed over the direct or alternating current they carry.

Shielding is a braid of tin-plated or cadmium-plated copper wire around the outside of the insulation. This braid is connected to ground through a crimped-on ground lead.

Shielding is used to prevent electromagnetic radiation from the wire, but in the wiring for some of the sensitive electronic components, the wires are shielded to keep electromagnetic radiation out of them. The shielding used on most of the components in the powerplant system of an aircraft may be grounded at both ends, but the shielding on the wires in some electronic circuits are grounded at only one end to prevent current flowing in the shielding and causing electromagnetic interference.

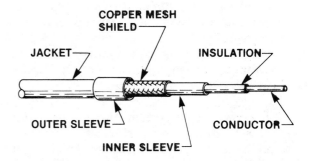

PREPARATION OF THE WIRE FOR THE ATTACHMENT OF THE GROUND WIRE.

(A)

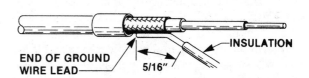

THE END OF THE GROUND WIRE IS SLIPPED BETWEEN THE INNER AND THE OUTER SLEEVE, IN DIRECT CONTACT WITH THE SHIELDING.

(B)

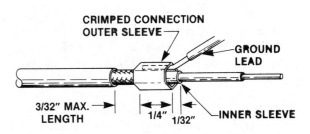

THE SLEEVES ARE CRIMPED WITH THE PROPER CRIMPING TOOL TO HOLD THE GROUND LEAD.

(C)

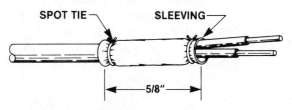

THE FINISHED CONNECTION IS INSULATED WITH A PIECE OF VINYL TUBING HELD IN PLACE WITH SPOT TIES OF LACING CORD.

(D)

Fig. 1A-27 Attachment of ground lead to a shielded wire, using a crimped-on sleeve.

II. ELECTRICAL SYSTEM REQUIREMENTS

One of the most important functions of an A&P technician is to maintain aircraft in an airworthy condition, and this basically amounts to inspecting and repairing the aircraft so it will continue to meet the requirements for certification under which it was manufactured.

The powerplant portion of the electrical system is primarily that portion of the system that produces the electrical energy that is used throughout the aircraft. And in order to know that this portion of the system remains airworthy, we have two documents we must refer to: Federal Aviation Regulations Part 23, which includes the certification requirements for most of the aircraft in the general aviation fleet, and Advisory Circular 43.13-1A, which contains approved methods of repairing certificated aircraft so they will retain their airworthiness. Both of these documents and the maintenance manual provided by the aircraft manufacturer should be consulted before any repair is made to the aircraft electrical system.

A. Generating System

The electrical power generating system for a certificated aircraft must have enough output capacity to adequately supply all of the electrical load that can be imposed upon it.

If it is probable that the electrical load can exceed the current rating of the generator or alternator, the flight crew must have some way to reduce the electrical load and keep it within the capacity of the generator.

When the aircraft battery is part of the electrical system, the generator or alternator must be able to maintain the battery in its fully charged condition except for temporary conditions in which heavy loads such as the engine starters or electrical landing gear loads take current from the battery.

There are several ways to prevent overloading the electrical system:

(1) Placards may be used to inform the pilot or flight engineer of the electrical loads that may be connected to the system at any one time.

(2) A load meter, calibrated in percent of the generator output, may be installed between the generator and the electrical bus to show the flight crew the portion of the generator output that is being used.

(3) There may be an ammeter between the battery and the system bus to show the flight crew when the battery is supplying current to the electrical system.

If the use of placards or monitoring the generator output is not practical, the electrical load can be limited to 80% of the rated generator output.

When two or more generators share the electrical load, the flight crew must be able to quickly disconnect enough of the electrical load from the system, if one of the generators fails, to operate within the output of one generator.

It is important that the flight crew be provided with a list of all of the electrical equipment installed in the aircraft and the amount of current each piece uses, so that in the event of the loss of a generator, they will know what equipment they can operate with only one generator. When preparing the electrical load chart, be sure that the aircraft battery is fully charged and the generator is producing the correct output voltage. Use an accurate ammeter and record the current drawn by each individual piece of equipment.

B. Circuit Protection

1. Fuses

FAR Part 23 requires that all aircraft certified under this part use some form of circuit protective device. These devices should be in all of the electrical circuits except the main circuit of starter motors and any circuit in which "no hazard is presented by their omission." Many aircraft use fuses for this type of circuit protection.

A fuse is a strip of metal of a specific size and shape and of a material that will melt in two when a specific amount of current flows through it. This strip of metal is enclosed in a glass tube with a metal cap on either end to make contact in the circuit. When excessive current flows, the metal link melts and opens the circuit. The circuit cannot be restored until the blown fuse is removed and replaced with a new one.

Some electrical loads such as electric motors and incandescent lamps have a high inrush of current when the switch is first closed, and then the current drops to a much lower operating level. To prevent this high inrush of current opening the circuit, a slow-blow fuse may be used. A slow-blow fuse has a fusible link that will carry the high initial current without melting, but if there is a sustained flow of current in excess of the operating current, the link will overheat and a coil spring will pull the link apart.

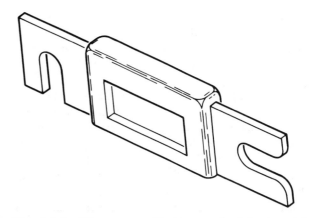

Fig. 2A-2 The current limiter used with some of the larger high-output DC generators is a special slow-blow fuse. It is usually mounted in the engine nacelle near the generator.

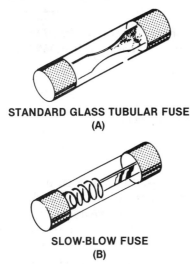

STANDARD GLASS TUBULAR FUSE
(A)

SLOW-BLOW FUSE
(B)

Fig. 2A-1 Typical aircraft fuses

Almost all fuses used in an aircraft electrical system are of the glass tubular type and are installed in panel-mounted screw-in or push-in-and-turn type holders. The current rating of the fuse is marked on the end cap, and the rating for the fuse required for the circuit must be marked on or adjacent to the fuse holder so the person replacing the fuse will know the size to install.

Federal Aviation Regulations Part 91, General Operating Rules, requires that all aircraft using fuses as the circuit protective devices carry "one spare set of fuses, or three spare fuses of each kind required."

Many of the large reciprocating-engine-powered aircraft do not have an automatic current limiter in their generator control system, but use a slow-blow fuse-type current limiter, such as the one in Fig. 2A-2, mounted in the engine nacelle. If the generator is forced to carry a load in excess of its rating, the current limiter fuse will blow and open the circuit. This device is not replaceable in flight.

2. Circuit breakers

Most modern aircraft use circuit breakers rather than fuses to protect their electrical circuits, because the circuit can be restored without having to remove and replace a component.

FAR Part 23 specifies that each resettable circuit protective device require a manual operation to restore service after the device has interrupted the circuit. This requirement prevents the use of automatic reset circuit breakers, such as are used in some commercial electric motors that open the circuit when they overheat and automatically restore the circuit when they cool down.

A second requirement is that the circuit protection device open the circuit regardless of the position of the operating control. This requirement is met by what is called "trip-free" circuit breakers that cannot override an open circuit. This type of circuit breaker makes it impossible to manually hold the circuit closed when excessive current is flowing.

There are three basic types of circuit breakers used in aircraft electrical systems. The push-to-reset type of circuit breaker cannot be used as a switch. It is normally in, and only when a circuit is overloaded does the breaker pop out. It may be pushed back in to restore the circuit.

The push-pull-type circuit breaker can be used as a switch as well as a circuit breaker. The knob is large enough to pull it out to open the circuit, and pushing the knob in restores the circuit.

When the circuit is opened by an excess of current, the knob will pop out enough that a white band is visible and is easy to see, even in a darkened cockpit.

The toggle-type circuit breaker is used in some of the higher current circuits, and it serves as a switch as well as a circuit breaker. When the circuit is opened by an overload of current, the handle will move partially toward the open circuit position. To restore the circuit, the handle must be moved all of the way to the open-circuit position, and then back to the closed-circuit position.

PUSH TO-RESET-TYPE CIRCUIT BREAKER
(A)

PUSH-PULL-TYPE CIRCUIT BREAKER
(B)

TOGGLE-TYPE CIRCUIT BREAKER
(C)

Fig. 2A-3 Aircraft circuit breakers

The circuit protection devices are installed primarily to protect the wiring, and the rating of the device is based on the amount of current that can be carried without overheating the insulation. The chart in Fig. 2A-4 gives the size of both circuit breakers and fuses that are recommended to protect wiring routed in a bundle.

COPPER WIRE AWG SIZE	CIRCUIT BREAKER AMP	FUSE AMP
22	5	5
20	7.5	5
18	10	10
16	15	10
14	20	15
12	25	20
10	35	30
8	50	50
6	80	70
4	100	70
2	125	100
1		150
0		150

Fig. 2A-4 Recommended circuit protection devices for various sizes of MIL-W-5086 copper wire

C. Control Devices

1. Switches

Toggle switches have been used for years as the standard type of switch for use in aircraft cockpits, but in recent years the rocker switch is becoming more accepted as the standard, and electrically the two types of switches are the same. When the switch is closed, the two contacts are held together to complete the circuit, and when the operating control, the toggle or the rocker, is moved toward the open position, the contacts snap apart.

Switches are rated according to both the voltage and the current they can control, and a typical aircraft switch may be rated at five amps 125 volts, or 35 amps at 24 volts. When the switch is used in a circuit that has a high inrush of current, or it the load it controls is inductive, the switch must be derated so that it can carry all of the load without damage.

Incandescent lamps and electric heaters have a very low resistance when they are cold, and the current is high when the switch is first closed; but as soon as the current heats up the filament or the heater element, the resistance goes up and the current drops. When a switch is installed in a DC lamp circuit or a heater circuit, it must be derated by the factors shown in Fig. 2A-5.

Electric motors, relays, and solenoids all have coils or wire in them, and the inductance of these coils tries to keep the current flowing as the switch contacts are separating. To prevent this induced current causing arcing across the contacts as they open, the switch must be derated, that is, the amount of current it can carry must be decreased.

TYPE OF LOAD	DC VOLTAGE	DERATING FACTOR
LAMP	24 12	8 5
RELAY OR SOLENOID	24 12	4 2
HEATER	24 12	2 1
MOTOR	24 12	3 2

Fig. 2A-5   Derating factors for aircraft switches

2.   Relays and solenoids

The powerplant electrical systems uses many relays and solenoids to control the heavy current that is used in many of the powerplant circuits.

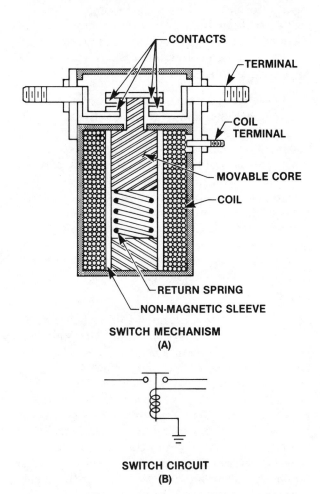

Fig. 2A-6   Typical solenoid switch

Relays and solenoids are very similar in their function, but their operating principles are somewhat different. A solenoid is a type of magnetically operated switch that uses a movable iron core such as we see in Fig. 2A-6. When the coil is energized, it produces a magnetic field that pulls the movable core into the center of the coil. This closes the contacts and completes the electrical circuit between the terminal. When the coil is de-energized, the return spring forces the contacts apart and breaks the electrical circuit.

A relay is similar to a solenoid except that the core is stationary. When the coil is energized, the armature is pulled down toward the coil core and the contacts complete the circuit between the terminals.

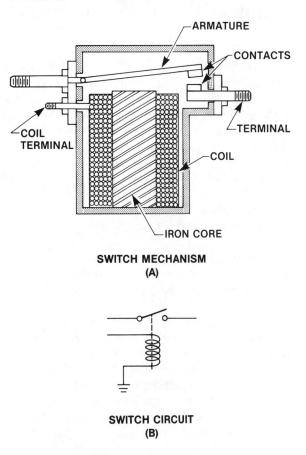

Fig. 2A-7   Typical relay-type switch

When replacing a relay or a solenoid in an aircraft circuit, be sure to use the particular make and model specified in the aircraft manufacturer's service and parts manual, because while many of the units look much alike, they may be quite different in their electrical and mechanical characteristics.

Section 5-II

Powerplant Instrument Systems

I. THE EVOLUTION OF POWERPLANT INSTRUMENTS

The first airplanes that flew had no powerplant instruments. The engines ran for so short a time that instruments were not needed. As the engines became more dependable, tachometers were used to give the pilot an idea of the power the engine was developing, and oil pressure gages were carried to warn of an impending engine failure. Later, oil or water temperature gages were added, and this group became the standard engine instrumentation. Even today these are all that are required in our simplest airplanes.

At the peak of their development, the large reciprocating engine-powered airliners had a large instrument panel full of engine-monitoring instruments that the flight engineer had to watch during flight. But turbine engines do not require as many instruments as a reciprocating engine, and the flight engineer's panel on a modern jet transport is much smaller than that for a reciprocating engine airplane. Much of the panel is taken up with the controls for the electrical system and the environmental control system, and the instruments relating directly to the powerplant are few.

Powerplant instrumentation did not change a great deal in the years between 1920 and 1970. It is true that as airplanes became larger, more of the direct-reading instruments were replaced with remote-indicating electrical instruments that merely transmitted the position of the pointer shaft from the instruments located in the engine compartment and displayed it on a standard instrument dial in the cockpit.

Today, powerplant instrumentation is experiencing a quantum leap in technology, with mechanical systems being replaced with digital electronic systems. Round instruments are being replaced with vertical instruments or with instruments having digital readouts. And transducers (devices that convert mechanical movement into electrical signals) having no moving parts feed their information into computers that are made primarily of solid-state integrated circuit chips. These computers display their output in discrete numbers or in the form of a bar graph.

We have mentioned digital electronics in the Communications and Navigation Systems portion of this Integrated Training Program, but we should emphasize here that digital electronics must not be confused with *digital display instruments*. A digital display instrument is one that presents its information in the form of discrete numbers rather than by the position of a pointer over a dial. Pointer and dial instruments give us an analogy of the pressure, temperature, or whatever other information the instrument is giving us, and these are called analog instruments.

Digital *electronics*, on the other hand, is the branch of electronics that deals with only two electrical conditions in a circuit; on or off or high or low. Pressure, temperature, mechanical movement, or any other parameter that must be measured is converted by a transducer into a series of these two conditions and this string of bits, as they are called, is carried into a computer. Here the bits are processed and then directed into the indicator, where they operate a display of either liquid crystals or light emitting diodes (LED's) to give the pilot the information the transducer sensed.

Fig. 1B-1   Powerplant instrument group for a twin-engine general aviation airplane

The accuracy, sensitivity, dependability, and low cost of digital electronic instrumentation will make this the main form of engine instruments that will be used in the very near future.

II. PRESSURE MEASURING INSTRUMENTS

A. Types of Pressure

There are several ways we can classify aircraft instruments. They can be classified according to their system and, in this regard, all of the instruments discussed in this book are powerplant instruments. We can also classify them according to their operating mechanism, but in this text we will classify them according to the parameters they measure. And, because we will use this classification, all of the instruments that measure pressure will be discussed together.

Pressure is a measure of the amount of force acting on a given unit of area, and in order to correctly define pressure, we must specify the point from where it is measured, or referenced. Absolute pressure is measured from zero pressure, or a vacuum. Gage pressure is measured from the existing atmospheric pressure and is actually the amount of pressure that is added to a fluid by a pump. Differential pressure is the difference between two pressures.

1. Absolute pressure

The most familiar absolute pressure gage used in aircraft instrumentation is the altimeter.

The altitude flown by an airplane is indicated by an absolute pressure gage that measures the pressure caused by the weight of the air above the instrument. An adjustable scale calibrated in either inches of mercury or millibars allows the pilot to set the reference from which the pressure measurement is made. If the pressure is measured from the existing sea level pressure, the altitude is called "indicated altitude." If the altimeter is adjusted to the standard sea level pressure of 29.92 inches of mercury or 1013.2 millibars, the altitude is called "pressure altitude." The pointers on the instrument indicate the number of feet above the reference the instrument is located.

Manifold pressure gage

The most widely used absolute pressure gage in powerplant instrumentation is the manifold pressure gage used with reciprocating engines. This gage, almost always calibrated in inches of mercury, shows the pilot the amount of absolute pressure inside the intake manifold of the engine. This is the amount of pressure available to push the fuel-air mixture into the cylinders.

The amount of power developed by a reciprocating engine is determined by the RPM of the engine and the amount of pressure pushing down on the piston inside the cylinder. There is no practical way to measure the pressure that is actually inside the cylinder during the power stroke, but fortunately there is a definite relationship between the absolute pressure of the air pushing the mixture into the cylinder and the pressure that is developed when the mixture burns.

The tachometer shows the RPM of the engine which is directly related to the number of power strokes per minute, and the manifold pressure is related to the cylinder pressure. And so, by knowing and being able to control these two variables, the pilot can adjust the engine controls to produce the power he needs.

Most of the manifold pressure gages installed with normally aspirated engines use a simple evacuated phosphor bronze capsule, called an aneroid capsule, to measure the absolute pressure. The case of the instrument is sealed airtight and is connected to the intake manifold with a piece of aluminum tubing or flexible instrument hose. Pressure in the instrument case below atmospheric pressure allows the capsule to expand, and this expansion is carried by the link to the

rocking shaft which rotates the sector gear. This in turn drives the pinion gear. The pointer pressed onto the pinion gear shaft moves over the dial on which the scale is marked.

DIAL FOR A DUAL MANIFOLD PRESSURE GAGE FOR A TURBOCHARGED ENGINE

(A)

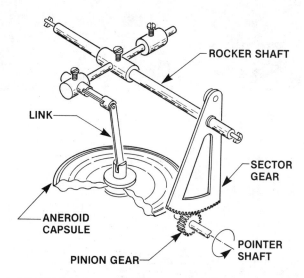

MECHANISM OF A MANIFOLD PRESSURE GAGE USING AN EVACUATED ANEROID CAPSULE

(B)

Fig. 2B-1   Manifold pressure gage

The manifold pressure gage used with normally aspirated engines usually has a range of from about ten inches of mercury up to slightly above 30 inches. Instruments for turbocharged engines have a range in excess of 40 inches. Most of the World War II vintage military and transport airplanes used manifold pressure gages that measured up to 70 inches of mercury, and many of the fighters had instruments with an upper range of 110 inches of mercury.

One of the problems with manifold pressure systems is the accumulation of moisture in the instrument line, and some of the larger aircraft have a purge valve connected into the line at the back of the instrument, with a knob sticking through the instrument panel near the manifold pressure gage. The pilot can purge the line of moisture by depressing the knob when the engine is idling and the manifold pressure is below atmospheric pressure. Pressing the knob pushes the ball off of its seat so that atmospheric pressure can enter the manifold pressure line and the low pressure inside the induction system will draw the water into the engine. Holding the valve open for just a few seconds will purge the line of any moisture.

Some of the smaller aircraft have a calibrated hole in the manifold pressure gage line that allows a flow of air into the intake manifold to keep the line purged of moisture. This constitutes a manifold air leak, but it is a calibrated leak, and the system is designed to compensate for it.

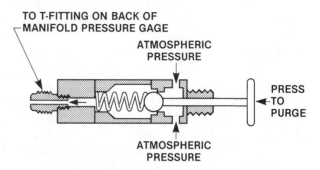

Fig. 2B-2   Manifold pressure purge valve

The larger, high-range manifold pressure gages use a differential bellows arrangement, such as the one we see in Fig. 2B-3. One of the

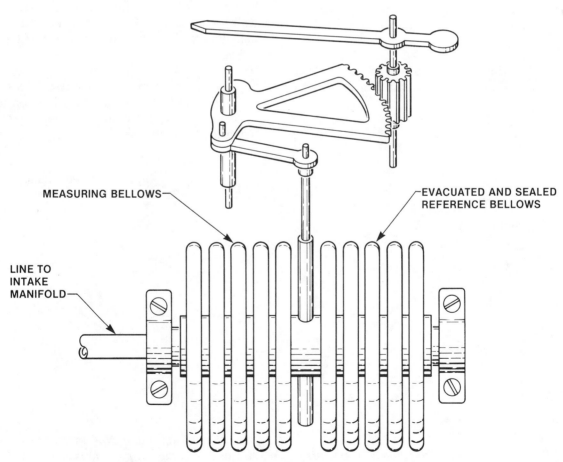

Fig. 2B-3   Mechanism for a differential-bellows-type manifold pressure gage

bellows is evacuated and sealed, and the other bellows connects to the intake manifold. As the pressure inside the intake manifold changes, the measuring bellows expands or contracts and presses against the constant opposition of the reference bellows. The movement of the bellows is transmitted to the sector gear through the link, and movement of the sector gear rotates the pinion which moves the pointer across the calibrated scale on the instrument dial.

2. Gage pressure

When an oil pump or a fuel pump adds pressure to a fluid, the resulting pressure can be expressed as absolute, differential, or gage, depending upon the way we will use it. Most of the time, oil pressure is measured in terms of gage pressure, since we are interested only in the pressure the pump has added to the oil. Atmospheric pressure pushes the oil into the pump, but it also opposes the oil as it flows into the bearings.

Oil pressure and other gage pressure above about ten pounds per square inch are usually measured with a bourdon tube-type instrument. The mechanism of this type gage is seen in Fig. 2B-4.

The bourdon tube is made of a thin-wall bronze tube flattened to an elliptical cross section, and is formed into a semi-circular shape. The fluid whose pressure we are measuring is taken into the bourdon tube at the end that is anchored into the instrument case. When the pressure of the fluid inside the bourdon tube increases, it attempts to change the cross-sectional shape of the tube from elliptical to round, since a tube with a circular cross section can hold more fluid than one with an elliptical shape.

As the cross section changes, the semi-circular tube tries to straighten out, and the movement of the unrestrained end of the tube pulls on the link which rocks the sector and rotates the pinion. The pointer is attached to the pinion shaft, and it moves across the dial to indicate the amount of pressure inside the bourdon tube.

Bourdon tubes in instruments used to measure low pressures may be in the form of a helix of several turns, rather than in the semi-circular shape shown. A helical tube provides more movement of the free end for a small pressure change than the simpler tube provides. High pressure instruments such as those used to measure hydraulic pressure have a tube with a relatively thick wall. Bourdon tube instruments are quite rugged and simple, but they are not particularly sensitive.

A diaphram-type instrument with a mechanism such as we see in Fig. 2B-5 is often used to measure low gage pressures, either above or below atmospheric pressure.

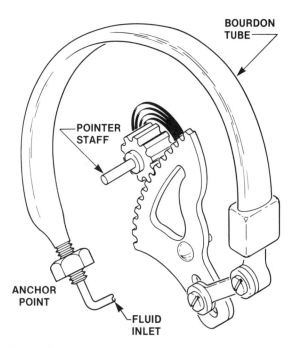

Fig. 2B-4 Bourdon tube instrument mechanism for measuring gage pressure

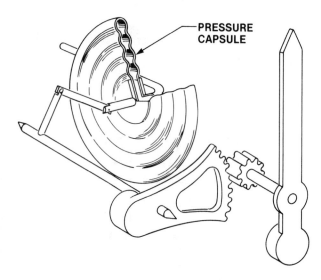

Fig. 2B-5 Diaphragm-type instrument for measuring low values of gage pressure or differential pressure

The pressure to be measured is taken into the inside of the diaphragm, which causes it to expand or contract as the pressure changes. The movement of the diaphragm is carried by a link to the rocking shaft which rocks the sector gear back and forth to rotate the pinion gear. The pinion moves the pointer over the instrument dial.

Temperature measurement with a pressure gage

The oil temperature on most of the smaller aircraft is measured with a pressure gage. A bourdon tube-type gage-pressure instrument is connected by a capillary tube to a bulb that is installed inside the engine, usually in the oil strainer. The bulb, capillary tube, and the bourdon tube are sealed as a unit and are filled with methyl chloride, which is a highly volatile liquid. The vapor pressure of the methyl chloride is measured with the bourdon tube, and the higher its temperature, the higher the vapor pressure.

Care must be taken when working with the capillary tube that it does not get kinked or broken. It is a small copper tube and is often encased in a copper braid for physical protection against damage from rough handling.

COURTESY OF SCOTT AVIATION CORP.

Fig. 2B-6 Pressure-type temperature measuring instrument for measuring the oil temperature in a typical small aircraft engine.

3. Differential pressure

We often have need to know the difference between two pressures, and this difference is called differential pressure. The differential pressure gage most familiar to us in aviation is the airspeed indicator. This instrument measures the difference between the pressure of the air rammed into a forward-facing pitot tube and the pressure of the still, or static, air surrounding the airplane. The difference is given to us in units of knots, miles per hour, or in kilometers per hour.

In powerplant operation there are several different pairs of pressure whose differences we must measure.

a. Differential fuel pressure

In the operation of large reciprocating engines using pressure carburetors, we must know the difference between the pressure of the fuel at the carburetor inlet and the pressure of the air at the carburetor upper deck. We find this pressure with a differential pressure gage.

We also need to know the fuel pressure across the injector nozzles in a continuous-flow fuel injection system. If the engine is normally aspirated, we can use a simple gage-pressure instrument, but in turbocharged engines, the pressure in the induction system is higher than atmospheric pressure, and turbocharger pressure is applied to the vent of the injector nozzle to assure that fuel is forced into the cylinders under all operating conditions. In order to know the amount of pressure that is forcing the fuel through the nozzle orifices, we can measure the pressure drop across the nozzle. We find this pressure with a differential pressure gage, measuring the pressure difference between the fuel pressure at the distribution manifold and the air pressure in the shroud around the nozzle. The gage, rather than being calibrated in units of pressure, is calibrated in units of flow. (See Fig. 2B-7 on page 26.)

b. Engine pressure ratio

One of the most important instruments used with axial-flow turbine engines is the engine pressure ratio, or EPR, indicator.

Centrifugal-flow engines often use RPM as an indicator of the thrust being produced, but

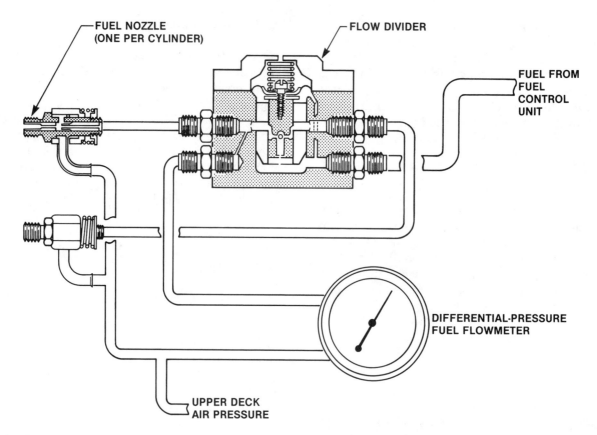

Fig. 2B-7   Differential pressure gage used to measure fuel flow in a turbocharged fuel-injection engine

axial-flow engines more effectively use the engine pressure ratio as the primary indicator of thrust.

This instrument, a differential pressure gage, measures the ratio between the turbine discharge total pressure and the equivalent of the compressor inlet total pressure. These pressures are normally called P_{T2} and P_{T5} or P_{T7} (P_{T5} if the engine is a single-spool, and P_7 if it is a two-spool engine.) See Fig. 2B-8 on page 27.

The compressor inlet total pressure must be corrected for inlet duct loss, and the instrument must be calibrated for each individual type of installation. The pressure pickup for the compressor inlet total pressure is not usually at the face of the compressor, but it is as near the engine inlet as is practical, and corrections based on flight tests are automatically biased into the indicating system so it will indicate the correct pressure ratio.

4. *Pressure switches*

We often need to inform the pilot or flight engineer of a condition in the engine that is likely to indicate an inpending problem. Almost all of the larger aircraft have an annunciator panel where warning lights are grouped so any problem, regardless of the system, can be easily noted. In such fluid systems as the fuel, oil, hydraulic, or pneumatic systems, the warning lights are turned on by pressure switch similar to the fuel pressure warning switch we have in Fig. 2B-9.

Fuel pressure is sensed at port P_1, and ambient air pressure is sensed at port P_2. You will notice that these two pressures act on the opposite sides of the diaphragm assembly, and when a preset differential pressure exists across the diaphragm, the actuator arm will close the microswitch and turn on the warning light on the instrument panel.

III.   TEMPERATURE MEASURING INSTRUMENTS

Pressure and temperature are the two most important parameters to be measured in power-

288

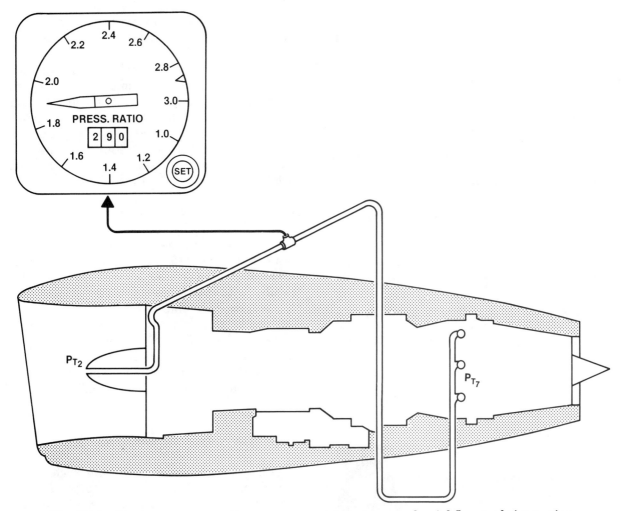

Fig. 2B-8 *Engine pressure ratio measurement in a two-spool axial-flow turbojet engine*

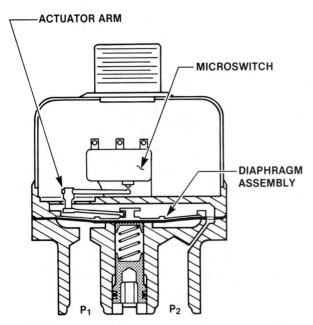

Fig. 2B-9 *Electrical pressure switch for sensing differential pressure*

plant operating systems. Pressures indicate the availability of energy in a system to force oil or fuel to flow, and temperature reflects the amount of heat energy that is in the engine. Too little heat can be bad in some instances, and too much can be bad in others.

There are two basic ways temperature can be measured in an aircraft engine; electrically and non-electrically.

A. Non-electrical Temperature Measurement

Temperature can be measured non-electrically by measuring the physical changes that take place in matter when it absorbs or gives off heat. These changes exist in solids, liquids, and in gases.

We have already mentioned that we measure the oil temperature in most small engines by

27

measuring the pressure of the vapors of methyl chloride. This liquid is sealed in a bulb and capillary tube attached to a bourdon tube and, as the methyl chloride in the tube gets hot, its vapor pressure increases and the bourdon tube measures this change in pressure and indicates it on a dial calibrated in terms of temperature.

The outside air temperature gage we see in Fig. 3B-1 measures the temperature of the outside air by using the changes in the dimensions of two dissimilar metals as the temperature of the outside air changes. A bimetallic strip made of two different types of metal welded together is twisted into a helix, and one end is attached to the pointer and the other end is anchored to the outside end of a protective metal tube. As the temperature of the air changes, the two metals expand and contract in different amounts, and the helical strip either tries to straighten out or tries to twist tighter. This twisting moves the pointer over the dial to indicate the outside air temperature in both degrees Fahrenheit and degrees Celsius.

Fig. 3B-1 The pointer on this outside air temperature gage is moved by the differential expansion of a bimetallic strip.

We are all familiar with mercury thermometers, but because of the difficulty in reading them and the ease with which they can be broken, they find little or no practical use in powerplant instrumentation. They can be made highly accurate, however, and they are often used as a master in-

dicator to calibrate the more rugged and easier to read mechanical or electrical instruments.

The mercury thermometer is made of a glass tube with a thick wall and an extremely tiny but uniformly even bore. A bulb of mercury is made onto the bottom of the tube and as the temperature of the mercury increases, it expands and moves up in the tube. Calibration marks are either etched into the glass tube or marked on a metal plate on which the tube is mounted.

B. Electrical Measurement of Temperature

As with almost every other type of measurement used in aircraft instrumentation, the measurement of temperature is now being done electrically, with much greater ease and accuracy than was ever possible with non-electrical means. This can be done in two ways: by measuring the change in resistance of a metal caused by a change in its temperature and by measuring the amount of voltage generated when a thermocouple junction made up of two dissimilar metals is heated.

1. Resistance-type thermometers

Most of the lower temperatures such as outside air temperature and oil temperature are measured with resistance-type thermometers, usually of the ratiometer type.

The circuit in Fig. 3B-2 is typical of the ratiometer used in an oil temperature indicator. The pointer of the instrument is attached to a small permanent magnet that is free to rotate inside two coils as shown in Fig. 3B-3. This magnet is influenced by the magnetic fields of the coils. The temperature sensing bulb is made of a coil of small diameter nickel wire wound on a strip of mica and is mounted inside a thin-wall stainless steel tube immersed in the oil inside the engine. The resistance of the wire in the bulb changes as its temperature changes, increasing as the temperature increases.

When the temperature of the oil is low, the resistance of the bulb is low, and most of the current flowing through the instrument will pass through resistor resistors R_2 and R_3, the low-end coil, and the bulb. The magnetic field from the low-end coil will pull the magnet and the pointer over so the pointer will indicate the temperature on the low side of the scale.

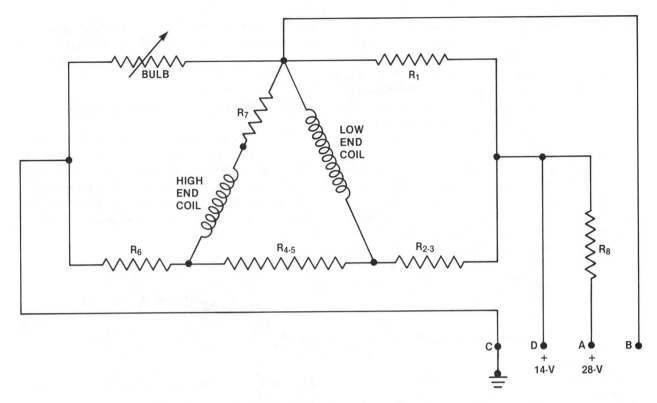

Fig. 3B-2   Electrical circuit of a ratiometer-type resistance thermometer

As the oil temperature increases, the resistance of the bulb increases and more current flows through resistors R_1, R_7, and R_6, and the high-end coil. Less current flows through the bulb, and the magnetic field of the high-end coil pulls the magnet and the pointer over so that it indicates on the high side of the dial.

You can easily see from the way this indicator works that if the bulb is shorted, all of the current will flow through the low-end coil and the pointer will be driven off scale on the low side. But, if the bulb is disconnected from the circuit, or if it is open, all of the current will have to pass through the high-end coil and the pointer will be off scale on the high side of the dial.

Ratiometer-type thermometers are powered from the aircraft electrical system and most of them are made so they will operate in either a 14-or a 28-volt aircraft. If the instrument is installed in a 14-volt system, the positive lead is connected to pin D in the connector plug, and pin C is connected to the aircraft ground. If the aircraft has a 28-volt electrical system, the power comes in through pin A and the voltage is dropped across resistor R_8, so the same voltage reaches the coils as it does when the instrument is installed in a 14-volt system.

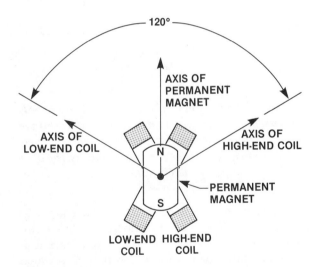

Fig. 3B-3   Physical placement of the low-end and the high-end coils in a ratiometer-type resistance thermometer

2.   Thermocouple instrument systems

One of the main characteristics and advantages of thermocouple-type temperature measuring instruments is their complete independence of the electrical system of the aircraft. Thermocouple-type instruments are used to measure cylinder head temperature (CHT), turbine inlet temperature (TIT), and exhaust gas temperature (EGT) on reciprocating engine-powered aircraft. On turbine engine-powered aircraft they are used to measure the exhaust gas temperature (EGT), turbine inlet temperature (TIT), or intermediate turbine temperature (ITT). Regardless of the parameter they measure, these instruments work on the same principle.

When the junction of wires made of two dissimilar metals is heated, current will flow from the junction through one of the wires, through the coil of the measuring instrument, and back to the junction. The amount of this current is determined by two factors: by the resistance of the circuit and by the temperature difference between the hot, or measuring junction, and the cold, or reference, junction.

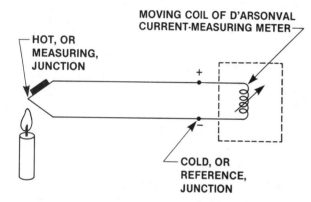

Fig. 3B-4 *The amount of deflection of the needle in a thermocouple-type instrument is determined by the temperature difference between the measuring and the reference junctions and by the electrical resistance of the circuit.*

The two metals used in the thermocouple for measuring cylinder head temperature are either copper and constantan or iron and constantan. Both combinations work essentially the same, but the copper and constantan combination is normally used for temperatures up to about 300 degrees Celsius, and iron and constantan thermocouples are good up to more than 400 degrees C. For measuring the higher temperatures such as those found in the exhaust of both reciprocating and turbine engines, chromel and alumel are used for thermocouples. This combination is good for measuring temperatures as high as 1,000° C.

The complete thermocouple system consists of a probe which is held at the point at which the

temperature is to be measured, leads between the probe and the indicator, and the indicator itself. Some systems require a resistor to adjust the total resistance of the circuit.

a. Cylinder head temperature indicators

Many reciprocating engine installations require the pilot to be able to monitor the temperature of the cylinder head that has been proven by flight tests to run the hottest. The temperature pickup may be either a special gasket under the spark plug, or a bayonet-type pickup, such as the one in Fig. 3B-5.

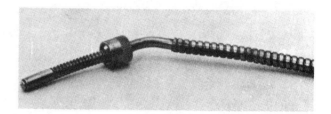

Fig. 3B-5 *Bayonet-type cylinder head temperature probe*

Leads, made of either iron and constantan or copper and constantan, are installed between the probe and the instrument. The resistance of the leads is critical. Most of the smaller aircraft with a short distance between the instrument panel and the engine use leads with two ohms of resistance, and the larger, multi-engine aircraft use eight-ohm leads. If there is not enough resistance in the leads, a spool of constantan wire is available to connect into the negative (the constantan) lead to bring the circuit resistance up to exactly the value needed.

If the leads are too long for the installation, rather than cutting them off to make a neat installation which would have too low a resistance, the leads should be coiled up and securely fastened in the aircraft so it will not be in the way, and yet the lead will have the correct resistance.

The cylinder head temperature indicator used with a thermocouple system is a D'Arsonval-type current-measuring meter. The constantan lead connects to the negative terminal of the meter, and the iron or copper lead connects to the positive terminal. These instruments should all be marked with the type of thermocouples they are to use and with the resistance of the lead and probe combination.

Since these instruments operate from the current generated in the thermocouple, rather than from the aircraft electrical system, they operate all of the time. The instrument should be checked on a pre-flight inspection by observing its indication before the engine is started and comparing the temperature shown on the cylinder head temperature gage with that shown on the outside air temperature gage.

There are two types of probes; those that screw into bosses welded to the exhaust stack and the more commonly used type that fits tightly in a hole drilled in the stack and is held in place with a stainless steel clamp which is part of the probe. The clamp encircles the stack and holds the probe so it will form a gas-tight seal. It is important that exhaust gases do not leak out of the stack around the probe, as such a leak will cause severe damage to both the probe and the exhaust stack, as well as creating a fire hazard in the engine compartment.

COURTESY OF ALCOR, INC.

Fig. 3B-6   Cylinder head temperature gage

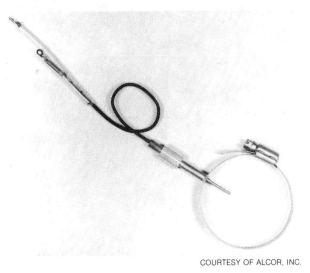

COURTESY OF ALCOR, INC.

Fig. 3B-7   Exhaust gas temperature probe for a reciprocating engine

b. *Exhaust gas temperature gage for reciprocating engines*

Powerplant engineers have known for years that there is a direct correlation between the temperature of the exhaust gas leaving the cylinders of a reciprocating engine and the efficiency of the combustion within the cylinder. In recent years, there has been a great deal of emphasis placed on the EGT as a means of properly adjusting the fuel-air mixture for efficient operation, and in the Fuel Metering portion of this Integrated Training Program we discuss the way the EGT varies with the different fuel-air mixture ratios. In this portion of the text, however, we are concerned with the mechanism that is used to monitor and measure the temperature.

Probes for measuring the temperature of the gas as it leaves the cylinders are mounted in holes drilled through the exhaust stacks at locations specified by the engine manufacturers.

There are two types of EGT systems used in reciprocating engine-powered aircraft. The simpler systems have only one probe and it is installed in the cylinder recommended by the engine manufacturer. More efficient installations have one probe for each of the cylinders and a selector switch that allows the pilot to scan all of the readings and leave the instrument connected to the probe in the hottest cylinder. Different cylinders will run hottest under different operating conditions.

The exhaust gas temperature indications for reciprocating engines are not precise, but the information we need is *relative*. As the mixture ratio is leaned from a relatively rich mixture, the exhaust gas temperature will increase until it peaks, and then continued leaning will cause the temperature of the exhaust gases to decrease. The temperature at the peak will be different with different power settings, and the pilot is concerned with the mixture that will produce a peak temperature. Then he can adjust the mixture control to provide the correct amount of temperature drop on one side or the other of peak.

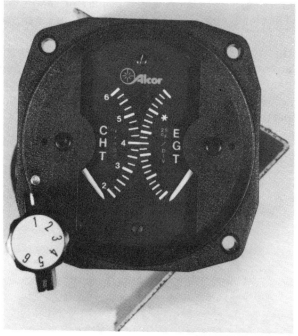

Fig. 3B-8 Combination exhaust gas temperature and cylinder head temperature gage, with a selector switch to measure the temperature of any six cylinders

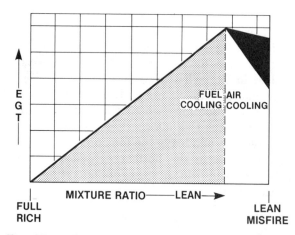

Fig. 3B-9 The temperature of the exhaust gas relates directly to the fuel-air mixture ratio being burned.

You will notice that the EGT shown in Fig. 3B-8 has no numbers on its dial, but there is an asterisk about 4/5 of the way up the scale, and there is a movable red line that may be adjusted with the screw in the front of the instrument glass. This asterisk is used as a limit mark, and it is not a mark that should be reached in normal leaning. The graduations on this instrument are 25 degrees Farenheit apart and knowing the increment between the marks is all the pilot needs in order to find the best operating fuel-air mixture.

When the airplane is at its cruise altitude and is trimmed for cruise flight with the engine power adjusted, the pilot can scan the cylinders to find the cylinder with the highest exhaust gas temperature, and then he can lean the mixture until the EGT peaks. He then enriches the mixture until he sees a 25 degree drop in the EGT. This procedure is pretty well standard, but it may vary with particular aircraft, and the recommendations found in the aircraft flight manual must be followed in detail.

While it is not important that the pilot have specific numbers on the EGT gage, it is important that the limit be such that the pilot will never lean the mixture when the engine is producing power in excess of that recommended for leaning. This could cause an excessive EGT which could result in overheating the cylinders and will cause serious damage to the engine.

The turbine inlet temperature (TIT) indicator used on turbocharged engines is almost the same as an EGT system, except that the probe is installed near the inlet of the turbocharger, and the indicator is calibrated with the actual temperatures on the dial. It is important that the exhaust gas at the inlet of the turbine never exceed the temperature specified in the aircraft service manual. This is usually somewhere around 1,600 degrees Fahrenheit.

The manufacturers of the EGT instruments make a calibration unit such as the one seen in Fig. 3B-10. The tester can be connected to the thermocouple lead and adjusted to produce a voltage in the lead that is equivalent to that produced when the thermocouple is sensing 1,550° or 1,600° Farenheit, depending upon the instrument being calibrated. With this voltage on the thermocouple, a resistor inside the instrument can be adjusted until the indicator needle is opposite the asterisk on the dial.

A more accurate calibration of the entire system uses a heater that is supplied with the tester. The probe is removed from the exhaust stack and placed in the cavity of the heater, and a heating current is supplied from the tester. A thermocouple inside the heater is used to provide a voltage to the instrument in the tester to indicate the temperature of the heater, and the instrument in the aircraft should be adjusted to read the same as the indicator in the tester.

c. Exhaust gas temperature system for turbine engines

The temperature of the exhaust gases in a turbine engine is one of the more important parameters that must be monitored. This is especially true during engine start-up, when an overtemperature condition in the hot section can ruin the engine in a matter of seconds.

The indicator in the cockpit may be labeled Exhaust Gas Temperature (EGT), Turbine Inlet Temperature (TIT), Turbine Gas Temperature (TGT), or Intermediate Turbine Temperature (ITT). These names indicate the location at which the temperature is being measured. TIT is the temperature of the gas forward of the turbine wheels, ITT is the temperature measured between the turbine stages, and EGT or TGT is measured after the gases leave the turbine.

The most important temperature is that at the inlet of the turbine, just forward of the first-stage turbine nozzle, However, because of the difficulties in measuring the temperature at this location, many engine manufacturers measure the temperature of the gases as they leave the turbine. There is a definite correlation between

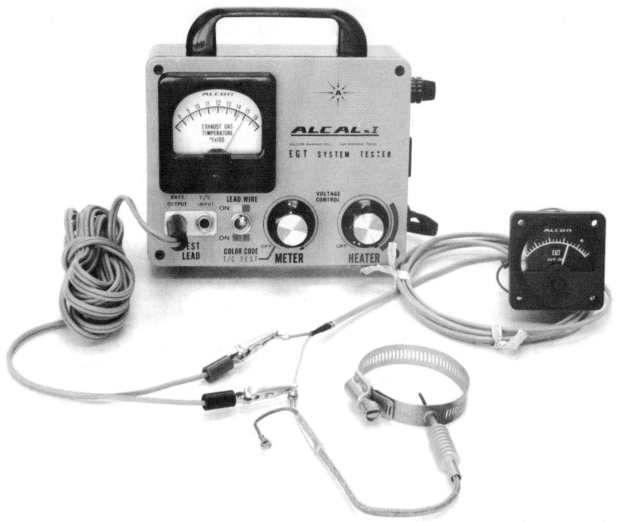

COURTESY OF ALCOR, INC.

Fig. 3B-10 Exhaust gas temperature system tester

the temperatures on the two sides of the turbine, and the turbine inlet temperature can be kept within safe operating limits by limiting the EGT to a specified value.

The complete EGT system for a turbine engine consists of: the probes that sense the temperature of the exhaust gas, the harness that surrounds the engine tail pipe and serves as a connection for all of the probes, extension wires that carry the current from the probes into the cockpit, resistors to adjust the resistance of the thermocouples to the value required for the system, and the indicating instrument in the aircraft instrument panel. The probes are mounted in the tail pipe and are connected in parallel so that their output is averaged.

Some EGT systems use as their indicator a special form of direct current measuring D'Arsonval meter movement very similar to the one used in reciprocating engine EGT systems. But some of the other systems feed the output of the thermocouples into an electronic circuit where the DC voltage from the thermocouples is converted into pulsating DC which is fed into a servo-type instrument. This type of indicator can give the pilot either an analog or a digital readout and, in many instances, both types. In some of the more

Fig. 3B-11 *Turbine gas temperature indicator. This is also called an exhaust gas temperature gage.*

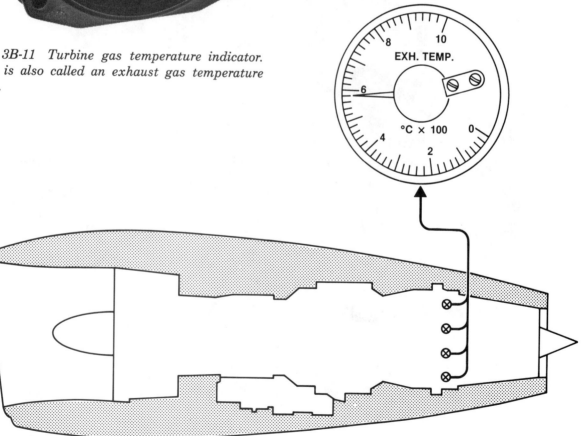

Fig. 3B-12 *Method of measuring the exhaust gas temperature in a two-spool, axial-flow turbojet engine*

Fig. 3B-13 Harness assembly for the exhaust gas temperature probes in a turbine engine

modern instruments, DC output from the thermocouples is processed so it can be converted into digital electronic pulses which actuate light emitting diode digits on the face of the indicator. We see this type of display on the indicators in Fig. 3B-11.

Since it is so vital that accurate EGT information be furnished the pilot or fight engineer, several manufacturers of test equipment have portable testers that can be used by an A&P technician to calibrate the EGT system. The tester shown in Fig. 3B-14 allows the technician to check the accuracy of the engine torque pressure, the tachometer, and the EGT system on a particular turboprop engine while it is installed in the aircraft.

Heater probes are clamped around the EGT probes in the engine and are heated by electric current. The temperature of the probe is measured with a highly accurate electronic circuit and is displayed in digital form in the tester. This allows you to compare the actual temperature of the probe with the indication on the cockpit EGT gage. The circuit can be recalibrated if the indication is not within tolerance.

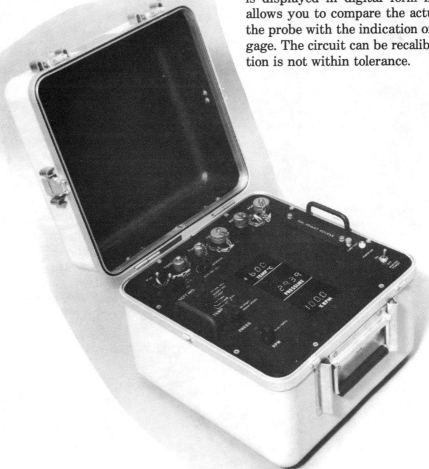

Fig. 3B-14 Tester for calibrating the EGT, the tachometer, and the torque pressure indicating systems in a turboprop engine.

IV. MECHANICAL MEASUREMENT

A. Tachometers

It is important for the flight crew to know the rotational speed of both reciprocating and turbine engines, and the instrument for this is the tachometer. As with aircraft temperature measuring instruments, there are both electrical and non-electrical tachometers.

1. Non-electrical tachometers

Almost all of the small general aviation aircraft use non-electrical magnetic-drag tachometers. The mechanism in these instuments is the same as that used in an automobile speedometer.

COURTESY OF AC SPARK PLUG DIV., G.M. CORP.

Fig. 4B-1   Typical magnetic drag tachometer for a light general aviation aircraft

A twisted steel cable housed inside a metal casing is installed in the aircraft, with two ends of the cable swaged to a square cross section. One end is held in the tachometer drive in the engine, and the other end is inserted into a hole in the rotating permanent magnet inside the instrument. The cable drives the magnet so that it spins in the instrument at one-half the engine speed. (See Fig. 4B-2 below.)

An aluminim cup fits close over the spinning magnet but it does not touch it. As the magnet spins, its lines of flux cut across the aluminum cup and induces a voltage in it. This voltage causes current (eddy current) to flow in the aluminum, and this eddy current produces its own magnetic field that opposes the field that caused it. The two fields produce a torque that rotates the drag cup against the restraint of a calibrated hairspring. The faster the magnet spins, the greater the eddy current and the greater its magnetic field, and the more drag cup will be rotated. The drag cup is supported in a brass bushing by a steel shaft onto which the pointer is pressed. When the engine is not running, the restraining hairspring hold the drag up over so the pointer indicates zero RPM on the dial. (See Fig. 4B-3 on page 37.

Magnetic drag tachometers are calibrated at the speed at which the engine operates most of the time. The calibration is done by driving the magnet at one-half of this speed, and adjusting the restraint of the hairspring so the pointer will indicate twice the speed of the magnet.

A series of drum-type counter wheels is driven by a worm gear from the spinning magnet to indicate the number of hours the tachometer has operated. This method of measuring time is accurate only at the speed for which the tachometer is calibrated, and this speed is stamped on

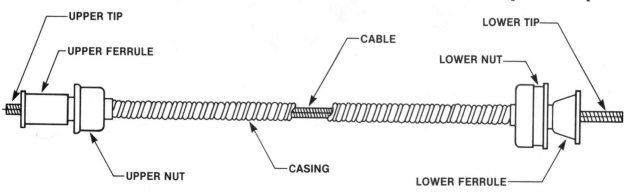

Fig. 4B-2   Cable for connecting a magnetic drag tachometer to an aircraft engine

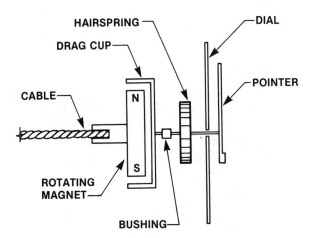

HAIRSPRING · **DRAG CUP** · **CABLE** · **ROTATING MAGNET** · **BUSHING** · **DIAL** · **POINTER** · **N** · **S**

Fig. 4B-3 Mechanism in a magnetic drag tacho-meter

the case of the instrument. While this is, at best, an approximation of the engine operating time, it is considered by the FAA to be accurate enough to use in determining the operating time of the engine.

Magnetic drag tachometers have a reputation for inaccuracy, and because of the great amount of power developed for takeoff by the modern lightweight aircraft engine, these engines operate at as high a speed as they can possibly tolerate. If the tachometer is as little as five percent slow, an engine that is red-lined at 2600 RPM would actually be turning up 2730 RPM, and this could cause serious damage to both the engine and the propeller. These tachometers are sealed and do not permit any change in calibration, so if the instrument is found to be out of calibration, it must be replaced.

Because of the critical nature of engine RPM, and the bad reputation of magnetic drag tachometers, the tachometer should be checked before making any adjustment to a propeller governor or before taking drastic action to remedy a power-loss problem in an engine.

The most accurate and easiest way of checking the calibration of an installed tachometer is to use some form of electronic strobe light tachometer. These instruments use an accurate electronic oscillator to drive a strobe light that is shone on the propeller. The frequency of the oscillator is adjusted until the propeller appears to stand still (the light flashes on the propeller each time one of the blades is in the same position in its rotation), and the actual RPM of the propeller is then shown on the tester.

2. Electric tachometers

a. Electronic tachometers

One popular tachometer used in many of the larger general aviation aircraft is an electronic tachometer that, instead of using any kind of mechanical drive from the engine, uses electrical pulses from the magneto to produce the signal that indicates the engine RPM.

The magnetos used with electronic tachometers have an extra set of breaker points that are totally separate from the electrical circuit of the magneto. Current from the aircraft electrical system flows through these breaker points and they chop it up into pulsating direct current. Electronic circuitry inside the instrument converts the frequency of these pulses, which is determined by the RPM of the engine into direct current. This moves the pointer of a D'Arsonval meter across the dial that is calibrated in engine RPM.

b. AC electrical tachometers

(1) Reciprocating engine tachometers

Some of the early electric tachometers used a permanent magnet generator driven by the engine to produce a voltage that was proportional to the engine speed. The generator voltage was measured by a D'Arsonval-type voltmeter which was calibrated in engine RPM. Some systems used AC generators and others used DC generators. This type of system was used for many years, but it had the serious drawback that its accuracy was determined by the strength of the permanent magnet in the generator, and engine heat, vibration, and age all work on the magnet to cause it to lose strength. This made the system inaccurate.

Modern AC tachometers do not depend upon the voltage produced by the tachometer generator, but rather on the *frequency* of the AC the generator produces, and this frequency is directly related to the engine speed.

The tachometer generator is driven by the engine at one-half engine speed, and it is a permanent magnet-type three-phase AC generator. Its output voltage is somewhere in the neighborhood of one volt for each one hundred RPM but this voltage is not at all critical.

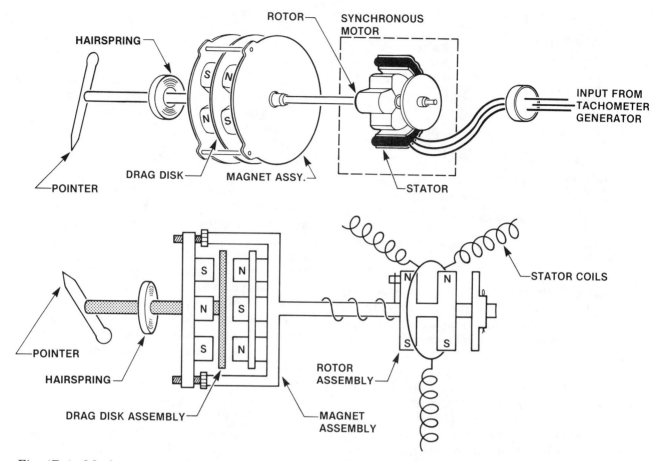

Fig. 4B-4   Mechanism in an AC tachometer generator. The synchronous motor turns at exactly the same speed as the rotor in the tachometer generator.

The indicator used with this type of system consists of a three-phase synchronous AC motor that is driven by the output of the tachometer generator, and it turns at exactly the same speed as the magnet in the generator. This synchronous motor drives a magnet assembly that causes an aluminum magnetic drag disc to rotate against the restraint of a calibrated hairspring. The magnetic drag operation is identical in principle to that used in the simple mechanically driven magnetic drag tachometer. (See Fig. 4B-4.)

This type of tachometer is highly accurate and is quite rugged, and because of this, it is the type of tachometer that is used when accuracy is more important than a low first cost. This type of tachometer is used with both reciprocating engines and turbine engines.

(2)   Turbine engine tachometers

Two-spool turbine engines use two tachometers, one to measure the speed of the low-

pressure compressor, and this is called N_1. The other tachometer measures the speed of the high-pressure compressor, and this is called N_2. These systems are similar in operation to the AC electric tachometer used with reciprocating engines, except that the indicators are calibrated in terms of percent of engine RPM, and they have a vernier scale. The indicators shown in Fig. 4B-5 show that the low-pressure compressor (N_1) is turning at 97% of its rated RPM and the high-pressure compressor (N_2) is turning at 94% of its rated speed. (See Fig. 4B-5 on page 40.)

(3)   Helicopter tachometers

Helicopters use a dual tachometer of the same type that is used in reciprocating engine airplanes. One indicator needle shows the engine's RPM, and the other needle shows the RPM of the main rotor. The ratio of the gears that drive the two generators is such that when the rotor clutch is fully engaged, the needle for the rotor RPM is directly in line with the needle for the

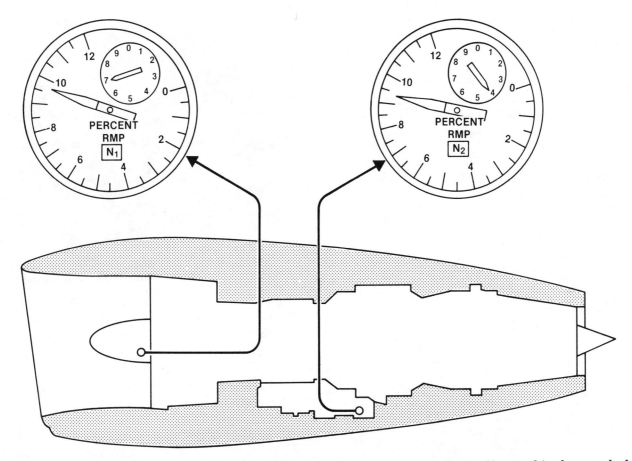

Fig. 4B-5   Method of measuring RPM in a twin-spool, axial-flow turbine engine. N_1 speed is the speed of the low-pressure compressor, and N_2 speed is the speed of the high-pressure compressor.

engine RPM, and the needles are said to be "married." Any time the rotor clutch is not fully engaged, the needles will be split and the pilot will be aware of the condition.

(4)   Synchroscopes

Some multi-engine airplanes use a dual tachometer with a built-in synchroscope, a small differential AC motor that has two stator windings. One winding is excited by the output of the right tachometer generator, and the other from the left tach generator. A small segmented wheel is attached to the rotor shaft and is visible through a hole in the tachometer dial, Fig. 4B-6. When the two engines are turning at exactly the same speed, the output from the two generators cancels each other and the rotor does not turn. But when one engine gets faster or slower than the other, the rotor turns at one-half of the difference between the speeds of the two engines, and this shows the pilot which of the propeller governors should be adjusted to bring the engines back into synchronization.

Fig. 4B-6   A dual AC electric tachometer with a synchroscope. The segmented disk of the synchroscope turns at one-half of the difference between the speeds of the two engines.

B. Fuel Flow Instruments

1. Pressure type

The fuel flow indication used for fuel-injected horizontally opposed engines is actually a measure of the pressure drop across the fuel injection nozzles. This is not an especially accurate way of measuring the fuel flow, since a restricted injector nozzle decreases the fuel flow, but it causes an increased pressure drop. This condition will show as an increase in the fuel flow, which is exactly the opposite of what is really happening.

Normally aspirated engines use a gage-pressure indicator connected as we see in Fig. 4B-7, and turbocharged engines measure flow with a differential pressure gage connected as seen in Fig. 2B-7 on page 26.

Fig. 4B-8 Fuel flow indicator. This is actually a pressure gage that measures the pressure drop across the injector nozzles of a fuel-injected engine.

2. Volume-flow-type remote indicating flow meters

Most of the large airplanes from World War II through the end of the reciprocating engine airliners used a flowmeter that consisted of a movable vane in the fuel line between the engine-drive fuel pump and the carburetor. In Fig. 4B-9 we see the way this vane moves proportional to the rate of fuel flow. The rotor of an Autosyn transmitter is connected to the movable vane in-

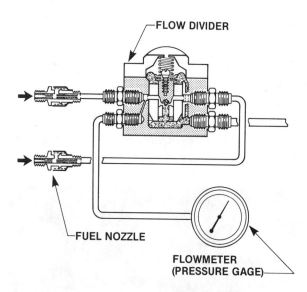

Fig. 4B-7 Pressure-type flowmeter indicator for a normally aspirated fuel-injected engine

The indication given the pilot is seen in Fig. 4B-8, with the inner marks on the dial representing the percent of power the engine is developing. To adjust the fuel flow, the pilot first uses the tachometer and manifold presure to get the desired percent of power output from the engine. Then he can adjust the mixture control to get a fuel flow that causes the needle on the indicator to align with the mark for the percent of power the engine is producing. This is an approximation, and the exhaust gas temperature gage should be monitored to be sure that this mixture is correct for the power being produced.

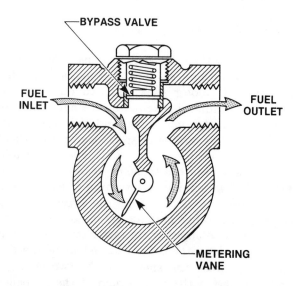

Fig. 4B-9 Flowmeter transmitter for a volume-flow remote indicating fuel flowmeter system

302

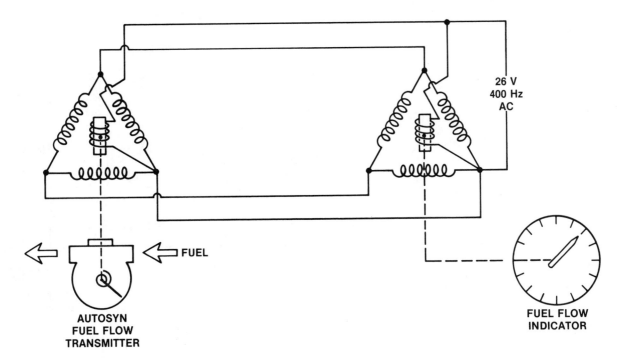

Fig. 4B-10   Electrical circuit for an Autosyn-type remote indicating fuel flowmeter

side the housing, and as the fuel flows and moves this vane, the rotor moves. The movement of the rotor is transmitted electrically to the indicator, and its rotor moves the exact same amount.

A pointer is pressed onto the shaft of the indicator rotor and its movement follows the movement of the transmitter vane. The electrical operation of an Autosyn system is described in the Position and Warning Systems portion of the Airframe section of this Integrated Training Program.

3.   Turbine engine mass-flow measurement

The fuel flow meters we have just described measure the volume of the fuel that flows through the transmitter, but the engine is concerned not with the volume of the fuel, but with its mass. Some of the simpler flowmeter indicators have the pounds of fuel flow marked on their dial, but this is only an approximation based on the normal density of the fuel, and there is no compensation for changes in fuel density caused by changes in fuel temperature.

For turbine engines, it is important that the mass of the fuel flowing into the engine be known, and a mass-flow type of fuel flowmeter is used. This type of instrument is shown in Fig. 4B-11. The impeller is rotated at a constant speed by a

three-phase AC motor, and as the fuel passes through the impeller, its rotation imparts a rotary, or swirling motion to the fuel. As this swirling fuel passes through the turbine, it tries to rotate it, but since the turbine is restrained by the calibrated restraining springs, it can deflect but not rotate. The amount the turbine deflects is determined by both the volume of the flow and the density of the fuel.

A permanent magnet is mounted on the end of the transmitter shaft, and this serves as the rotor of a remote indicating transmitter, which is connected electrically with an indicator on the flight engineer's panel. This shows the actual number of pounds of fuel that passes through the transmitter in a given period of time.

4.   Fuel flow computer

Digital electronics and microprocessors are revolutionizing all forms of instrumentation and control systems. The basic premise is that if any parameter can be measured, it can be controlled, and if it can be controlled, it can be controlled automatically.

In Fig. 4B-12, we have the instrument panel display of a fuel flow computer. Turbine-type flow transmitters are mounted in the fuel lines leading to the fuel injection systems in each engine.

41

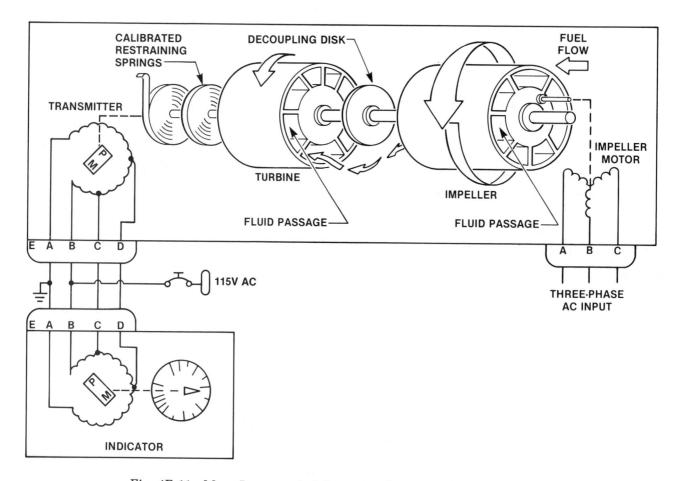

Fig. 4B-11   Mass-flow-type fuel flowmeter for use with turbine engines

COURTESY OF FUELTRON

Fig. 4B-12   A fuel flow computer shows the rate of fuel consumption by each engine, the amount of fuel remaining on board, and the time the fuel will last at the indicated flow rate.

These turbines produce a pulsed output that is proportional to the flow of fuel to the engines, and these pulses are manipulated by the digital circuitry inside the computer to give the pilot an indication of the amount of fuel on board, the flow rate of both of the engines, and the time of fuel remaining at the present flow rate.

C.   The Measurement of Torque

The power produced by an engine-propeller combination is determined by the RPM of the propeller and the amount of torque that drives the propeller. But, because of the difficulty in measuring the torque, most of the smaller propeller-driven airplanes use the RPM and the manifold pressure as an indirect measure of the amount of power the engine is producing.

Some of the large radial engines were equipped with a torque nose that gave the flight engineer a measure of the torque reaction between the propeller and the engine. Rather than being calibrated in terms of torque, the indicators used with this system were usually calibrated in pounds per square inch of brake mean effective pressure (BMEP), and a chart using BMEP and RPM was used to find the amount of power the engine was producing.

Torque is an important measurement in modern turboprop engines, because the engine must never be allowed to produce more torque than the airframe can absorb. The amount of power a turboprop engine can produce is limited basically by the temperature of the gases at the turbine, but there may be a lower limit on the amount of power the engine is allowed to produce because of airframe restrictions.

In Fig. 4-13, we see a curve that shows the limits of a typical turboprop airplane. The engine is capable of producing any power below the temperature limit curve, and from sea level up to the altitude represented by the vertical line, the amount of power allowed is limited by the airframe. The horizontal line is the flat-rated torque. The engine must be held to a value of RPM and torque that will not exceed this power. At altitudes above that represented by the intersection of the flat-rated torque line and the temperature limit curve, the engine is temperature limited and the power must be reduced to keep the turbine sections from being subjected to excessive temperatures.

The power developed by a turboprop powerplant may be found by the formula:

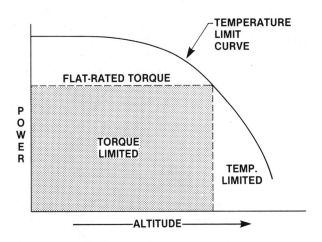

Fig. 4B-13   Power limits for a turboprop engine

Horsepower = K × RPM × Torque (foot-pounds)

The value of K is determined for each powerplant and is a constant. The RPM and torque are values read from instruments in the airplane. The red line, or upper limit of operation, on a torque meter is usually the maximum torque that is allowed when the engine is operating at 100% RPM.

Torque produced by a turboprop engine is found by measuring the torque reaction between the propeller and the engine, and on one turboprop engine it is found by measuring the amount of twist in a torsion shaft.

In Fig. 4B-14, we see the basic principle of the way this shaft twists in proportion to the amount of restraint the propeller puts on the tur-

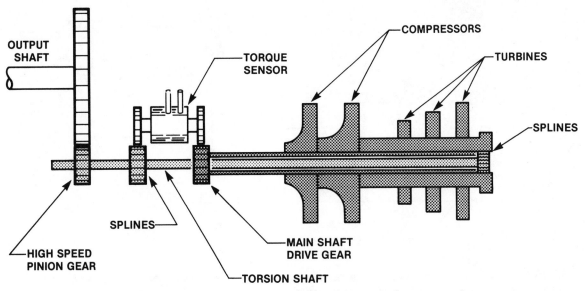

Fig. 4B-14   Method of measuring torque in a turboprop engine

bine shaft. The torsion shaft is splined into the back end of the main turbine shaft and into the shaft for the high-speed pinion gear. As the turbine drives the pinion through the torsion shaft, it twists; the more torque, the more twist.

The torque sensor is mounted inside the gear case and it consists of a helical gear set, driven by two gears that mesh with the main shaft drive gear and the high-speed pinion gear. There is no relative movement between these two gears when the torsion shaft is not twisted, but when the engine is running and the propeller is opposing the rotation of the turbine, the torsion shaft twists. As the two gears on the torque sensor sense the twisting, the helical gear set moves out and increases the spring pressure on the pilot valve.

Engine lube oil flows through the metering orfice and into the gear case, and the pressure on

the downstream side of the orifice is sensed as the torque pressure. As the torque produced by the engine increases, the male helical gear moves out of the female gear, and this increases the spring force on the pilot valve. The oil pressure must build up to a higher value to force the pilot valve off of its seat, and the oil pressure shown on the gage increases. The gage, while actually reading oil pressure, is calibrated in foot-pounds or inch-pounds of torque.

Because the pressure inside the gear case opposes the oil flowing through the pilot valve, this pressure is sensed by the differential pressure gage that reads the pressure drop across the pilot valve. This pressure is proportional to the amount the helical gears have moved, and this movement is proportional to the amount the torsional shaft has twisted. The amount of twist is proportional to the torque the engine is producing.

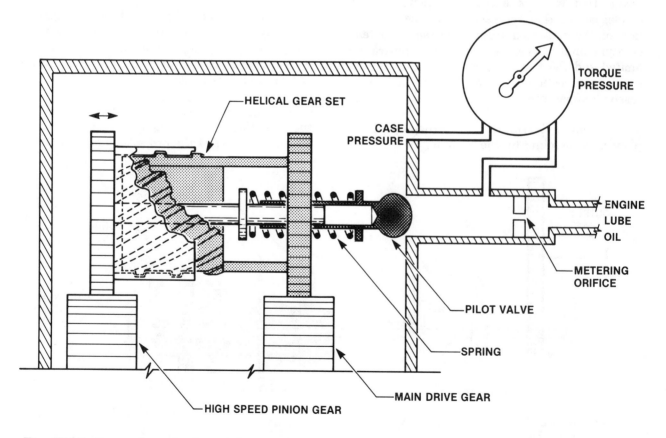

Fig. 4B-15 Internal mechanism of the torque sensor that produces an oil pressure that is proportional to the amount of torque produced by the engine

V. INSTRUMENT INSTALLATION AND MARKING

A. Installation

The engine instruments are often mounted on the right side of the instrument panel, as we see in Fig. 5B-1, or across the top of the instrument panel, as we see in Fig. 5B-2. A convenient layout for a twin-engine turboprop airplane is shown in Fig. 5B-3 where the engine instruments are in two vertical columns just to the left of the engine control console. These instruments are all convenient to the pilot, with the left instruments referring to the left engine and those in the right column to the right engine. Across the top of the instrument panel is a cluster of warning lights, called the annunciator panel.

COURTESY OF CESSNA

Fig. 5B-1   The engine instruments for this twin-engine airplane are mounted in the right-hand instrument panel.

COURTESY OF CESSNA

Fig. 5B-2   The engine instruments for this twin-engine airplane are mounted across the top of the center instrument panel.

Fig. 5B-3   The engine instruments for this airplane are mounted in two vertical rows between the left instrument panel and the electronics panel in the center. Across the top of the center panel is the annunciator panel with all of the warning lights grouped together.

Aircraft instrument panel space is becoming more valuable all the time as new radar and exotic navigation equipment is being developed and installed, and this crowding of the panel has made vertical instruments more popular. Vertical instruments are not new, however; purely mechanical vertical instruments were used as early as 1929 in the Army C-3, which was a military version of the Ford Trimotor airliner (Fig. 5B-4).

In the early generation of turbojet airplanes, servo-type vertical tape instruments became popular. In a tape instrument, the parameter to be measured is sensed by a transmitter in the engine compartment, and this information is carried into the instrument where a servo motor moves a tape up or down to display the measurement as a white or a colored band against a scale on the side of the tape.

The newest generation of vertical instruments is seen in Fig. 5B-5. There are no moving parts in these instruments. The bars are made up of light emitting diodes (LED's) that are il-

Fig. 5B-4   Vertical instruments are nothing new. They were used as early as 1929 in this Army C-3 transport, the military version of the Ford Trimotor airliner.

308

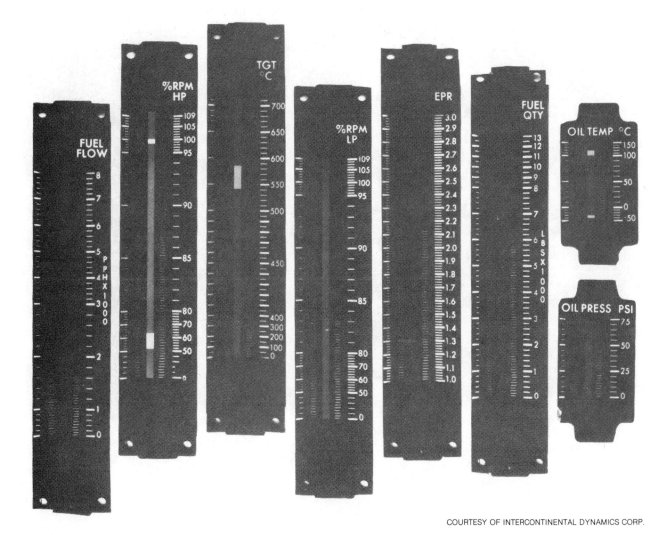

Fig. 5B-5 Modern vertical instruments for a turbojet airplane. The vertical indications are made by light-emitting diodes (LEDs) that are driven by digital electronic circuits.

luminated according to signals from digital electronic circuits inside the instrument.

There are two ways in which round-dial instruments are mounted in the instrument panel, and in Fig. 5B-6 we see turbine gas temperature gages in each of the two types of cases. The case on the left is mounted in the panel with four 6-32 machine screws through the holes in the corners of the instrument bezel. This type of instrument case can be mounted either in front of the panel or behind the panel. Spring-mounted nut plates may be clipped into the four mounting holes in the instrument case if the instrument is mounted behind the panel, or they may be clipped into the holes in the instrument panel when the instrument is front mounted. When these nut plates are used to mount the instrument, you do not have to

hold a wrench on the nut to prevent it turning when the instrument is being installed or removed.

The round case of the instrument on the right in Fig. 5B-6 is mounted in a clamp attached to the back of the instrument panel. The wiring or tubing that connects to the instrument is attached, and the instrument is slipped into the clamp. A single screw in the panel below the instrument is tightened to hold the instrument in place.

Some aircraft whose powerplant instruments are all clamp-mounted have the instruments oriented in their mounting holes so that all of the pointers are horizontal when the parameters they are measuring are normal. The flight engineer sees a solid line of pointers when everything is as

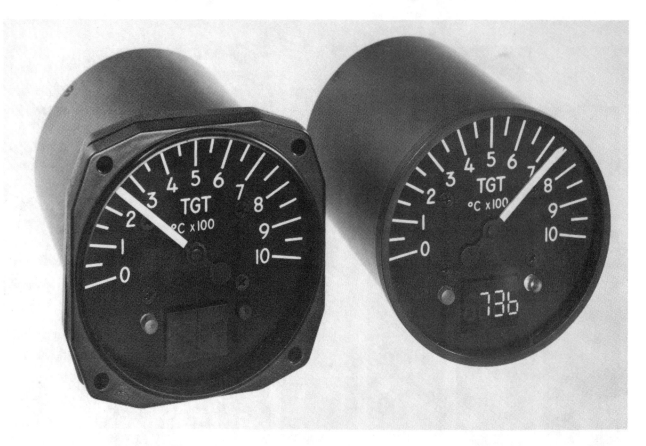

Fig. 5B-6   Methods of mounting aircraft instruments. The instrument on the left is held in place with four 6-32 machine screws. The instrument on the right is held in a clamp that is attached to the back of the instrument panel.

it should be, but if any indicator shows an abnormal reading, the pointer will be out of the straight line, which calls immediate attention to this instrument.

B.   Range Marking

Many of the powerplant instruments require range marking, to allow the pilot or flight engineeer to know whether the indication shown on the instrument is safe or dangerous.

Most of these markings are painted on the instrument dial itself, but some ex-military aircraft that are now in the civilian fleet have the range marks put on the instrument glass. When the marks are on the glass, there is a possibility that the glass could slip and the markings would not be in their correct position relative to the dial. For this reason, instruments with the range marks on the glass must have a slip mark. This is a small mark of white paint, usually at the bottom of the

instrument glass, with part of the mark on the glass and part on the instrument bezel. If the glass slips, the two parts of the mark will be out of aligment and the pilot or flight engineer will know that the range marks are not correct.

The following range marks are typical for powerplant instruments:

1. Carburetor air temperature

   a. Red radial line—Maximum permissible carburetor inlet air temperature recommended by the engine manufacturer

   b. Green arc—Normal operating range for trouble-free operation. The upper limit is at the maximum carburetor inlet air temperature, and the lower limit is at the point where icing may be anticipated. An additional green arc may be required in the temperature range below the icing range.

c. Yellow arc—Range indicating where icing is most likely to occur.

2. Cylinder head temperature

   a. Red radial line—Maximum permissible cylinder head temperature.

   b. Green arc—the upper end of this arc is at the maximum permissible temperature for continuous operation, and the lower end is the minimum recommended for continuous operation.

   c. Yellow arc—This arc extends from the maximum temperature for continuous operation to the maximum permissible temperature.

3. Manifold pressure

   a. Red radial line—This marks the maximum permissible absolute manifold pressure for wet or dry operation, whichever is greater.

   b. Green arc—the upper end of this arc is the maximum permissible manifold absolute pressure for continuous operation, and the bottom of the arc is the minimum pressure selected by the aircraft manufacturer for cruise power.

   c. Yellow arc—This arc extends from the maximum pressure for continuous operation to the maximum permissible pressure.

4. Fuel pressure

   a. Red radial line—Maximum and minimum permissible pressures established as the engine operating limits.

   b. Green arc—Normal operating range

   c. Yellow arc—Caution ranges, indicating any portential hazard in the fuel system.

5. Oil pressure

   a. Red radial line—Maximum and minimum permissible pressure established as the engine operating limits.

   b. Green arc—Normal operating range

   c. Yellow arc—Caution ranges, indicating any potential hazard due to overpressure during a cold start of low pressure during idle.

6. Oil temperature

   a. Red radial line—Maximum and minimum permissible temperatures established as the engine operating limits.

   b. Green arc—Normal operating range

   c. Yellow arc—Caution ranges indicating any potential hazard from overheating, or from low temperature causing a high viscosity.

7. Tachometer, reciprocating engine

   a. Red radial line—Maximum permissible RPM

   b. Green arc—the top of this arc is the maximum permissible RPM for continuous operation, and the bottom of the arc is the minimum recommended RPM for continuous operation except in the restricted ranges, if any.

   c. Yellow arc—This extends from the maximum RPM for continuous operation to the maximum RPM allowed.

   d. Red arc—This is a restricted range of operation, and it should be avoided except to pass through it as rapidly as possible. This type of range is usually caused by resonance of the engine and propeller combination.

8. Tachometer, turbine engine

   a. Red radial line—Maximum permissible RPM

   b. Green arc—This arc extends from the maximum RPM for continuous operation to the minimum RPM recommended for continuous operation.

   c. Yellow arc—This extends from the maximum RPM for continuous operation to the maximum RPM.

9. Tachometer, turboshaft helicopter

   a. Red radial line—Maximum permissible RPM

10. Dual tachometer, helicopter

   a. Red radial line, Engine—Maximum permissible RPM

   b. Red radial line, Rotor—Maximum and minimum rotor RPM for power-off operational conditions.

c. Green arc, Engine—This arc extends from the maximum continuous RPM to the minimum RPM recommended for continuous operating power except in restricted ranges, if any.

d. Green arc, Rotor—This extends from the maximum to the minimum RPM for the normal operating range.

e. Yellow arc, Engine—This is the precautionary range for such restrictions as altitude limits.

11. Torque

a. Red radial line, Maximum permissible torque pressure for wet or dry operation, whichever is greater.

b. Green arc—This extends from the maximum torque pressure for continuous operation to the minimum torque pressure recommended.

c. Yellow arc—This extends form the maximum torque pressure for continuous operation to the maximum permissible torque pressure.

12. Exhaust gas temperature, turbine engines

a. Red radial line—Maximum permissible gas temperature for wet or dry operation, whichever is greater.

b. Green arc—This extends from the maximum permissible temperature for continuous operation to the minimum temperature recommended by the engine manufacturer.

c. Yellow arc—This extends from the maximum temperature for continuous operation to the maximum permissible gas temperature.

I. FIRE DETECTION SYSTEMS

The powerplant area in an aircraft is a natural fire hazard area. Fuel and oil are available in large quantities and are often under pressure. The exhaust system encloses high temperature gases and flames. Vibration and the lightweight construction of the engine components make fluid and exhaust leaks possible and, to compound these problems, the high-velocity air flowing through some of the engine compartment areas can carry explosive fumes into areas of high temperature.

A. Fire Zones

The powerplant area has been divided into fire zones based on the volume of airflow and the smoothness of the airflow. These classifications allow us to match the type of detection and extinguishing system to the fire conditions.

Class-A fire zones have large quantities of air flowing past regular arrangements of similarly shaped obstructions. The power section of a reciprocating engine where the air flows over the cylinders is a class-A fire zone.

Class-B fire zones have large quantities of air flowing past aerodynamically clean obstructions. Heat exchanger ducts and exhaust manifold shrouds constitute class-B fire zones in a reciprocating engine installation. Also considered class-B fire zones are turbine engine compartments in which the engine surfaces are aerodynamically smooth and the inside of the airframe structure housing the engine is covered with a smooth fireproof liner.

Class-C fire zones have a relatively small airflow through them. The compartment behind the firewall of a reciprocating engine is considered to be a class-C fire zone.

Class-D fire zones are areas that have little or no airflow. Wheel wells and the inside of a wing structure are class-D fire zones.

Class-X fire zones are areas in the powerplant portion of an aircraft that have large volumes of air flowing through them at an irregular rate. These are the most difficult of all areas to protect from fire, and the amount of extinguishing agent required for adequate protection of a class-X fire zone is normally twice that required for other zones.

B. Types of Fire Detection Systems

There are a number of fire detection systems that are able to detect the presence of a fire before it is visible to the flight crew, but there are a few basic types that are popular for detecting fires or overheated conditions in the powerplant portion of a modern aircraft.

1. Thermal-switch fire detection system

This is a spot-type fire detection system that uses a number of thermally actuated switches, such as the one in Fig. 1C-1, to warn of a fire. The detector, or thermal switch, is mounted inside a stainless steel housing that expands and elongates when it gets hot.

All of the detectors are connected in parallel with each other, and the combination is in series with the warning light. If any detector reaches the temperature to which it is adjusted, it will complete the circuit to ground and turn on the warning light, indicating a fire or an overheat condition. A detector may be adjusted by heating its case to the required temperature and turning the adjusted screw in or out until the contacts just close.

The entire circuit can be tested by closing the test switch which actuates the test relay and grounds the end of the conductor that ties all of the detectors together. This turns on the warning light.

The warning light burns at full brilliance in the daytime, but when the navigation lights are

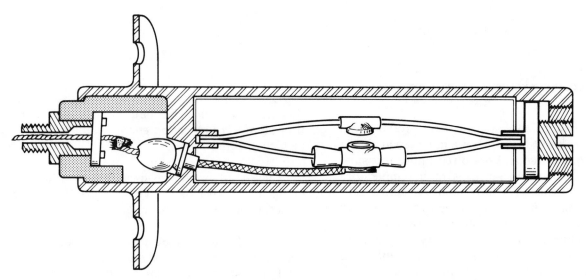

INTERNAL DETAILS OF A THERMOSWITCH FIRE DETECT'
(A)

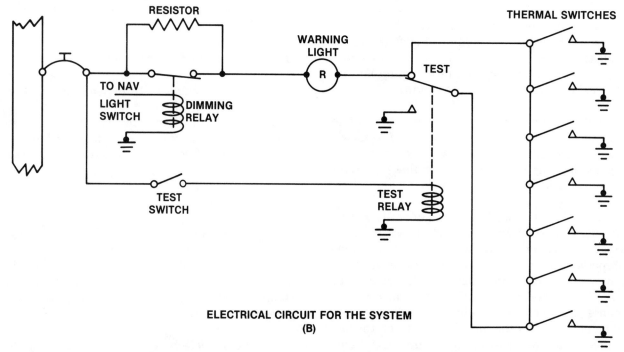

ELECTRICAL CIRCUIT FOR THE SYSTEM
(B)

Fig. 1C-1   Thermoswitch-type fire detection system

turned on, the dimming relay is actuated and the warning light burns at a reduced brilliance.

2. *Thermocouple-type fire detection system*

The thermal-switch-type fire detection system will actuate to warn of a fire any time any one of the detectors reaches the temperature for which it is adjusted; but the thermocouple-type system is actuated to warn of a fire only when the temperature of the area protected by the active thermocouples rises at a rate faster than the temperature rises at the reference thermocouple.

In Fig. 1C-2, we see the circuit of this system. The active thermocouples are placed around the engine compartment at locations where a fire is most likely to start. However, the reference thermocouple is installed at a location that is not likely to be involved in the initial flame of a fire, but in an area that normally reaches the same temperature as the area in which the active thermocouples are mounted.

Remember, in our discussion of thermocouple instruments such as the cylinder head temperature gage and exhaust gas temperature gages, that the current that flows in a thermocouple circuit is proportional to the temperature *difference* between the active, or measuring, junction and the reference junction. When the engine is started and the temperature of the engine compartment rises relatively uniformly, all of the thermocouples will be at essentially the same temperature and no current will flow in the thermocouple circuit.

But if a fire breaks out in the engine compartment, the temperature of one or more of the active thermocouples will suddenly rise much higher than the temperature of the reference thermocouple, and a voltage difference will be produced between the thermocouples that will cause current to flow. This current will energize the sensitive relay, and when its contacts close, the slave relay is energized. The contacts for the slave relay are connected to the red warning light, and when the slave relay closes, the warning light comes on.

To test this system, close the test switch and current will flow through the heater element in the thermal test unit. This unit has an enclosed thermocouple that is in series with both the active and the reference thermocouples, and when the test thermocouple is heated, a voltage is produced which causes current to flow in the circuit and close the sensitive relay. This in turn closes the slave relay which turns the warning light on.

3. *Continuous loop fire detection system*

This system works on the same basic principle as the spot-type fire detection system, except that instead of using individual thermal switches

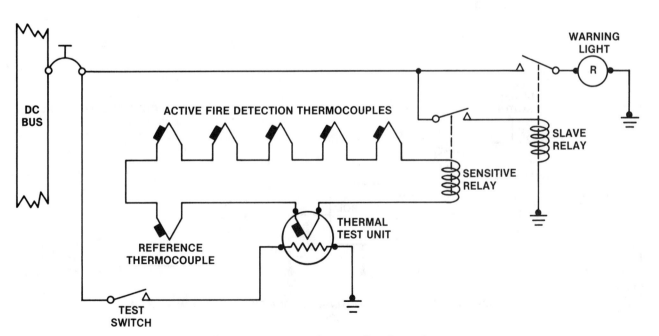

Fig. 1C-2 Thermocouple-type fire detection system

that physically close when a preset temperature is reached, the continuous loop-type of system has as its sensor a long inconel tube. This tube is arranged around the area it is protecting in such a way that it will sense any fire or overheat condition in the compartment.

There are two types of sensing elements in use. One of them consists of an inconel tube in which two sensing wires are embedded in a special ceramic insulating core. This ceramic has the characteristic that its resistance changes with the temperature. At low temperature it acts as an insulator, but as its temperature increases, as it would if there were a fire or an overheat condition in the engine compartment, the resistance will decrease and current will flow between the two sensing wires, and the alarm will sound. This system senses the average temperature of the loop as well at the temperature of any spot in the system.

The other type of sensing element uses a single conductor supported in the inconel tube by ceramic beads. Covering all of the beads and making contact with both the center conductor and the inconel tube is a eutectic salt whose electrical conductivity is determined by its temperature. At normal temperature, the eutectic salt acts as an insulator, but when a fire occurs in the compartment in which the sensor is installed, the resistance of the salt suddenly drops and current flows from the control circuit to ground, and the warning bell sounds.

In Fig. 1C-3, we see the two types of sensing elements and the basic electrical circuit for this type of system. When the resistance between the sensor and ground is decreased because of a high temperature, the control circuit senses this drop in resistance and provides a ground for the warning light and the fire bell. The pilot has a fire bell disable switch that allows him to silence the bell so it will not be distracting.

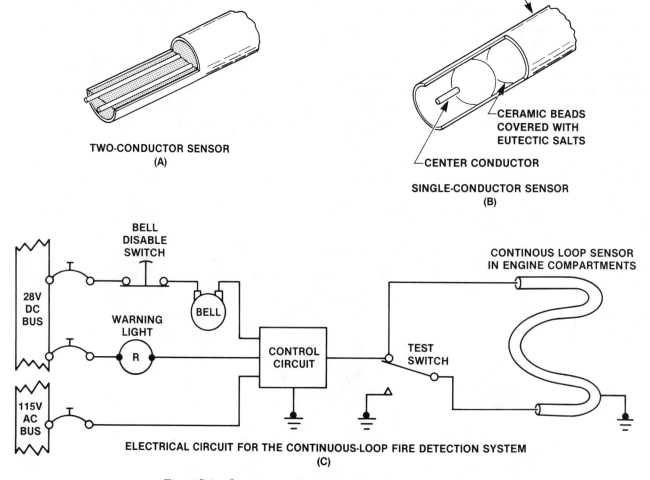

TWO-CONDUCTOR SENSOR
(A)

SINGLE-CONDUCTOR SENSOR
(B)

ELECTRICAL CIRCUIT FOR THE CONTINUOUS-LOOP FIRE DETECTION SYSTEM
(C)

Fig. 1C-3 Continuous-loop-type fire detection system

4. Pressure-type sensor-responder fire detection system

This fire detection system activates the fire warning signal when any portion of the sensor reaches a temperature that signals a fire condition. It also activities when a large portion of the element is exposed to a lower temperature, as would happen when the compartment in which it is installed is overheated enough to cause structural damage, or to a temperature that precedes a fire.

The sensor in this system is a sealed, gas-filled tube in which there is an element that absorbs the gas at a low temperature, but releases it when it is heated. The tube is connected to a pressure switch that closes when the gas pressure reaches a preset value.

If the temperature in the engine compartment rises to the value for which the pressure switch is set, the switch will close and activate the fire warning bell and turn on the fire warning light. This system also has a bell disable switch that allows the flight crew to turn off the bell so it will not be distracting.

To test this system, the test switch is depressed and held. This passes low-voltage AC through the metal tubing that houses the sensitive element. This current heats the sensor and produces the fire warning signal. When the test switch is released, the current stops, the element cools down, and the fire warning stops.

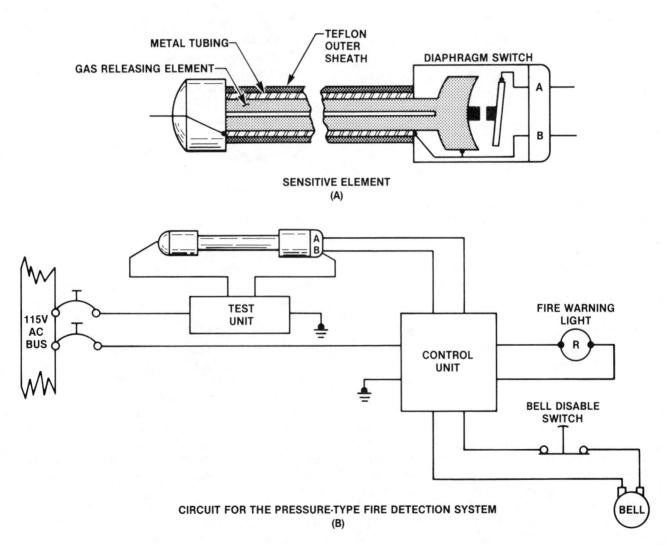

Fig. 1C-4 *Pressure-type sensor-responder type of fire detection system*

II. FIRE EXTINGUISHING SYSTEMS

A. Types of Fires

The National Fire Protection Association has classified fires according to their fuel, and the fire extinguishing agents must be matched to the class of fire they are used to extinguish.

Class-A fires are those that have ordinary combustible materials such as paper, cloth, or wood as their fuel. These are not normally the class of fire that occurs in an aircraft powerplant area.

The Class-B fire has flammable liquids as its fuel, and this is the most common type of fire that occurs in the powerplant area of an aircraft.

Class-C fires involves energized electrical equipment. The fire fighting procedures used with this type of fire must disconnect the electrical power from the fire zone.

B. Fire Extinguishing Agents

There are three requirements for a fire; there must be fuel that can be burned, there must be oxygen to unite with the fuel, and there must be enough heat to raise the temperature of the fuel enough that it can chemically combine with the oxygen. To extinguish a fire, any one of the three requirements may be taken away, and the fire will go out. Removing the fuel, blanketing the fire with an inert material to exclude the oxygen, or lowering the temperature of the burning fuel enough that the chemical combination can no longer occur are all used to extinguish aircraft fires, and actually most successful systems use more than one of these methods.

1. Water

Water cools the burning fuel below its combustion temperature, but it is not normally used to combat fires in engine compartments, because most of these fires contain either flammable liquids or energized electrical equipment. Gasoline, turbine fuel, and oil are lighter than water and will float on the water and spread the fire. The stream of water is electrically conductive, and a fire fighter can be seriously injured if the stream

of water being used to extinguish a fire connects him to the electrical power.

2. Dry powder

Dry powder used as a fire extinghishing agent is similar to sodium bicarbonate, and it releases carbon dioxide gas when it is heated. This CO_2 gas extinguishes the fire by blanketing it and excluding the oxygen it needs. Dry powder is excellent for some classes of fires, but it is not usually used for aircraft engine fires because of the damage the powder can cause to some of the engine components. The other two types of fire extinguishing agents are so much better than either water or dry powder for powerplant fires that they are used exclusively.

3. Inert gas

Carbon dioxide, CO_2, and liquid nitrogen LN_2, are the two inert gases that are used successfully in combating aircraft engine fires. CO_2 systems were the first fire extinguishing systems installed in engine nacelles, and the liquid nitrogen systems are one of the newest developments in installed fire protection systems.

a. Carbon dioxide

Carbon dioxide is such a handy and effective fire extinguishing agent that it is used for portable fire extinguishers in aircraft. It was also used as an installed engine fire protection system from the middle 1930's until the more stringent fire extinguisher requirements of the large jet airplanes demanded even more effective extinguishing agents.

Carbon dioxide is a colorless, odorless, gas that is about one and one-half times as heavy as the air. It is liquified by compressing and cooling it, and then it is stored in steel cylinders. When it is released into the atmosphere, it expands and cools to a temperature of about $-110°$ F, and becomes a white solid that resembles snow. When it changes from this solid into a gas, it does not pass through the liquid state, but goes directly from the solid snow to CO_2 gas.

Reciprocating engine airplanes that have an installed CO_2 fire extinguisher system have one or more CO_2 cylinders in the fuselage connected by aluminum tubing to the engine nacelles. Here they discharge through perforated tubes into the accessory section of the engines.

Installed CO_2 fire extinguisher systems have two blowout disks on the side of the fuselage near the storage bottles so they are easily seen by the flight crew on their walk-around inspection. If the CO_2 bottles have been discharged because of an overheat condition in the storage area or because of excessive pressure, the red disk will blow out. But when the system is discharged by normal system actuation, the yellow disk will be blown out. The disk are replaced when full cylinders are installed.

b. *Liquid nitrogen*

Liquid nitrogen held at a temperature of $-320°$ F in a Dewar flask (a special double-wall storage container that insulates the material from ambient heat) is effective in extinguishing a fire by blanketing it and excluding the needed oxygen. Because of the difficulty in handling this liquid, it is not generally used in civilian aircraft, but its effectiveness as a fire extinguishing agent and as an inerting agent for purging air from the space above the fuel in fuel tanks makes it an agent that is likely to find wide use in the future.

4. *Halogenated hydrocarbon agents*

It is not competely clear just how a halogenated hydrocarbon extinguishes a fire, but basically it interferes with the chemical combination that takes place between the oxygen and the fuel. This method of fire extinguighing is highly effective and is currently the most widely used method of fire control for aircraft engine compartments.

One of the first fire extinguishing agents carried on board aircraft for fire protection was carbon tetrachloride, which is a member of the halogenated hydrocarbon family. It was carried in a small hand-pump-type fire extinguisher, and was very effective in extinguishing engine fires. Carbon tetrachloride is no longer allowed to be used as a fire extinguishing agent because of its extremely harmful effects on the human body and the fact that when it comes in contact with a fire it changes into a lethal gas, phosgene.

There are a number of compounds that are used in high-rate discharge (HRD) fire extinguishing systems to combat powerplant system fires. These agents have Halon numbers, and are often spoken of by the generic term "freon."

Halon 1001 is the identification for methyl bromide which is also called "MB." It is relatively low in cost and is readily available, but it is corrosive to aluminum and magnesium.

Halon 1202 is lightweight and non-corrosive to aluminum steel, or brass, but it is relatively toxic and is high in cost.

Halon 1301 is one of the most effective of the modern fire extinguishing agents. It is non-toxic and non-corrosive and is far more effective that most of the other agents. It is a liquified gas whose freezing point is $-270°$ F. For use in aircraft powerplant fire extinguishing systems, it is stored in spherical high-rate-discharge bottles.

C. *Complete Fire Protection System*

The fire protection system used on the Boeing 727 is typical of those found on most of the high-performance jet aircraft. The engine compartments are monitored by the fire detection sensors, and if a fire is sensed, a red light inside the engine fire switch comes on and the fire alarm bell rings. When the warning light comes on, the pilot pulls the switch. This arms the fire bottle discharge switch, trips the generator field relay, and shuts off the fuel to the engine, the hydraulic fluid to the pump, and the engine bleed air, and deactivates the engine-driven hydraulic pump low-pressure lights.

When the fire switch is pulled, the bottle discharge switch is uncovered, and when the pilot has determined that a fire does actually exist in the engine compartment, he can press the bottle discharge switch and hold it for one second. This discharges one of the high-rate-discharge bottles of fire extinguishing agent into the engine compartment. If the fire warning light does not go out within thirty seconds, indicating that the fire has been extinguished, the pilot can move the bottle transfer switch to its other position to select the other bottle of fire extinguishing agent and again push the bottle discharge switch. When the fire extinguisher bottles have been discharged, or when their pressure is low, the appropriate bottle discharge light comes on.

The controls for the fire detection system are on the copilot's side of the panel. They consist of

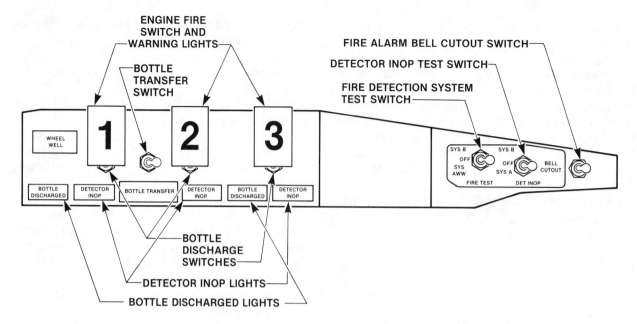

CONTROLS MOUNTED ABOVE THE MAIN INSTRUMENT PANEL
(A)

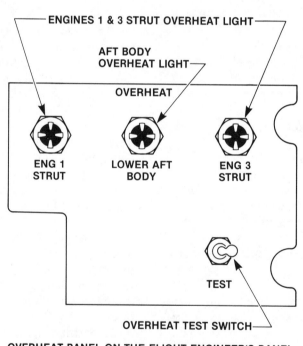

OVERHEAT PANEL ON THE FLIGHT ENGINEER'S PANEL
(B)

Fig. 1C-5   Controls for the fire protection system on a Boeing 727 jet transport airplane

the fire alarm bell disable switch, the fire detection system test switch, and the detector inop test switch. The fire alarm bell disable switch silences the fire bell when it has been started by a fire alarm in the engines or in one of the wheel wells. The fire detection system test switch checks the continuity of the detectors and the operation of the warning system. The detector inop test switch tests the circuits that activate the "Detector Inop" lights and, if the systems are functioning properly, will momentarily illuminate the detector inop lights.

320

There is a small panel on the upper portion of the flight engineer's panel that has overheat warning lights to warn of an overheated condition of the engine struts for the side-mounted number one and three engines and for the area adjacent to the engine bleed air ducts on the lower portion of the aft body. A test switch is also on this panel to test the lights.

In Fig. 1C-6, we see the complete fire extinguishing system for the powerplant areas of a Boeing 727. The system is protected by two high-rate-discharge bottles of fire extinguishing agent, and these bottles are sealed with a metal seal. When the fire switch is pulled, the bottle discharge circuit is armed, and as soon as the bottle discharge switch is pressed, a powder charge is ignited with an electric squib. The charge blows a knife into the seal and dumps the contents of the bottle into the fire manifold. Pulling the fire switch opened the engine selector valve and the extinguishing agent flows to the correct engine.

Each of the two agent bottles have a pressure gage to show the pressure of the contents, and an electrical pressure switch is mounted on each bottle to turn on a bottle discharge light on the instrument panel when the pressure on the agent bottle is low.

Each bottle has a thermal fuse that will melt and release the contents if the bottle is subject to a high temperature, and if a bottle is emptied in this way, a red blowout disk on the side of the fuselage is blown out. This can be seen on a walk-around inspection so the flight crew will know the reason for the discharge. If the bottles are discharged by normal operation of the system, the yellow disk will be blown out.

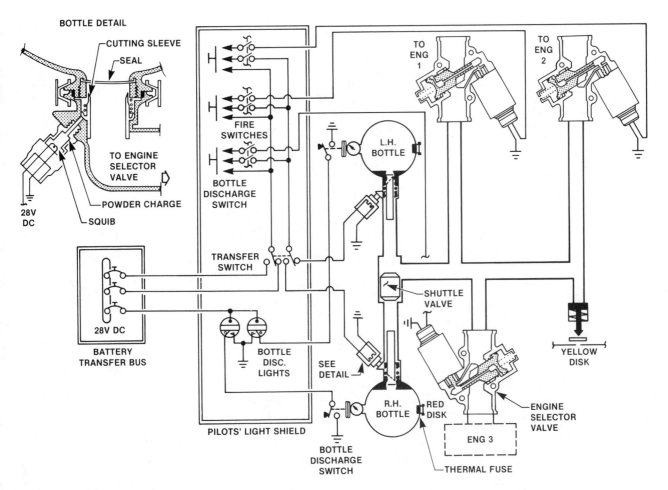

Fig. 1C-6   Fire extinguishing system for a Boeing 727 jet transport airplane

I. RECIPROCATING ENGINE FUEL METERING SYSTEMS

A. *Transformation of Energy*

1. *Source of energy*

The internal combustion engine used in our modern aircraft is a form of heat engine; that is, it is a device which changes heat energy into mechanical energy. And the heat energy used in this engine comes from our chief source of energy, the sun. Solar energy has been radiated to the earth where it was changed into chemical energy in vegetation and plant life, then into animals that ate the vegetation. During some great up-heaval that took place billions of years ago, the plants and animals were buried beneath millions of tons of earth, and heat and pressure turned them into fossil fuel, the petroleum products we use in our heat engines. Petroleum is an organic chemical, a compound of hydrogen and carbon, of the family known as hydrocarbons.

2. *Conversion of heat energy into power*

a. *Chemistry of combustion*

In order to release heat energy from the fuel, a chemical reaction must take place. And for this to occur, the hydrocarbon fuel must be brought into contact with a source of oxygen and the temperature of the fuel raised to its kindling point. When this happens, the oxygen will combine with the fuel, and oxidation, or burning, occurs.

Each molecule of aviation gasoline, our most widely used fuel for aircraft reciprocating engines, is made up of eight atoms of carbon and eighteen atoms of hydrogen, and this may be written as the chemical formula C_8H_{18}. In order to release all of the energy in the fuel, we must burn all of it; that is, all of the fuel combined with oxygen. Two atoms of oxygen must unite to form one molecule of oxygen gas, and for complete burning, we must have two molecules of gasoline and 25 molecules of oxygen gas. Looking at this as a chemical equation, we see the following has taken place:

$$2\ C_8H_{18} + 25\ O_2 \rightarrow 16\ CO_2 + 18\ H_2O + \text{HEAT}$$

Three products are formed when aviation gasoline is burned: sixteen molecules of carbon dioxide (CO_2), 18 molecules of water (H_2O), and, along with the formation of these two new chemical compounds, the very thing we want most, *heat*.

You can see from this chemical equation that a definite amount of oxygen is needed to unite with the gasoline to produce water and carbon dioxide and not have any leftovers, either from the fuel or from the oxygen.

Air is a physical mixture made up of several gases, principally nitrogen and oxygen, and since nitrogen is an inert gas, it does not enter into the chemical reaction. Fifteen pounds of air is needed to unite with one pound of gasoline to completely combine all of the gasoline and oxygen. If there is more air than is needed, oxygen will be left over after the burning is completed, and if there is

1

more fuel in the mixture than is required for the oxygen, free carbon will be left; this usually shows up as black smoke or soot.

The mixture ratio of fifteen pounds of air to one pound of gasoline is known as a stoichiometric mixture, which is a chemically correct mixture in which all of the chemical elements are used and none are left over. A 15:1 air-fuel ratio may also be expressed as a fuel-air ratio of 0.067 (1/15 = 0.067).

Combustion can occur with a mixture as rich as 8:1 (0.125) or as lean as 18:1 (0.055), but the maximum amount of heat energy is released with the stoichiometric mixture of 0.067. In a lean mixture there is less fuel and therefore less heat energy. But if the mixture is overly rich, there is not enough oxygen and some of the fuel will not be burned, and so a smaller amount of heat energy will be released.

It would seem that since the most heat energy is released from the fuel with a fuel-air mixture of 0.067, this ratio would be used to produce the most power. This is not actually the case, however. The design of the engine induction system and the valve timing requires a mixture that is slightly richer than chemically perfect in order to produce the maximum power. Maximum power is normally considered to be produced with a mixture of approximately 0.083 or 12:1.

Aircraft engines are built as light as possible and because so much power is produced in such a lightweight structure, they are highly susceptible to damage from too much heat. Cylinder head

temperature and oil temperature are used to indicate the operating temperatures of air-cooled aircraft engines, but both are too slow to provide much information about the amount of heat being released from the fuel. So, in recent years EGT systems have been developed to measure the temperature of the exhaust gas to indicate the efficiency of the combustion inside the cylinder. A thermocouple probe is inserted into an exhaust stack where it can measure the temperature of the flow of exhaust gases as they come out of the cylinder.

There is a direct relationship between the temperature of the exhaust gas and the mixture ratio being burned. In Fig. 1-1, we see that as the mixture ratio is leaned, the exhaust gas temperature rises until it reaches a peak, and then it drops off. This same relationship exists regardless of the amount of power the engine is developing. The actual temperature depends upon the amount of fuel being burned, but the peak temperature will always be reached with the same fuel-air ratio, and so the peak can be used as a reference for adjusting the fuel-air mixture ratio. In Fig. 1-2, we see a typical exhaust gas temperature gage. Some indicators have the actual temperature shown on the scale, but others such as this one, have an asterisk or a red line about 4/5 the way up the scale. Many EGT indicators begin their indication at 1,200° F and go up to

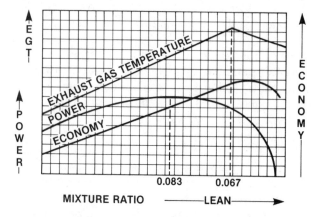

Fig. 1-1 Relationship between fuel-air mixture ratio, engine power, engine economy, and exhaust gas temperature.

Fig. 1-2 Exhaust gas temperature gage for a reciprocating engine. Some instruments have the degrees marked on the dial, and others have only an asterisk about 4/5 of the way up the scale.

2

1,700° F, and if they are not calibrated with numbers, they are adjusted so the pointer is opposite the asterisk when the probe is sensing 1,600° F.

To properly adjust the mixture ratio, the airplane is trimmed up for cruise flight and the power is adjusted with the propeller pitch control and the throttle to get the power required for cruise flight. The mixture is then leaned until the EGT of the hottest cylinder peaks. Then, the mixture control is moved back toward rich until the EGT drops about 25° F. This procedure is typical, but it may be different for different airplanes, and the recommendations of the aircraft manufacturer must be followed in detail.

b. Thermal efficiency

Aviation gasoline has a nominal heat energy content of 20,000 British thermal units per pound, and one Btu of heat energy will produce 778 foot-pounds of work.

When an airplane engine burns 12 gallons of aviation gasoline per hour, enough heat energy is released from the fuel to produce 566 horsepower. We find this by knowing that aviation gasoline has a nominal weight of six pounds per gallon, and so when the engine burns 12 gallons of fuel per hour it will release $12 \times 6 \times 20,000$ or 1,440,000 Btus of heat energy per hour, and this amount of energy will do $1,440,000 \times 778$ or 1,120,300,000 foot-pounds of work per hour. Since one horsepower is equal to 33,000 foot-pounds of work done in one minute, this fuel will release enough heat to produce 565.8 horsepower.

However, an aircraft engine burning 12 gallons of aviation gasoline per hour does not produce nearly this much power. A typical reciprocating engine burning fuel at this rate will produce only about 135 brake horsepower.

The brake thermal efficiency of the engine (the ratio of the amount of brake horsepower produced by the engine to the amount of horsepower in the fuel used to produce the power) is 135/566 or 23.8%. This is about typical for an aircraft reciprocating engine and is one reason much research remains to be done to make reciprocating engines fuel efficient.

c. Specific fuel consumption

We seldom use thermal efficiency to rate or to compare the performance of aircraft engines. Instead, we use a measure called specific fuel consumption. This is the number of pounds of fuel burned per hour for each horsepower developed, and when we use brake horsepower, we find the brake specific fuel consumption. The engine used in our example for thermal efficiency has a brake specific fuel consumption of 0.53 pounds of fuel per brake horsepower, per hour. We found this by using the formula:

$$\frac{\text{Pounds of fuel burned per hour}}{\text{Brake horsepower produced}} = \frac{72}{135} = 0.53$$

d. Production of Power

The horsepower produced by an aircraft engine may be computed by using the formula:

$$\text{Horsepower} = \frac{P\,L\,A\,N\,K}{33,000}$$

P = Brake mean effective pressure. This is the average pressure inside the cylinder during the power stroke.

L = Length of the stroke in feet

A = Area of the piston head in square inches

N = Number of power strokes in one minute. This is the RPM of the engine divided by two, since only every other stroke is a power stroke.

K = Number of cylinders in the engine

When we multiply the amount of pressure inside the cylinder, expressed in pounds per square inch, by the area of the piston head, we find the number of pounds of force that is pushing down on the piston. Then when we multiply this by the length of the stroke, which is the distance the piston moves during each power stroke, we have the number of foot-pounds of work done on each power stroke. The number of foot-pounds per stroke multiplied by the total number of power strokes per minute gives us the number of foot-pounds of work done each minute. And when we divide by the constant 33,000, we have the number of horsepower the engine is developing.

The pilot has no control over the area of the piston, the length of the stroke, or the number of cylinders in the engine, but he does have control over the pressure inside the cylinder and over the number of power strokes produced each minute. The throttle controls the cylinder pressure, and the propeller governor controls the RPM, or the number of power strokes per minute.

The pilot has two instruments to monitor the power the engine is producing. The tachometer shows the RPMs which relate directly to the number of power strokes per minute, and the manifold pressure gage relates to the pressure inside the cylinders.

Manifold pressure is measured in inches of mercury and is the absolute pressure inside the induction system of the engine. It is not the cylinder pressure, but since cylinder pressure is so difficult to measure, and since the pressure inside the induction system relates directly to this pressure, it is adequate for engine monitoring and control to use manifold pressure. The pickup for the manifold pressure gage may be at any convient point between the throttle valve and the intake valve in any of the cylinders.

e. Factors affecting engine power

(1) Density altitude

Density altitude is the altitude in standard air that corresponds to the existing air density, and it is an easy way to visualize the relative density of the air the engine is breathing. When the temperature of the air increases, it becomes less dense and the density altitude increases. The engine operates as though it were at a higher altitude.

The combination of fuel and air takes place in an engine on the basis of the weight (actually the mass) of the air and not its volume. So, as the air becomes less dense—that is, as the mass for its volume decreases—less oxygen is available in the cylinder to unite with the fuel.

(2) Humidity

An aircraft engine will produce less power on a humid day than it will on a dry day, because water vapor has only about 5/8 the mass of an equal amount of dry air. And since part of the air inside the cylinder is displaced by water vapor,

the charge inside the cylinder becomes less dense and there is less oxygen to unite with the fuel.

(3) Carburetor air temperature

When the air entering the carburetor is heated, it expands and the same mass of air occupies a greater volume. Its density decreases, and there are fewer molecules of oxygen for a given volume of air.

(4) Exhaust back pressure

An aircraft reciprocating engine is an air pump, and when the exhaust valve opens and the piston moves outward to force the burned gases out of the cylinder, the pressure of the air on the outside of the exhaust valve prevents some of the gases leaving. These trapped gases dilute the new charge in the cylinder so there will be less oxygen in the fuel-air charge. As the airplane goes up in altitude, the atmospheric pressure decreases and the diluting effect becomes less.

(5) Supercharging

Air is a physical mixture of gases in which the percentage of oxygen remains relatively constant, and if we increase the pressure of the air entering the engine, more oxygen will be available to combine with the fuel and more heat energy can be released.

As an airplane moves through the air, its forward motion rams air into the carburetor. This increases the power put out by the engine by about five percent. The power can be further increased by compressing the air with a mechanical compressor called a supercharger. Sea level power can be maintained to a high altitude by the supercharger increasing the intake air pressure to compensate for the decrease in air density with altitude.

(6) Compression ratio

The more the fuel-air mixture is compressed before it is ignited, the higher the cylinder pressure and temperature will be after combustion occurs. And the higher the temperature for a given amount of fuel and air, the lower will be the specific fuel consumption. There is, however, a definite practical upper limit to which we can raise the pressure and temperature in a cylinder, and this limit is imposed by detonation.

4

(7) Detonation

Detonation is a condition of uncontrolled burning which occurs inside a cylinder when the fuel-air mixture reaches its critical pressure and temperature. Under normal conditions, the fuel-air mixture is compressed and ignited by the two spark plugs, and as it burns, the flame fronts move across the face of the piston from both sides. Ahead of the flame front, the mixture is heated and further compressed.

With normal combustion, the cylinder pressure rises smoothly until it peaks about 20 degrees after the piston passes over the top of its stroke. This gives a smooth push to the piston. But, if for any reason the fuel-air mixture reaches its critical pressure and temperature, it will explode rather than burn, and instead of pushing smoothly on the piston, the pressure inside the cylinder will rise almost instantaneously and apply a sharp blow to the piston as we see in Fig. 1-3.

The pressure shock waves caused by the explosion travel at sonic speed and produce a ping, or knock. (This is easily heard in automobile engines but, because of other noises, it is not generally heard in an aircraft engine.) The rapid rise in pressure and temperature imposes extreme loads on such internal parts of an engine as the connecting rods, bearings, valves, piston heads, and combustion chamber walls, and it often leads to complete destruction of the engine.

(8) Preignition

Normal combustion inside an aircraft engine cylinder is started at an accurately controlled time in the operational cycle by two spark plugs, but if any flake of carbon or a feather-edge on a valve is heated to incandescence, it will ignite the fuel-air mixture before the correct time and the mixture will burn as the piston is moving outward. This longer burning period heats and compresses the fuel-air charge until it reaches its

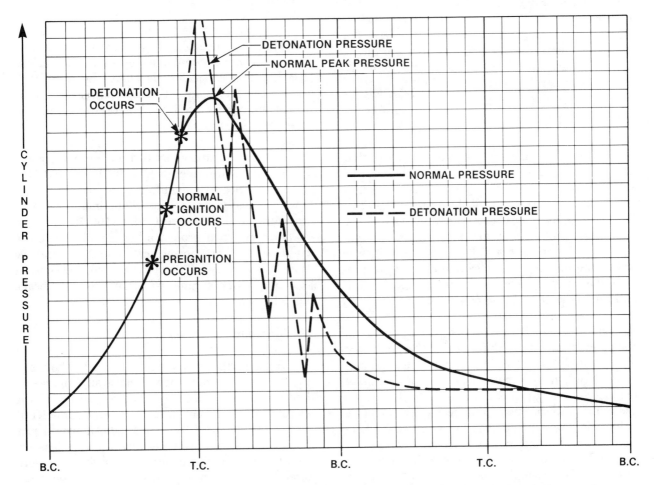

Fig. 1-3 Relationship between cylinder pressure and piston pistion in a reciprocating engine

5

critical pressure and temperature and then it detonates. The highest temperature in the cylinder will be concentrated at the same location each time the detonation occurs, and holes can be burned in pistons, and cylinder heads can be blown off by detonation.

Since detonation is caused by excessive temperature, its most effective control is to keep the cylinder temperature below the critical temperature of the fuel. And prevention of preignition is one way to keep the temperature within the allowable range.

(9) Fuel-air mixture ratio

The proper rate of fuel burning depends to a great extent upon the chemical composition of the gases inside the cylinder. If the mixture is correct for complete combustion, the flame will travel across the piston at a fast rate. But if the mixture is leaner than stoichiometric, the molecules of fuel will be more widely scattered and the flame will travel slower. The mixture will still be burning when it goes out the exhaust valve. This slow burning will not only decrease the amount of power the engine can develop, but it will also cause excessive heating, which is likely to lead to preignition and detonation.

(10) Ignition timing

All aircraft engines certificated in the United States must have dual ignition which requires each cylinder to have two spark plugs. Safety, in the event of failure of one of the ignition systems, is only one reason for this requirement. One of the most important reasons for using dual ignition is to assure that the flame will start on both sides of the piston and travel across it uniformly, to prevent the rapid and excessive buildup of pressure and temperature that can cause detonation.

Some engines have a diluted fuel-air charge in the cylinder near the exhaust valve, because they are not able to completely scavenge the exhaust gases. And, it is a common practice for these engines to use staggered timing; that is, to time the ignition so the spark plug nearest the exhaust valve will fire a few degrees of crankshaft rotation before the one on the other side of the cylinder. This early firing allows the flame to progress evenly across the piston and meet near the center of the piston head.

If an engine operates at high power with one of the spark plugs fouled, or if any ignition malfunction prevents one spark plug firing, the flame will travel completely across the piston head from one side, heating and compressing the fuel-air charge ahead of it as it goes. This type of single-spark operation will very likely cause detonation.

f. Engine performance charts

Before an engine is presented to the FAA for certification, the manufacturer must perform a series of tests and compile charts that show the relationship between manifold pressure, RPM, fuel consumption, and horsepower, both at sea level and at altitude.

(1) Sea level performance

In Fig. 1-4, we have a typical performance curve taken from an Operator's manual for a popular four-cylinder, fuel injected aircraft engine.

The scale across the bottom of the chart is the absolute manifold pressue, in inches of mercury, and the diagonal lines represent the various RPM's. The vertical scale along the right-hand side of the chart shows the horsepower the engine is developing for each combination of RPM and manifold pressure.

Let us use this chart to find the horsepower the engine is developing at 2100 RPM and 23.6 inches of mercury manifold pressure. Project a line vertically from the 23.6 inch manifold pressure until it intersects the 2100 RPM curve at point B. Then project a horizontal line to the right to the horsepower scale (point C), and we find that at this RPM and manifold pressure, at sea level, the engine will produce 112 brake horsepower.

(2) Altitude performance

As the airplane goes up in altitude, the maximum available horsepower naturally decreases because there is less oxygen in the air. If the engine were to be able to maintain the same RPM and manifold pressure at altitude, it would produce more horsepower because there is less exhaust back pressure for the engine to work against when it pushes the exhaust gases from the cylinders.

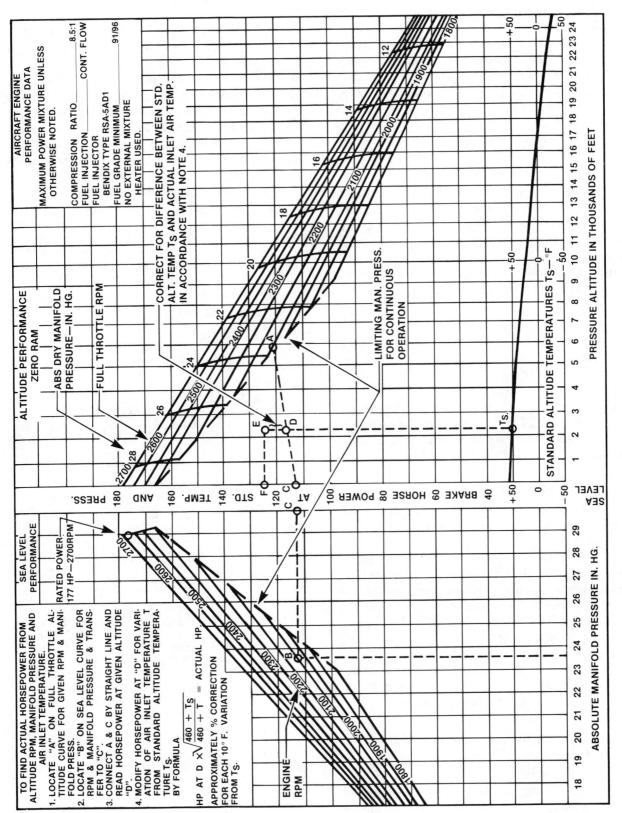

Fig. 1-4 Typical sea level and altitude performance chart

7

Locate the point of intersection of the 23.6 inches of manifold pressure and 2100 RPM on the altitude side of the chart. This is point A. Now transfer the sea level horsepower to the altitude chart at point C, and draw a line between points A and C. This line shows us the change in horsepower with altitude for a fixed manifold pressure and RPM combination. At the bottom of the altitude chart we have the pressure altitude scale, and we can draw a vertical line along the pressure altitude we are flying until it crosses the horsepower line we just drew in.

In our example, if we are flying at a pressure altitude of 2,300 feet, we would project a horizontal line to the left from point D and find that the engine will develop 118 brake horsepower if the air temperature is standard for this altitude. But if the air temperature is not standard, we must put in a correction.

The actual correction involves finding the square root of the ratio of the standard absolute temperature to the actual absolute temperature. But, for practical purposes, we can add one percent of the uncorrected brake horsepower for every ten degrees Fahrenheit that the actual temperature is below the standard temperature for that altitude, or we can subtract the same amount for temperatures above standard.

We can find the standard temperature for any altitude by using the bottom curve on the altitude chart: Project a horizontal line to the left from the intersection of the standard altitude temperature curve and the vertical line representing the pressure altitude we are flying. In our example, we are flying at a pressure altitude of 2,300 feet, and the standard temperature for this altitude is +50° F. Using our rule of thumb, and knowing from reading the outside air temperature gage that the outside air temperature is −10° F., we will add six percent of the brake horsepower to 118 and get a corrected horsepower of 125 brake horsepower. This is point F.

(3) *Full-throttle versus propeller load power curves*

Another useful set of curves furnished by the engine manufacturer is the power curves shown in Fig. 1-5. These curves relate the RPM, brake horsepower, and specific fuel consumption, both under dynamometer conditions with full throttle and with the load imposed by the propeller.

First, let's look at the curve relating RPM with brake horsepower. Both lines increase with RPM, showing that as the RPM increases, the brake horsepower increases. The two curves meet at 2700 RPM and 168 brake horsepower. This is the maximum power the engine can develop. At this power the BSFC is 0.59, which means that the engine will burn 0.59 pound of fuel per hour for every horsepower it produces. Because aviation gasoline has a nominal weight of six pounds per gallon, the engine will burn 16.52 gallons of fuel per hour at full throttle.

When the airplane is cruising at 65% power, the engine is developing 109.2 brake horsepower and its fuel consumption is 9.28 gallons per hour. (The BSFC for this power is 0.51, which gives us 55.92 pounds of fuel per hour, and this is 9.28 gallons per hour).

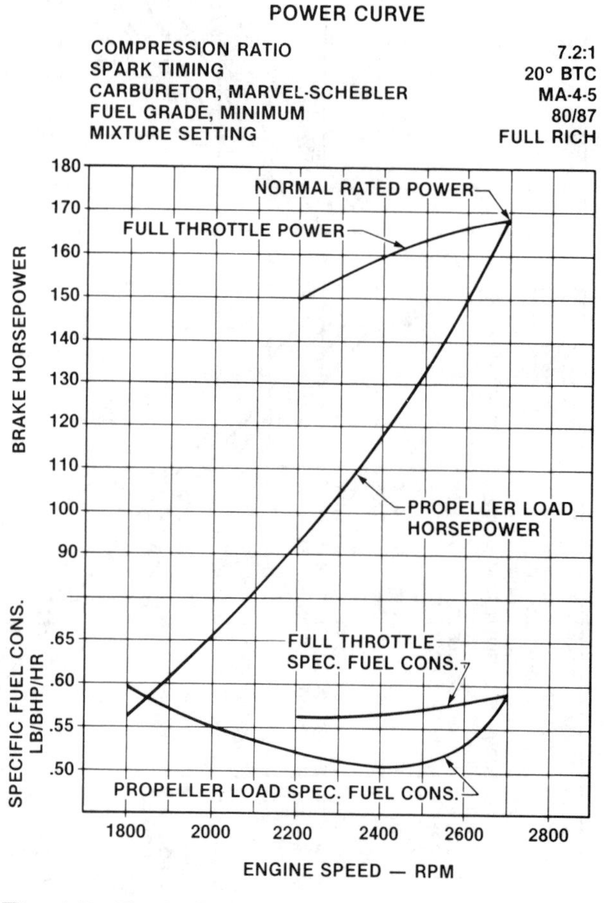

POWER CURVE

COMPRESSION RATIO	7.2:1
SPARK TIMING	20° BTC
CARBURETOR, MARVEL-SCHEBLER	MA-4-5
FUEL GRADE, MINIMUM	80/87
MIXTURE SETTING	FULL RICH

Fig. 1-5 Typical power curve for a four-cylinder reciprocating aircraft engine.

8

B. Reciprocating Engine Fuels

1. Requirements for aviation fuel

Selecting a fuel suitable for use in a particular aircraft is more involved than simply deciding whether we should use "regular" or "premium." The specifications for aviation fuels have been laid down by the petroleum industry and accepted and approved by the Federal Aviation Administration. Every certificated aircraft has in its Type Certificate Data Sheets a list of the fuel that is approved for its use. The use of improper fuel may cause engine failure, or it can at least reduce the power output of the engine below that required for the aircraft.

When a refinery produces a fuel for aviation use, it must consider two basic factors: the chemical and the physical characteristics of the fuel. Chemically, the fuel must have a high heat energy content, and it must be free from any constituents that will form acids or gums. It must have a high boiling point and a low freezing point. Its vapor pressure must be low enough that it will readily ignite from the spark plug, yet it must not be so low that it is hazardous for normal handling. Physically, the fuel must be free from contaminants, it must be easy to filter, and it must be pumpable at very low temperatures.

2. Aviation gasoline

Aviation gasoline is the most widely used fuel for reciprocating aircraft engines, but with gasoline prices increasing at such a fast rate, a good deal of study is being made of alternate fuels of aircraft. Alcohol fuels have a good probability of becoming the long-range substitute, but much research is needed before alcohol can replace gasoline.

One of the major contenders as a substitute for aviation gasoline is automobile gasoline which, while high in price, is not nearly as costly as aviation gasoline. Several studies have been made, and are being made, regarding the suitability of automobile gasoline for aviation use. And there seem to be some good arguments for adopting it—within limits, of course. But, because of the liability involved and because of the lack of control the FAA may have over the production and uniformity of automotive gasoline, both the engine manufacturers and the FAA have resisted all efforts to make automotive gasoline a legal substitute for aviation gasoline in certificated aircraft.

Aviation gasoline is a hydrocarbon fuel refined from crude oil. The crude petroleum is distilled, and the fractions, as they boil off, are condensed to form the various petroleum products. Gasoline produced in this manner is called straight-run gasoline, and it makes up the largest amount of the gasoline used in aircraft. Some of the heavier fractions which are unsuitable for use as aviation gasoline are further treated by a process known as cracking. Here the hydrocarbon is heated under pressure with a catalyst to break it down into products having high volatility which are suitable for use in gasoline.

a. Gasoline blends

(1) Paraffin series

This is the most stable series of hydrocarbons. They are clean burning and have a high heat energy content for their weight, but because they are so light, their heat energy content for a unit volume is low. The paraffin series all have very low boiling points.

(2) Cycloparaffin series

This is another stable series, sometimes called the naphthalene series. These products have a lower heat energy content per unit weight than the paraffins, but they have a higher boiling point.

(3) Aromatic series

These products tend to dissolve or swell rubber fuel lines, rubber tank liners, and diaphragms. They have a high freezing point and a high density, and they produce a good deal of smoke when they burn, but the anti-detonation characteristics of this series is very good.

(4) Olefin series

These are the most unstable of the hydrocarbons used in gasoline, and they combine with themselves through a process known as polymerization to produce a gum-like residue. On their credit side, they are clean burning and have a high boiling point and low freezing point. Because of their unstable nature, they are not found in natural petroleum products, but are formed in the cracking process.

(5) Blended gasolines

Because no one hydrocarbon series produces all of the desirable characteristics wanted in an aviation fuel, aviation gasoline is a blend of the various hydrocarbon series.

There are also in aviation gasoline such undesirable constituents as sulfur compounds, which combine with other elements to form acids which promote corrosion and damage the fuel pumps, valves, metering systems, and even the engine itself.

Gums and varnishes form from the combustion of gasoline and cause the piston rings and valves to stick. Some of these gums form during storage, especially if the fuel is exposed to sunlight or to elevated temperatures. And it is the sulfur in this gum which gives old gasoline its characteristic sour odor.

b. Gasoline ratings

(1) Heat energy content

Aviation gasoline is required to have a minimum of 18,700 Btu per pound, but its nominal rating is 20,000 Btu per pound.

(2) Reid vapor pressure

Liquid gasoline does not readily combine with oxygen, so in order to burn, it must be vaporized, or evaporated. A liquid evaporates when the pressure of the escaping gases is greater than the pressure of the air above the liquid, so a liquid may be made to evaporate by either lowering the pressure above it or by raising its temperature. The amount of pressure required to hold the vapors in a liquid is known as its vapor pressure, and it is expressed in pound per square inch at a specific temperature.

Vapor pressure is measured in a Reid vapor pressure bomb. The fuel to be rated is enclosed in a container where its temperature can be accurately controlled, and the pressure of the vapors above the liquid is measured at the test temperature. The allowable range of Reid vapor pressure for aviation gasoline is from 5.5 to 7.0 psi at 100° F.

If the vapor pressure of aviation gasoline is too low, the fuel will not vaporize properly and this will cause hard starting, especially in cold weather. But, if the vapor pressure is too high, the fuel will "boil" in the lines of the fuel system. Fuel vapors released in the lines by this boiling have a tendency to collect in high points and cause a vapor lock. A bubble of fuel vapor in the line, because of its compressibility, resists the flow of fuel from the tank to the carburetor or to the fuel pump.

(3) Critical pressure and temperature

When the fuel-air mixture in a cylinder reaches a certain pressure and temperature, it will explode rather than burn evenly, and this explosion is known as detonation.

In order to get the maximum amount of power and the lowest specific fuel consumption from an aircraft engine, the cylinder pressures must be raised as high as possible and this is usually done by increasing the compression ratio. But the maximum compression ratio is limited by the critical pressure of the fuel.

(a) Octane rating

In order to rate the fuel according to its critical pressure, a comparative rating system has been established. The detonation characteristics of two hydrocarbon fuels are used as references and a variable compression ratio test engine is used to establish the detonation characteristics. Iso-octane (C_8H_{18}), a member of the paraffin series of hydrocarbons, has a high critical pressure and desirable anti-detonation characteristics and is assigned a rating of 100. Normal heptane (C_7H_{16}), on the other hand, has a very low critical pressure and undesirable detonation characteristics, and so it is assigned a rating of zero.

For the rating test, the fuel is run in the test engine and the compression ratio is raised until a definite condition of detonation is produced. The test fuel is then switched out and a metering system is put into operation which feeds a mixture of iso-octane and normal heptane into the engine. The ratio of the octane and heptane is varied until the same detonation characteristics are obtained as were obtained with the fuel under test. If the blend of reference fuels is, for example, 80% octane and 20% heptane, the fuel is given an octane rating of 80.

The fuel-air mixture ratio determines the detonation characteristics of a fuel and, for a period of time, aviation gasoline was given a dual rating based on the mixture ratio. A test was run on the fuel using a rich mixture ratio, such as would be used for takeoff, and the octane rating was determined. Then the test was repeated, this time using a lean mixture ratio as would be used for cruise flight. The fuel was given a rating such as 80/87 which means that with a rich mixture, the fuel has a rating of 87 octane. With a lean mixture, however, its rating is 80 octane. This dual rating system has been superseded with a fuel grade rating, and the same fuel is now called grade-80 aviation gasoline.

(b) Performance number

When aircraft engines grew in size and power output, fuels were demanded that had anti-detonation characteristics that were better than those of iso-octane. In order to rate these fuels, various amounts of tetrethyl lead was added to the reference fuel to increase its critical pressure, and ratings greater than 100 were created. These ratings were not called octane numbers, but are performance numbers.

c. Gasoline additives

In order to get better anti-detonation characteristics from a particular aviation gasoline, tetraethyl lead, a heavy, oily, poisonous liquid is added. Grade 80 gasoline is allowed to have a maximum of 0.5 milliliter per U.S. gallon, and grade 100 gasoline is allowed to have as much as 4.6 milliliters per gallon. Tetraethyl lead allows engines to develop more power without detonation, but using a fuel with a lead content higher than the engine is designed to accommodate leads to problems of spark plug lead fouling and sticking valves.

In recent years, the economics of aviation fuel production have caused the petroleum industry to try to phase out the production of grade 80 gasoline, but the higher lead content of the grade 100 fuel makes it impractical for use in engines designed for the lower lead content grade 80 gasoline. To accommodate the lower lead engines and at the same time have a fuel with an octane rating high enough for the high compression engines, the petroleum industry has brought out a fuel called grade 100-LL. This low-lead 100-octane fuel has a maximum of two milliliters of lead per gallon, and it seems to be a workable compromise. The two milliliters of lead provides enough lubrication of parts requiring the lead, and at the same time its lead content is low enough that spark plug fouling and valve sticking is not a major problem.

Tetraethyl lead has a lower volatility than gasoline and under conditions of low power output or of uneven fuel-air distribution, some spark plugs may have their electrodes bridged over by a conductive lead oxide which completely shorts out the spark plug. A scavenging agent, ethylene dibromide, is added to the fuel to combine with the lead oxide and form lead bromide. This is more volatile than the oxide and it passes out the exhaust as a gas.

d. Fuel grades

Aviation gasoline is manufactured in several grades, depending upon the octane or perfomance number and the amount of tetraethyl lead it contains. The various grades are dyed for identification.

Grade 80 aviation gasoline was formerly called 80/87 gasoline, and it is dyed red. It may contain up to 0.5 milliliter of tetraethyl lead per gallon.

There was at one time a 91/97 aviation gasoline that was dyed blue, but this grade of fuel has been phased out. In its place, grade 100-LL or 100-octane low-lead gasoline is available, and it is dyed blue. This grade of fuel is allowed to have up to two milliliters of lead per gallon.

The popular 100-octane fuel, which was formerly called 100/130, is dyed green and it may have up to 4.6 milliliters of lead per gallon. This was the main fuel for the military and the airlines before jet aircraft took over in both of these areas. But now, of course, this vast market for 100-octane aviation gasoline has disappeared and the petroleum industry is trying to fill the need for this fuel with 100-LL with its reduced lead content.

Large, high-powered engines such as the Pratt and Whitney R-4360 require a fuel with better anti-detonation characteristics than the 100/130 fuel had, and the 115/145 aviation gasoline was brought out. This fuel is dyed purple.

3. Fuel contamination

Fuel contaminants clog fuel strainers and this has come to be the cause of many aircraft accidents.

a. Types of contaminants

(1) Solid particles

Sand and dust blown into the storage tanks or into the aircraft tanks during the fueling operation, as well as rust from unclean storage tanks, are common types of solid particles which clog fuel strainers and restrict the flow of fuel.

(2) Surficants

These partially soluble compounds are by-products of the fuel processing or may be caused by fuel additives. They tend to adhere to other contaminants and cause them to drop out of the fuel and settle to the bottom of the tank as sludge.

(3) Water

Though water has always been present in aviation fuel, it is now considered to be a major source of contamination, since modern airplanes fly at altitudes where the temperature is low enough to cause entrained, or dissolved, water to condense out of the fuel and form free water. This freed water can freeze and clog the fuel screens.

(4) Microorganisms

Airborne bacteria gather in the fuel, where they remain dormant until they come into contact with free water. The bacteria then grow at a prodigious rate as they live in the water and feed on the hydrocarbon fuel and on some of the surficant contaminants. The scum which they form holds water against the walls of the fuel tanks and causes corrosion.

b. Detection of contaminants

Draining a sample of fuel from the main strainers has long been considered an acceptable method of assuring that the fuel system is clean. But tests on several designs of aircraft have shown that this cursory sampling is not adequate to assure that no contamination exists.

In one test reported to the FAA, three gallons of water were added to a half-full fuel tank, and after time was allowed for this water to settle, it was necessary to drain ten ounces of fuel before any water appeared at the stainer. In another airplane, one gallon of water was poured into a half-full fuel tank, and more than a quart of water had to be drained before the water appeared at the strainer. The tank sumps had to be drained before all of the water was eliminated from the system.

A commercial water test kit is available to test for water in aircraft fuel. This kit contains a small glass jar and a supply of capsules containing a grayish-white powder. A 100-cc sample of fuel is taken from the tank or from the fuel truck and put into the jar, and a capsule of powder is dumped into it. The lid is screwed on, and the contents are shaken for about ten seconds. If the powder changes color from gray-white to pink or purple, the fuel has a water content of more than 30 parts per million, and the fuel is not considered to be safe for use. This test is fail-safe, meaning that any error in performing the test will cause an unsafe indication to be given.

c. Protection against contamination

All fuel tanks are required to have their discharge protected by an eight- to 16-mesh finger screen. Downstream of this finger screen is the main fuel strainer, which usually is either a fine wire mesh or a paper-type element.

Each fuel tank is normally equipped with a quick drain valve from which a sample of fuel may be taken on a preflight inspection. When draining the main strainer, some fuel should flow with the fuel tank selector set for each tank individually, because draining fuel when the selector valve is on the Both position will not necessarily drain all of the water that has collected in the fuel lines.

d. Importance of the proper grade of fuel

Aircraft engines are designed to operate with a specific grade of fuel and will not operate efficiently or safely if an improper grade of fuel is supplied to the engine.

The required grade of fuel must be placarded on the filler cap of the aircraft fuel tanks, and it is important to know that the required grade is being pumped into the tanks. The various grades of aviation gasoline are dyed for identification, and turbine fuel has a distinctive color and odor to distinguish it from gasoline.

If an improper grade of fuel has been inadvertently used, you should do the following:

IF THE ENGINE HAS NOT BEEN OPERATED

1. Drain all of the improperly filled tanks.

2. Flush out all of the fuel lines.

3. Refill the tanks with the proper grade of fuel.

IF THE ENGINE HAS BEEN OPERATED

1. Perform a compression check of all cylinders.

2. Inspect all of the cylinders with a borescope.

3. Drain the oil and inspect the oil screens.

4. Drain the entire fuel system, including all of the tanks and the carburetor.

5. Flush the entire system with the proper grade of fuel.

6. Fill the tanks with the proper grade of fuel.

7. Perform a complete engine run-up check.

4. Fuel handling

Aircraft maintenance technicians are often required to fuel aircraft and to maintain the fueling equipment. Each type of bulk fuel storage facility is protected from static electricity discharges and from contamination as much as is practical, and it is the responsibility of the operator of these facilities to assure that the proper grade of fuel is put into the fuel truck and that the truck is electrically grounded to the bulk fuel facility when it is being filled. All of the fuel filters should be cleaned before pumping, and all water traps must be carefully checked for any indication of water.

When the aircraft is fueled from a tank truck, it is the responsibility of the truck driver to posi-

tion it well ahead of the aircraft and to be sure that the brakes are set, so there will be no possibility of the truck rolling into the aircraft. The sumps on the truck storage tanks should be checked and a record made of the purity of the fuel. A fully charged fire extinguisher should be mounted on the truck ready for instant use if the need should arise, and static bonding wires should be attached between the aircraft and the truck, with a ground connected between the truck and the earth. A ladder or stand should be used if needed, and a wing mat should be put in place to prevent damage to the aircraft.

The fuel nozzle must be free of any loose dirt which could fall into the fuel tank, and when inserting the nozzle into the tank, take special care to not damage the light metal of which the tank is made. Be sure that the end of the nozzle doesn't strike the bottom of the tank. When the fueling operation is completed, replace the nozzle cover and secure the tank cap. Remove the wing mat and return all of the equipment to the truck and roll the hose and bonding wire back onto their storage reels.

5. Fire protection

All fueling operations must be done under conditions which allow a minimum possibility of fire. Never refuel an aircraft in a hangar, and defueling, as well, must be done in the open. Electrical equipment that is not absolutely necessary for the fueling opertion should not be turned on, and fueling must not be done where radar is operating, as enough electrical energy can be absorbed by the aircraft to cause a spark to jump and ignite the fuel vapors.

If a fire should break out, it can be extinguished either with a dry powder or with a carbon dioxide fire extinguisher. Soda-acid or any water-type fire extinguishers should not be used, because fuel is lighter than water and it will float away, spreading the fire.

C. Aircraft Float Carburetors

1. Principles of fuel metering

In order for an engine to develop its power most efficiently, the fuel must be mixed with exactly the correct weight of air. The volume of this mixture must be controllable by the pilot, and it must be uniformly distributed to all of the cylinders.

The mixture ratio between the fuel and the air must be variable in order to provide for either full power or for economy as the operating conditions require. And provision must be made to compensate the mixture ratio for variations in the air density caused by the changing temperature and altitude.

Absolute dependability is essential for an aircraft fuel metering system, and the system must operate efficiently under conditions of moisture, dust, vibration, and engine heat.

Modern aircraft engines are being operated with cylinder pressures so high that any mismanagement of the fuel-air mixture ratio can cause detonation that can destroy an engine in a very few seconds.

The fuel metering systems used with aircraft engines have evolved from a very simple drip-type system in which liquid gasoline dripped into a hot portion of the cooling water jacket and vaporized. Then the vapors were drawn into the cylinders.

The float carburetor that followed this primitive system has remained basically the same for the last sixty years or so. Float carburetors are simple and dependable, but they have several limitations that are primarily caused by their non-uniform mixture distribution and their susceptibility to carburetor icing. These problems have been solved to a great extent by the pressure carburetor and the fuel injection systems. We will discuss the float carburetor, the pressure carburetor, and two different types of fuel injection systems in some detail.

2. Systems of an aircraft float carburetor

a. Main metering system

This system provides a uniform fuel-air mixture that remains essentially constant as the airflow through the engine varies.

(1) Production of the pressure drop

All of the air that enters into the combustion process inside the cylinders of an aircraft engine must pass through the venturi in the carburetor.

A venturi is a specially shaped restrictor in the main air passage in the carburetor that converts the kinetic energy in the airflow into a pressure differential that is the heart of the fuel metering done by a float carburetor.

Energy, as we know, exists in two forms: potential, and kinetic. In the flow of a fluid, such as that of the air entering an engine, the kinetic energy relates to its velocity, and the potential energy relates to its pressure. And, according to the law of conservation of energy, we can neither create nor destroy energy, but we can change its form. If energy is neither added to nor taken away from the stream of air flowing into the engine, as the air passes through the venturi and speeds up because of the decreased area, the kinetic energy increases. But in order to keep the total energy constant, the potential energy, or the pressure of the air, must decrease. The main discharge nozzle in Fig. 1-6 is placed at the point in the venturi where the air velocity is the greatest and the pressure is the lowest.

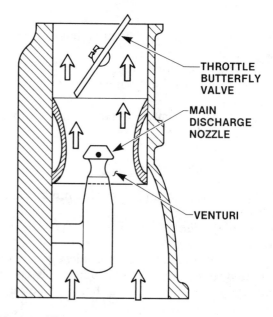

Fig. 1-6 *The venturi in a float carburetor decreases the air pressure at the main discharge nozzle by accelerating the air as it passes through the restriction.*

(2) Fuel metering forces

Fuel from the aircraft fuel tank is delivered to the carburetor by the aircraft fuel system, and inside the carburetor it first passes through a fine mesh wire screen, and then into the float bowl as we see in Fig. 1-7.

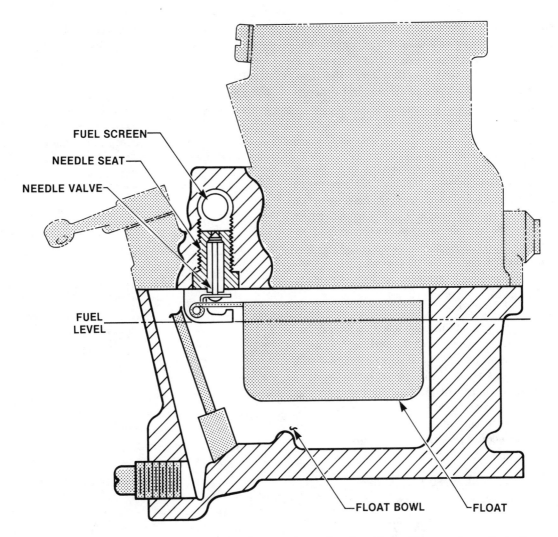

Fig. 1-7 The fuel level in a float carburetor is maintained by a float and needle valve.

A float-actuated needle valve maintains the fuel in the float bowl at a specific level which is just below the outlet of the discharge nozzle. As the fuel is used from the bowl, the float drops down and opens the needle valve, allowing more fuel to flow in and restore the level.

The fuel is metered by a balance of forces. One force is that required to lift the fuel in the discharge nozzle from its static level up to its outlet. This is called the fuel metering head. The other force is caused by the pressure differential between the atmospheric pressure inside the float bowl (P_1), and the lowered pressure at the discharge nozzle (P_2), caused by the increased velocity of the air as it flows through the venturi.

The fuel metering head remains relatively constant throughout the entire range of engine

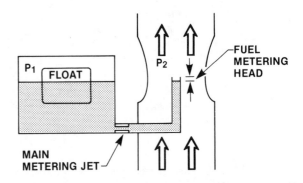

Fig. 1-8 The metering force in a float carburetor is determined by the pressure difference between air in the float chamber and that at the main discharge nozzle and by the distance the fuel must be lifted from its normal level to the lip of the discharge nozzle.

15

operation, but the air pressure differential varies with the volume of air taken into the engine. The more air, the greater the differential. The maximum amount of fuel that can flow from the float bowl is determined by the size of the main metering jet.

(3) Airbleed

One of the disadvantages of this type of arrangement is the uneven fuel-air mixture ratio that results as the airflow changes. The reason for this change in mixture ratio is easy to see when we consider the simple system we have in Fig. 1-8. The pressure differential between that at the discharge nozzle and that in the float bowl increases as the airflow increases, and as more fuel flows, the mixture becomes richer.

But, if we put an unrestricted airbleed between the main discharge nozzle and the ambient air, as we have in Fig. 1-9, the mixture will become leaner as the airflow through the venturi increases. The amount of air flowing through the airbleed will increase faster than the amount of fuel flow through the main metering jet, and the mixture will become progressively leaner.

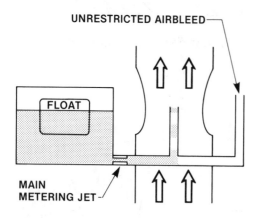

Fig. 1-9 An unrestricted airbleed allows the fuel-air mixture to become leaner as the airflow through the venturi increases.

We can put a jet, or restrictor, in the airbleed passage that will allow just enough air to enter the fuel passage to cancel the enriching tendency, and the fuel-air mixture will remain essentially constant as the airflow changes. The size of the airbleed is critical. Exactly enough air must be admitted to the fuel on the way to the discharge nozzle to keep the fuel-air mixture ratio constant.

When the airflow through the venturi is low, the pressure differential between the air at the discharge nozzle and that in the float bowl is relatively small, and there is a corresponding small flow of fuel through the main metering jet and air through the airbleed orifice. As the airflow increases, the pressure differential increases, and the flow of fuel and the flow of air both increase in an essentially constant ratio.

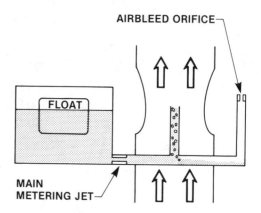

Fig. 1-10 By the proper choice of airbleed orifice and main metering jet sizes, the fuel-air mixture ratio may be maintained relatively constant as the air volume through the carburetor changes.

Another function of the airbleed is to aid in the atomization of the fuel. It introduces air into the stream of fuel, breaking it into tiny bubbles, or an emulsion of air and fuel. This emulsion is less dense than the liquid fuel and is brought up to the lip of the discharge nozzle more easily. And the larger surface area of the emulsion allows it to vaporize much more readily than if it were liquid fuel.

(4) Airflow limiter

All of the air that combines with the fuel in an aircraft reciprocating engine must pass through the carburetor venturi. The throttle butterfly valve in Fig. 1-6, located downstream of the venturi, controls the amount of air that flows into the engine under all conditions other than full throttle. When the throttle is wide open, the venturi becomes the airflow limiting device. This makes the size of the venturi critical, and it is chosen to provide the proper air velocity and the correct pressure drop for the volume of air needed by the engine.

The amount of air entering the engine at full throttle is determined by the pressure drop across the venturi, and increasing the pressure at the carburetor air inlet by ramming it in by forward speed or by turbocharging, increases the airflow into the engine. Any obstruction such as a clogged air filter will produce a corresponding power decrease because it restricts the airflow into the engine.

b. Mixture control systems

There are two ways to control the amount of fuel that is metered into air in a float carburetor: by varying the pressure drop across a fixed size orifice, and by varying the size of the orfice while maintaining a constant pressure differential across it. Both methods are used to control the fuel-air ratio in aircraft float-type carburetors.

(1) Back-suction mixture control

The back-suction mixture control we see in Fig. 1-11 varies the pressure in the float chamber between atmospheric pressure and a pressure slightly below atmospheric. This pressure variation is accomplished by using a control valve located in the float chamber vent line.

The float chamber is vented to the low pressure area near the venturi through a back-suction channel. This lowers the pressure in the float bowl. When the mixture control is in the Rich position, the vent valve is open and the pressure in the float bowl is raised to essentially the atmospheric pressure, and a differential pressure exists across the main metering jet. This causes fuel to flow out of the discharge nozzle. When the mixture control is moved to Lean, it closes the vent valve, and pressure in the float chamber is decreased to a pressure that is essentially the same as that of the discharge nozzle. This decreased pressure differential decreases the flow of fuel.

(2) Variable-orifice mixture control

A more widely used method of varying the fuel-air mixture ratio is by controlling the size of the opening in the fuel passage between the float bowl and the discharge nozzle. The float chamber has an unrestricted vent to maintain atmospheric pressure on the fuel in the float bowl, and either a needle valve or a step-cut rotary valve, such as the one we see in Fig. 1-12, is in series with the main metering jet. When the valve is completely closed, no fuel can flow and the engine cannot run. This is the Idle Cutoff position.

When the valve is open, fuel flows to the discharge nozzle and is metered by the mixture

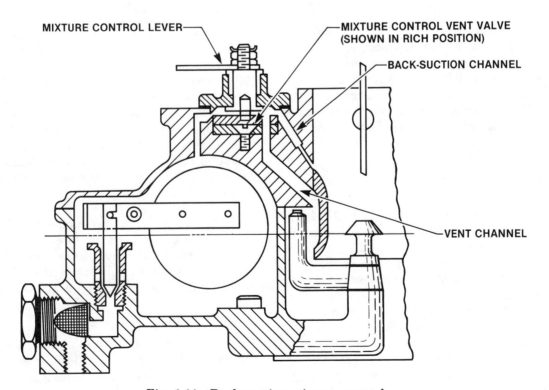

Fig. 1-11 Back-suction mixture control

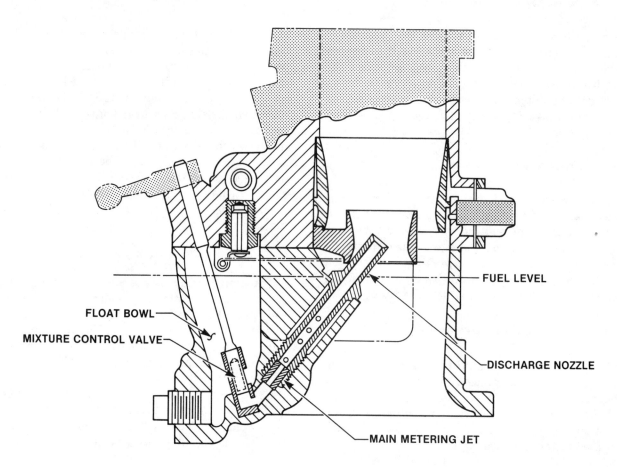

FLOAT BOWL

MIXTURE CONTROL VALVE

FUEL LEVEL

DISCHARGE NOZZLE

MAIN METERING JET

Fig. 1-12   Variable orfice mixture control

control valve as long as the area of the opening of the valve is smaller than the area of the main metering jet. When the mixture control is fully open or is in the Full Rich position, the area of the opening of the mixture control is larger than the area of the main metering jet, and the jet limits the amount of fuel which can flow to the discharge nozzle.

c. Idle system

At engine speeds below about 1000 RPM, the airflow through the venturi is not sufficient to produce a pressure drop at the main discharge nozzle great enough to pull fuel from the float bowl, and so an auxiliary system is provided. The throttle butterfly valve restricts the air which flows into the engine, and during idling, it is almost completely closed, and the only air which flows into the engine must pass around the edge of the disk. The velocity of the air at this point is naturally quite high, and the pressure at the edge of the valve is low.

In the wall of the throttle body where the butterfly valve almost touches, there are two or three small holes, or idle discharge ports. These ports are connected by an idle emulsion tube to a supply of fuel between the float bowl and the discharge nozzle. This emulsion tube contains the idle metering jet and the idle airbleed. The upper idle discharge port is fitted with a tapered needle valve to control the amount of fuel-air emulsion allowed to flow from the discharge ports when the throttle valve is closed.

The idle RPM of the engine is adjusted by controlling the amount the throttle valve is held open by the throttle stop screw, Fig. 1-14. The idle mixture which determines the efficiency of burning is adjusted to get the smoothest idling, or the lowest manifold pressure for the RPM.

As the throttle valve is opened, the edge of the butterfly valve moves down over the lower idle discharge holes and provides a smooth transition between idle RPM and the speed that pro-

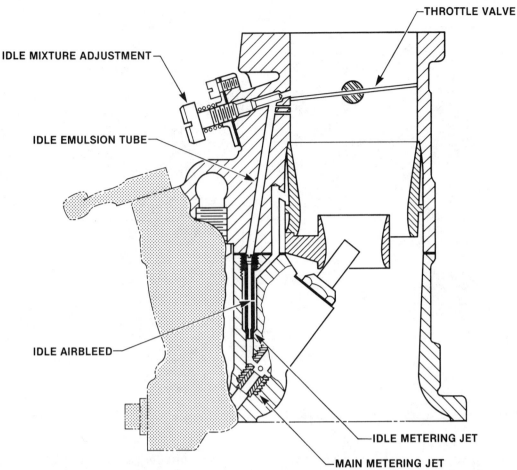

Fig. 1-13 Typical idling system in a float carburetor

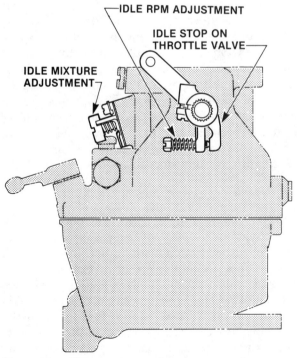

Fig. 1-14 Idle system controls on a float carburetor

vides enough airflow to pull fuel from the main discharge nozzle.

d. Acceleration system

Between the time the idle system loses its effectiveness and the time there is sufficient airflow for the main metering system to operate, there is a tendency for the engine to develop a "flat spot," or a point where there is insufficient fuel for continued acceleration. The acceleration system is installed to overcome this condition.

(1) Acceleration well

The acceleration system may be as simple as the acceleration well we see in Fig. 1-15. In this simple system, an enlarged annular chamber around the main discharge nozzle at the main airbleed junction stores a supply of fuel during idling. And this fuel is readily available between the airbleed and the discharge nozzle to produce

19

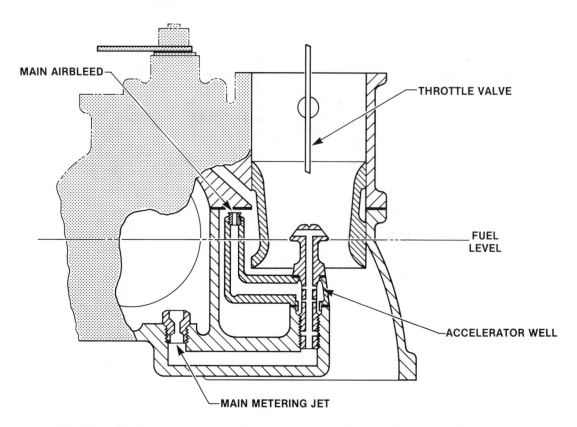

MAIN AIRBLEED

THROTTLE VALVE

FUEL LEVEL

ACCELERATOR WELL

MAIN METERING JET

Fig. 1-15 Accelerator well-type acceleration system used on small float carburetors.

a rich mixture when the throttle is suddenly opened.

(2) Accelerator pump

In Fig. 1-16, we see a typical accelerator pump which uses a leather packing held against the walls of the pump chamber by a coiled spring. The pump is actuated by a linkage from the throttle. When the throttle is closed, the piston moves upward, filling the pump chamber with fuel from the float bowl through the pump inlet check valve. When the throttle is opened, the piston moves downward, closing the inlet check valve and forcing the fuel out past the pump discharge check valve into the airstream through the accelerator pump discharge nozzle.

The piston is mounted on a spring-loaded telescoping shaft. When the throttle is opened, fuel is unable to discharge immediately because of the restriction of the nozzle, so the shaft telescopes and compresses the spring, and the spring pressure sustains the discharge, providing a rich mixture all during the transition period.

e. Power enrichment system

Aircraft engines are designed to produce a maximum amount of power consistent with their weight. But since they are not designed to dissipate all of the heat the fuel is capable of releasing, provisions must be made to remove some of this heat. This is done by enriching the fuel-air mixture at full throttle. The additional fuel absorbs this heat as it changes into a vapor. Power enrichment systems are often called economizer systems, because they allow the engine to operate with a relatively lean and economical mixture for all conditions other than full power.

(1) Needle-type enrichment system

One of the simplest types of power enrichment systems uses an enrichment metering jet in parallel with the main metering jet. In series with this jet is a needle valve that is operated by the throttle. When the engine is operating at all conditions other than full throttle, the spring holds the needle on its seat and no fuel flows to the economizer jet. But when the throttle is wide open, the economizer needle is pulled out of the valve and additional fuel flows to the discharge nozzle.

20

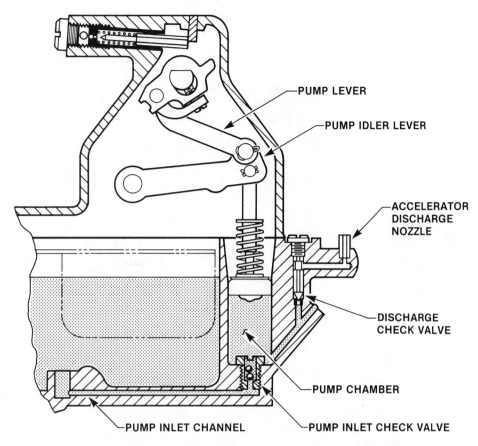

Fig. 1-16   Movable piston-type accelerator pump used on a float carburetor

PUMP LEVER

PUMP IDLER LEVER

ACCELERATOR
DISCHARGE
NOZZLE

DISCHARGE
CHECK VALVE

PUMP CHAMBER

PUMP INLET CHANNEL

PUMP INLET CHECK VALVE

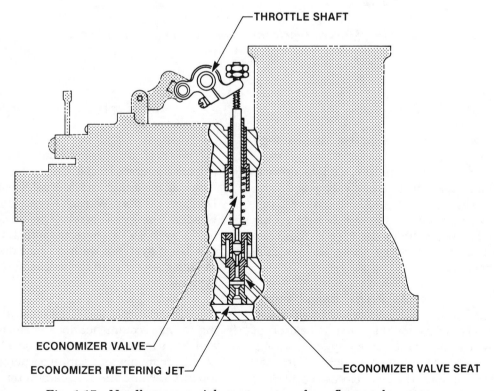

THROTTLE SHAFT

ECONOMIZER VALVE

ECONOMIZER METERING JET

ECONOMIZER VALVE SEAT

Fig. 1-17   Needle-type enrichment system for a float carburetor

21

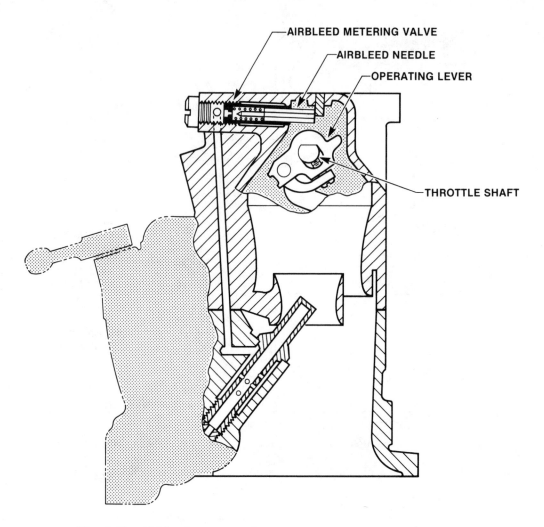

Fig. 1-18 Airbleed-type enrichment system for a float carburetor

(2) Airbleed enrichment system

When we increase the air velocity through the main venturi, we get an increased pressure drop that enriches the mixture, and to prevent this enrichment, an airbleed of a very precise size is used between the float bowl and the discharge nozzle. If we increase the size of this airbleed, we lean the mixture, and if we decrease it, more fuel is pulled from the discharge nozzle and the mixture becomes richer.

The air for the airbleed comes from the float chamber and passes through the airbleed metering valve. The needle for this valve is held off of its seat by a spring and is closed by an operating lever attached to the throttle shaft. When the throttle is wide open, the lever closes the airbleed valve and enriches the fuel-air mixture.

3. Service and maintenance of aircraft float carburetors

Most of the float carburetors used on certificated aircraft in the United States are manufactured by either the Marvel-Schebler/Tillotson Division of Borg-Warner Corporation or by the Bendix Energy Controls Division. Carburetors made by both of these manufacturers have the five basic systems we have just discussed, even though they differ in the mechanical details of the systems.

All carburetor servicing must be done according to the recommendations of the manufacturer, and both manufacturers have available a set of service manuals and approved parts lists for the aircraft technician. The carburetor, as with any other aircraft appliance, must be maintained in the configuration and with the parts list by which they were certificated, and the use of any part other than the ones specified will render the carburetor unairworthy and will compromise its functions.

a. Inspection

(1) Preflight

On each preflight inspection, the pilot should determine that no fuel is leaking from the carburetor. Leaking fuel is indicated by fuel dye stains on the carburetor body or in the cowling below the carburetor.

(2) One-hundred-hour or annual inspection

A much more comprehensive inspection of the entire fuel system is performed on the one-hundred-hour or annual inspection. The carburetor gets special attention as the fuel bowl is drained of any sediment, and the fuel strainer is removed and cleaned. The air filter is cleaned or replaced. Paper-type filters are usually replaced, and the cloth-covered wire screens are washed in solvent and blown out with compressed air.

All controls are checked to be sure there is no indication of binding or loosening, with particular attention paid to the throttle shaft and its connections. Looseness of the throttle shaft in its bushings allows an air leak which destroys the calibration of the carburetor.

b. Overhaul

There is normally no particular number of operational hours between overhauls specified by the carburetor manufacturer, but good operating practice dictates that at the time of engine overhaul, the carburetor should also be completely overhauled. Operating beyond this time can cause poor fuel metering, which could lead to detonation and subsequent damage to a freshly overhauled engine.

(1) Disassembly

Disassembe the carburetor in a clean work area where all of the parts may be laid out systematically. Have the latest carburetor service and overhaul instructions and follow them in detail.

(2) Cleaning

After the carburetor has been disassembled and a preliminary inspection made of all of the parts for wear or for breakage, clean the entire unit. First, wash all of the parts in a solvent such as varsol or Stoddard solvent to remove the grease and dirt. Dry the parts with compressed air, and then immerse them in a decarbonizing solution. Carburetor decarbonizer is normally of the water-seal or glycerine-seal type, in which the water or glycerine floats on top of the active ingredients to prevent evaporation of the highly volatile solvents. When placing the part in a decarbonizer, be sure it is completely covered by the active agent and is all below the seal layer.

A word of caution regarding some of the commercially available decarbonizers: Some are more active than others and will attack certain metals used in carburetors, so be sure to check the instructions before using them. Use extreme care that the decarbonizer does not get on your skin or in your eyes. If this should happen, wash the affected areas with running water, and if you get any in your eyes, get medical attention immediately. After the parts have been decarbonized for the appropriate time, remove them, rinse them in hot water and dry them throughly with compressed air.

(3) Inspection

After all of the parts are clean, carefully inspect them for any indication of wear or damage. Needle valves and seats should be closely examined for indications of grooves or scratches. The needle and seat in some models of Bendix carburetors may be lapped together to faciliate complete sealing, but all of the Marvel-Schebler carburetors require the float valve and seat assembly to be replaced with new units at each overhaul. All parts subject to wear should be checked against a table of limits or else replaced. Be sure to follow the manufacturer's overhaul instructions in detail.

(4) Parts replacement

The throttle shaft bushings are subject to the most wear and will likely have to be replaced. Bendix recommends about a 0.003-inch loose fit for the throttle shaft in the bushings. Drive the old bushings out, clean out the holes in the throttle body, and press the new bushings in with an

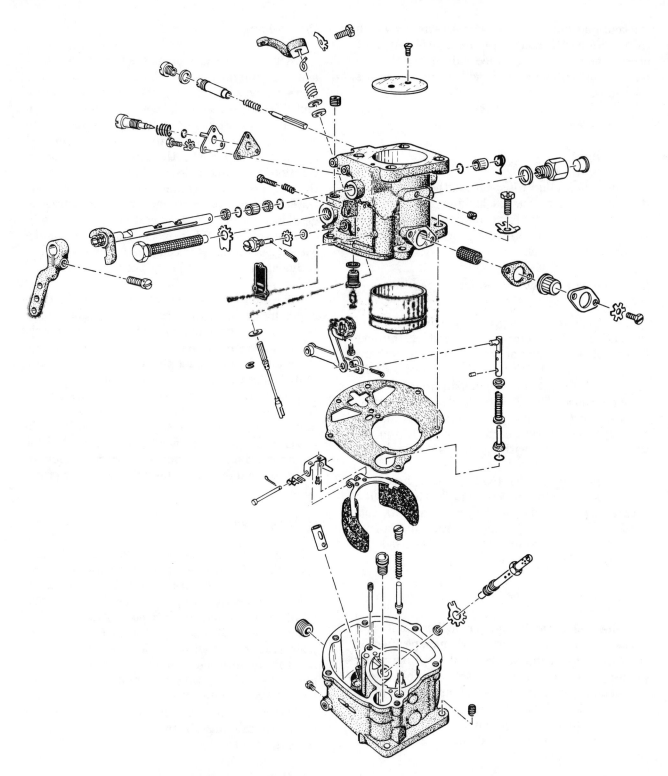

Fig. 1-19 Exploded view of typical aircraft float carburetor.

arbor press and the proper tool. When the bushings are in, line-ream them to assure a uniform fit between both bushings and the throttle shaft.

When reaming the bushings, always turn the reamer in its cutting direction, as it will be quickly dulled if it is turned in the opposite direction.

Some carburetors use Oilite bronze bushings held in place with Loctite compound. If this is the type of bushings used, be sure they are installed according to the manufacturer's instructions and be sure to use the proper alignment tool.

(5) Reassembly

When reassembling carburetors, many operations require the use of special tools, and it is poor economy to use anything other than the proper tool for some of these specialized jobs, as the use of an incorrect tool may damage a part costing far more than the tool.

Most carburetor bodies are made of aluminum alloy castings, and when installing straight plugs or jets, put a drop or two of engine oil on the threads. When installing tapered plugs, insert the plug into the casting for one thread, then put a very small amount of thread lubricant on the second thread of the plug. When you screw the plug into the hole it will squeeze the lubricant between the threads and prevent galling. Use extreme care to prevent any of the lubricant getting inside the carburetor, since it is insoluble in gasoline and it may plug up the jets or passages.

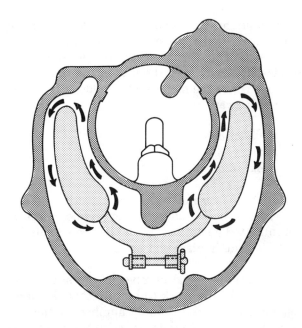

Fig. 1-21 *After the float level has been adjusted, a test fixture is placed over the float, and a gage is passed all around the float to be sure it does not hang up at any position.*

ly around the floats without binding. When the float level and side clearance are both correct, check the float drop as specified in the overhaul manual.

Some Bendix carburetors have the float mounted in the fuel bowl, and the *fuel level* must be checked on these carburetors. Level the carburetor body on a flow bench and put the test fuel into the bowl under the recommended pressure. Adjust the level the fuel rises in the bowl before it

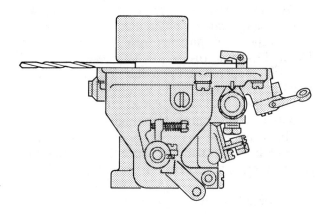

Fig. 1-20 *Method of measuring the correct float level*

Install the needle valve and float and adjust the float level in the way the manufacturer recommends. Most carburetors have the level adjusted by adding or removing shims between the needle seat and the throttle body. After getting the correct float level, check the side clearance of the float in the float bowl. This usually requires a special cut-away float bowl to check to see that a drill rod gage of specified size will pass complete-

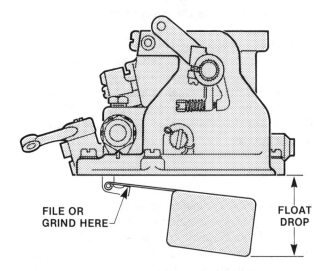

FILE OR GRIND HERE

FLOAT DROP

Fig. 1-22 *Method of measuring the float drop*

is shut off by the needle valve. Add or remove shims from under the needle seat until the fuel level is correct. When checking this level, be sure to measure away from the wall of the float bowl, because where the fuel wets the walls, it is higher than it is away from the walls.

c. Installation

(1) Mounting

After the carburetor has been overhauled and is ready to be installed on the engine, check it carefully to be sure that all of the proper safeties and seals are in place and are intact.

Record the serial number in the aircraft records and double check to be sure that the parts list number or the assembly number for the carburetor is correct for the particular aircraft on which it is to be installed. Mount the carburetor on the engine, using a new gasket, and torque the holddown nuts, as specified by the engine manufacturer.

Connect all of the controls and make sure that the cockpit controls move through their complete travel and the stop on the carburetor is reached before the stop in the cockpit. In the case of multi-engine aircraft, the controls for all engines must be aligned in both their Full Forward and Full Rearward positions. The carburetor air scoop should be installed with particular attention paid to the fully open and fully closed position of the carburetor heat air valve. The air filter should be checked to see that there is no leak around the filter element where unfiltered air could be drawn into the carburetor.

Before connecting the fuel line, it is good practice to open the fuel valve and drain a small amount of fuel out through the line to get rid of any foreign matter which might have accumulated while the carburetor was removed. This will also assure that there is adequate flow of fuel to the carburetor.

(2) Adjustment

After an overhauled carburetor is mounted on the engine, the engine must be run and the idling RPM and manifold pressure adjustments made. Before attempting to adjust anything on a carburetor, make sure that all of the other systems of the engine are functioning properly.

When the engine is properly warmed up, run it up to full power and check for the proper static RPM and manifold pressure. Check the magnetos to be sure that both ignition systems are operating properly and that the magneto drops are well within the allowable limits.

When you know that the engine is functioning properly, pull the throttle back and hold the RPM at the desired speed with the throttle. While holding the RPM constant, adjust the idle mixture until the engine runs smoothest. And if the engine is equipped with a manifold pressure gage, the operator in the cockpit should signal when the manifold pressure is at its lowest value. This indicates that the mixture is properly set.

After making this adjustment, run the engine up to near Full Throttle, to clear the spark plugs of any fouling that might have occurred during this prolonged idling. Return the throttle to the desired RPM and adjust the idle speed screw until it just contacts the throttle stop. Increase the RPM and bring the throttle back against its stop, to be sure there is no slippage in the linkage and that the engine idles at the same RPM each time the throttle is retarded.

A final check of the correctness of the idle adjustments is made by pulling the mixture control back to the cutoff position with the engine idling. The RPM should rise somewhere between 10 and 35 RPM, depending upon the engine, before it stops running. This slight rise before a complete dropoff shows that the idle mixture is properly adjusted. A clean dropoff without any preceding rise indicates that the mixture is too lean. But if the rise is more than that specified for the particular engine, the mixture is too rich.

After all of the proper checks indicate that the carburetor and engine are functioning properly, a final inspection must be made of the installation. Check for indication of fuel leaks, loose rigging, missing safeties, or any incomplete work. Signing off the work in the engine records completes the installation procedure.

d. Servicing

(1) Troubleshooting

For some reason, the carburetor seems to be a mysterious component and comes away with more than its share of blame for performance deterioration. Fortunately, though, careful trouble-shooting will eliminate the carburetor as the offender more often than not. Carburetors seldom change their calibration or operating characteristics without some outside influence, and these changes usually build up over such a time period that a warning is given.

Before assuming a carburetor to be at fault when troubleshooting an engine problem, isolate the carburetor.

a. Be sure both ignition systems check out and have the proper RPM drop when checking each system separately.

b. Be sure the engine develops the proper RPM and manifold pressure when making a Full-Throttle static runup.

c. Be sure that all cylinders have the proper compression.

d. Be sure the carburetor air filter is clean and the proper filter element is installed.

e. Be sure that the proper fuel pressure is supplied to the carburetor.

f. Be sure the fuel filter is clean.

g. Be sure the proper propeller is installed.

h. If the propeller is a constant-speed propeller, be sure the governor is properly adjusted.

i. Be sure the muffler is not causing an excessive amount of back pressure in the exhaust system.

j. If the performance deterioration has been noticed as a drop in RPM, be sure the tachometer is accurate.

After all other possibilities have been eliminated and the fault persists, the carburetor may be legitimately suspected. If it is at all possible, before opening a carburetor, exchange the suspected carburetor with a similar one known to be good, and then run the engine. If the trouble is still present, the carburetor is not at fault; but if the trouble no longer exists, the carburetor is pretty well proven to be the culprit.

Float carburetor servicing logically divides itself into three levels, and the level attempted by an A&P technician should be governed by his experience and by the availability of the proper replacement parts and the required service equipment.

(2) Level I carburetor servicing

The first level of service may be considered as all that can be done with the carburetor on the aircraft. It consists of checking the control linkages for proper travel, freedom of movement, and proper contact with the stops. The fuel lines and air hoses are checked for kinks or distortion or indication of leakage. The main fuel filter and carburetor strainer are checked on this level of service to assure the proper amount of fuel is being delivered to the carburetor. The air filter may be checked for the proper installation, and to be sure that there is no air leaking around the filter. The carburetor heat valve should be checked for proper travel and for any leakage of warm air into the carburetor when the control is in the Cold position.

The idle RPM and mixture can be adjusted to provide the smoothest running engine at the closed throttle speed recommended by the aircraft manufacturer.

While this level of service requires a knowledge of engine and carburetor operation and the same professional integrity as any other aircraft maintenance, it can be performed with a very minimum of stock and equipment. Replacement air filters and fuel filter gaskets are the only parts required, and no special tools are needed for this level of servicing.

(3) Level II carburetor servicing

Maintenance facilities which perform work beyond that normally considered as line maintenance do what might be considered as a second level of servicing. This level requires that the carburetor be removed from the aircraft and opened for closer inspection and replacement of parts. This level of maintenance should never be attempted without the proper replacement parts available.

Level II servicing consists essentially of removing the carburetor from the engine and thoroughly cleaning and inspecting its exterior.

If the carburetor has been flooding, the needle and seat may be replaced with new parts having the proper part numbers, and the float level adjusted. All interior passages should be inspected with an otoscope while the carburetor is open, and any obstructions should be blown out with clean compressed air. The carburetor is reassembled, using new gaskets. It is safety wired and reinstalled on the engine, where it is given the normal run-up tests, final adjustments, and safeties.

(4) Level III carburetor servicing

Level III servicing is major overhaul. The carburetor is completely disassembled, and all parts are cleaned with the recommended carburetor cleaner and rinsed and dried with compressed air. All passages are carefully inspected with an otoscope, and the carburetor is reassembled. All parts recommended by the manufacturer and included in the carburetor overhaul kit are replaced. The carburetor manufacturer's recommendations must be followed to smallest detail, and no alterations should ever be made with hopes of "improving" on the manufacturer's engineering and experience.

D. Pressure Carburetors

1. Characteristics of pressure carburetors

The float carburetors have several notable limitations; among them are:

Susceptibility to icing

Uneven fuel-air mixture distribution

They are critical with respect to altitude

Metering is a function of the volume of air, not its mass

Most of these limitations are overcome in the pressure carburetor. The pressure carburetor uses a closed fuel system; that is, one that is not open to the atmosphere at any point from the tank to the discharge nozzle, as it is in the float bowl of a float carburetor. Fuel leaves the tank under pressure from the boost, or auxiliary, pump and goes through the filter and the engine pump to the carburetor. Here the fuel is metered and is directed to the discharge nozzle.

Pressure of the fuel delivered to the metering jet is controlled by the volume and the density of the air flowing into the engine. In this way, the fuel flow becomes a function of the mass airflow.

The Bendix PS7BD carburetor shown in Fig. 1-23 is typical of the modern pressure carburetors used on light reciprocating engine aircraft and is the unit we will discuss in this text.

a. Air metering force

Air flows into the engine, passing first through the air inlet filter and then into the carburetor throttle body. Then through the venturi, past the throttle valve and discharge nozzle, into the intake manifold. As the air enters the carburetor body, some of it flows into the channel around the venturi where its pressure increases due to its decrease in velocity. This impact pressure is directed into chamber A of the regulator unit, and any change in the pressure of the air entering the carburetor changes the pressure in chamber A.

Air flowing through the venturi produces a low pressure that is proportional to the velocity of the air entering the induction system. This low pressure is directed into chamber B of the regulator, where it operates on the opposite side of the diaphragm from the impact air pressure. These two forces work together to move the inner regulator diaphragm proportional to the volume of the air entering the engine.

The mass, or weight, of the air is a function of the air density, and to modify the effect of air volume to reflect its density, an automatic mixture control is placed in the vent line between the two air chambers. When the air density is decreased, the automatic mixture control opens the vent, decreasing the pressure drop across the diaphragm, and lowers the metering force. (See Fig. 1-24, page 30.)

b. Fuel metering force

Fuel enters the carburetor from the engine pump under a pressure from approximately nine to 14 pounds per square inch, and passes through a fine mesh wire screen into chamber E, on its way to the poppet valve.

The amount the poppet valve opens is determined by the balance between the air metering force and the regulated fuel pressure. The air

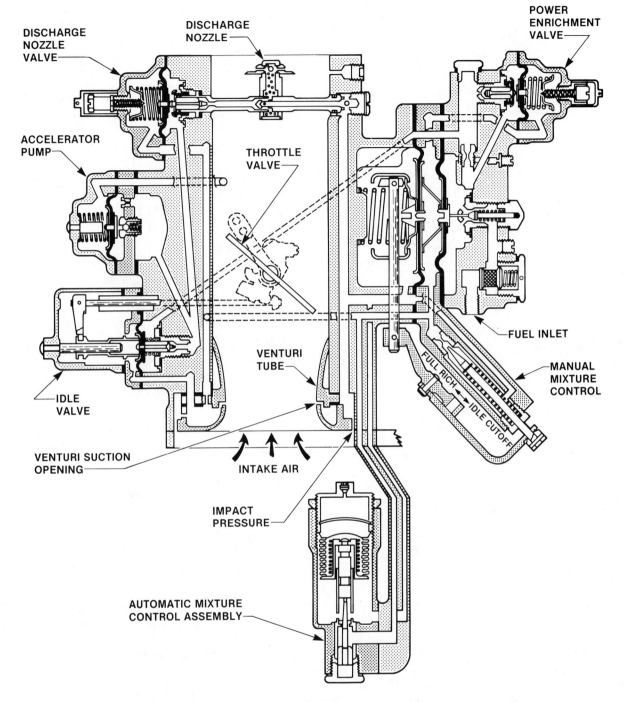

Fig. 1-23   Bendix PS7BD pressure carburetor

metering force moves the diaphragm to the right, and opens the poppet valve.

When the poppet valve opens, the fuel flows from chamber E into chamber D and exerts a force on the outer regulator diaphragm, moving it back enough to allow the spring to close the poppet valve. The fuel pressure in chamber D is, in

this way, regulated so that it is proportional to the mass of the air flowing into the engine.

When the engine is idling without enough airflow to produce a steady air metering force, the large coil spring in chamber A forces the diaphragm over and opens the poppet valve to provide the fuel pressure required for idling.

29

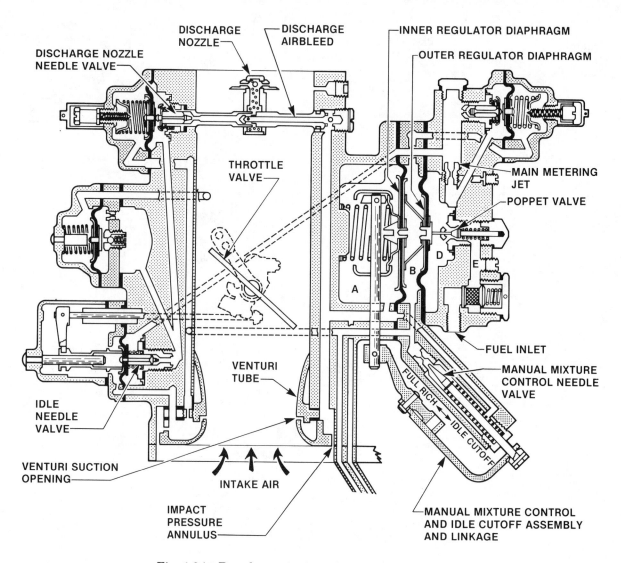

Fig. 1-24   Regulator section of a pressure carburetor

Fuel from chamber D, regulated but unmetered, flows through the main metering jet and through the idle needle valve. For all conditions other than idle, this valve is off its seat enough that its opening is larger than that of the main metering jet, and so no metering is done by the idle valve. Fuel flows from the idle valve to the discharge valve and the discharge nozzle where air from the impact annulus is mixed with the fuel in the nozzle to produce a spray for better vaporization.

The spring-loaded diaphragm type discharge valve provides a fast and efficient cutoff when the mixture control is placed in the Idle Cutoff position. When the fuel pressure drops low enough, a spring forces the needle valve onto its seat, stopping all flow of fuel from the nozzle. This valve

also provides a constant pressure downstream of the metering jet; so the variable pressure from the regulator will force a flow through the jet proportional to the mass airflow.

c.   Mixture control system

As altitude increases, the air density becomes less, and unless a correction is made for this, the mixture will become richer and the engine will lose power. To maintain an essentially constant fuel-air mixture, the pilot must decrease the weight of the fuel flowing to the discharge nozzle. This is done by decreasing the pressure differential across the air metering diaphragm (the inner regulator diaphragm) by opening the bleed between the two chambers.

When the pilot wishes to stop the engine, he pulls the mixture control to the Idle Cutoff position. The mixture control needle valve is pulled back so that the pressures in chambers A and B are essentially equalized. When the control is in this position, the idle spring is depressed by the release contact lever, and its force is removed from the diaphragm, closing the poppet valve, and shutting off all of the flow to the metering sections.

An automatic mixture control relieves the pilot of the necessity of regulating the mixture as altitude changes. A brass bellows, filled with helium and attached to a reverse-tapered needle, varies the flow of air between chambers A and B as the air density changes and an inert oil in the bellows damps out vibrations. This automatic mixture control, by varying the amount of air bleed between the two chambers, maintains a

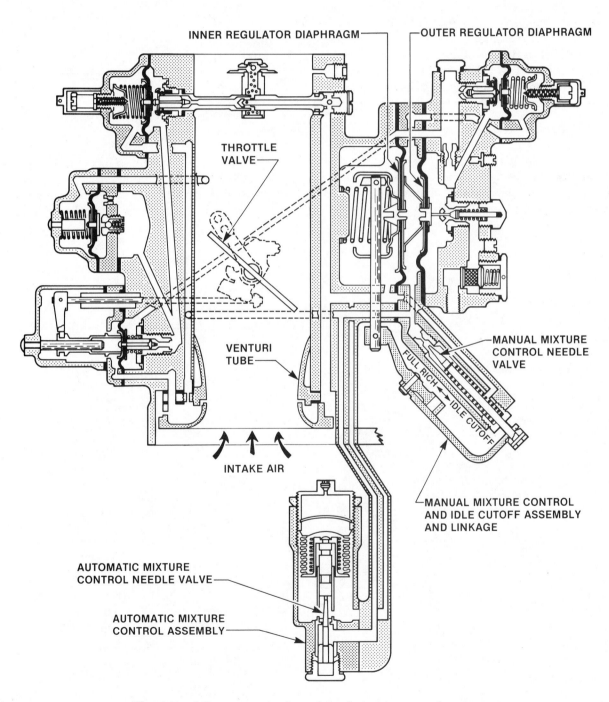

Fig. 1-25   Mixture control section of a pressure carburetor

31

pressure differential across the air diaphragm appropriate for any air density.

d. Idle system

The idle system of the Bendix PS carburetor controls both the idle air and the idle fuel. All of the air which flows into the engine during idle must flow around the almost closed throttle valve, and the amount this valve is held away from closing is controlled by the idle speed adjustment, which is an adjustable stop on the throttle shaft extension. The amount of fuel allowed to flow during idling is regulated by the amount the idle fuel valve is held off of its seat by the control rod, as it contacts a yoke on the throttle shaft extension. We see this in Fig. 1-26.

e. Acceleration system

A single-diaphragm pump is used on most carburetors of this type to provide a momentarily rich mixture at the main discharge nozzle when the throttle is suddenly opened.

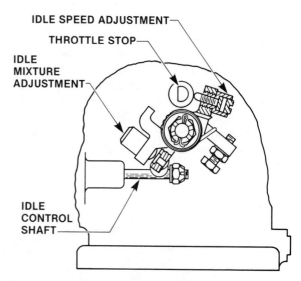

Fig. 1-26 Idle speed and mixture adjustments on a pressure carburetor

The accelerator pump is located between the idle valve and the discharge nozzle, with one side of the diaphragm vented to the manifold pressure downstream of the throttle. The other side of the

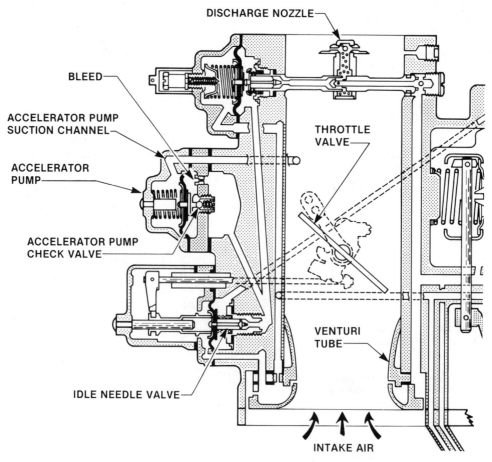

Fig. 1-27 Acceleration system for a pressure carburetor

diaphgram is in the fuel line between the main metering jet and the discharge nozzle. The coil spring in the air side compresses when the manifold pressure is low, and fuel fills the pump. When the throttle is opened, the manifold pressure increases and the spring pushes the diaphragm over. This forces the fuel out of the discharge nozzle, momentarily enriching the mixture.

Some pumps, such as the one in Fig. 1-27, have a divider in the fuel chamber with a combination check valve and relief valve and a bleed. The valve allows a rapid discharge of fuel when the throttle is first opened, but it soon seats, and a lesser but sustained flow of fuel discharges through the pump bleed. When the throttle is suddenly closed, the decrease in manifold pressure causes a rapid movement of the pump diaphragm and the check valve closes to prevent the pump starving the discharge nozzle. This would cause the mixture to go momentarily lean.

f. Power enrichment system

(1) Manually controlled power enrichment valve

A double-step idle valve is used on some pressure carburetors. For operation up to approximately 65% power, the needle in the orifice limits the flow of fuel, but at powers above 65%, the pressure differential across the diaphragm is great enough to pull the needle completely out of the orifice, and the fuel flow is limited by the main metering jet.

(2) Airflow power enrichment valve

A spring-loaded valve is located in the fuel passage parallel with the main metering jet, Fig. 1-29. When this valve is closed, the main metering jet limits the flow, providing a lean mixture for cruise. At the higher power settings, the venturi air pressure and unmetered fuel pressure are great enough to overcome the spring force and open the valve, enriching the mixture.

2. Pressure carburetor installation, servicing, and maintenance

A pressure carburetor is a precision piece of equipment that requires the use of a flow bench for any major maintenance or calibration. The manufacturer has established a network of authorized service facilities equipped to handle any maintenance with a minimum of down time for the aircraft. Installation, inspection, and troubleshooting are, however, a routine part of the work of the aviation maintenance technician.

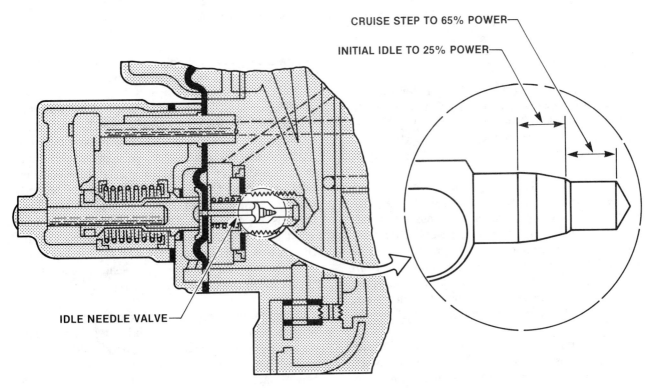

CRUISE STEP TO 65% POWER

INITIAL IDLE TO 25% POWER

IDLE NEEDLE VALVE

Fig. 1-28 Manually controlled power enrichment valve

33

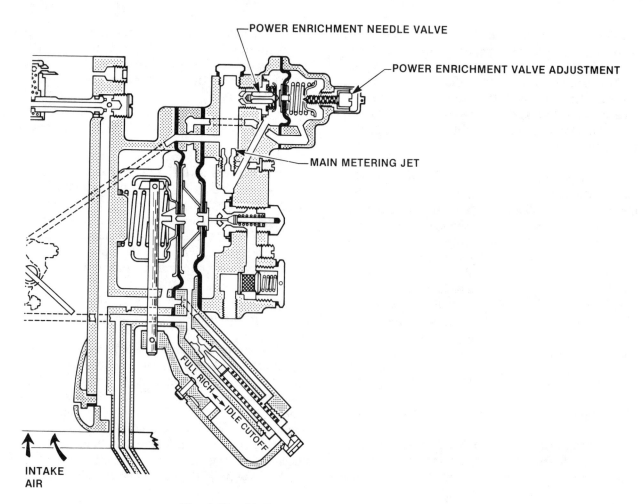

POWER ENRICHMENT NEEDLE VALVE

POWER ENRICHMENT VALVE ADJUSTMENT

MAIN METERING JET

FULL RICH

IDLE CUTOFF

INTAKE
AIR

Fig. 1-29 Airflow-type power enrichment valve

a. Installation

When a pressure carburetor is in storage, it is filled with a preservative oil, and it must be conditioned, or "soaked," before the fuel diaphragms have the exact pliability they had when the carburetor was calibrated.

Remove the shipping plugs and drain all of the preservative oil from the regulator unit. Install the carburetor, using a new gasket, and attach the fuel line as soon as practical. Then turn the fuel supply on and place the mixture control on the Rich position. Remove the pipe plug from the bottom of the regulator unit, and turn the boost pump on until clear, oil-free fuel flows out of the regulator and the discharge nozzle.

At this point, place the mixture control in the Idle Cutoff position and turn the boost pump off. Allow the carburetor to stand with the fuel in it for as long a period as practical. Eight hours is

recommended, to completely condition the diaphragms to their calibrated condition.

Then connect the fuel pressure gage line, the vapor vent return line, and the throttle and mixture control linkages. When screwing any tapered fittings into the carburetor, be very careful that only the correct amount of thread lubricant is used, and that this is placed only on the second thread from the end of the male fitting. The threads on both the fitting and in the housing must be perfectly clean and free of burrs. Apply the thread lubricant and tighten the fitting only enough to assure a leak-tight seal, as excessive tightening may crack the housing.

The throttle and mixture control must operate freely throughout their full range of travel, and the stops on the carburetor should be contacted before the stop on the cockpit control. Springback in the control system is your assurance that the carburetor control is fully actuated.

Install the air filter and check the alternate air mechanism to determine that in both the direct and alternate position, the valve allows an unrestricted flow of air.

b. Operation

Start the engine and allow it to warm up until the oil temperature is in the proper operating range. Check the magnetos to be sure that both ignition systems are operating properly and the engine develops full power. The engine fuel pump pressure must be within plus or minus one pound of the inlet pressure recommended for the particular carburetor. If an adjustable pitch propeller is used, place its control in Low Pitch, and turn the electric fuel boost pump On.

c. Idle adjustment

Adjust the throttle to approximately 600 RPM and slowly pull the mixture control to the Idle Cutoff position. Watch the tachometer as you lean the mixture—the engine should pick up approximately ten RPM before it begins to cut out. Be sure to return the mixture control to the Full Rich position before the engine cuts out completely. A rise of more than ten RPM indicates that the idle mixture is too rich, and an immediate decrease in RPM without the momentary rise is indicative of too lean an idling mixture. If the engine is equipped with a manifold pressure gage, it may be used along with the tachometer. The correct mixture is indicated when the manifold pressure holds steady as the mixture control is moved toward Idle Cutoff and then rises as the RPM decreases. If the pressure drops and then rises, the idle mixture is too rich. If it increases immediately, the mixture is too lean.

Adjust the idle mixture by moving the idle adjustment screw one or two notches in or out as needed, and repeat the check procedure until you have the proper indication. When adjusting the idling, clear the engine by momentarily opening the throttle to approximately 2000 RPM each time a change is made in the setting. After making the proper mixture adjustment, set the idling RPM and adjust the throttle stop. Check to see that the engine returns to the same RPM each time the throttle is pulled back.

d. Enrichment valve adjustment

Carburetors with the manual enrichment valve we saw in Fig. 1-28 must have the enrichment valve setting checked after the idle adjustments are made. This valve has a cruise step on the idle valve, and any repositioning for idle may affect the position from which the enrichment system operates. Put the proper adjusting gage directly over the throttle stop as we see in Fig. 1-30 and open the throttle until the lever contacts the gage. Hold the throttle in this position and adjust the enrichment valve adjusting screw until it just contacts the control rod. Tighten the locknut and safety. If the carburetor is equipped with an airflow-type enrichment system, the opening of the enrichment valve is accomplished by the airflow through the carburetor, and this adjustment must be made on a flow bench. It is not a field adjustment.

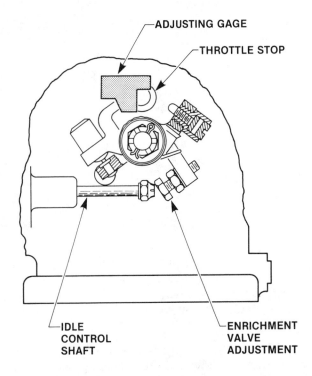

Fig. 1-30 Method of adjusting the manual power enrichment valve

e. Metered fuel pressure adjustment

The basis of metering used by the Bendix pressure carburetors is a varying pressure differential across the fixed metering jet. This differential is achieved by holding the pressure of the discharge fuel constant with a constant pressure discharge valve. A small amount of adjustment in the field is permissible on certain models of this carburetor by adjusting the tension on the discharge valve spring.

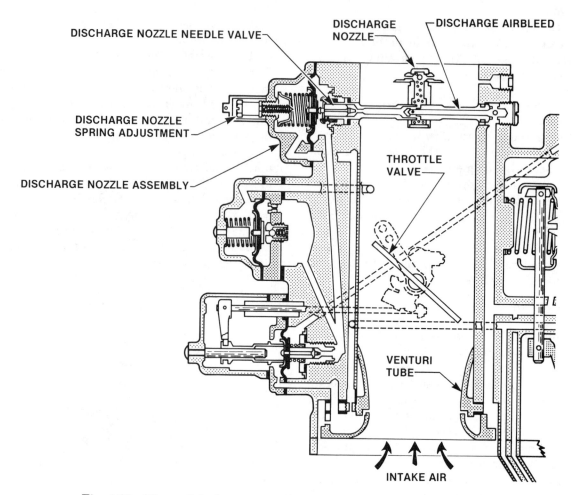

DISCHARGE NOZZLE NEEDLE VALVE

DISCHARGE NOZZLE

DISCHARGE AIRBLEED

DISCHARGE NOZZLE SPRING ADJUSTMENT

DISCHARGE NOZZLE ASSEMBLY

THROTTLE VALVE

VENTURI TUBE

INTAKE AIR

Fig. 1-31   Metered fuel pressure adjustment on a pressure carburetor

This adjustment may be made on only those carburetors that have discharge nozzle adjustment limiters. Carburetors without these adjusters could be adjusted lean enough to damage the engine.

If the engine operates rough in the RPM range between idle and cruise, it is possible that the discharge pressure could need adjusting. Before making this adjustment, it is especially important that the engine be checked for proper operation of all systems that could affect its operation. Engine oil temperature and cylinder head temperature should be in the proper operating range, and the idle RPM and mixture adjustment must be correct and both ignition systems operating properly.

Set the engine speed between 1200 and 1500 RPM. Pull the mixture control back, and note the amount of rise before the RPM starts to drop off. The mixture control must be returned to the Full Rich position before the engine stops.

If the RPM rises, the discharge nozzle setting is on the rich side of the best-power mixture. If there is no rise, the setting is at or leaner than the best-power mixture. The off-idle mixture may be enriched by screwing the discharge nozzle adjustment counterclockwise or it may be leaned by screwing it clockwise. Make this adjustment by turning the nozzle adjustment one notch at a time until the desired performance is reached.

After making the off-idle check, the engine must be cleared and the idling checked and readjusted as necessary. Any adjustment of the idle mixture will necessitate readjustment of the manual enrichment valve.

E.   Bendix Fuel Injection System

Uneven fuel-air mixture distribution, which is a major problem with conventional carburetors, has become of increasing importance as the horsepower requirements of modern aircraft engines have continued to go up. The modern aircraft engine operates with such high cylinder pressures that an inadvertently lean mixture on an individual cylinder can cause detonation that can destroy the engine.

Some of the larger reciprocating engines have, in the past, used a direct fuel injection system with a master control unit. This was similar to a large pressure carburetor that metered the correct amount of fuel into two multi-cylinder piston pumps which forced the fuel directly into the combustion chambers as timed, high-pressure spurts.

Timed injection systems, because of their complexity, have been superseded by the simpler, yet effective, continuous-flow systems.

1. Bendix RSA fuel injection system

The Bendix Energy Controls Division has produced two very successful and popular fuel metering systems, the RS system and the RSA system. The RS system uses a side-by-side regulator and a flow of servo fuel to regulate the flow control valve. The more recent system, the RSA, does not use servo fuel, and this is the system we will discuss.

The Bendix RSA fuel injection system is a continuous-flow system which meters the fuel by a pressure drop that is proportional to the airflow through the venturi.

a. Air metering force

The air metering force is similar to that discussed with the Bendix PS pressure carbure-

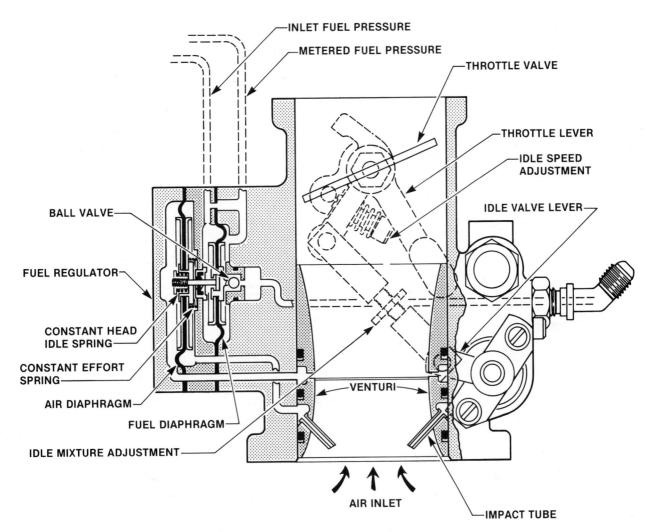

Fig. 1-32 Regulator unit of a Bendix RSA fuel injection system

tor. The impact tubes in the inlet of the throttle body sense the pressure of the air entering the engine, and the venturi senses its velocity. These two forces move the air diaphragm proportional to the amount of air taken into the engine.

b. Fuel metering force

Fuel from the engine-driven fuel pump enters the fuel control through the strainer and the mixture control valve. Some of this fuel acts on the fuel diaphragm to cause it to close the ball valve. Fuel for the engine operation flows through the main metering jet and the throttle fuel valve, into the metered fuel chamber of the regulator. This metered fuel opens the ball valve.

c. Metered fuel flow

The actual metering is done by the pressure drop across the orifices in the fuel control. In this system, the metering force is determined by the position of the ball valve in its seat. The inlet pressure is held relatively constant by the engine-driven fuel pump, and the outlet pressure is controlled by the balance between the fuel and the air metering forces. When the throttle is opened, the total air metering force increases. This moves the

air diaphragm to the left, which opens the ball valve and lowers the pressure downstream of the orifice in the metered fuel chamber. The pressure in the inlet fuel pressure chamber is greater than the metered fuel pressure by the amount of pressure dropped across the fuel control, and this tends to close the valve. The balance between the air and the fuel forces therefore holds the valve off its seat a stabilized amount for any given airflow.

d. Flow divider

After the fuel leaves the regulator, it flows through a flexible hose to the flow divider, Fig. 1-34, located on top of the engine in a central location. The injector nozzles are connected to the flow divider by 1/8-inch stainless steel tubing. A pressure gage in the cockpit reads the pressure at the outlet of the flow divider. This is actually the pressure drop across the injector nozzles and is directly proportional to the fuel flow through the nozzles. For all flow conditions other than idle, the restriction of the nozzles causes a pressure to build up in the metered fuel lines which influences the fuel metering force. Under idle flow conditions, the opposition caused by the nozzles is so small that the metering would be erratic, and to

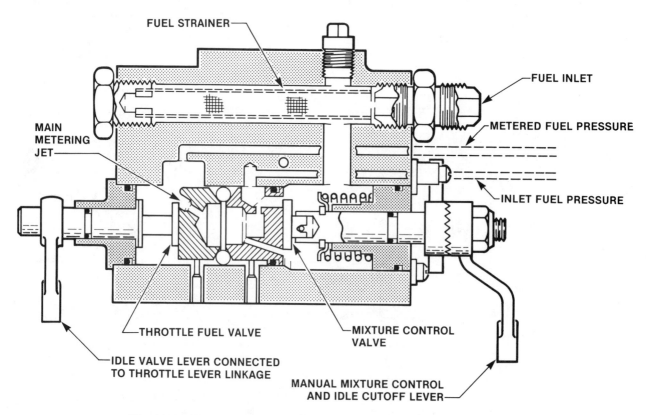

Fig. 1-33 Fuel control unit of a Bendix RSA fuel injection system

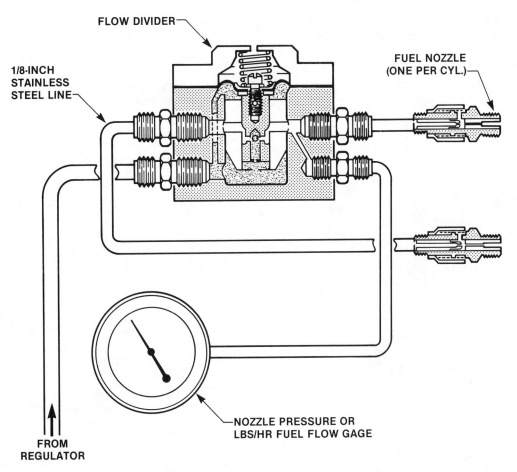

Fig. 1-34   Flow divider for a Bendix RSA fuel injection system

prevent this, a coil spring holds the flow divider valve closed until metered fuel pressure becomes sufficient to off-seat it.

For idle fuel flow, the flow divider opens only partially and thus serves the double function of creating the downstream pressure for the fuel control, and dividing the fuel to the cylinders for these extremely low flow conditions. When the mixture control is placed in the Idle Cutoff position, the flow divider cuts off the fuel flow and leaves the fuel lines to the injector nozzles full of fuel.

e. Injection nozzles

This system uses air bleed-type fuel nozzles, which screw into the cylinder head near the intake ports. Each nozzle consists of a brass body which incorporates a metering orifice, an air bleed hole, and an emulsion chamber. Around this body is a fine mesh metal screen and a pressed steel shroud. These nozzles are calibrated to flow within plus or minus two percent of each other,

and are interchangeable between engines, and between cylinders. An identification mark is stamped on one of the hex flats of the nozzle opposite the air bleed hole, and when installing a nozzle in a horizontal plane, the air bleed hole should be positioned as near the top as practical, in order to minimize fuel bleeding from the opening immediately after the engine is shut down.

f. Idle system

When there is a low airflow through the engine such as is encountered during idling, the air metering forces are not high enough to open the ball valve to provide adequate fuel flow for idling. The air diaphragm is between two springs which hold the ball valve off its seat until the airflow becomes sufficient. The constant head spring we see in Fig. 1-36 pushes against the air diaphragm and forces the ball valve off its seat. This maintains a constant head of pressure across the fuel control. As the airflow increases, the air diaphragm moves over, compressing the constant head spring until the diaphragm bushing makes solid contact with the ball valve shaft.

Beyond this point, the ball valve acts as though it is directly connected to the diaphragm.

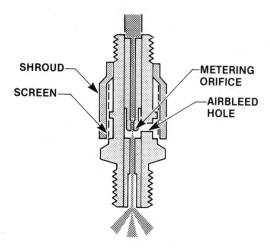

CUTAWAY VIEW OF THE NOZZLE
(A)

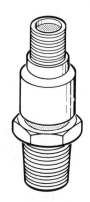

NOZZLE FOR A NORMALLY ASPIRATED ENGINE
(B)

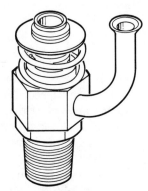

NOZZLE FOR A TURBOCHARGED ENGINE
(C)

Fig. 1-35 Fuel nozzles for a Bendix RSA fuel injection system

A smooth transition between idle and cruise RPM is provided by the use of the constant effort spring working between the air diaphragm and the housing. This spring, in effect, preloads the air diaphragm, giving it an initial loaded position from which to work.

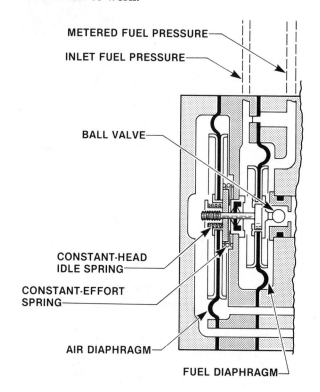

Fig. 1-36 Idling springs in a Bendix RSA fuel injection system

As with all fuel metering systems, this system controls the idle RPM by limiting the amount of air allowed to flow past the throttle valve, and controls the idle mixture by the amount of fuel allowed to flow to the discharge nozzles.

A spring-loaded screw, Fig. 1-37, contacts a stop on the throttle body to limit the amount the throttle air valve can close. An adjustable-length rod connects the throttle air valve to the throttle fuel valve and controls the amount the throttle fuel valve remains open. Adjustment of this length determines the idle mixture ratio and ultimately the idle manifold pressure.

g. *Manual mixture control*

A spring-loaded, flat, plate-type valve in the fuel control is moved by a linkage from the cockpit to regulate the amount of fuel that can flow to the main metering jet. When the mixture control is placed in the Idle Cutoff position, the passage

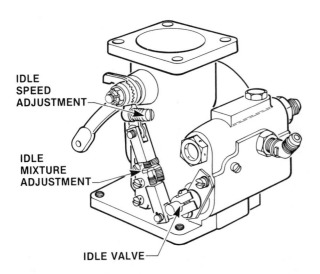

IDLE
SPEED
ADJUSTMENT

IDLE
MIXTURE
ADJUSTMENT

IDLE VALVE

CHANGING THE LENGTH OF THE CONNECTING ROD
BETWEEN THE THROTTLE AIR VALVE AND THE THROT-
TLE FUEL VALVE ADJUSTS THE IDLE MIXTURE RATIO.

(A)

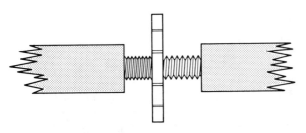

THE LENGTH OF THE ROD IS ADJUSTED BY THE DIF-
FERENTIAL THREADS IN THE IDLE MIXTURE ADJUST-
MENT LINKAGE.

(B)

Fig. 1-37 Idle speed adjustment for a Bendix
RSA fuel injection system

the bellows causes it to expand. This moves the needle and increases the bleed across the air diaphragm, which decreases the air metering force and leans the mixture.

i. Installation and servicing

The Bendix RSA fuel injector mounts on the engine in the location normally occupied by the carburetor, and the flow divider mounts on the crankcase above the cylinders. The injector connects to the flow divider with a flexible line, and the flow divider connects to the individual nozzles with stainless steel tubing. The fuel flow indicator also connects to the flow divider.

The aircraft fuel system required for this injector is quite standard, using a boost pump of either the plunger or centrifugal type to supply fuel pressure to the injector for starting.

Fuel flows through the main strainer, through or parallel with an engine-driven fuel pump, into the main fuel control unit. The inlet fuel pressure is not critical, but it must be within the range specified by the airframe manufacturer.

Service required by this injection system is typical for fuel metering systems. The strainer should be cleaned after the first twenty-five hours of operation, and every fifty hours after that. All nuts and screws should be checked for security, and the entire installation checked for indication of fuel stain. The throttle and mixture control linkage should be checked for tightness and freedom of operation.

j. Starting procedure

To start an engine equipped with this injection system, first place the mixture control in the Idle Cutoff position and open the throttle about 1/8 of the way. Turn on the master switch and the boost pump. Move the mixture control to the Full Rich position until there is an indication of flow on the flow meter, and then return the mixture control to Idle Cutoff. Turn the ignition On and engage the starter. As soon as the engine starts, move the mixture control to the Full Rich position.

Fuel injected engines have the reputation of being difficult to start when they are hot, and this is largely due to the fact that the high temperature in the engine nacelle causes the fuel lines to

to the main metering jet is completely closed and no fuel can flow to the jet. In the Rich position, the opening afforded by the mixture control is larger than the metering jet, and the jet limits the flow. In any intermediate position, the opening is smaller than the main jet, and the mixture control becomes the flow-limiting device. (See Fig. 1-33, page 38.)

h. Automatic mixture control

A reverse-tapered needle attached to a bellows, Fig. 1-38, varies the airbleed between the two air chambers of the regulator. The bellows contains helium to sense density changes and a small amount of inert oil to dampen vibrations. As the air density decreases, from an increase in either altitude or temperature, the pressure inside

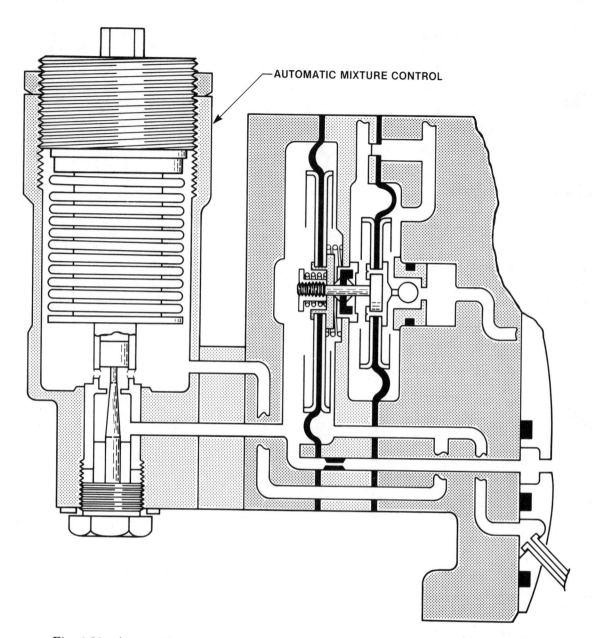

Fig. 1-38 Automatic mixture control for the Bendix RSA fuel injection system

be full of vapor rather than liquid fuel. The lines must be purged of all vapor before an effective start can be made.

k. Idle speed and mixture adjustment

Before attempting to make any adjustment to the fuel metering system, you must determine that the engine is operating properly. Both magnetos and all of the spark plugs must be doing their proper job, and the engine should develop its proper static RPM and must have the proper oil or cylinder head temperature.

Turn the aircraft cross wind and close the throttle to the desired idle speed, usually somewhere near 650 RPM. Slowly and carefully move the mixture control toward the Idle Cutoff position and observe the tachometer as you lean. The RPM should increase approximately 25 to 50 RPM as the mixture control is pulled back. Return the mixture control to Rich before the engine dies. An increase of more than this indicates that the mixture is too rich, and you should shorten the idle mixture control linkage to decrease the fuel flow. This is done by screwing the coarse threads of the adjustment screw into the rod.

After making this adjustment, momentarily open the throttle to approximately 2000 RPM to clear the spark plugs of any fouling that may have occurred during idling. If, when the throttle is returned to the idle position, the RPM has changed, it must be corrected, and a final check made of the mixture before the idle adjustments may be considered completed.

F. Teledyne-Continental Fuel Injection System

The Teledyne-Continental fuel injection system meters its fuel as a function of the engine RPM, and does not use airflow as a metering force. A special engine-driven fuel pump which is an integral part of the system, produces the fuel metering pressure.

1. System components

a. Injection pump

The pump is the heart of this fuel injection system. It is basically a vane-type, constant-displacement pump, with special features that allow it to produce an output pressure that varies with the engine speed.

If a passage containing an orifice bypasses the pump mechanism, Fig. 1-39, the output pressure will vary according to the speed of the pump, and the size of the orifice will determine the pressure for any given speed. If its size is increased, the output pressure will decrease.

This system works well for flows in the cruise and high power range, but when the fuel flow is low, as it is during idling, the orifice does not produce enough restriction to maintain a constant output pressure, and so an adjustable pressure relief valve is installed in the line. During idle, the output pressure is determined by the setting of the relief valve, and the orifice has no effect. At the high-power end of operation, the relief valve is off its seat, and the pressure is determined by the orifice.

All fuel injection systems must have vapor-free fuel in their main metering section, and provision is made in the pump to remove all vapor from the fuel and return it to the tank.

Fuel enters the pump through a chamber where the vapor is swirled out of the liquid and collects in the top. Some fuel from the pump outlet returns to the fuel tank through a venturi, or jet pump, arrangement on top of this chamber. This produces a low pressure which attracts the vapors and returns them to the tank.

A final feature in this pump is a by-pass check valve around the pump, so fuel from the air-

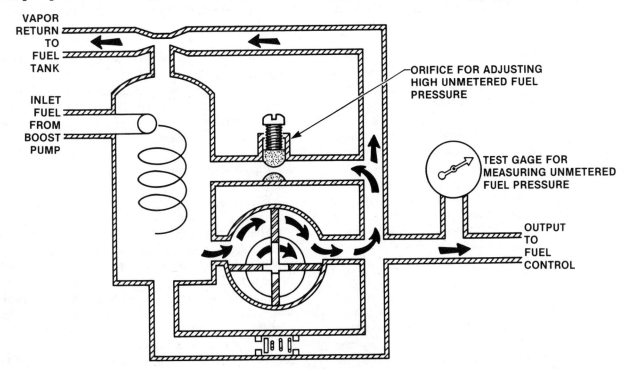

Fig. 1-39 Teledyne-Continental fuel pump showing the high, unmetered fuel pressure adjustment

43

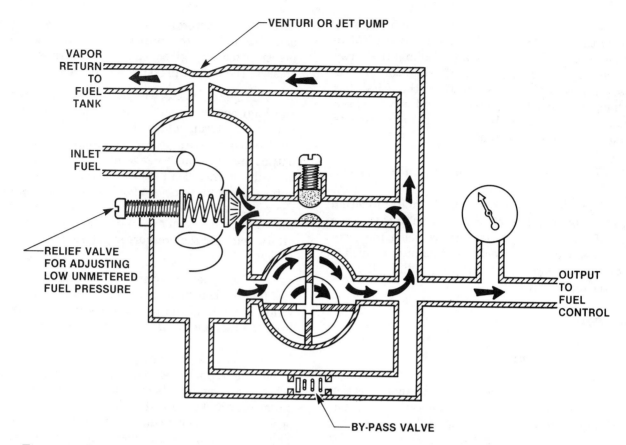

Fig. 1-40 Teledyne-Continental fuel pump showing the low, unmetered fuel pressure adjustment

craft boost pump may flow to the fuel control for starting. As soon as the engine pump pressure becomes higher than that of the boost pump, the valve closes and the engine pump takes over.

Turbocharged engines have a unique problem during acceleration. If the fuel flow increases before the turbocharger has time to build up to speed and increase the airflow proportionately, the engine may falter from an overly rich mixture. Pumps for these engines have the simple orifice replaced with a variable restrictor controlled by an aneroid valve.

An evacuated bellows is surrounded by upper deck pressure which is actually the turbocharger discharge pressure. This bellows moves a valve which controls the size of the orifice, varying the output fuel pressure proportional to the inlet air pressure.

When the throttle is suddenly opened and the engine speed increases, rather than immediately supplying an increased fuel pressure to the control, the aneroid holds the orifice open until the turbocharger speed builds up and increases the

air pressure into the engine. As the inlet air pressure increases, the orifice becomes smaller, and the fuel pressure, and therefore the fuel flow, increase.

The drive shaft of this fuel pump has a loose coupling to take care of any slight misalignment between the pump and the engine drive, and there is also a shear section in the coupling that will break to protect the pump and the engine if the pump should ever seize.

b. Fuel control unit

The diagram in Fig. 1-42 shows the fuel control used with this type of fuel injection system. Fuel leaves the engine-driven fuel pump at a pressure that is proportional to the engine speed, modified by the turbocharger discharge pressure. It then flows through the fuel control filter into the manual mixture control valve. This valve differs from that used in the Bendix fuel injection system, as it acts as a variable selector valve rather than a shutoff valve. When the mixture control is in the Idle Cutoff position, all fuel is bypassed back to the pump, and none flows to the

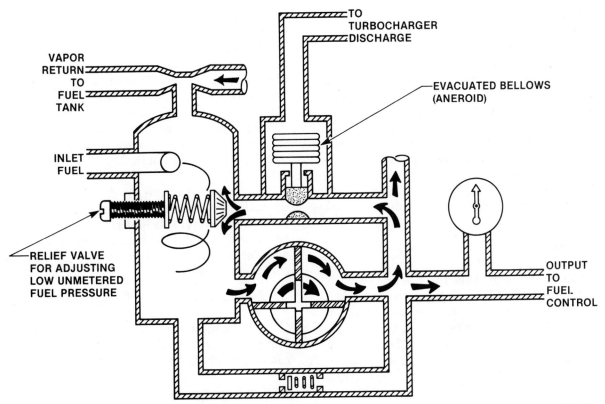

Fig. 1-41 *Teledyne-Continental fuel pump used with a turbocharged engine. The evacuated bellows senses the upper deck (turbocharger discharge) pressure to adjust the high, unmetered fuel pressure.*

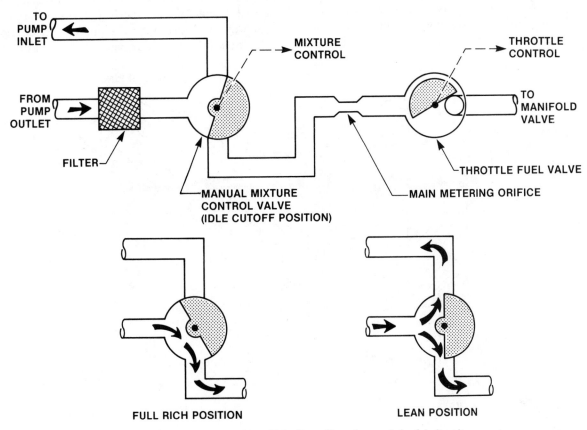

Fig. 1-42 *Fuel control unit for a Teledyne-Continental fuel injection system*

45

engine. In the Full Rich position, all of the fuel flows to the engine. Any intermediate position drops the pressure upstream of the metering orifices by routing some of the fuel back to the pump and some of it to the engine. A metering plug with a precision orifice limits the maximum amount of fuel that can flow into the engine under Full Throttle, Full Rich conditions. The throttle in the cockpit controls both the throttle air valve, similar to that used in a carburetor, and the throttle fuel valve, which is essentially a variable orifice which determines the amount of fuel that is allowed to flow to the engine for any given fuel pressure.

c. Fuel manifold valve

After the fuel leaves the throttle fuel valve, it flows through a flexible fuel line to the manifold valve, Fig. 1-43, which is usually located on top of the engine. This valve serves two basic functions: It distributes the fuel evenly to all of the cylinders, and it provides a positive shut off of the fuel when the mixture control is put in the Idle Cutoff position.

When the pressure rises, the diaphragm lifts the valve off its seat, but a spring-loaded poppet valve inside the cutoff valve stays on its seat until the fuel pressure has opened the valve completely so it can do no metering. Then the poppet opens and allows fuel to flow to the nozzles. The opposition caused by this valve provides a constant pressure downstream of the jets for metering at idle. Above idle, the valve is fully open, and offers no opposition. When the mixture control is placed in its Idle Cutoff position, the fuel pressure drops and the spring closes the manifold valve to provide a positive shutoff of the fuel to the nozzles. If the poppet should become plugged or sticky, erratic or rough idling will result.

The chamber above the diaphragm is vented to static atmosphere to allow unrestricted movement of the valve. It is important that ram air not be allowed into this chamber, so the vent must be open and positioned either toward the side of the engine or to the rear.

d. Injector lines

The nozzles are connected to the manifold valve by six stainless steel lines, 1/8-inch in diameter and all the same length.

e. Injector nozzles

Possibly the simplest, yet some of the most important components in a fuel injection system are the injector nozzles, Fig. 1-44. One nozzle in each cylinder provides the point of discharge for the fuel into the induction system. Fuel flows

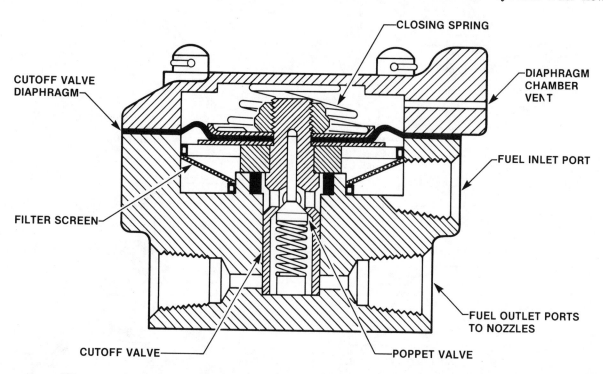

Fig. 1-43   Fuel manifold valve for a Teledyne-Continental fuel injection system

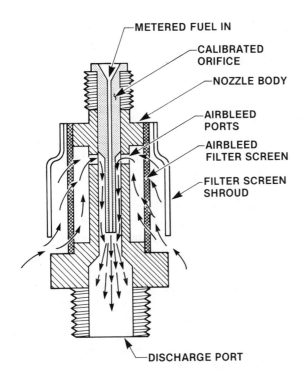

METERED FUEL IN

CALIBRATED ORIFICE

NOZZLE BODY

AIRBLEED PORTS

AIRBLEED FILTER SCREEN

FILTER SCREEN SHROUD

DISCHARGE PORT

Fig. 1-44 Fuel injector nozzle for a Teledyne-Continental fuel injection system

from the manifold valve into the nozzle, through a calibrated orifice into the intake valve chamber of the cylinder head. Air is drawn through a screen into the injector nozzle where it mixes with the fuel and forms an emulsion to aid in vaporization.

Plugged nozzles can be a source of rough operation and should be one of the first components checked when troubleshooting the system. Nozzles must be cleaned by soaking them in lacquer thinner or acetone and blowing them out with clean compressed air in the direction opposite the regular flow.

Wires, drills, or other cleaning devices should *never* be used to remove obstructions from the orifice. If soaking does not clear out the obstruction, replace the nozzle. The metal shroud that is pressed over the nozzle to protect the screen should never be removed.

There are three sizes of nozzle orifices in use with this system; each is identified by a letter stamped on the nozzle. The A-size nozzle will flow a given amount of fuel for a given pressure. A set of B-size nozzles flows one-half gallon more fuel per hour with the same pressure, and the C-size nozzles flow yet another half gallon per hour. When the engine was calibrated in its factory run-

in, the proper size nozzles were installed and this size should be kept in the engine from then on.

Different models of engines require different styles of nozzles. Some engines require long nozzles to inject the fuel farther into the intake chamber; others function best with short nozzles. The nozzles differ in appearance, but their function is the same. Nozzles used with turbocharged engines have their shrouds vented to the turbocharger side of the throttle valve. If this were not done, increasing the manifold pressure above atmospheric would blow fuel out of the bleed holes.

2. System operation

a. Starting

Fuel injected engines have a reputation of being difficult to start. This reputation is founded, however, on ignorance rather than fact. Starting an engine equipped with a Teledyne-Continental fuel injection system consists mainly of the normal pre-start settings of the switches and controls. Turn the fuel on, crack the throttle approximately one-eighth of the way, and place the mixture control in the Full Rich position and the boost pump on High. When fuel flow is indicated on the flowmeter, engage the starter. The engine will start and the boost pump may be turned off.

Starting a hot engine is sometimes more difficult. If the proper procedure is not used, the engine can progress from a starved to a flooded condition without starting. When a hot engine is shut down, the heat inside the cowling may cause the fuel to boil out of the injector lines. All of the lines in the system will be filled with fuel vapors instead of liquid fuel, and the engine will not be able to get enough fuel to start. In the process of trying to start, the engine will pass from the starved condition to a condition in which too much fuel is supplied for the amount of air the starter can pull into the cylinders. The engine is then flooded, and it cannot be started until it is cleared.

To successfully start a hot engine, it is first necessary to remove all of the fuel vapors and get liquid fuel into the lines. The mixture control should be placed in the Idle Cutoff position and the throttle opened wide so that the boost pump can operate at full pressure. Turn the boost pump on high and allow the fuel to circulate through the pump for about fifteen or twenty seconds. Then turn the boost pump off and place the mixture control in Full Rich. Close the throttle to the correct position for starting and engage the starter. The lines are now full of liquid fuel, and a normal start can be made.

b. Inspection

At least each one hundred hours of operation, all of the components in a fuel injection system should be checked for security and integrity of mounting. All lines should be checked to see that they are not chafed or leaking, as would be indicated by the presence of fuel dye stains. All moving parts should be checked for wear, and the ends of the throttle link rods lubricated with a drop of engine oil.

Clogged vents are a source of potential trouble and should be checked on each inspection. The vent in the relief valve chamber of a pump for a naturally aspirated engine should be checked, as well as the vent in the manifold valve. Be sure when checking these vents that you do not use a wire or anything else that could damage the diaphragm. A small piece of waxed or wet thread will go into the vent hole if it is open.

Injector nozzles do not require much service, but every 300 hours of operation they should be removed from the engine, soaked in acetone or lacquer thinner, and blown out. A clogged nozzle can usually be located by the presence of fuel dye stain on the cylinder head around the nozzle. Since fuel cannot be drawn into the cylinder through a plugged nozzle, it will escape through the air bleed hole.

c. Adjustments

At each one-hundred-hour inspection, the low and high unmetered fuel pressure and the high metered fuel pressure must be checked. An accurate fuel pressure gage is connected into the fuel line between the pump and the fuel control. The engine is started, warmed up, and the idle speed properly set. The pressure indicated on the test gage should be that specified by the latest Continental service bulletin for the engine being inspected. If it is not within the specified limits, the relief valve must be adjusted to get this pressure.

After the correct pressure is obtained, the idle mixture is checked. The mixture control is pulled slowly to the Idle Cutoff position and the tachometer watched. There should be an increase of approximately 50 RPM before the engine starts to die. If the rise is less than this, the mixture is too lean, and the idle mixture linkage should be shortened a bit to enrich the mixture. If the RPM increases more than is allowed, the mixture is too rich and the rod should be lengthened by turning the idle mixture adjustment to the left. This leans the mixture.

With the low unmetered fuel pressure and the idle mixture and RPM adjusted, the high unmetered fuel pressure should be checked. The engine is operated at full throttle, maximum RPM, and the test gage read. This pressure should be within the range specified in the service bulletin. If it is not, the orifice should be adjusted to bring this pressure to the value desired. At this high RPM, every precaution must be taken to operate the engine for the absolsute minimum time on the ground.

After the pressures are checked and the test gage is removed from the system, a flight test should be made to check the high metered fuel pressure on the airplane's fuel flow gage. This reading must be made at full throttle, maximum RPM, after the engine has had about ten or fifteen minutes at cruise power to stabilize.

d. Flow matching

Sometimes on a multi-engine installation, there is found to be a difference in the mixture control position and fuel flow indications for the same power settings of the engines. This can be caused by improper nozzle sizes or an improper manifold valve. If, for a given power setting, an excessive fuel flow (pressure) indication is required, it may be caused by a set of nozzles of too small a size in the engine. The installation of a set with a higher letter, such as a C for a B or a B for an A, would provide the proper flow with a lower pressure indication on the flowmeter. Mismatched flows may also be corrected by changing the manifold valves.

e. Rough operation

Fuel injected engines are apt to become flooded, so provisions are made in the induction system to drain some of the excessive fuel when this occurs. Manifold drains are simply holes in the intake system to allow the excess fuel to drain out.

These holes are calibrated air leaks, so for proper engine operation they must be open. If they become plugged, the engine will operate excessively rich and run rough, so on each inspection these drain holes should be checked to assure that they are clear. At idle, air should be drawn through them.

Turbocharged engines have a sniffle valve in their drain opening. These valves are spring loaded to the open position when the pressure on both sides is the same. When the manifold pressure is either above or below atmospheric, the valve will close. These valves are checked by assuring that they are closed at idle. If they do not close, the engine will get too much air and will idle rough.

f. Flowmeter fluctuation

The fuel flowmeter used with a fuel injected engine is nothing more than a pressure gage used to measure the pressure drop across the injector nozzles. If air gets into the line between the manifold valve and the gage, the gage will fluctuate. The line must be purged of all air and filled with some type of light oil such as kerosene. Air which causes the fluctuation may enter the system at any point, so a careful check must be made of all connections, especially around the fuel filter to be sure no air leak is present.

II. TURBINE ENGINE FUEL METERING SYSTEMS

A. Principles of Turbine Engine Fuel Systems

The fuel system functions to supply a precise amount of fuel to the engine in all conditions of ground and air operations. The system must be free of dangerous operational characteristics such as a vapor lock. This is a condition that restricts fuel flow through units designed to handle liquids rather than gases.

Further, it must be possible to increase and decrease power at will to obtain the thrust required for any operating condition. In a gas turbine engine, this is accomplished by a device called a fuel control, which meters fuel to the combustion chamber. The pilot selects a fuel flow condition by a power lever, causing the fuel control to automatically schedule fuel flow as per ambient and mass airflow conditions. These automatic features prevent rich or lean flameout and overtemperature or overspeed conditions from occurring.

Lean die-out occurs as one might expect; a lessening of fuel in the mixture occurs until combustion is no longer supported. A rich blowout occurs when the force of fuel flow during low airflow conditions interrupts the normal burning process near the fuel distribution nozzle. This can cause unstable combustion, and the flame will blow away from the nozzle and out of the combustor. At this point combustion ceases and a relight (restart) procedure is required. It is generally stated that maximum flame speed for supporting combustion using hydrocarbon fuel must be less than 0.4 Mach.

1. Gas turbine fuels

Jet fuels are liquid hydrocarbons similar to kerosene with gasoline blended in. This mixture is designed to freely oxidize (mix with oxygen) at combustion flow rates and temperatures. The blending of gasoline reduces the kerosene fuel's tendency to become too viscous at high altitudes. This is a problem which ground powerplants do not encounter; they use heavier No. 2 fuel oil, diesel fuel, and the like without blending.

The oxides which are formed by combustion in a gas turbine engine are mostly gases. This is another quality designed into jet fuels which keeps solid particles to a minimum, solids that would impinge on turbine nozzle vanes and turbine blades, causing erosion.

The following jet fuels are most commonly used in commercial and general aviation:

Jet-A: A fuel of heavy kerosene base, flash point 110-150° F, freezing point −40° F, 18,600 Btu/lb; similar to Navy JP-5 fuel.

Jet-A1: A fuel identical to Jet-A except for a freezing point of −58° F. Similar to NATO fuel JP-8 with additives.

Jet-B: A fuel of heavy gasoline base, flash point 0° F, freezing point −76° F, 18,400 Btu/lb; similar to military JP-4 fuel.

Jet-A, Jet-A1, and Jet-B are the primary commercial fuels and are interchangeable for use in most gas turbine engines. Military JP-4 and JP-5 are often suitable alternate fuels. Aviation grades 80-145 octane reciprocating engine fuels are often emergency alternate fuels for most engines.

For the approved fuel and fuel additives used to service a turbine engine, the technician should check the aircraft operator's manual or the type certificate data sheet file.

a. Fuel additives

The most common fuel additives protect the fuel as anti-icing and anti-microbiological agents. Anti-icing additives keep entrained water from freeze-up without the use of fuel heat, except at very low temperatures. The manufacturer's manual will state this low value where fuel heat must be applied. Microbiocidal agents kill microbes, fungi, and bacteria which form a slime and in some cases a matted waste in fuel systems.

Most often the additives are pre-mixed in the fuel by the distributor. If they are not, the service person must add the agents when fueling the aircraft. A popular brand of a combined anti-icing and anti-microbiological mixture is called PRIST. It is designed to be added during servicing. The type and amount, however, must be determined to maintain the airworthiness of the fuel system in the existing climatic conditions.

2. Thrust specific fuel consumption (TSFC)

TSFC is a ratio of fuel consumption to the engine thrust. This ratio is usually included in the engine specifications and affords a means of comparing the fuel consumption or economy of one engine to another regardless of its thrust rating. Specifically it is the amount of fuel in pounds consumed by an engine while producing one pound of thrust for one hour.

$$\text{TSFC} = \frac{\text{Total weight of fuel consumed}}{\text{Pounds of thrust}}$$

EXAMPLE 1:

An engine with TSFC of .49 lb/lbt/hr, has a thrust rating of 3,500 lbs. How many pounds of fuel will it consume per hour?

$$\text{TSFC} = \frac{W_f}{F_n(\text{or } F_g)}$$

$W_f = 3,500 \text{ lbt} \times .49 \text{ lb/lbt} \div \text{hr}$

$W_f = 3,500 \text{ lbt} \times .49 \dfrac{\text{lb}}{\text{lbt}} \times \dfrac{1}{\text{hr}}$

$W_f = 1,715 \text{ lb/hr}$

where:

W_f = fuel flow in pph
$F_n(\text{or } F_g)$ = thrust in pounds
TSFC = .49 lb/lbt/hr

A comparison of an early model turbojet, the Westinghouse J-34, with a modern turbofan, the Garrett-AiResearch TFE-731, best indicates the current state of the art in engine fuel efficiency.

Engine	Thrust (lbs) (F_g)	TSFC	W_f (pph)
J-34	3,250	1.06	3,445
TFE-731	3,500	0.49	1,715

Another way in which TSFC is used can be seen in the statistics of a General Electric CF-6 high by-pass turbofan engine. See Table 1.

EXAMPLE 2:

Using the specifications of the G.E. CF-6 engine, calculate the maximum continuous fuel consumption in pounds per hour.

If: $\text{TSFC} = W_f / F_g$

Then: $W_f = \text{TSFC} \times F_g$

and: $W_f = 0.385 \dfrac{\text{lb/hr}}{\text{Lbt}} \times 46,200 \text{ lbt}$

$W_f = 17,787 \text{ lb/hr}$

EXAMPLE 3:

Now calculate fuel consumption in PPH at maximum cruise at altitude.

If: $W_f = \text{TSFC} \times F_n$

Then: $W_f = 0.654 \times 10,800$

$W_f = 7,063 \text{ lb/hr}$

If we consider that maximum continuous thrust on the ground and maximum cruise power in flight are the same power setting, note that even though TSFC is higher in flight, actual fuel consumption is only 40% of the fuel used on the ground. This shows the fuel economy a gas turbine engine experiences in flight.

The reason TSFC is higher in flight is that in order to keep thrust up at altitude where the air is rarefied and mass airflow is lower, more fuel is needed per pound of thrust to keep thrust at an acceptable level.

B. Fuel Controls

The fuel control is the most commonly used fuel metering device. It is an engine driven ac-

Ground Performance:

Thrust Setting	Mach Number	Ambient Temperature	Gross Thrust Lb	TSFC Lb/Hr/Lbt	Fuel Flow Lb/Hr
Takeoff (sea level)	0	Standard	50,200	0.394	19,779
Maximum Continuous	0	Standard	46,200	0.385	17,787
75% Takeoff	0	Standard	37,600	0.371	13,579
Flight Idle	0	Standard	5,190		2,320
Ground Idle	0	Standard	1,740		1,490

Altitude Performance:

Thrust Setting	Mach Number	Ambient Temperature	Net Thrust Lb	TSFC Lb/Hr/Lb	Fuel Flow Lb/Hr
Max Climb	0.85	Standard	11,500	0.664	7,636
Max Cruise	0.85	Standard	10,800	0.654	7,063

Table 1   Performance statistics of a General Electric CF-6 turbofan engine

cessory which can be operated by mechanical, hydraulic, electrical or pneumatic forces in various combinations. The purpose of the fuel control is to maintain a correct air-to-fuel mixture ratio of 15:1 by weight. This ratio represents weight of combustor primary air to weight of fuel. Sometimes this is expressed as a fuel-air ratio of 0.067. All fuels require a certain proportion of air for complete burning, but at rich or lean mixtures the fuel will also burn, but not completely. The ideal proportion for air and jet fuels is 15:1 and it is called the stoichiometric (chemically correct) mixture.

When the pilot moves the fuel control power lever forward, fuel flow is increased. This increase in fuel flow creates increased gas expansion in the combustor which in turn raises the level of power in the engine. For the turbojet and turbofan, that means a thrust increase. For the turboprop and turboshaft, it means an increase in power to the output drive shaft. This could mean a speed increase at a given propeller load or a stabilized speed at an increasing blade angle and load.

On a single compressor engine the fuel control is driven directly by the accessory gearbox and indirectly from the compressor. On the dual- and triple-spool engines, the fuel control is normally driven at high compressor speed.

The professional technician should become very familiar with the maintenance manuals and

with all the fuel control functions on the aircraft he maintains, in order to troubleshoot intelligently and prevent inadvertent component change, causing unnecessary expense.

Many signals are sent to the fuel controls for their automatic controlling of the air-fuel ratio. A list of the most common signals are as follows:

1. Engine speed signal (N_c)—given to the fuel control by a direct drive to the engine gearbox through a flyweight governor within the control; used for steady state, acceleration, and deceleration fuel scheduling (acceleration of most gas turbine engines is in the range of 5-10 seconds from idle to full power).

2. Inlet Pressure (P_{t_2})—a total pressure signal transmitted to a bellows from a probe in the engine inlet, used to give the control a sense of aircraft speed and altitude as ram conditions in the inlet change.

3. Compressor discharge pressure (P_{s_4})—a static air pressure signal sent to a bellows within the control; used to give the control an indication of mass airflow at that point in the engine.

4. Burner Can Pressure (P_b)—a static pressure signal sent to the control from within the combustion liner. There is a linear relationship between Burner Pressure and weight of

airflow at this point in the engine; that is, if Burner Pressure increases 10%, the mass airflow has increased by 10% and the burner bellows will schedule 10% more fuel to maintain the correct air-fuel ratio. The quick response this signal gives makes it valuable in preventing stalls, flameouts, and overtemperature conditions.

5. Inlet temperature (T_{t_2})—a total temperature signal from the engine inlet to the control; a sensor connected by a capillary tube to the control. It is filled with a heat sensitive fluid or gas which expands and contracts as a function of inlet temperature. This signal provides the control with an airflow density value against which a fuel schedule can be established.

1. Simplified fuel control schematic

Fig. 2-1 is a simplified schematic of a gas turbine fuel control. It functions to meter fuel as follows:

a. Fuel metering section

(1) Movement of the shutoff lever to the open position allows fuel to flow out to the engine. The manual shutoff lever is needed because the minimum flow stop prevents the metering valve ever completely closing. This makes the full rearward position of the power lever the idle position; against the minimum flow stop and prevents the power lever from becoming a shutoff lever. The shutoff lever in this illustration also provides the function of ensuring the correct working pressure buildup within the control for accurate metering before allowing fuel to flow to the engine. This lever, then, prevents roughly metered fuel entering the engine before its correct time.

(2) Fuel from the supply system is then pumped by the main fuel pump to the main metering valve. As fuel flows through the orifice created by the taper of the valve, a pressure drop occurs. Fuel, from that point of the metering valve to the fuel nozzle, is referred to as metered fuel. Fuel in this instance is metered by weight rather than volume, because Btu/lb is constant, regardless of fuel temperature, while Btu per unit volume is not.

(3) Fuel now passes out to the engine in a correctly metered condition to the combustor.

(4) The principle involved in metering fuel by weight is mathematically expressed as:

$$W_f = KA \sqrt{\Delta P}$$

where: W_f = weight of fuel flow in lb per hr
  K = constant for a particular fuel control
  A = area of orifice
  ΔP = pressure differential across orifice

If only one orifice size were needed, there would be no variable in this formula as pressure drop would remain constant. But, aircraft engines obviously must change power settings. This is accomplished by moving the power lever. By advancing the power lever, the orifice area will subsequently increase. This action creates a mathematical variable. If it were not for the differential pressure regulating valve, the pressure differential across the metering orifice would change to create a second variable. This arrangement allows for what is referred to as a linear relationship between orifice size and weight of fuel flow. With only one variable—orifice size—such a relationship exists; if the opening changes, the weight of fuel flow changes proportionally.

A constantly changing fuel bypass is the means by which pressure differential is maintained at a constant value, regardless of orifice size. By directing metered fuel to the spring side of the differential pressure regulator diaphragm, the pressure differential will always return to the value of the spring tension. This spring tension being a constant value, the differential pressure will also be constant.

To understand this concept more fully, consider that the fuel pump always delivers fuel in excess of the fuel control's needs. The by-pass valve continually returns excess fuel back to the pump inlet.

EXAMPLE:

The unmetered fuel pressure is 500 psig, metered fuel pressure is 420 psig, and spring force is 80 psi. At this point, there is 500 psig on both sides of the pressure regulator diaphragm. The by-pass valve will be in a state of equilibrium

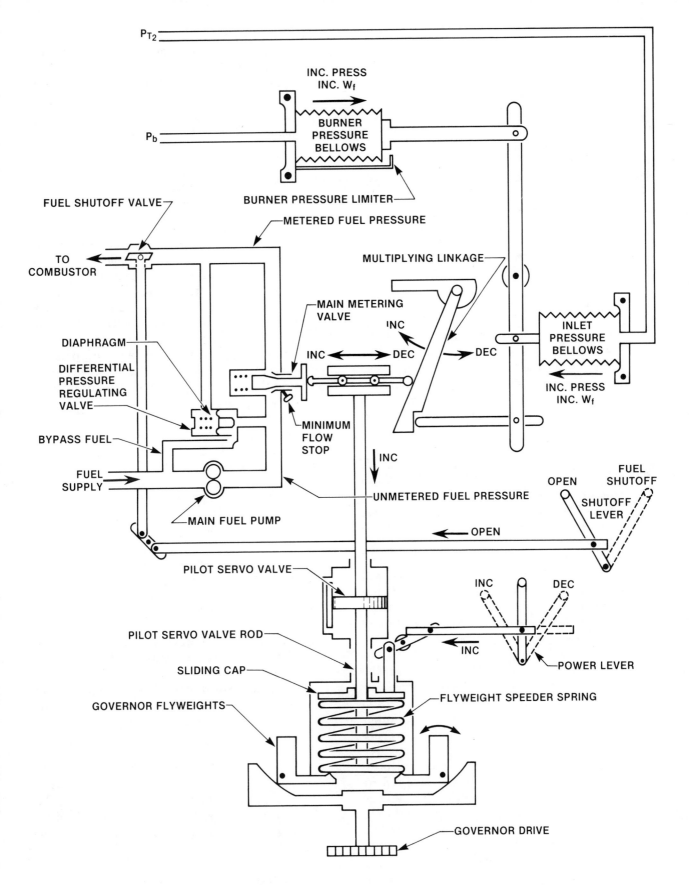

Fig. 2-1 Simplified diagram of a turbine engine fuel control unit

53

and will be bypassing fuel which is surplus to the needs of the engine.

Now, if the pilot moves the power lever forward, the orifice size at the metering valve will increase and the metered pressure downstream will also increase. Let us say that metered pressure increases to 440 psig. This action will create a total pressure of 520 psig on the spring side of the regulator diaphragm, pushing the by-pass valve toward closed. This reduced bypassing will now cause unmetered pressure to rise until 80 psig is re-established for the new orifice size. As previously stated, the differential pressure will always come to the setting of the pressure regulator spring as equilibrium occurs.

b. *Computing section*

(1) During engine operation, movement of the power lever causes the spring cap to slide down the pilot servo valve rod and compress the flyweight speeder spring. In doing so, the spring base forces the flyweights in at the top, creating an underspeed condition. The pilot servo valve prevents sudden movement as its fluid is displaced bottom to top. As the pilot servo valve rod moves down, the roller, riding on the inclined plane formed by the multiplying linkage, moves to the left and forces the metering valve to the left against its spring, allowing increased fuel to flow to the engine. With increased fuel flow, the engine will speed up and drive the governor faster. The new flyweight force will come to equilibrium with the speeder spring force as the flyweights return toward an upright position. They are now in position to act at the next speed change.

(2) On many engines, static pressure in the burner can is a useful measure of mass airflow. If mass airflow is known, fuel-air ratio can be more carefully controlled. As burner pressure (P_b) increases, the burner pressure bellows expand to the right. Excessive movement is restricted by the burner pressure limiter. If the pilot servo valve rod remains stationary, the multiplying linkage will force the roller to the left, opening the metering valve to match fuel flow to the increased mass airflow. This condition could occur in an aircraft nosedown condition, which would increase airspeed, inlet ram air, and engine mass airflow.

(3) An increase in inlet pressure would also cause the inlet pressure bellows to expand, forcing the multiplying linkage to the left and causing the metering valve to open wider.

2. *Bendix DP-L2 hydropneumatic fuel control unit*

This hydropneumatic fuel control is installed on the JT15D turbofan and the PT6 turboprop, manufactured by United Aircraft of Canada.

Fuel is supplied to the fuel control unit at pump pressure (P_1) which is applied to the entrance of the metering valve. The metering valve, in conjunction with the metering head regulator valve system, serves to establish fuel flow. The fuel pressure immediately downstream of the metering valve is known as P_2. The by-pass valve maintains an essentially constant fuel pressure differential ($P_1 - P_2$) across the metering valve, assuring that fuel flow is a function of orifice area only. (See Fig. 2-2.)

a. *Fuel flow sequence*

(1) *Component functions*

1. Fuel Inlet — from fuel storage tank.

2. Filter — coarse screen.

3. Gear Pump — pump discharge referred to as P_1 fuel.

4. Filter — fine mesh.

5. Relief Valve — prevents excessive (P_1) fuel pump discharge pressure build-up.

6. Metering Head Regulator — bypasses unwanted (P_0) fuel and establishes a constant fuel pressure differential ($P_1 - P_2$) across the metering valve.

7. Fuel Temperature Bimetallic Disks — automatically compensates for specific gravity changes with fuel temperature changes. Can also be manually adjusted for use of differing specific gravity values of various jet fuels.

8. Metering Valve — meters P_2 fuel to the fuel nozzles. Positioned by the torque tube con-

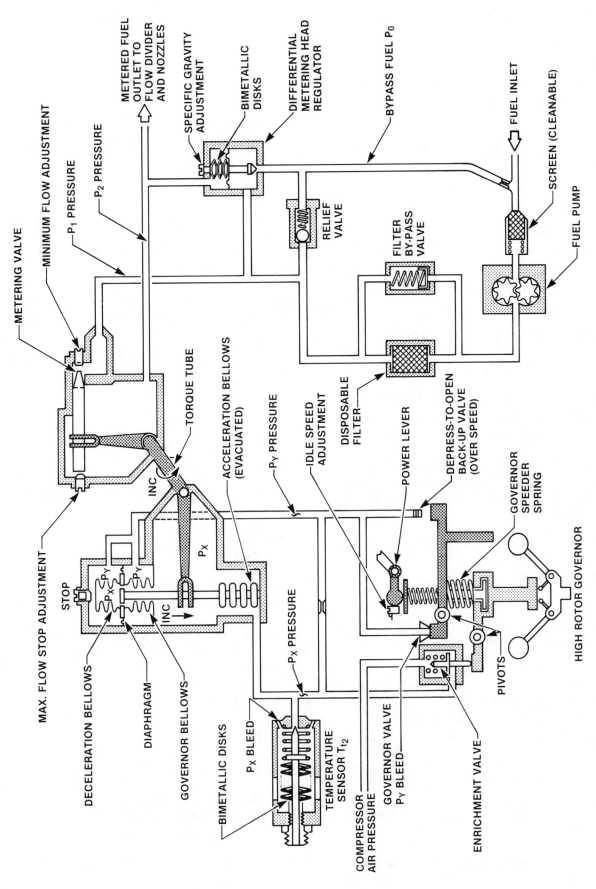

MAX. FLOW STOP ADJUSTMENT

STOP

DECELERATION BELLOWS

DIAPHRAGM

GOVERNOR BELLOWS

BIMETALLIC DISKS

P_X BLEED

TEMPERATURE SENSOR T_{t2}

COMPRESSOR AIR PRESSURE

GOVERNOR VALVE P_Y BLEED

ENRICHMENT VALVE

PIVOTS

HIGH ROTOR GOVERNOR

GOVERNOR SPEEDER SPRING

DEPRESS-TO-OPEN BACK-UP VALVE (OVER SPEED)

POWER LEVER

DISPOSABLE FILTER

IDLE SPEED ADJUSTMENT

P_Y PRESSURE

ACCELERATION BELLOWS (EVACUATED)

TORQUE TUBE

METERING VALVE

MINIMUM FLOW ADJUSTMENT

P_1 PRESSURE

P_2 PRESSURE

METERED FUEL OUTLET TO FLOW DIVIDER AND NOZZLES

SPECIFIC GRAVITY ADJUSTMENT

BIMETALLIC DISKS

DIFFERENTIAL METERING HEAD REGULATOR

BYPASS FUEL P_0

FUEL INLET

SCREEN (CLEANABLE)

FUEL PUMP

FILTER BY-PASS VALVE

RELIEF VALVE

P_X

P_X P_Y

P_Y

INC

INC

P_X PRESSURE

Fig. 2-2   Bendix DP-L2 hydropneumatic fuel control used on a Pratt and Whitney of Canada JT-15 turbine engine

necting the bellows unit to the metering valve.

9. Minimum Flow Adjustment — prevents metering valve from completely closing on deceleration.

10. Maximum Flow Stop Adjustment — sets maximum rotor speed for trimming of engine.

11. Bellows Unit

Governor — Receives differential air pressures P_x and P_y to position the torque tube and change fuel schedule and engine speed.

Deceleration — Expands to its stop when P_y air pressure decreases to cause a reduction in engine speed.

12. Temperature Sensor — bimetallic disks sensing engine inlet temperature T_2 to control P_x air pressure to the bellows unit.

13. Enrichment Valve — receives P_c compressor air pressure and controls P_x and P_y pressure to the bellows unit.

14. High Speed Governor — flyweights throw outward under centrifugal loading of increased engine speed. This action modifies P_y air pressure.

15. Power Lever — exerts a direct force to position the governor.

b. *Operation of control*

1. Unmetered fuel (P_1) is supplied to the fuel control by the fuel pump.

2. P_2 pressure drops across the metering valve orifice in the same manner as was discussed earlier in the simplified fuel control diagram. P_1 then becomes P_2 pressure which flows out to the engine and also influences the operation of the differential pressure regulating valve, here referred to as the Metering Head Regulator.

3. The fuel which bypasses back to the fuel pump is labeled P_0.

4. The air section is operated by compressor discharge air P_c; when modified this air becomes P_x and P_y air which position the main metering valve.

5. When the power lever is advanced:

a. The flyweights droop in, the speeder spring force being greater than the flyweight force.

b. The governor valve closes off the P_y bleed.

c. The enrichment valve moves toward closed, reducing P_c airflow. (Not as much air pressure is required when P_y bleeds are closed)

d. P_x and P_y air pressures equalize on the surfaces of the governor.

e. P_x air contracts the acceleration bellows and the governor bellows rod is forced downward. The diaphragm allows this movement.

f. The torque tube rotates counterclockwise and the main metering valve moves open.

g. The flyweights move back outward as engine speed increases and the governor valve opens to bleed P_y air.

h. The enrichment valve reopens and P_x air increases over P_y air value.

i. Reduced P_y value allows the governor bellows and rod to move back up.

j. The torque tube rotates clockwise to decrease fuel flow and engine speed stabilizes.

6. When the power lever is retarded:

a. The flyweights move outward—speeder spring force now being less than flyweight force under high engine speed.

b. The governor valve opens dumping P_y air. The backup valve is also depressed, dumping additional P_y air.

c. The enrichment valve opens, allowing increased P_x airflow.

d. P_x air expands the governor and deceleration bellows to its stop. The governor rod also moves up and the main metering valve moves towards closed.

e. P_x air decreases with engine speed decrease but the acceleration bellows holds the governor rod up.

f. As engine speed slows, the flyweights come back in, closing the P_y bleed at the governor valve and the backup valve.

g. The enrichment valve moves toward closed and P_y air increases in relation to P_x value.

h. The deceleration bellows moves downward. The metering valve moves slightly open and engine speed stabilizes.

7. When ambient temperature goes up for any fixed power lever position:

The T_{t_2} sensor expands to reduce P_x bleed and this in turn causes fuel flow to increase to keep a constant idle thrust. It also decreases off-idle stall tendencies by increasing the RPM vector for angle-of-attack purposes. Above idle this sensor loses its authority.

3. Electro-hydromechanical fuel controls

This type of fuel control has not been as widely used as the hydromechanical or hydropneumatic control. In recent years, the electro-hydromechanical control has been incorporated on some of the newer engine designs for both commercial and general aviation. In essence, the electro-hydromechanical unit is actually a basic hydromechanical control with the addition of an electronic sensor circuit. This circuit is powered by the aircraft bus and operates by receiving engine operating parameter signals from systems such as exhaust gas temperature and engine RPM.

a. Example system (Rolls-Royce RB-211)

An analysis of Fig. 2-3 will show that the control amplifier receives a signal from turbine gas temperature (TGT) and two compressor speed signals (N_1 and N_2).

This control operates on a hydromechanical schedule until near full engine power. Then the electronic circuit starts to function as a fuel limiting device.

The pressure regulator in this installation is similar to the pressure regulating valve in the simplified hydromechanical fuel control diagram, except that in this system, the bypassing of fuel occurs at the fuel pump outlet rather than the fuel control. Near full power, when predetermined TGT and compressor speed values are reached, the pressure regulator reduces fuel flow to the spray nozzles by returning increased amounts of fuel to the fuel pump inlet. The fuel flow regulator in this control acts as a hydromechanical control, receiving signals from the high speed compressor (N_3) gas path air pressure (P_1, P_2, P_3) and power lever position.

b. Example system (Garrett-AiResearch ATF-3)

The ATF-3 is one of the new generation business jet, turbofan engines. It is a three-spool design and incorporates an electro-hydromechanical fuel control system, sometimes referred to as an electronic fuel control because of the major role of the electronic circuits. Analysis of the schematic in Fig. 2-4 will reveal that the electronic computer receives the following inputs:

1. Fan speed

2. N_2, low-pressure compressor speed

3. N_3, high-pressure compressor speed

4. T_{t_2}, inlet total temperature

5. T_{t_8}, high-pressure turbine inlet temperature

6. P_{t_2}, inlet total pressure

7. 28-volt DC power

8. Permanent magnet generator

9. Power lever angle

10. Inlet guide vane position

The electronic portion of the fuel control analyzes the input data and sends a command to position the inlet guide vanes and to schedule fuel flow at the hydromechanical portion of the fuel control unit.

57

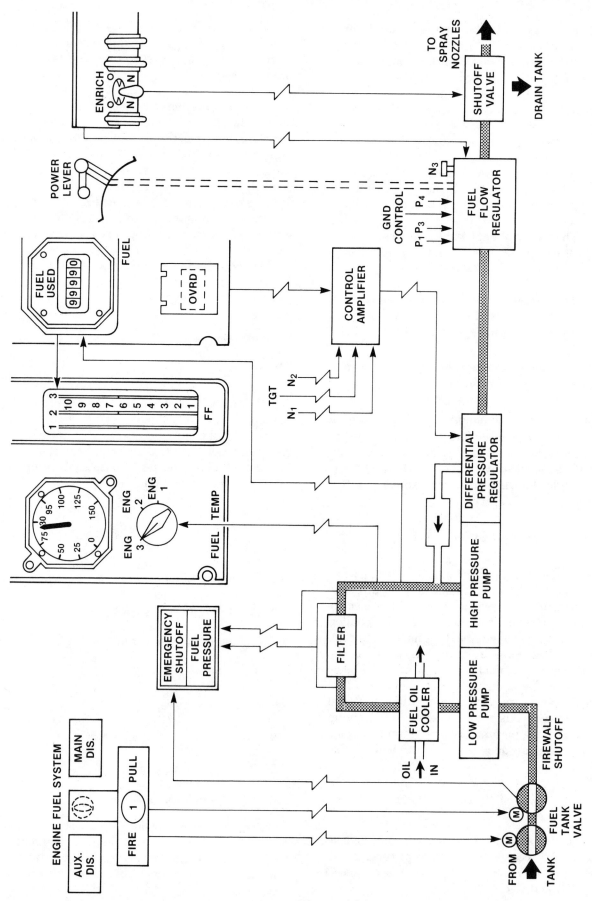

Fig. 2-3   Electro-hydromechanical fuel system used on the Rolls Royce RB-211 turbofan engine

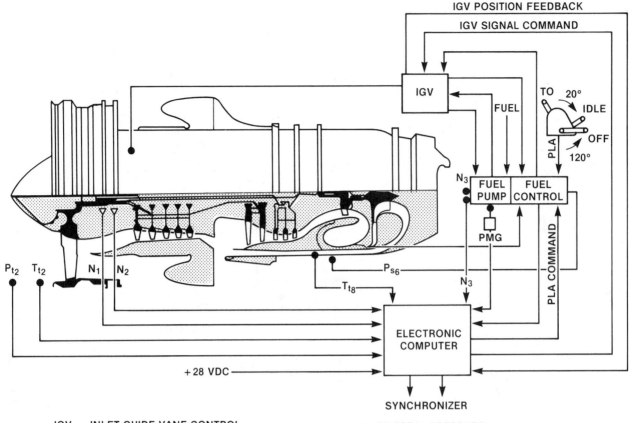

Fig. 2-4 *Electro-hydromechanical fuel system used on the Garrett-AiResearch TFE-731 and ATF-3 turbofan engines*

IGV = INLET GUIDE VANE CONTROL

N_1 = FAN SPOOL SPEED

N_2 = LOW PRESSURE SPOOL SPEED

N_3 = HIGH PRESSURE SPOOL SPEED

PLA = POWER LEVER ANGLE

T_{t2} = INLET TOTAL TEMPERATURE

P_{t2} = INLET TOTAL PRESSURE

W_f = FUEL FLOW

P_{s6} = HIGH PRESSURE COMPRESSOR DISCHARGE STATIC PRESSURE

T_{t8} = HIGH PRESSURE TURBINE INLET TEMPERATURE

PMG = PERMANENT MAGNET GENERATOR

Manufacturer's information states that this is a full authority (full time) system and more accurately schedules fuel flow than a comparable hydromechanical unit. It also provides the engine with overtemperature and overspeed protection and stall-free rapid acceleration by continually monitoring turbine inlet temperature.

4. Auxiliary power unit fuel control (APU)

Auxiliary gas turbine engines are widely used for supplying electrical power and air when the main engines are not operating. The same type of gas turbine is used in ground power units (GPU).

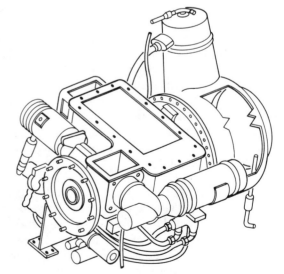

Fig. 2-5 *Aircraft auxiliary power unit, an APU*

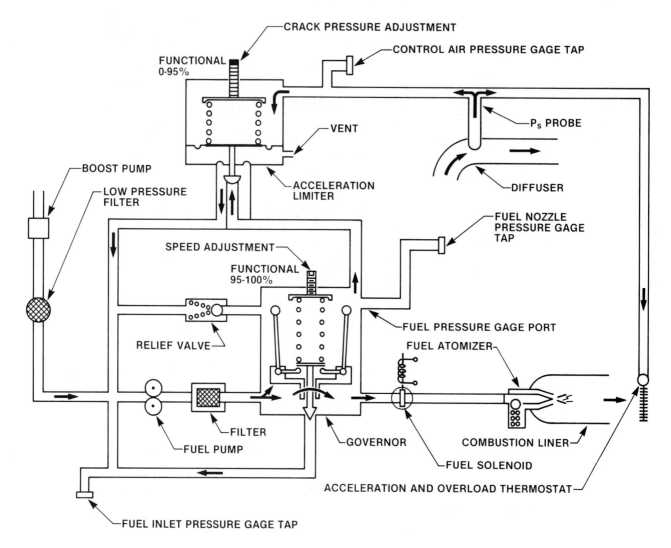

Fig. 2-6 APU fuel system. This is a hydromechanical control, with a pneumatic temperature limiter.

a. Example system (Garrett-AiResearch GTP-30)

Auxiliary gas turbine fuel systems are fully automatic and do not require a power lever. After actuation of a start switch, the fuel system functions to provide the correct amount of fuel for smooth acceleration to the rated speed. Thereafter, the fuel system schedules fuel to maintain a constant engine speed under varying pneumatic bleed and electrical loads.

In this installation there is a controlled-leak acceleration and overload thermostat located in the exhaust stream. It is normally closed and is used to dump a portion of the diffuser pressure signal which goes to the bellows of the acceleration limiter. The thermostat expands with exhaust heat and relieves some of the diffuser signal

pressure into the exhaust stream. The acceleration limiter's function is to control a by-pass valve which returns excess fuel to the pump inlet to protect against engine overtemperature during acceleration from 0 to 95% RPM.

A flyweight governor also controls fuel by-pass to maintain a steady state speed condition by returning unwanted fuel to the pump inlet. This governor protects against overspeed by overriding the acceleration limiter from 95 to 100% RPM.

Sequence of fuel flow:

1. Fuel flows from the aircraft boost pump through the low pressure filter to the main fuel pump under low pressure.

2. Fuel is pumped to the governor under higher pressure and to the fuel atomizer.

3. Fuel bypasses back to the main pump inlet at the acceleration limiter if an overtemperature occurs when the acceleration overload thermostat expands to leak control air overload.

4. Fuel bypasses above 95% engine speed to limit speed as per the preset value of the speeder spring.

5. The relief valve limits pump pressure in the event of a malfunction in the bypass portion of the system.

6. The fuel solenoid opens and closes for starting and operation.

5. Fuel control adjustments

Fuel control maintenance is usually limited to removal and replacement of the fuel control unit and resetting of the control adjustments. The adjustments include: specific gravity, idle RPM, and maximum power on the usual hydromechanical or hydropneumatic control. Electronic control adjustments are beyond the scope of this text and will not be discussed.

a. Specific gravity adjustment

The specific gravity adjustment, Fig. 2-7A, is seen on many fuel controls. It is a means of resetting tension on the Differential Pressure Regulating Valve spring within the fuel control when an alternate fuel is used. On some controls this adjustment is eliminated and other compensating mechanisms within the control make the needed density correction.

It is not uncommon to have the maintenance manual call for a trim check when an alternate fuel is used to ensure proper scheduling.

If, however, a run-up is to be made for performance checks, the normal fuel should be used, because the Btu value as well as specific gravity may be different in an alternate fuel, and the validity of the performance check could be compromised.

b. Trimming adjustments

Trimming is a term applied to idle speed and and maximum thrust adjustment made during

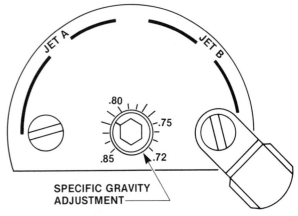

SPECIFIC GRAVITY ADJUSTMENT
(A)

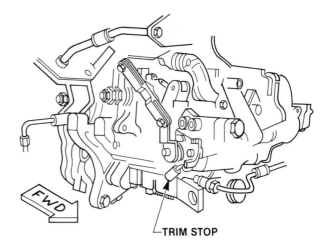

TRIM STOP ADJUSTMENT
(B)

Fig. 2-7 Trim adjustments on a turbine engine fuel control

performance testing. Idle speed adjustment is very similar from one engine to the next. It sets engine idling speed to the manufacturer's best economy and performance range. Idle speed is used during periods of operation when thrust is not required. Depending on the particular engine, a maximum percent RPM adjustment or maximum EPR (Engine Pressure Ratio) adjustment assures the correct maximum thrust is being produced by the engine. This thrust value is referred to as manufacturer's guaranteed rated thrust, sometimes shortened to rated thrust.

Most manufacturer's recommend that in order to stabilize cams, springs, and linkages within the fuel control, all final adjustments must be made in the increase direction. If an over-adjustment occurs, the procedure would be repeated by decreasing trim to below target values. Then increase trim back to the desired values.

Another important part of the trim procedure is to check for power lever cushion or springback. The technician, before and after the trim run, moves the power lever full forward and releases it. The distance the lever springs back is measured against prescribed tolerances. On an airliner, for instance, this distance might measure one inch or more. If out of limits, the technician will adjust the aircraft control system which attaches to the fuel control connection. This springback insures that the pilot will not only be able to obtain takeoff power but also additional power lever travel for emergencies. If the springback is correct, the fuel control will reach its internal stops before the cockpit quadrant reaches its forward stop. The springback, then, is the stretch in the aircraft linkage system which ensures the pilot's command over the engine's entire power range.

The trim check is accomplished whenever engine thrust is suspect and after such maintenance tasks as prescribed by the manufacturer. A few of these might include trimming after engine change, fuel control change, or loss of cushion. Trimming is otherwise required due to deterioration of engine efficiencies, as service time takes its toll, and also aircraft linkages stretch to cause misalignment in the cockpit to engine control systems.

In conjunction with the trim check, the idle speed is set and then an acceleration check is usually accomplished as an additional check on engine performance. After the trim check is completed, a mark is placed on the cockpit power lever quadrant at the takeoff trim position. The power lever is then advanced from idle position to takeoff position and the time is measured against a published tolerance. The time, even for a large gas turbine engine, is quite low, in the range of 5 to 10 seconds.

c. Part power trim

In order to save wear on the engine and also to save fuel, both EPR rated engines and speed rated engines are typically trimmed at less than takeoff power. This procedure usually involves placing a physical obstruction in the path of the fuel control lever linkage called a part power trim

stop. The power lever is advanced to hit the stop during trimming, and the trimming adjustment is made at this position with the engine operating (Part power position is approximately the maximum continuous fuel schedule). After trimming, the stop is removed in order to perform a takeoff power check.

d. Engine pressure ratio (EPR)

Many aircraft utilize an EPR gage for thrust indication. This gage conveniently reads the ratio of turbine discharge pressure divided by compressor inlet pressure and automatically corrects for changing ambient conditions.

During trimming procedures, the EPR system is checked for accuracy by the technician who compares the cockpit gage reading to his precision test gages.

EXAMPLE:

If turbine discharge pressure is measured to be 59.2 inches of mercury when the engine is correctly trimmed and engine inlet pressure reads 30.0 inches of mercury. The engine pressure ratio is calculated as:

$$EPR = \frac{\text{Turbine Discharge Pressure}}{\text{Compressor Inlet Pressure}} = \frac{59.2}{30.0} = 1.97$$

If the cockpit indicator reads within the accepted tolerance of 1.97 the EPR system and the engine are performing satisfactorily.

e. Trimming EPR rated engines

On many axial flow engines, there exists a better relationship between internal engine pressures and thrust than between compressor speed and thrust. In fact on some dual-spool engines the last 10% of the RPM range can change thrust by as much as 30%.

The trim setscrew arrangement for the EPR rated engine is similar to Fig. 2-7B. By turning the maximum trim adjustment, fuel flow and thrust will be increased or decreased. This in turn will affect the relationship between compressor inlet pressure and turbine discharge pressure. That is, as the latter goes up with fuel flow increase, EPR will also go up for a given compressor inlet condition.

A maximum EPR adjustment is provided on a variety of small and large engines. By turning the maximum trim setscrew, fuel flow and thrust are increased or decreased.

The trim procedure is as follows:

With finely calibrated gages, the technician measures inlet conditions of barometric pressure and ambient temperature before running the engine. He then installs the necessary fittings to measure turbine discharge pressure with an absolute pressure gage, and with the engine running, trims the engine to a value prescribed by a manufacturer's graph.

A caution to observe here is not to use the temperature at the control tower; there could be a measurable difference between that reading and the on-site reading. Also do not use the aircraft Outside Air Temperature (OAT) gage, because it could be heat soaked from the sun and again be

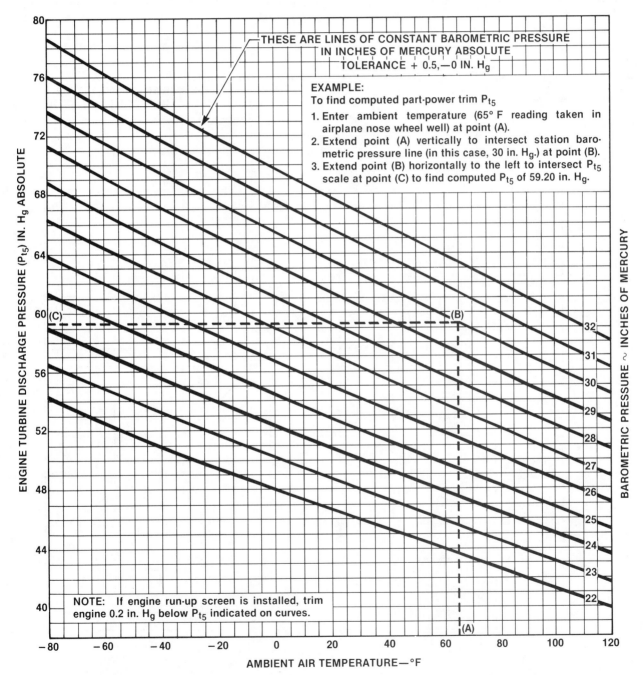

THESE ARE LINES OF CONSTANT BAROMETRIC PRESSURE IN INCHES OF MERCURY ABSOLUTE
TOLERANCE + 0.5,—0 IN. H_g

EXAMPLE:
To find computed part-power trim P_{t5}
1. Enter ambient temperature (65° F reading taken in airplane nose wheel well) at point (A).
2. Extend point (A) vertically to intersect station barometric pressure line (in this case, 30 in. H_g.) at point (B).
3. Extend point (B) horizontally to the left to intersect P_{t5} scale at point (C) to find computed P_{t5} of 59.20 in. H_g.

NOTE: If engine run-up screen is installed, trim engine 0.2 in. H_g below P_{t5} indicated on curves.

ENGINE TURBINE DISCHARGE PRESSURE (P_{t5}) IN. H_g ABSOLUTE

BAROMETRIC PRESSURE ∼ INCHES OF MERCURY

AMBIENT AIR TEMPERATURE—°F

Fig. 2-8   Part-power trim chart for a Pratt and Whitney JT-12 turbojet engine

385

significantly different from a thermometer reading taken at the engine. An industry-wide procedure is to hang the thermometer in the shade of the nose-wheel well.

Manufacturers also commonly recommend that two or more accelerations be made to a high power setting, in order to properly preload the internal mechanisms within the fuel control and ensure a higher degree of repeatability before the actual acceleration to trim power setting is made.

Fig. 2-8 shows a typical EPR rated business jet's part-power engine trim curve. After calculating the necessary turbine discharge pressure for an outside air temperature of 65° F and inlet pressure of 30 inches of mercury, if the engine is not producing 59.20 inches of mercury turbine discharge pressure, the technician will up-trim the engine to this value. The other engine instruments such as EGT, fuel flow, etc., are then checked, and if they are within allowable limits, the engine is considered to be airworthy from the standpoint of thrust output. (See page 63.)

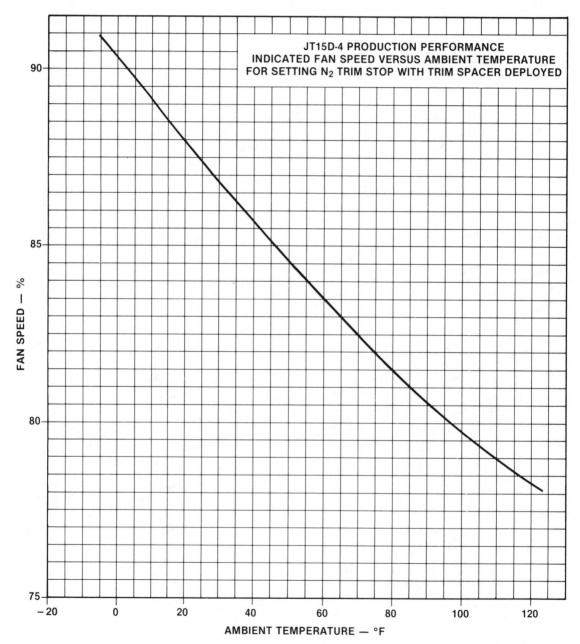

Fig. 2-9   Part-power trim chart for a Pratt and Whitney of Canada JT-15 turbofan engine

Note that Fig. 2-8 utilizes barometric pressure (P_{am}) rather than compressor inlet pressure (P_{t_2}). This is common practice because there is no convenient way to measure P_{t_2} on most aircraft. The graph, of course, is corrected for this and the engine is actually trimmed to an EPR of P_{t_5} divided by P_{t_2}.

f. Trimming speed-rated engine

A maximum speed trim is generally performed on turbofans of the dual-spool configuration where there exists a direct relationship between fan speed (N_1) and thrust. Referring to Fig. 2-7, counterclockwise rotation of the maximum adjustment will increase fuel flow to the engine, increase fan speed and consequently engine thrust output. Adjusting the idle setscrew will reset idle speed to the manufacturer's recommended percent RPM value. The amount of adjustment will depend on the parameters of an engine performance curve such as Fig. 2-9 on page 64.

The maximum trim determination in this case is dependent on ambient temperatures. The JT15 engine is speed rated. That is, its thrust is determined by a comparison to fan speed. High pressure compressor speed, turbine temperature, and other engine operational parameters must fall within an allowable range when rated thrust is obtained. If they do not, one of two conditions probably exists; a turbo-machinery malfunction, or a fuel scheduling malfunction. Dirty compressors and damaged compressors can cause aerodynamic problems. Damaged hot sections can cause thermodynamic problems. Poor fuel scheduling can cause anything from flameout to overtemperatures. What occurs is that in order to obtain the required thrust, some part of the engine is being over-taxed and this shows up on one of the condition monitoring instruments in the cockpit. The cause of the problem must then be found and corrected before the engine is released for service.

To ensure accuracy of readings, many operators utilize a portable precision tachometer slaved into the percent RPM indicating system. This gives the trim technician an immediate indication of condition of the cockpit instrument and allows him to more completely assess the condition of the engine.

The trim procedure is as follows:

1. With a precision thermometer reading taken at the aircraft, plot the fan (N_1) speed for the ambient temperature on Fig. 2-10.

2. Deploy trim stop.

3. Run aircraft, advancing power lever until linkage hits trim stop.

4. Adjust maximum setting and record gage readings — check against manufacturer's limits.

5. Retract trim stop and advance power lever to value shown in Fig. 2-10. Check N_1 speed, N_2 speed and T_5. (See page 66.)

g. Data plate speed check

A performance check that is quite often completed along with the trim check is the data plate speed check. On many gas turbine engines a small metal plate is attached at time of manufacture.

The plate is stamped with the engine speed at which a known thrust value in the part power range was obtained at final test. No two engines would necessarily be stamped with the same speed, because production tolerances of engine parts would vary the speed-to-thrust relationship on nearly every engine in some measurable way.

The purpose of this check is to compare future engine performance against the as new performance data. For example, consider that the data plate is stamped "87.25% N_2 Speed at 1.61 EPR, 59°F". As a ground performance check on engine condition when the engine accumulates service time (cycles and hours), the data plate speed check can be accomplished. The procedure in this instance would be to operate the engine at 1.61 EPR and observe the N_2 tachometer speed, then compare that speed to the as new speed on the data plate. If the tolerance is +2.0% and the engine being tested goes to 89.5% N_2 speed, the engine is out of limits for the data plate speed check. This check is an assessment of engine performance and not necessarily a pass or fail type check. If the engine is otherwise performing satisfactorily, the operator might decide to keep the engine in service even though it is no doubt consuming more fuel than when new, to give the required 1.61 EPR. It might also tell the operator

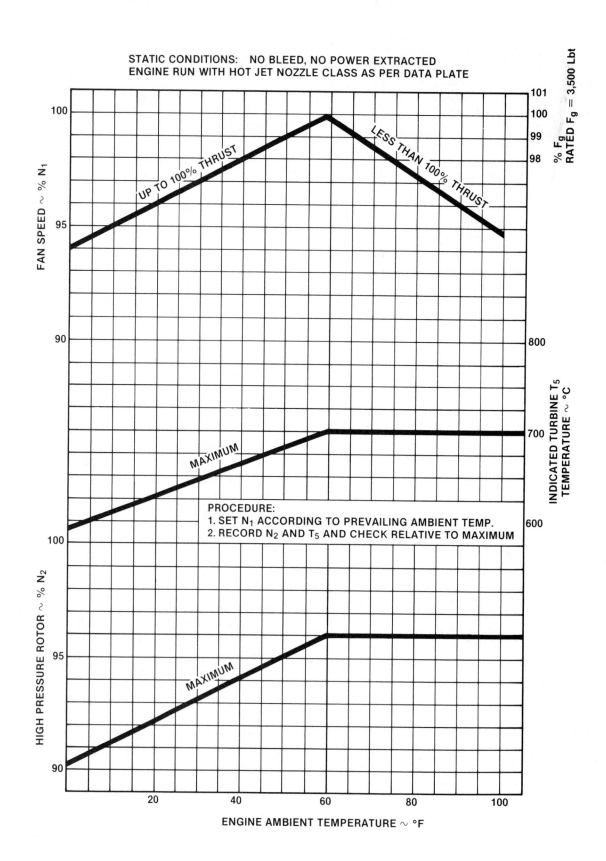

Fig. 2-10 Engine performance check limits for a Pratt and Whitney of Canada JT-15 turbofan engine

Fig. 2-11 Test equipment for trimming a turbo-jet engine

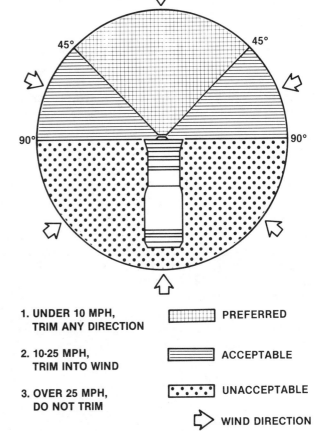

1. UNDER 10 MPH, TRIM ANY DIRECTION

2. 10-25 MPH, TRIM INTO WIND

3. OVER 25 MPH, DO NOT TRIM

PREFERRED

ACCEPTABLE

UNACCEPTABLE

WIND DIRECTION

Fig. 2-12 Wind direction and velocity limits to be observed when trimming a Pratt and Whitney JT-12 turbojet engine.

that the compressor is contaminated and requires field cleaning or that the hot section is deteriorating and should be scheduled for repair at the next inspection interval. The test run will seldom if ever be accomplished on a Standard Day, so the observed speed on the tachometer indicator will have to be corrected as per manufacturer's charts before it can be compared with the data plate speed.

The discussion above described an EPR rated engine, but the data plate speed check is equally applicable to the turbofan, speed rated engine, and the turboprop engine which is rated in either torque psi units or horsepower units.

h. Trim restrictions

There are always ambient condition restrictions on trimming which must be closely followed. Typical wind direction and velocity restrictions are shown in Fig. 2-12. Excessive wind in the direction of the tailpipe will cause a false high turbine discharge pressure and subsequent low trim. Excessive wind in the inlet will cause a false high inlet pressure and a subsequent low trim.

These situations occur as follows: wind up the tail pipe causes a back pressure and a false

high turbine discharge pressure for which the technician would compensate by down-trimming. Then later in calm air, turbine discharge pressure would be low and consequently EPR would be low. Wind into the inlet causes a false high compressor inlet pressure which generally affects compression to a greater degree than it affects the cockpit EPR reading. The tendency here is for higher inlet pressure to lower the cockpit EPR reading but because the compressor magnifies the false high inlet pressure by increased compression, it causes a rise in turbine discharge pressure. This causes what appears to be a higher EPR reading or over-trimmed engine. The technician compensates for this by down-trimming and then later in calm air the engine is under-trimmed. Wind is also seldom steady and when gusting it causes erratic gage readings which makes trimming difficult.

Other trim restrictions such as moisture content (rain), low temperature, and moisture (icing) will also be found in the aircraft operations manual and must be adhered to or a false trim will result.

The same restrictions apply to the speed rated engines. Even though the trim instrumentation is not as directly affected, the engine performance will be affected, and an incorrect trim could result.

i. Trim danger zones

An inherent danger to personnel and equipment exists in the area of operating turbine engines. Maintenance technicians should be constantly alert to keep out of the danger areas when approaching or leaving an operating jet engine. This can be a major problem when engines are being trimmed. Loose articles of clothing, microphone chords, mechanic wipe rags, tools etc., should be carefully secured so they won't be ingested by the engine. If possible, use an intercom system between the cockpit and the ground crew to minimize the danger present while working on operating jet engines, or use hand signals such as those shown in Fig. 2-13 on page 69.

j. Ear protection from noise

A large turbojet or turbofan engine operating at trim power settings can generate sound levels up to 160 decibels at the aircraft. Smaller engines of all types can generate sound levels up to approximately 130 decibels. This noise intensity is sufficient to cause either temporary or permanent hearing loss if adequate ear protection is not used by ground personnel.

The most effective ear protection device is the muff type which fits over the entire ear and defends against noise to the ear opening and also to the bone structure behind the ear.

The Federal Standard for noise protection is outlined in the Occupational Safety and Health Administration Standard, 36th Federal Register, 105, Section 19-10.95.

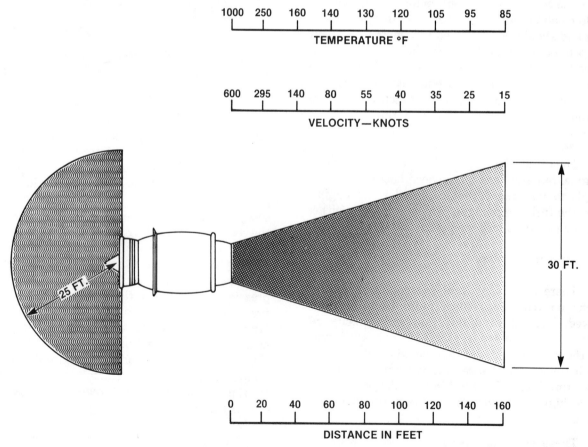

Fig. 2-14 Inlet and exhaust jet wake danger areas for a JT-15 turbofan engine

EMERGENCY SIGNALS

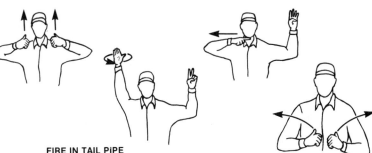

PERSONNEL IN DANGER
(for any reason)
REDUCE THRUST AND SHUT DOWN ENGINE(S).
(A) Draw right forefinger across throat. When necessary for multi-engine aircraft. Use a numerical finger signal (or point) with the left hand to designate which engine should be shutdown.
(B) As soon as the signal is observed, cross both arms high above the face. The sequence of signals may be reversed, if more expedient.

FIRE IN TAIL PiPE
TURN ENGINE OVER WITH STARTER.
(A) With the fingers of both hands curled and both thumbs extended up, make a gesture pointing upward.
(B) As soon as the signal is observed, use circular motion with right hand and arm extended over the head (as for an engine start). When necessary for multi-engine aircraft, use a numerical finger signal (or point) with the left hand to designate the affected engine.

FIRE IN ACCESSORY SECTION.
SHUTDOWN ENGINE AND EVACUATE AIRCRAFT.
(A) Draw right forefinger across throat. when necessary for multi-engine aircraft, use a numerical finger signal (or point) with the left hand to designate which engine should be shutdown.
(B) As soon as the signal is observed, extend both thumbs upward, then out. Repeat, if necessary.

GENERAL SIGNALS

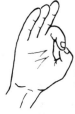

AFFIRMATIVE CONDITION SATISFACTORY, OK, TRIM GOOD, ETC.
Hold up thumb and forefinger touching at the tips to form the letter "O".

NEGATIVE CONDITION UNSATISFACTORY NO GOOD, ETC.
With the fingers curled and thumb extended, point the thumb downward toward the ground.

ADJUST UP
(higher)
With the fingers extended and palm facing up, move hand up (and down), vertically, as if coaxing upward.

ADJUST DOWN
(lower)
With the fingers extended and palm facing down, move hand down (and up), vertically, as if coaxing downward.

SLIGHT ADJUSTMENT
Hold up thumb and forefinger, slightly apart (either simultaneously) with the other hand when calling for an up or down adjustment, or with the same hand, immediately following the adjustment signal.

SHORTEN ADJUSTMENT
(as when adjusting linkage)
Hold up thumb and forefinger somewhat apart (other fingers curled), then bring thumb and forefinger together in a slow, closing motion.

LENGTHEN ADJUSTMENT
(as when adjusting linkage)
Hold up thumb and forefinger pressed together (other fingers curled), then separate thumb and forefinger in a slow, opening motion.

NUMERICAL READING
(of any instrument or to report a numerical value of any type)
Hold up appropriate number of fingers of either one or both hands, as necessary, in numerical sequence, i.e., 5, then 7 = 57.

ENGINE OPERATING SIGNALS

CONNECT EXTERNAL POWER SOURCE.
Insert extended forefinger of right hand into cupped fist of left hand.

DISCONNECT EXTERNAL POWER SOURCE
Withdraw extended forefinger of right hand from cupped fist of left hand.

START ENGINE
Circular motion with right hand and arm extended over the head. When necessary for multi-engine aircraft, use a numerical finger signal (or point) with the left hand to designate which engine should be started.*

*NOTE: To use an "ALL CLEAR TO START" signal, pilot or engine operator initiates the signal from the aircraft cockpit. Ground crewman repeats the signal to indicate "ALL CLEAR TO START ENGINE."

Fig. 2-13 Common hand signals for use with turbojet aircraft

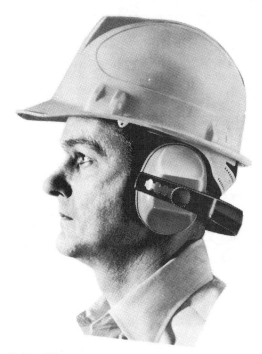

Fig. 2-15 Hearing protectors must be worn when working around turbojet aircraft

k. Flat rating

Most commercial engines and some business jet engines are referred to as flat rated. This refers to the flat shape of the full thrust curve. The flat portion of the curve ends at the point on the ambient temperature scale at which the thrust starts to drop below 100%. Fig. 2-16 shows this concept on a speed-rated engine. Analysis of the curve will reveal that a fan speed of 96% corresponds to 100% thrust on this engine, and that this value can be obtained at any ambient temperature up to 90° F. That is, by moving the power lever more and more forward, the pilot can obtain rated thrust at any temperature up to 90° F. After 90° F, more forward movement of the power lever is not permitted because it might result in an engine overtemperature.

When ambient temperature exceeds the flat rating of the engine, 100% thrust can no longer be obtained. This being the case, the aircraft's gross weight might need to be adjusted, or at the very least, runway takeoff roll will increase and the flight crew will need to consider this.

Some engines are flat rated only to 59° F; others, to as high as 100° F. This fuel system design depends largely on the needs of the operator and the prevailing local temperature condi-

tions. Generally, flat rating is thought of as enabling the engine to produce a constant rated thrust over a wide range of ambient temperatures without working the engine harder than necessary, in the interest of prolonging engine service life.

For example, an engine rated at 3,500 lbt at 59° F might be rerated to 3,350 lbt at 90° F. The aircraft user perhaps does not need to utilize 3,500 lbt. nor the maximum gross weight of the aircraft, and would like to benefit from increased engine service life and lower fuel consumption by operating at 3,350 lbt maximum. So then, flat rating is a current way of rerating an engine to a lower rated thrust than it would have at Standard Day temperature and then be able to use that lower rated thrust over a wider temperature range to suit the needs of the aircraft user.

C. Water Injection

The principle of latent heat of vaporization applies to the water injection process in the gas turbine engine. That is, injection of fluid into the gas path causes a heat transfer. When the fluid evaporates, heat in the air will be transferred into the fluid droplets, cooling and making the gas flow more dense. This occurs because the fluid takes up less space than air. Also compression increases as the air cools, which then allows a more dense mixture to occupy the same space within the engine, resulting in a thrust increase.

The reciprocating engine by comparison also uses a water injection system, with a water-alcohol mixture to cool its cylinders at high power settings. However, no additional thrust results. The injection mixture merely allows a leaner fuel-air mixture to be used.

Water injection, then, in a gas turbine engine is a means of augmenting engine thrust. Augmentation occurs in two ways: First, addition of water to air in the compressor increases mass flow. Second, water cools the combustion gases which allows additional fuel to be used. So it can be stated that water injection causes an increase in mass airflow, compression, and expansion without exceeding maximum temperature limits during takeoff. These increases in engine parameters result in a thrust increase in the range of 10 to 15%. Inlet water injection is designed for use at ambient temperature above 40° F. Below this temperature, icing is likely to occur in the water injection system and in the engine inlet.

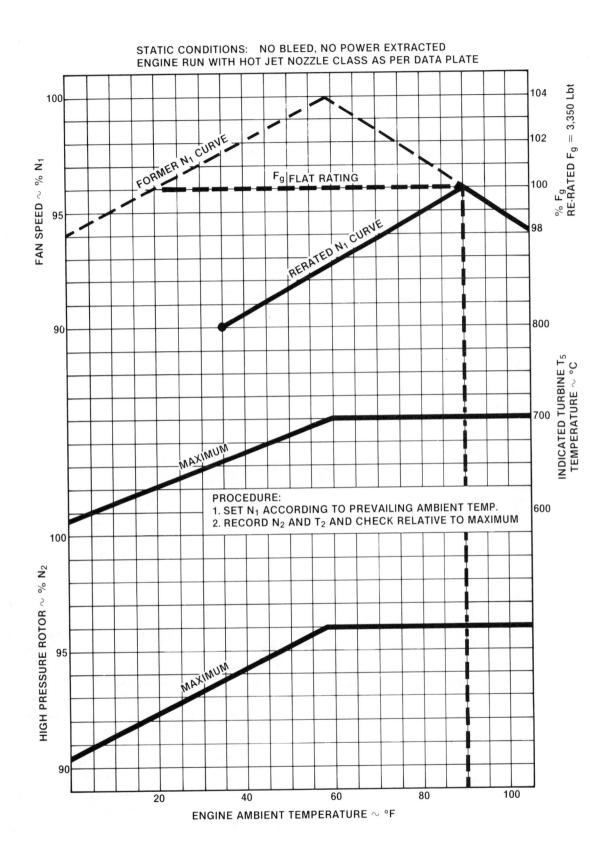

STATIC CONDITIONS: NO BLEED, NO POWER EXTRACTED
ENGINE RUN WITH HOT JET NOZZLE CLASS AS PER DATA PLATE

FORMER N₁ CURVE

Fg FLAT RATING

RERATED N₁ CURVE

MAXIMUM

PROCEDURE:
1. SET N₁ ACCORDING TO PREVAILING AMBIENT TEMP.
2. RECORD N₂ AND T₂ AND CHECK RELATIVE TO MAXIMUM

MAXIMUM

Fig. 2-16   Engine performance limits for flat rating a turbofan engine

71

Since available thrust quite often determines allowable aircraft takeoff weight, water injection is used almost exclusively at takeoff power settings. For instance, the older Boeing 707 aircraft carries approximately 300 gallons of water injection fluid per engine, using up the entire supply in a three minute takeoff and climb. This would equal an air-water ratio of approximately 5 to 1 based on a mass airflow of 160 lb/min and a water injection rate of 100 gal/min (850 lb/min).

Many airlines operating with older model aircraft which do not have the more modern turbofan or turboprop engines, frequently utilize water injection. This allows the operators to use higher aircraft takeoff weights in hotter ambient conditions and also when operating from higher airfield elevations.

1. Water injection fluid

Pure demineralized or distilled water is the most common water injection fluid. Ordinary tap water is not used, because it has high mineral solid content which can cause severe turbine distress when the minerals impinge on the turbine blades. Pure water is also widely used because it produces a greater cooling effect than a mixture of water and methyl or ethyl alcohol. Airliners can take advantage of this and not worry about altitude freeze-up by using the complete supply of water at takeoff. Aircraft, such as helicopters and turboprops, which make frequent takeoffs and landings, are forced to use a water-alcohol mixture to protect against freeze-up.

The following table shows the heat absorption or vaporization effect of the most common injection fluids.

FLUID	HEAT OF VAPORIZATION Btu/POUND	HEATING VALVE Btu/POUND
WATER	970 at 212° F	0
METHYL ALCOHOL	481 at 148° F	9,000 (APPROX)
ETHYL ALCOHOL	396 at 173° F	12,000 (APPROX)

Fig. 2-17 Characteristics of various types of water injection fluids

Even though water has no heating value, it has been determined that because of its heat absorption capability, more thrust can be obtained by injecting a given volume of water into the engine than on equal mixture of water and alcohol. Another way to think about this is that, although alcohol can be used as fuel after it is used as a coolant, the thrust augmentation factor per unit volume in a water-alcohol mixture is less than that of pure water.

2. Water injection system

A typical water injection system is shown in Fig. 2-18. Notice that it contains two independent injection nozzles, one to spray water into the compressor inlet and the other to spray into the diffuser. Compressor injection increases density and mass airflow and also cools the mixture, allowing increased fuel flow. The addition of fuel increases mass airflow to the turbine wheel relative to mass airflow being handled by the compressor and results in more mass flow and pressure left over in the tail pipe for thrust.

There is a limit to the amount of fluid injection any compressor or combustor can efficiently utilize. In the system shown, full thrust augmentation, when required, necessitated the use of both. On other installations it is common to see injection at only one location, either the compressor or diffuser.

When the ambient temperature is low, however, only the diffuser injection system can be used. Below 40° F at takeoff RPM, there is a danger of ice formation. At low ambient temperatures thrust is usually high enough without water injection for almost any aircraft gross takeoff weight. This water-injection system is controlled by a cockpit switch which arms the circuit and makes flow into both manifolds possible. When closed, the cockpit switch allows electrical current to flow to the fuel control microswitch. As the power lever reaches takeoff power, the microswitch is depressed and the water pump valve will be powered to open. This will allow compressor bleed air to flow through the air-driven water pump which supplies water under a pressure of 200 to 300 psig to the dual manifold. If compressor flow is not needed, a cockpit switch is installed to deactivate its flow valve. The pressure sensing tube to the fuel control is present to alert the control to reset fuel flow higher when water is flowing. This system is not generally needed if water-alcohol is used, because combustion of the alcohol keeps turbine inlet temperature at its required value.

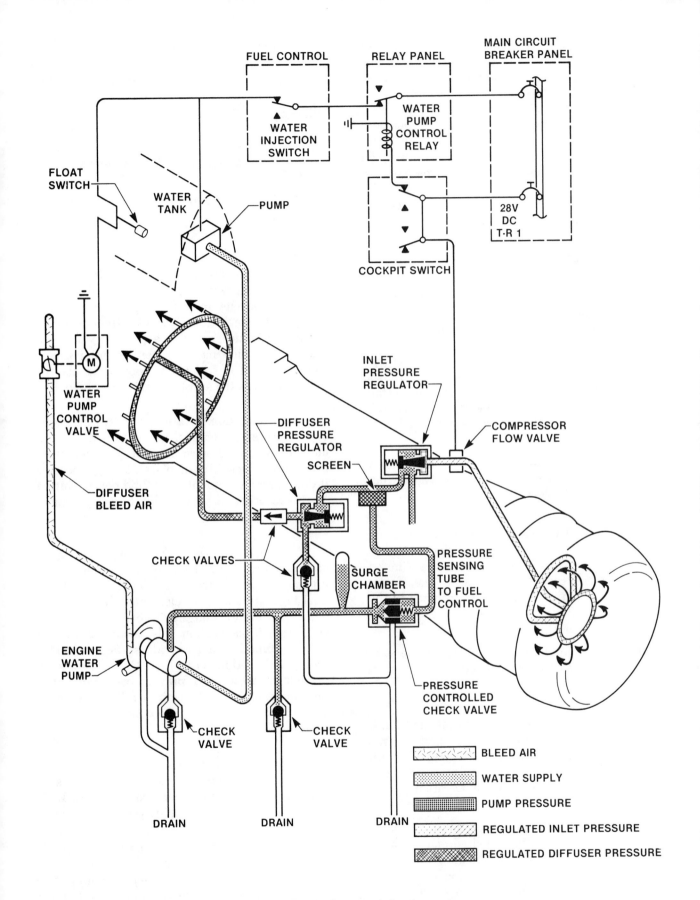

Fig. 2-18 Typical water injection system

73

A tank float level circuit will cut off power to the pump when the tank is empty and will also prevent the system from operating if the circuit is activated with an unserviced water supply. When the water injection system is not in use, the check valve at the diffuser prevents high temperature air from backing up into the water system. Drains are present to drain the lines when the system is not in use, preventing freeze-up. In some installations a bleed air system allows the pilot to purge the system of water after terminating water injection.

D. FAA Engine Power Ratings

Axial flow compressor engines, both turbojet and turbofan, are thrust rated in terms of either EPR or fan speed in the following categories: takeoff, maximum continuous, maximum climb, maximum cruise, and idle. For certification purposes, the manufacturer demonstrates to the FAA that the engine will perform at certain thrust levels for specified time intervals and still maintain its airworthiness and service life for the user.

These ratings can usually be found on the engine Type Certificate Data Sheets. The ratings are classified as follows:

Takeoff Wet — This rating represents the maximum thrust available while in water injection and is time limited. It is used only during takeoff operation. Engines are trimmed to this rating.

Takeoff Dry — Limits on this rating are the same as takeoff wet, but without water injection.

Max. Continuous — This rating has no time limit but is to be used only during emergency situations at the discretion of the pilot. For example, during one-engine-out cruise operation.

Max. Climb — Maximum climb power settings are not time limited and are to be used for normal climb to cruising altitude or when changing altitudes. This rating is sometimes the same as maximum continuous.

Max. Cruise — This rating is designed to be used for any time period during normal cruise, at the discretion of the pilot.

Idle — This power setting is not actually a power rating, but rather the lowest usable thrust setting for either ground or flight operations.

E. Fuel System Components and Accessories

1. Main fuel pump

The main fuel pump is an engine driven accessory. As such, when the engine speeds up, the pump also speeds up and delivers more fuel. The pump is designed to deliver a continuous supply of fuel to the fuel control at a quantity which is in excess of engine needs. After metering the required amount of fuel to the combustor, the fuel control returns the surplus fuel to the pump inlet.

Main fuel pumps are generally spur gear types with single or dual elements and often include a centrifugal boost element. The gear pump is classed as a positive displacement type because it delivers a fixed quantity of fluid per revolution. In this respect it is very similar to a gear type oil pump.

A typical positive displacement pump is shown in Fig. 2-19. The boost element is geared-up to produce the required inlet pressure to the dual high pressure gear elements. Dual elements with shear sections on the drive shaft are designed so that if one section fails, the other will continue to function and provide sufficient fuel for cruise and landing operation. Check valves are present in the outlet to prevent fuel recirculation into an inoperative element. A relief valve is incorporated to provide protection to the fuel system.

Pumps of this type produce the high pressure needed for atomization of fuel in the combustor. Many large gear pumps can produce up to 1,500 psi and a volume up to 30,000 pounds per hour.

On some smaller engines the fuel pump provides a mounting base for the fuel control. On others the fuel pump is not a separate unit at all but rather the pump housing is incorporated into the base of the fuel control.

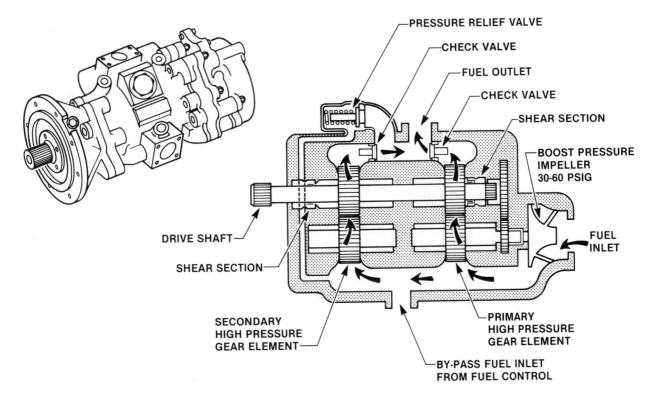

Fig. 2-19 *Engine-driven fuel pump for a turbine engine*

2. Fuel heater

Some engines use only the lubrication system oil cooler for heat transfer to the fuel, while others incorporate a separate fuel heater. Fuel heat is supplied to prevent ice crystal formation from entrained water in the fuel supply. When ice forms, fuel filter clogging occurs which can cause the filter to bypass. This condition allows unfiltered fuel to flow to downstream components. In severe cases, icing can cause flow interruption and engine flameout as ice forms again in the downstream components such as in the fuel control.

On engines where icing is critical, a pressure switch is often installed in the filter by-pass. If the filter ices, the pressure drop will cause a light to come on in the cockpit.

Fuel heat is designed to be used when the fuel temperature approaches 32° F. Fuel heat is either automatically activated three to five degrees F above the freezing point of water, or it is selected by a toggle switch in the cockpit. In this system, fuel on its way to the engine low-pressure filter passes through the cores of the heater assembly. The solenoid allows bleed air to pass over the cores to warm the fuel.

Typical operational restrictions are as follows:

1. Operate for one minute prior to takeoff and operate for one minute in every 30 during flight. Excessive heating can cause vapor lock or heat damage to the fuel control.

2. Do not operate during takeoff, approach, or go-around because of flameout possibilities from vaporization during these critical flight regimes.

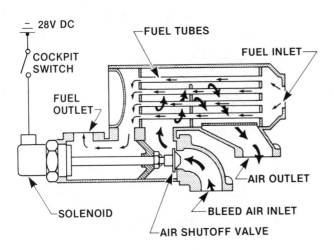

Fig. 2-20 *Fuel heater for use with a turbojet engine*

On some installations the cycle time is automatically controlled by an electric timer and gate valve arrangement. To check on the operation, as the system cycles on, the technician can observe gage indications such as EPR and oil temperature and the fuel filter light as follows:

1. EPR will usually drop due to compression loss as bleed air flows.

2. Oil temperature will usually rise as fuel temperature rises within the oil cooler.

3. If a filter by-pass light is illuminated due to filter icing, the light should extinguish as the system cycles on.

4. If the fuel by-pass light remains on, the technician would suspect a solid contamination at the fuel filter rather than icing.

In some aircraft, a separate fuel heat system is not installed, because the fuel-oil cooler plus additives in the fuel give sufficient protection against fuel icing. If an air-oil cooler is used, a fuel heat system is generally used.

3. Fuel filters

Two levels of filtration are generally required in turbine engine fuel systems. A low pressure coarse-mesh filter is installed between the supply tank and the engine, and a fine mesh filter be-

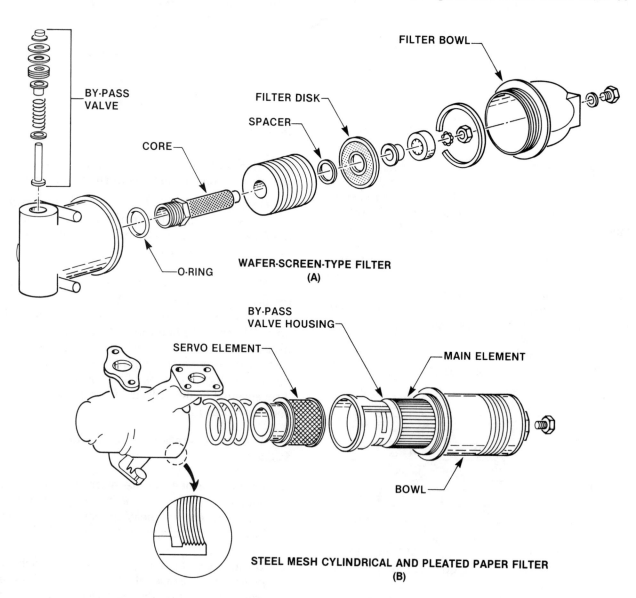

Fig. 2-21 *Fuel filters for use with a turbojet engine*

tween the fuel pump and the fuel control. The fine filter is necessary because the fuel control is a device with many minute passageways and fine tolerances.

Filters for this application have a micronic rating of 10 to 200 microns depending on the amount of contamination protection needed. (A micron is a metric linear measurement of one millionth of a meter, equal to 1/25,000 of an inch.)

Several varieties of filtering elements are used in fuel systems. The most common are: the steel mesh, the wafer screen, the steel mesh pleated screen, the steel mesh cylindrical screen, and cellulose fiber. The cellulose fiber filter has an equivalent micron rating. By way of explanation, a 35 micron filter of the wafer screen type will have square openings with a diameter of 35 microns and will prevent particles larger than 35 microns in diameter from passing through to the system. The cellulose element will filter out particles of relatively the same size.

Fuel filter checks are frequent inspection card items for the maintenance technician. If water or metal contaminants are present in the filter element or in the filter bowl, the technician must locate the source of the problem before returning the aircraft to service.

In Fig. 2-21A a wafer screen, bowl-type filter is illustrated containing a by-pass valve. This by-pass valve will open at 8 to 12 psid. Or in other words, it will open if the downstream pressure differs from the upstream pressure by 8 to 12 psi, as would occur if the filter starts to accumulate ice crystals or solid contaminants.

In Fig. 2-21B, notice that the filter contains two elements. The pleated mesh element filters the main system fuel on its way to the combustor and has a filtration rating of 40 microns. A by-pass valve relieves at 28 to 32 psid if this element clogs. The cylindrical mesh element is rated at 10 microns and filters fuel being routed to the fuel control to operate the servo mechanisms. The minimal flow in this part of the fuel system permits the use of a very fine filter. This fine filter is needed to protect the highly machined parts of the fuel control and protect against clogging of the numerous small fluid passageways.

a. Micron rating versus mesh size

Some fuel filters, rather than having a micron rating, are described in technical literature as having a certain mesh size or U.S. sieve number. An analysis of the chart of common filter sizes will show that there is a relationship between a micron rating and the mesh openings per linear inch. The mesh size is very close or in some cases exactly the same as the U.S. sieve number. For example, a 44 micron filter indicates that the opening size or the square mesh is 44 microns in diameter, or in English measurement equivalents, 0.0017 inch. Also, there are 323 square meshes per linear inch.

The difference in the two ratings, then, is that the micron rating deals with the diameter of the filter screen openings and the U.S. sieve number deals with the number of openings per linear inch.

SCREENS SIZE CHART			
OPENING IN MICRONS	US SIEVE #	OPENING IN INCHES	MESHES PER LINEAR INCH
10	---	.00039	1407.94
20	---	.00078	768.07
44	323	.0017	323.00
53	270	.0021	270.26
74	200	.0029	200.00
105	140	.0041	140.86
149	100	.0059	101.01
210	70	.0083	72.45
297	50	.017	52.36

Fig. 2-22 Comparison of opening sizes in fuel filters used with turbojet engines

4. Fuel nozzles (atomizing type)

This type of nozzle receives fuel from a manifold and delivers it to the combustor in a highly atomized, precisely patterned spray. The cone shaped, atomized spray pattern provides a large fuel surface area of very fine fuel droplets. This provides for optimum fuel-air mixing. The most desirable flame pattern occurs at higher compression ratios, and during starting and other off-design speeds, the lack of compression allows the flame length to increase. If the spray pattern is also slightly distorted, the flame, rather than being held centered in the liner, can touch the metal and cause a hot spot or even burn through. Another problem that occurs to distort the spray pattern is contaminant particles within the nozzle or carbon buildup outside the nozzle orifice which can cause what is termed hot streaking. That is, an unatomized stream of fuel is present which tends to cut through the cooling air blanket and impinge on the liner or on downstream components such as the turbine nozzle.

Some fuel nozzles are mounted on pads external to the engine to facilitate removal for inspection. Others are mounted internally and are only accessible when the combustion case is removed. The duplex nozzle shown in Fig. 2-23B is an externally mounted design: the simplex nozzle shown in Fig. 2-23A is an internally-mounted type.

a. The simplex design provides a single spray pattern and incorporates an internally fluted spin chamber to impart a swirling motion and reduce axial velocity of the fuel for better atomization as it exits the orifice. The internal check valve present in the simplex nozzle shown is there to prevent dribbling of fuel from the fuel manifold into the combustor after shutdown. Some fuel systems with simplex nozzles as the main fuel distributors incorporate a second smaller simplex nozzle called a primer or starting nozzle, which sprays a very fine atomized mist for improved light-off.

b. There are two common types of duplex fuel nozzles, the single-line and the dual-line.

The duplex nozzle shown, referred to as a single-line duplex type, receives its fuel at one inlet port and becomes a flow divider to distribute fuel through two spray orifices. Often, as shown, the center orifice, called the pilot or primary fuel, sprays at a wide angle during engine start and acceleration to idle. The outer orifice, referred to as main or secondary fuel, opens at a preset fuel pressure to flow along with the pilot fuel. Fuel of much higher volume and pressure, flowing from this outer orifice, causes the spray pattern to narrow so that the fuel will not impinge on the combustion liner at higher power settings.

The duplex nozzles also utilize spin chambers for each orifice. This arrangement provides an efficient fuel atomization and mixture residence time, as it is called, over a wide range of fuel pressures, and the high pressure supplied to create the spray pattern gives good resistance to fouling of the orifices from entrained contaminants.

The head of the fuel nozzle is generally also designed with air holes which provide some primary air for combustion, but these holes are mainly used for cooling and cleaning the nozzle head and spray orifices. At times of starting fuel flow only, the cooling airflow is also designed to prevent primary fuel from backflowing into the secondary orifice and carbonizing. A distortion of the orifice by carbon buildup around the head of the nozzle can distort the spray pattern. This buildup can be seen on a borescope check on some engines and if severe enough could require removal of the nozzles for cleaning. This could in some installations require an engine teardown to remove the carbon buildup. However, a recent development in decarbonizing allows the engine to be flushed with a special solution through the fuel manifolds. This purging under pressure loosens and removes the carbon and is a routine line maintenance procedure on some newer aircraft.

A second type of duplex nozzle—called a dual-line duplex—is discussed as part of the pressurizing and dump valve system in a subsequent paragraph.

5. *Fuel nozzle (vaporizing type)*

This type of fuel nozzle connects to a fuel manifold in an arrangement similar to the atomizing type. But, instead of delivering the fuel directly to the primary air in the combustor, as the atomizing type does, the vaporizing tube premixes the primary air and fuel. Combustor heat surrounding the nozzle causes the mixture to vaporize before exiting into the combustor flame zone.

Some vaporizers have only one outlet and are referred to as a cane-shaped vaporizer. Fig. 2-23D shows a dual outlet T-shaped vaporizer and is one of a set of eleven utilized in some models of the Lycoming T-53 turboshaft engine. Because vaporizing nozzles do not provide an effective spray pattern for starting, the T-53 incorporates an additional set of small atomizing type spray nozzles, which spray into the combustor during starting and then terminate flow on spool-up to idle. This system is generally referred to as a primer or starting fuel system. The Olympus engine in the Concorde SST also utilizes the vaporizing fuel nozzle and primer fuel nozzle systems but these systems are not in wide use throughout the industry.

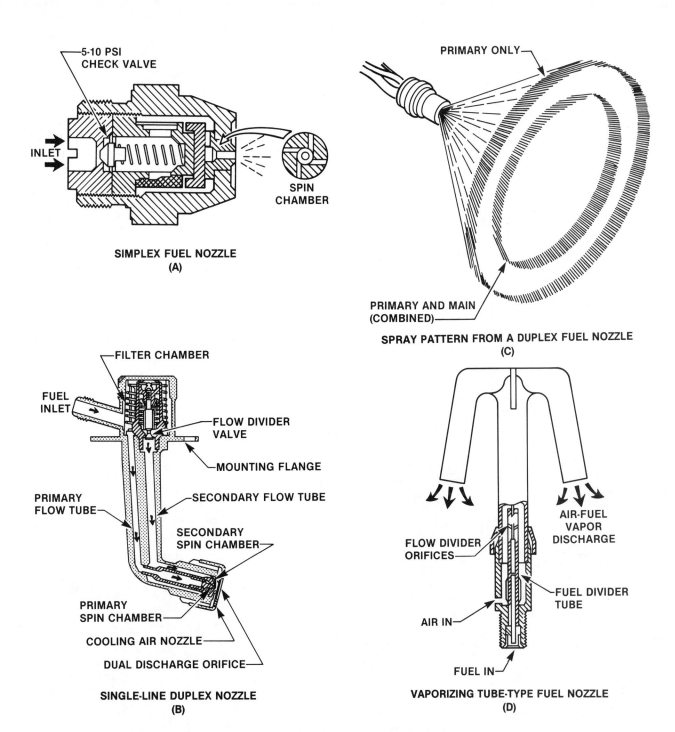

Fig. 2-23   *Turbine engine fuel nozzles*

6.   *Fuel pressurizing and dump valve*

A pressurizing and dump valve is often used along with a duplex fuel nozzle of the dual inlet line type. Rather than provide a flow divider in each nozzle, as with the single line duplex, this arrangement allows for one central flow divider called a pressurizing and dump valve. The term *pressurizing* refers to the fact that at a pre-set pres-sure, fuel flows into the main manifold as well as through the pilot manifold. The term *dump* refers to the valve's capability of dumping the entire fuel manifold after shutdown. Manifold dumping is a procedure which sharply cuts off combustion and also prevents fuel boiling as a result of re-sidual engine heat. This boiling would tend to leave solid deposits in the manifold which could clog the finely calibrated passageways.

79

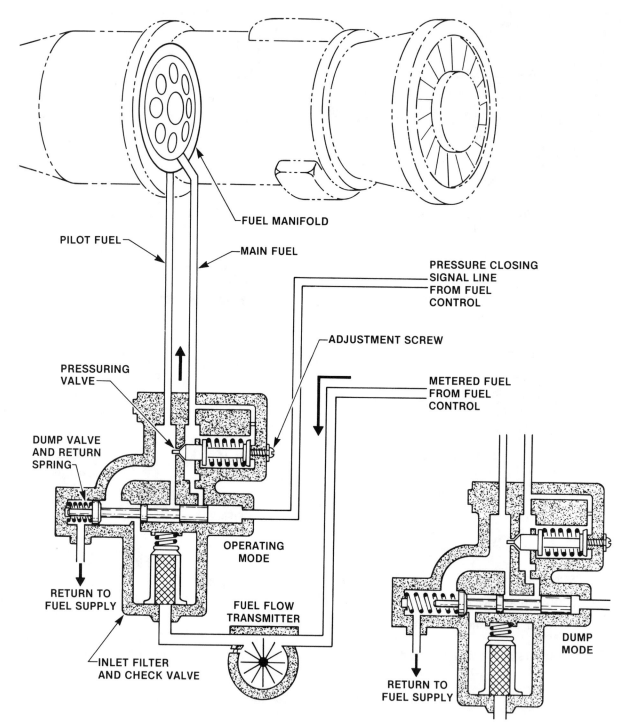

FUEL MANIFOLD

PILOT FUEL

MAIN FUEL

PRESSURE CLOSING
SIGNAL LINE
FROM FUEL
CONTROL

ADJUSTMENT SCREW

PRESSURING
VALVE

METERED FUEL
FROM FUEL
CONTROL

DUMP VALVE
AND RETURN
SPRING

OPERATING
MODE

RETURN TO
FUEL SUPPLY

FUEL FLOW
TRANSMITTER

INLET FILTER
AND CHECK VALVE

DUMP
MODE

RETURN TO
FUEL SUPPLY

Fig. 2-24   Pressurizing and dump valve installation

In Fig. 2-24, a pressure signal from the fuel control arrives at the pressurizing and dump valve when the power lever is opened for engine start. This pressure signal shifts the shuttle valve to the left, closing the dump port and opening the passageway to the manifolds. Metered fuel pressure builds at the inlet check valve until the spring tension is overcome and fuel is allowed to flow through the filter to the pilot manifold. At a speed slightly above idle, fuel pressure will be sufficient to overcome the pressurizing valve spring force, and fuel will also flow to the main manifold.

The compression on the pressurizing valve spring is normally adjustable as a line maintenance task. A valve opening too early can give an improper fuel spray pattern and hot starts or off-idle stalls. A late opening valve can cause slow acceleration problems. To delay the opening of the secondary manifold and eliminate hot start or off-idle stalls, the adjusting screw should be turned in to increase tension on the pressurizing valve spring. Conversely, to cause early fuel flow to the secondary manifold and enhance acceleration, the adjuster, should be turned outward.

To shut the engine off, the fuel lever is moved to Off. The pressure signal will be lost, and spring pressure will shift the shuttle valve back to the right, opening the dump valve port. At the same time, the inlet check valve will close, keeping the metered line flooded and ready for use on the next engine start.

Dump fuel, until recently, has been allowed to spill onto the ground or siphon from a storage container in flight. However, current FAA regulations prohibit this form of environmental pollution and several types of recycling systems have evolved. One such system returns fuel to the supply tank (Fig. 2-24), while another pushes fuel, which would have formerly been dumped, out of the fuel nozzles by introducing bleed air into the dump port. Still another plugs the dump port and fuel then drains out the lower fuel nozzles and finally out of the combustor case drain mentioned in subsequent paragraphs. This of course can only be accomplished where the fuel nozzle has no internal check valve to trap fuel within the head. There will no doubt be other systems designed to meet the FAA requirements and the needs of particular engines.

7. *Dump valve*

The dump valve, sometimes called a drip valve, is incorporated in the low point of fuel manifolds which utilize the simplex and single line duplex type of fuel nozzles. Its sole purpose is to drain the fuel manifold after engine shutdown. It is subject to the same environmental restrictions mentioned for the pressurization and dump valve.

8. *Combustor drain valve*

This valve is a mechanical device located in the low point of a combustion case. It is closed by

gas pressure within the combustor during engine operation and is opened by spring pressure when the engine is not in operation. This valve prevents fuel accumulation in the combustor after a false start, or any other time fuel might tend to puddle at the low point. Draining of fuel in this manner prevents such safety hazards as after-fires and hot starts. This drain also removes unatomized fuel which could burn near the lower turbine stator nozzles, causing serious local overheating during starting when cooling airflow is lowest.

As mentioned in the pressurization and dump valve discussion, if the dump line is capped off as an ecology control, the fuel manifolds will drain through the lower nozzles and fuel will exit the combustor via the mechanical drain valve into an aircraft drain receptical. A maintenance technician will periodically drain this receptical as a pollution control measure before it spills over onto the ramp.

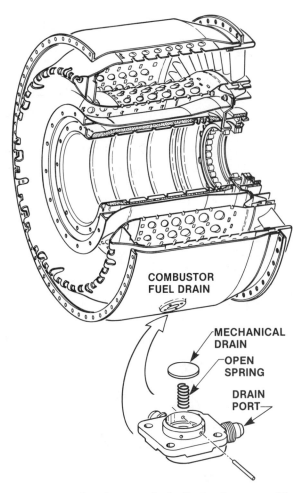

Fig. 2-25 Combustor fuel drain in a turbine engine

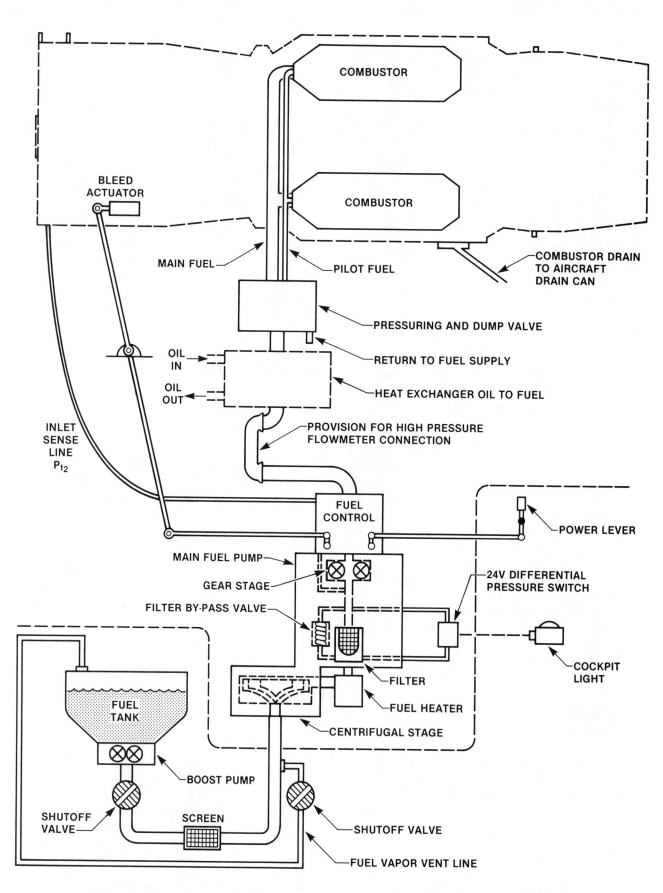

Fig. 2-26   Fuel system schematic for a Pratt & Whitney JT-12A turbojet engine

F. Example Fuel System (Pratt & Whitney JT-12)

The following is a complete fuel system schematic, in that it shows the relationship of component parts, one to another. This configuration is typical of the Rockwell Saberliner aircraft.

a. Fuel flow sequence

1. Storage Tank

2. Boost Pump — present to help eliminate vapor lock

3. Shutoff Valve — FAA regulation requires its presence

4. Screen — coarse mesh (200 micron)

5. Vapor Vent — present to help eliminate vapor lock

6. Centrifugal Stage — incorporated in the main fuel pump

7. Fuel Heater

8. Fine Filter — (20 micron)

9. Gear Stage — main fuel pump

10. Fuel Control — hydromechanical unit

11. Fuel Flowmeter

12. Heat Exchanger — (oil cooler)

13. Pressurization and Dump Valve

14. Pilot Fuel Manifold & Fuel Nozzle

15. Main Fuel Manifold & Fuel Nozzle

The 24V differential pressure switch is present to illuminate a cockpit warning light if the fuel pump filter is clogged and bypassing. The bleed actuator is a fuel operated compressor air bleed system.

G. Troubleshooting Fuel Systems

1. Engine motors over but doesn't start — no fuel when control shutoff valve is opened.		
Possible Cause	**Check For**	**Remedy**
a. No fuel to engine	fuel tank level	service
b. Improper rigging of shutoff valve (aircraft and engine)	full travel	re-rig linkage
c. Engine filters	clogging or icing	clean
d. Malfunctioning fuel pump	correct output pressure	adjust relief valve or replace pump
e. Malfunctioning fuel control	correct output pressure	replace control
2. Engine motors over but doesn't start — good fuel flow indication but no EGT		
Possible Cause	**Check For**	**Remedy**
a. Ignition system	weak or no spark	troubleshoot ignition system
b. P and D valve	dump valve stuck open	replace valve

3. Engine starts, but hangs up and will not self-accelerate to idle.

Possible Cause	Check For	Remedy
a. Starter	cutout speed—too early	troubleshoot starter
b. Fuel control	entrapped air preventing proper operation	bleed unit as per manual
c. Fuel control sensing lines	looseness causing loss of signal	tighten

4. Engine hot starts.

Possible Cause	Check For	Remedy
a. Fuel control	high fuel flow indication	replace control
b. P and D valve	1. partially open dump valve affecting fuel schedule and atomization at fuel nozzles	replace P and D valve
	2. partially open pressurizing valve—it should not be pressurized on start.	adjust, clean or replace as necessary

5. Engine unable to attain takeoff power.

Possible Cause	Check For	Remedy
a. Improper fuel control setting	trim or linkage problem	re-trim or re-rig
b. Fuel filters	partial clogging	clean
c. Fuel pump	correct output pressure	adjust relief valve or replace pump
d. Fuel control	correct output pressure and fuel flow	replace fuel control

6. Flame-out when applying takeoff power.

Possible Cause	Check For	Remedy
a. Fuel pump — one element sheared shaft	correct output pressure	replace pump
b. Fuel control	correct fuel flow indication	replace control

7. Engine is slow to accelerate/engine off idle stalls.

Possible Cause	Check For	Remedy
a. Same as 6b		
b. P and D valve	correct pressurizing valve spring adjustment	adjust spring tension or replace P and D

I. RECIPROCATING ENGINE INDUCTION SYSTEMS

A. Naturally Aspirated Engine Induction Systems

Aircraft reciprocating engines are air-breathing heat engines, which means that they must be able to take in a sufficient amount of air to combine with the fuel to release heat which is then converted into mechanical work. An induction system that allows the proper amount of clean air to be taken into the engine is one of the most important powerplant systems.

1. Carburetor heat systems

An engine equipped with a carburetor has the problem of carburetor ice. The fuel is sprayed into the air at the main discharge nozzle in the form of atomized liquid gasoline, and when it turns into gasoline vapor, it absorbs heat from the air. This loss of heat drops the temperature of the air enough to condense moisture out of it and form liquid water. The water then freezes and restricts the airflow into the engine.

To prevent the formation of carburetor ice, the FAA has this requirement: "Each reciprocating engine air induction system must have means to prevent and eliminate icing. Unless this is done by other means, it must be shown that in air free from visible moisture at a temperature of 30° F.—Each airplane with sea level engines using conventional venturi carburetors has a pre-heater that can provide a heat rise of 90° F with engines at 75 percent of maximum continuous

power. Each airplane with altitude engines using conventional venturi carburetors has a preheater that can provide a heat rise of 120° F with the engines at 75 percent of maximum power."—Federal Aviation Regulations Part 23, "Airworthiness Standards: Normal, Utility, and Acrobatic." This heat is provided by the exhaust system.

Normal inlet air is taken in from the front end of the cowling, where it can take advantage of the ram effect during normal flight to provide a slight increase in the amount of air the engine can take in.

Fig. 1A-1 *The carburetor air inlet for this aircraft is located between the two lights in the nose cowling.*

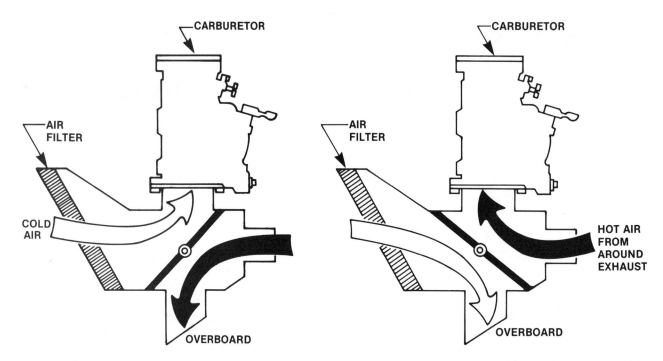

AIR FROM OUTSIDE OF THE AIRCRAFT IS TAKEN IN THROUGH THE FILTER AND DIRECTED INTO THE CARBURETOR. HEATED AIR FROM AROUND THE EXHAUST IS DUCTED OVERBOARD.

(A)

IN THE HOT POSITION, THE HOT, UNFILTERED AIR PASSES INTO THE CARBURETOR, AND THE FILTERED AIR IS DUCTED OVERBOARD.

(B)

Fig. 1A-2   Typical carburetor heat box

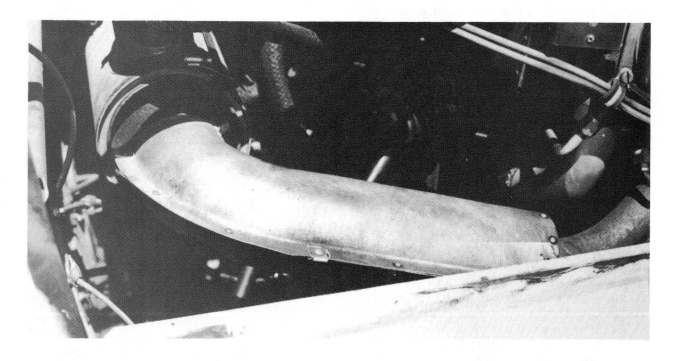

Fig. 1A-3   Air flowing through the thin sheet metal shroud around the exhaust pipe is heated before it is taken into the carburetor. This heated air prevents the formation of carburetor ice.

2

This air passes through a filter and then into the carburetor. When it is necessary to heat the inlet air, the air valve inside the carburetor heat box is moved, so the air that flows through the filter passes out of the heat box, and air is taken into the engine from around some part of the exhaust system.

The air that flows into the engine when the carburetor heat valve is in the Hot position is unfiltered, and for this reason carburetor heat should not be used when the aircraft is on the ground, especially when it is in a sandy or dusty location.

Engines equipped with fuel injection systems or with pressure carburetors that are not so highly susceptible to icing, may use air that has been heated by passing through the cylinder cooling fins instead of around the exhaust. Rather than calling this carburetor heat, we call it alternate air.

In Fig. 1A-4, we have a typical alternate air system. When the alternate air valve is in the Cold position, a butterfly valve allows air to flow through an opening in the nose cowling and through the air filter into the carburetor. Moving the alternate air valve in the cockpit to the Alternate Air position closes the butterfly valve which shuts off the flow of air through the filter. The low pressure caused by the pistons moving inward pulls open the spring-loaded alternate air door, and air that has passed through the cylinder cooling fins is taken in through the carburetor.

This spring-loaded alternate air door also opens if the aircraft flies into freezing rain, and the air filter ices over enough to shut off the flow of air into the engine. Rather than starving the engine of air, the door will automatically open and allow warm air to be drawn into the carburetor.

Some aircraft are equipped with carburetor air temperature gages to warn the pilot of impending danger from the formation of carburetor ice or of the equally serious problem of high inlet air temperature. In Fig. 1A-4, we see that the temperature probe is upstream of the carburetor air inlet. Icing is most apt to occur downstream of the fuel discharge nozzle, and so the temperature measured by the probe is not actually the temperature of the most vital part of the induction system. But, the aircraft manufacturer has determined by flight test the relationship between

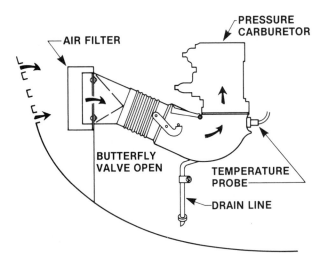

COLD AIR POSITION WITH THE BUTTERFLY VALVE OPEN, ALLOWING FILTERED AIR TO PASS INTO THE CARBURETOR.

(A)

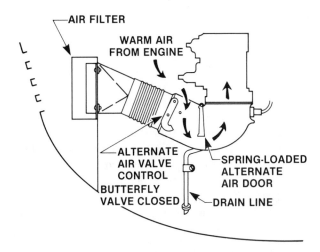

ALTERNATE AIR POSITION WITH THE BUTTERFLY VALVE CLOSED. WARM AIR FROM AROUND THE ENGINE FORCES THE SPRING-LOADED ALTERNATE AIR DOOR OPEN, AND THE WARM AIR FLOWS INTO THE CARBURETOR.

(B)

Fig. 1A-4 Alternate air system for an engine equipped with a pressure carburetor

the indication of the carburetor air temperature gage and the actual temperature at the discharge nozzle, and temperature limits are set to warn the pilot of any inlet air temperature that would allow ice to form. The temperature gages are usually marked with a low-temperature danger area that could indicate the danger of ice formation as well as a high-temperature area that warns of a possibility of detonation.

3

Some of the larger reciprocating engines have supplemented the carburetor heat or alternate air system with an alcohol deicing system that allows the pilot to spray alcohol into the inlet of the carburetor to remove ice and to assist the warm air in keeping the carburetor free from ice.

Radial engines have an advantage over opposed engines in their induction system, in that many of them have a diffuser section at the center of the engine into which the carburetor discharges its fuel-air mixture. Induction pipes of uniform length feed the mixture from this section to the intake valves of each of the cylinders. To assure complete vaporization of the fuel, some of these engines route a bit of the engine exhaust gas through a stainless steel tube that passes through the diffuser section just below the carburetor. This is called a "hot spot," and it helps vaporize the fuel for more uniform mixture distribution.

Modern horizontally opposed engines have a system that is far better for fuel vaporization. The carburetor is often mounted on the oil sump and the induction pipes pass through the oil. This serves the double function of cooling the oil and at the same time warming the induction air without the danger of heating it to the point that detonation could occur. The carburetor, when mounted in this position, is quite centrally located, and it can provide as even a fuel-air mixture to all cylinders as is possible with a horizontally opposed engine.

2. Induction air system filters

Any sand that gets into the engine will act as an abrasive and wear away the cylinder walls and piston rings. And silica from the sand will form a lead silicate contaminant on the nose core insulator of the spark plug that is extremely difficult to remove and is conductive when it is hot. This contaminant will short out the spark plug.

Many of the early air filters were made of screen wire that was filled with a fiber material called flock. These filters could be cleaned and serviced by washing them in varsol and, after they were clean, soaking them in a mixture of engine oil and preservative oil and allowing the excess oil to drain. Later, paper filters similar to those used in automobiles were adapted and approved for use in aircraft. Air passes through the porous paper filter element, but any dust and sand particles are trapped on its surface. Paper filters may be cleaned by blowing all of the dust out of them in the direction opposite the normal airflow, and then washing them in a mild soap and water solution and allowing them to dry. When servicing this type of filter by washing, be sure to follow any recommendations or restrictions imposed by the aircraft manufacturer.

The most effective filter today is a polyurethane foam filter impregnated with a glycol solution. The glycol gives these filters an affinity for dust, and at the time recommended for filter change, the foam element is removed and discarded and a new one is installed. It is not recommended that these filters be cleaned.

Fig. 1A-5 Foam-type carburetor inlet air filter. Rather than cleaning this type of filter, the foam sock is replaced with a new one.

B. Supercharged Engine Induction System

A naturally, or normally, aspirated engine is one whose cylinders are filled with air forced into them by atmospheric pressure. The intake valve opens as the piston is moving outward near the end of the exhaust stroke, and it remains open all through the inward movement of the piston on the intake stroke and until it has started back out on the compression stroke. The volumetric efficiency, which is the ratio between the volume of the charge of air in the cylinder to the actual volume of the cylinder, can never reach 100%. And, because of such factors as valve timing, the use of carburetor heat, or high density altitude, it will likely be considerably less than this.

In Fig. 1A-6, we have a power curve that illustrates the way the brake horsepower of a naturally aspirated engine decreases as the density altitude increases. As the air in the induction system becomes less dense, there is less oxygen to combine with the fuel, and the power decreases. The maximum power the engine can develop is determined by the maximum mass of air that can be taken into the cylinders.

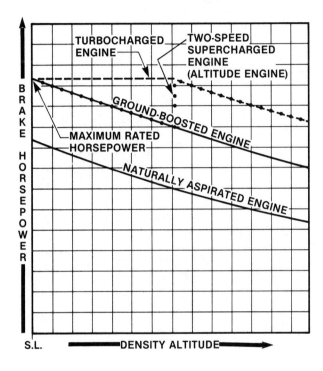

Fig. 1A-6 Relationship between power and density altitude for engines with various types of induction systems

Some engines are designed with a gear-driven air compressor, or impeller, to compress the fuel-air mixture after it leaves the carburetor and before it enters the intake valves at the engine cylinders. This geared supercharger, as it is called, can increase the power the engine produces at sea level, as well as the power it produces at all altitudes. The power will decrease as the density altitude increases, in the same way it does with a naturally aspirated engine. Engines whose sea level horsepower is boosted with an internal supercharger are called ground-boosted engines. We see the power curve for a ground-boosted engine in Fig. 1A-6.

Some of the large radial engines used up through World War II had two-speed superchargers that used an oil-operated clutch to drive the supercharger impeller through either a low-gear ratio of about eight to one or a high-gear ratio of about eleven to one.

Takeoff was made with the supercharger in low blower, and the engine acted in the same way as a ground-boosted engine. The power dropped off with altitude in the same way it does in a single-speed ground-boosted engine. When the aircraft reached a specified altitude, the throttle was pulled back somewhat, and the blower control shifted to high blower. The throttle advanced to get the manifold pressure that showed the engine was developing its rated horsepower. An engine equipped with this type of supercharger is called an altitude engine, and its power curve is also shown in Fig. 1A-6.

The step in the power produced when the supercharger is shifted from low to high blower is avoided by using an exhaust-driven turbosupercharger, or turbocharger, as it is commonly called. This is a centrifugal air compressor connected to the shaft of an exhaust gas-driven turbine. The amount of exhaust gas that flows through the turbine can be varied to produce a compressor speed that will maintain the manifold pressure needed to produce the rated engine power up to the critical altitude. This is the altitude at which full throttle and the waste gate fully closed is needed to produce the rated power.

1. Turbochargers

Gear-driven superchargers use a great deal of power for the amount of power increase they produce. And as early as 1918, a study was undertaken to find a method of increasing the manifold pressure with the least use of engine power. This study produced the turbosupercharger, which is an exhaust-driven turbine coupled directly to a centrifugal air compressor. This system provided a practical way of boosting engine power at altitude, and in World War II, almost all fighter planes and bombers were equipped with both gear-driven superchargers and turbosuperchargers.

Turbosuperchargers used with 1,000 to 2,000 horsepower engines are complex systems that use electronic controls to sense the manifold pressure and turbine speed to produce the power desired by the pilot.

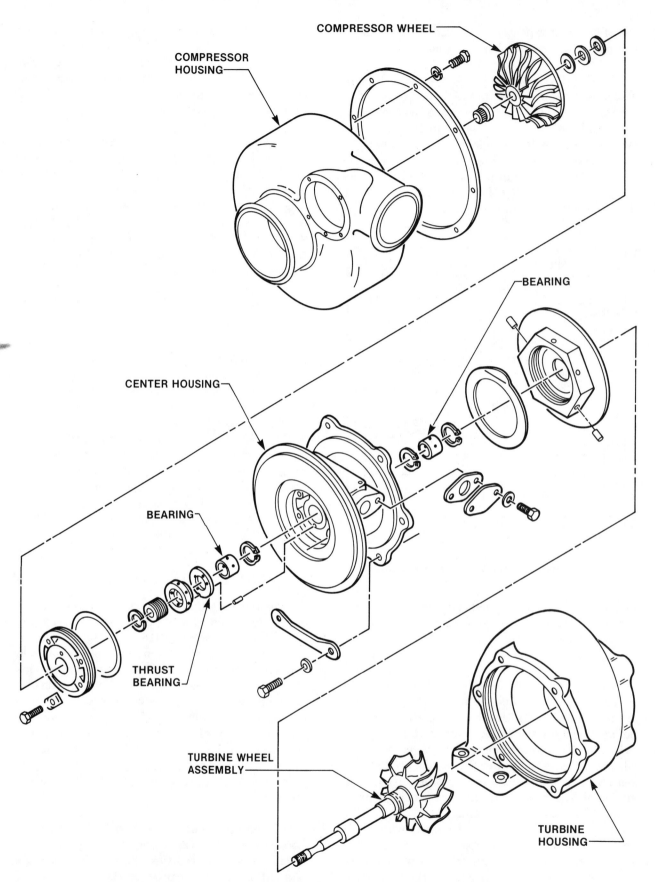

COMPRESSOR WHEEL

COMPRESSOR HOUSING

BEARING

CENTER HOUSING

BEARING

THRUST BEARING

TURBINE WHEEL ASSEMBLY

TURBINE HOUSING

Fig. 1A-7   Exploded view of a typical aircraft turbocharger

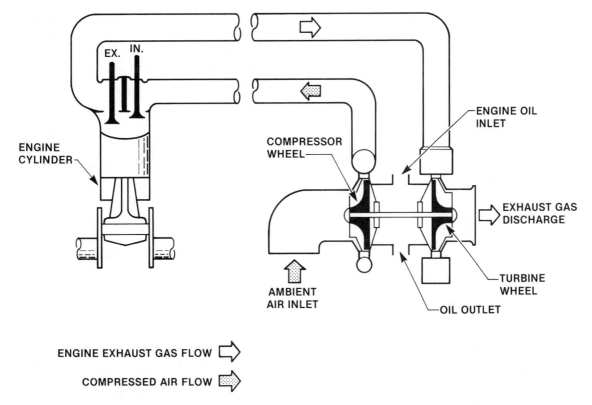

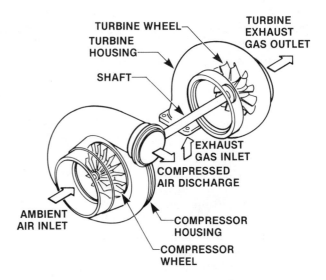

Fig. 1A-8 *Basic operating principle of a turbocharger in an aircraft engine induction system*

Turbochargers for general aviation aircraft are similar in principle to those used on the large military aircraft, but they are much smaller and far less complex.

In Fig. 1A-8, we see the basic way this system works. When the exhaust gas leaves the cylinder, it flows through the individual exhaust stacks into a collector and then to the periphery of the turbine wheel and out the center of the wheel. Connected to the same shaft as the turbine is a centrifugal air compressor. Air is taken from the normal air inlet of the aircraft, and it passes through the air filter and enters the center of the compressor. Centrifugal force then slings the air out from the compressor wheel, and it is collected at the periphery of the wheel and taken into the induction manifold of the engine.

Fig. 1A-9, gives a general idea of the basic physical appearance of two of the two turbocharger sections.

Between the compressor housing and the turbine housing, there is the center housing that contains two aluminum bearings that are free to rotate within their respective housings, and the shaft rotates in these bearings. Oil flows between the housing and the bearings, and also through

Fig. 1A-9 *Basic appearance of the turbine section and the compressor section of an aircraft engine turbocharger*

holes drilled in the bearings, so a continuous flow of oil is provided between the wheel shaft and the inside of the bearings. Lubricating oil from the engine oil system flows into the top of the center housing in Fig. 1A-10, and a scavenger pump inside the engine scavenges the oil from the turbo-

7

charger scavenger sump, out through the large opening in the bottom of the housing.

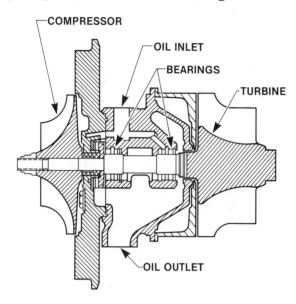

COMPRESSOR

OIL INLET

BEARINGS

TURBINE

OIL OUTLET

Fig. 1A-10 Turbocharger center housing showing the oil inlet, the bearings, and the oil outlet

About four to five gallons of oil per minute are pumped through the turbocharger center housing, not only to lubricate the bearings, but to take away heat from the housing. The turbine inlet temperature may get as high as 1,600° F, and the large flow of oil is needed through the bearings to keep them well below this temperature.

2. Turbocharger control systems

The simple system in Fig. 1A-8 explains the theory of operation of a turbocharger, but this is not a practical system. In this simple system, the turbine speed, and thus the speed of the compressor and the amount of air it compresses, are all a function of the amount the throttle is opened. And in this simple system it would be very easy to get an excess of manifold pressure which would cause detonation. Getting this excess manifold pressure is called overboosting the engine. The system could be designed so that under standard conditions the turbine could not turn fast enough to overboost the engine, but standard conditions do not exist throughout a flight, and if the engine is producing its rated power under high density altitude conditions, it would seriously overboost at low density altitude.

This limitation is overcome by using a controllable waste gate, as we see in Fig. 1A-11. When the waste gate is wide open, all of the ex-

haust goes out the tail pipe without passing through the turbocharger, and when the waste gate is fully closed, all of the exhaust must pass through the turbine to get to the tail pipe.

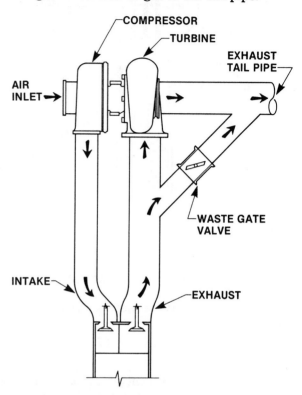

COMPRESSOR

TURBINE

EXHAUST TAIL PIPE

AIR INLET

WASTE GATE VALVE

INTAKE

EXHAUST

Fig. 1A-11 The turbocharger turbine speed is controlled by varying the amount of exhaust gases that are forced to flow through the turbine. When the waste gate valve is open, the exhaust gases bypass the turbocharger, but when the waste gate valve is closed, all of the gases must flow through the turbocharger.

The difference in the various turbocharger systems installed in general aviation aircraft have to do with the way the waste valve is controlled.

One of the simplest systems for turbocharger control uses a manual linkage between the engine throttle valve and the waste gate valve. For takeoff at low density altitude, the throttle is advanced to the takeoff position, and the engine develops full takeoff power with the waste gate fully open. As the airplane goes up in altitude, the engine power drops until full throttle will not produce the rated power. Then the throttle is advanced beyond its takeoff position, and the additional movement of the throttle begins to close the waste gate valve; the manifold pressure goes up to produce the rated horsepower. When the throttle is retarded, the waste gate opens, and then the fuel to the engine is decreased.

Some installations using this simple system have two controls: one control for the engine throttle and a separate control for the waste gate valve.

Another very simple turbocharger control system uses an adjustable, rather than controllable, waste gate valve and a pressure relief valve. We see such a system in Fig. 1A-12.

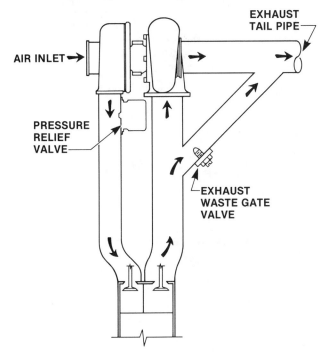

Fig. 1A-12 Turbocharger control system, using an adjustable waste gate valve and a pressure relief valve

This simple system is used on some of the newer generation aircraft that are designed for the simplest operation. An adjustable waste gate valve has a threaded bolt sticking out of the portion of the exhaust system that bypasses the turbocharger. Screwing this bolt in or out determines the amount of exhaust gas that is forced to flow through the turbocharger, and this, in turn, determines the amount of air that is forced into the induction system.

The waste gate valve is adjusted so the engine will produce its rated horsepower under sea level, standard-day conditions when the throttle is wide open. As the aircraft goes up in altitude and the air density decreases, the maximum manifold pressure will also decrease.

On takeoff when the outside air temperature is lower than standard, the pilot must monitor the manifold pressure gage to prevent overboosting the engine. But to protect the engine from inadvertent overboosting, the induction system is equipped with a pressure relief valve. This valve begins to offseat about one inch of manifold pressure below the maximum allowed, and by the time the maximum allowable manifold pressure is reached, the valve is off its seat enough to bleed off all of the pressure in excess of the maximum allowed.

More efficient control of the turbocharger pressure is maintained with an absolute pressure controller, often spoken of simply as an APC.

A waste gate valve that is opened by a spring and closed by engine oil pressure acting on a piston controls the flow of exhaust gas. When the valve is fully open, all of the exhaust gas flows out the tail pipe without passing through the turbocharger. But when the waste gate valve is closed, the exhaust gas must pass through the turbocharger turbine.

Oil flows from the engine lubricating system into the cylinder of the waste gate actuator through a capillary tube restrictor and out of the actuator through the APC and then into the engine oil sump. The APC is a variable restrictor in the oil return line from the waste gate actuator, and the size of its orifice is determined by the engine upper deck pressure, which is the turbocharger discharge pressure.

When the engine is started and is idling, the throttle is closed and the manifold pressure is low. The upper deck pressure is near atmospheric. The chamber in the APC in which the bellows is mounted is connected to the induction system, and it senses the upper deck pressure. When this pressure is low, the valve is closed.

The spring inside the waste gate actuator holds the waste gate valve open, but as engine oil flows into the actuator cylinder and out to the APC, which is closed, the pressure builds up and forces the piston over against the spring. This closes the waste gate so that all of the exhaust gases must flow through the turbine which spins the compressor.

At takeoff the throttle is opened, and the increased flow of exhaust gases increases the speed of the turbine, and both the upper deck pressure and the manifold pressure increase. Let's assume

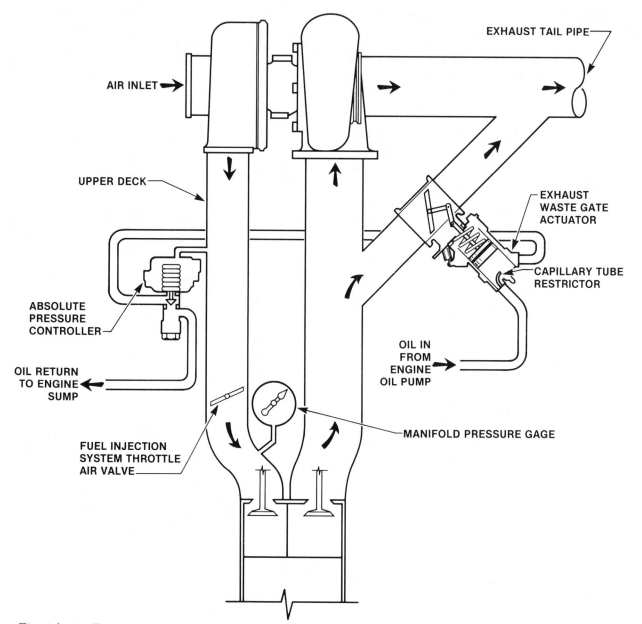

Fig. 1A-13   Turbocharger control system, using a hydraulically actuated waste gate actuator and an absolute pressure controller

that the controller is required to maintain a constant 32 inches of mercury manifold pressure. In order to do this, the upper deck pressure will have to be approximately 33 inches of mercury, because of the pressure drop across the fully open throttle valve. As the upper deck pressure builds up to 33 inches of mercury, the APC begins to open and drain the oil back into the engine sump. The oil can flow out of the waste gate actuator easier than it can flow in, because of the capillary tube restrictor on the oil inlet line, and as oil drains out of the actuator, the spring begins to open the waste gate so some of the exhaust gases

can flow out the tail pipe, bypassing the turbine rather than flowing through it. The turbine slows down a bit to hold the manifold pressure constant at 32 inches of mercury.

As the airplane goes up in altitude and the air becomes less dense, the upper deck pressure will decrease. This decrease is sensed by the APC, which closes a bit to restrict the oil flow from the waste gate actuator, and the oil pressure builds up and moves the waste gate valve toward its closed position to raise the upper deck pressure back high enough to produce 32 inches of manifold pressure.

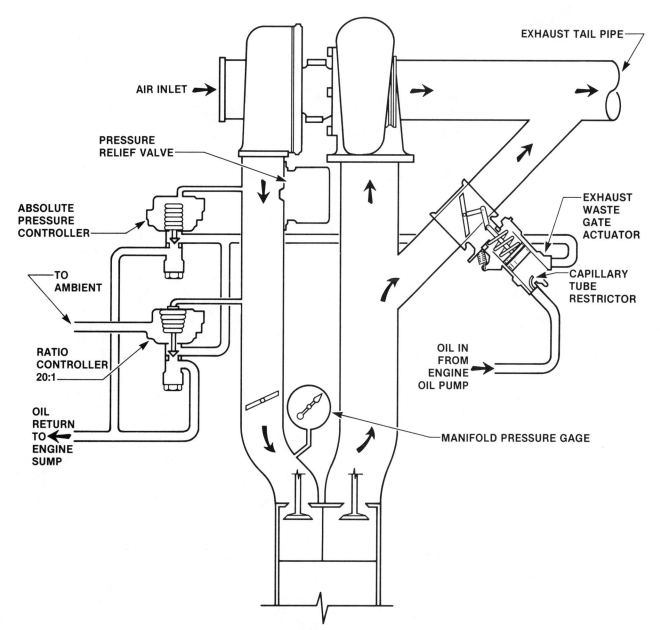

Fig. 1A-14 Turbocharger control system, using an absolute pressure controller and a ratio controller with a pressure relief valve as an emergency backup

As the aircraft continues to ascend, the upper deck pressure will decrease. The waste gate valve will continue to close until an altitude is reached when even with the throttle fully open and the waste gate valve fully closed, the turbocharger cannot maintain 33 inches of mercury upper deck pressure. This is the critical altitude of the engine, and the manifold pressure will decrease above this altitude.

The engine may be restricted to a maximum altitude at which it is allowed to maintain its maximum rated manifold pressure, and if this is the case, a stop may be installed on the waste gate to limit the amount it can close. This will prevent the engine reaching its maximum rated manifold pressure above its critical altitude, but if there are any leaks in either the induction system or the exhaust system, the engine will not be able to achieve its critical altitude.

To allow the engine to maintain a specific critical altitude and at the same time have enough capacity to compensate for some leakage, a more powerful turbine may be used and a ratio controller installed in parallel with the APC.

In Fig. 1A-14, we have a system that incorporates an absolute pressure controller, a ratio

11

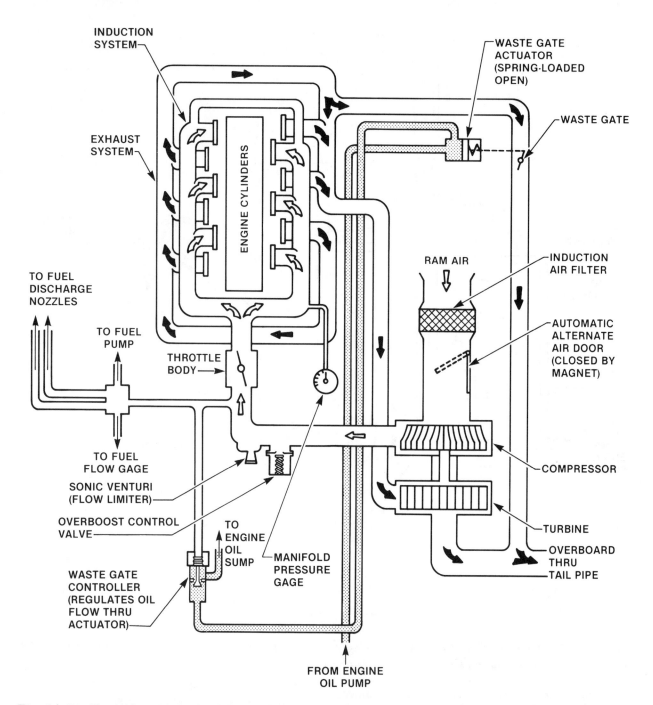

Fig. 1A-15 Typical turbocharger system for a general aviation aircraft. The turbocharger not only supplies the compressed air for the engine induction system, but it also provides air for cabin pressurization.

controller, and a pressure relief valve. The APC works in the same way we have just described, and for our explanation, we will assume that it will limit the maximum manifold pressure to 32 inches of mercury at 16,000 feet density altitude.

The ratio controller valve will remain seated as long as the upper deck pressure is no more than 2.0 times the ambient air pressure. At standard sea level pressure, the ratio controller will remain seated until the upper deck pressure reaches 59.84 inches of mercury (29.92 × 2.0 = 59.84). At 16,000 feet density altitude, the ambient pressure is approximately 16.2 inches of mercury and the ratio controller will prevent the manifold pressure exceeding 32.4 inches of mercury. Ascending above 16,000 feet will cause the ratio controller to reduce the upper deck pressure and the manifold pressure by bleeding some of the oil from the waste gate actuator to keep the manifold pressure no more than 2.0 times the ambient pressure. At 18,000 feet, the manifold pressure will be reduced to 29.9 inches of mercury, and at 24,000 feet, the maximum manifold pressure will be 23.2 inches of mercury.

The system we have in Fig. 1A-14 also has a pressure relief valve that is normally adjusted to offseat at about 1.5 inches of mercury above the maximum allowable upper deck pressure.

A final type of controller used with some turbocharged engines is the variable absolute pressure controller, which is often called a VAPC. This controller uses a cam, actuated by the engine throttle to maintain a constant upper deck pressure for each position of the throttle valve, rather than just for full throttle operation. The VAPC works in a manner similar to the APC with the bellows controlling the position of the valve, but in this controller, the engine throttle controls the position of the valve seat.

Fig. 1A-15 shows a typical turbocharger system for a general aviation aircraft. The upper deck pressure is used not only to supply air for the cylinders, but some of this pressure is taken off to serve as a reference pressure for all of the fuel discharge nozzles, as well as for the fuel pump and the fuel flow gage. The air for pressurizing the aircraft cabin is also taken from the turbocharger discharge, and it first passes through a sonic venturi which acts as a flow limiter. Air passing through the venturi reaches sonic speed and produces a shock wave that slows down all of the air passing through it and limits the amount of air that can flow into the cabin.

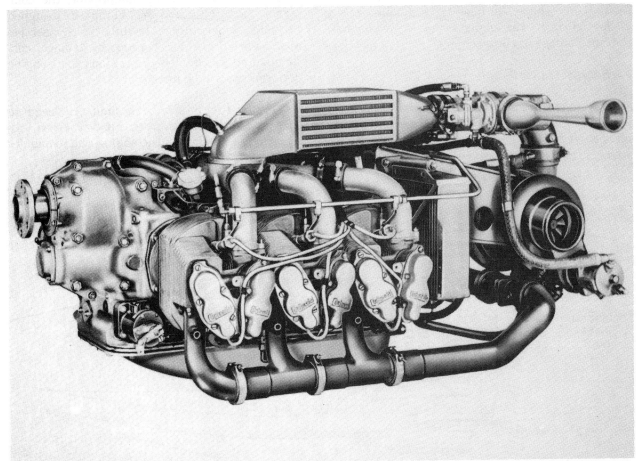

Fig. 1A-16 Turbocharged aircraft engine. The discharge from the turbocharger passes through an intercooler before it is distributed to the cylinders. This decreases the temperature of the induction air. The sonic venturi seen at the upper rear of the engine limits the amount of air that can flow into the cabin for cabin pressurization.

II. TURBINE ENGINE INDUCTION SYSTEMS

A. Turbine Engine Inlet Ducts

The air entrance, or flight inlet duct, is normally considered to be a part of the airframe, rather than part of the engine. Nevertheless it is usually identified as engine station one.

The air inlet to a turbine engine must furnish a uniform supply of air to the compressor so it can operate stall-free, and it must cause as little drag as possible. It takes only a small obstruction to the airflow inside the duct to cause a significant loss of efficiency, and if the inlet duct is to deliver its full volume of air with a minimum of turbulence, it must always be maintained as close to its original condition as possible. Any repairs to the inlet duct must allow it to retain its smooth aerodynamic shape. An inlet cover should be installed any time the engine is not operating to prevent damage or corrosion in this vital area.

1. Subsonic inlets

The inlet duct used on multi-engine subsonic aircraft, such as we find in the business and commercial jet aircraft fleet, is a fixed geometry duct whose diameter progressively increases from the front to back, as we see in Fig. 2A-1. A diverging duct is sometimes called an inlet diffuser because of the effect it has on the pressure of the air entering the engine. As air enters the inlet at ambient pressure, it begins to diffuse, or spread out, and by the time it arrives at the inlet to the compressor its pressure is slightly higher than the static pressure. Usually the air diffuses in the front portion of the duct, and then it progresses along at a fairly constant pressure past the engine inlet fairings and into the compressor. This allows the engine to receive the air with less turbulence and at a more uniform pressure.

This added pressure adds significantly to the mass airflow when the aircraft reaches its designed cruising speed. At this speed, the compressor reaches its optimum aerodynamic design point and produces the most compression for the best fuel economy. It is at this designed cruise speed that the flight inlet, the compressor, the combustor, the turbine, and the tail pipe are designed to match each other as a unit. If any section mismatches any other because of damage, contamination, or ambient conditions, the engine performance will be affected.

The inlet for a turbofan is similar in design to that for a turbojet except that it discharges only a portion of its air into the engine; the remainder passes only through the fan.

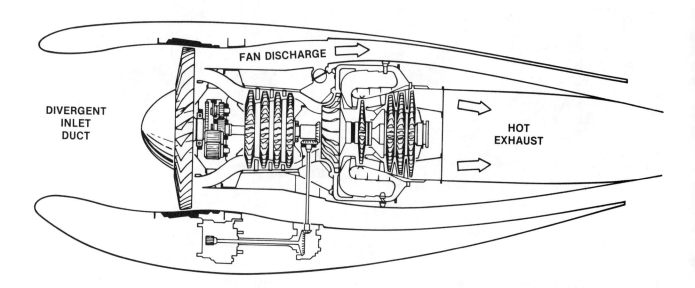

Fig. 2A-1 *The subsonic air inlet for a turbojet engine forms a diverging duct to increase the pressure of the air as it enters the fan or first-stage compressor.*

14

Fig. 2A-2 shows two common turbofan air-flow patterns: one is the short duct design used for high by-pass turbofans, and the other is the full duct used on low or medium by-pass turbofan engines. The long duct configuration reduces surface drag of the fan discharge air and enhances the thrust. A high by-pass engine cannot take advantage of this drag reduction concept, however, because the weight penalty caused by such a large diameter duct would cancel any gain.

When a turbine engine is operated on the ground, there is a negative pressure in the inlet because of the high velocity of the airflow. But, as the aircraft moves forward in flight, air rams into the inlet duct and ram recovery takes place. This ram pressure rise cancels the pressure drop inside the duct and the inlet pressure returns to ambient. Ram recovery is said to occur above about 160 miles per hour on most aircraft. From this point, the engine takes advantage of the increasing pressure in the inlet to create thrust with less expenditure of fuel.

2. Supersonic inlets

A variable geometry inlet duct is required on supersonic aircraft to slow the airflow at the face of the compressor to a subsonic speed, regardless of the aircraft speed. Most compressors require subsonic inlet airflow if shock waves are to be kept from the rotating airfoils.

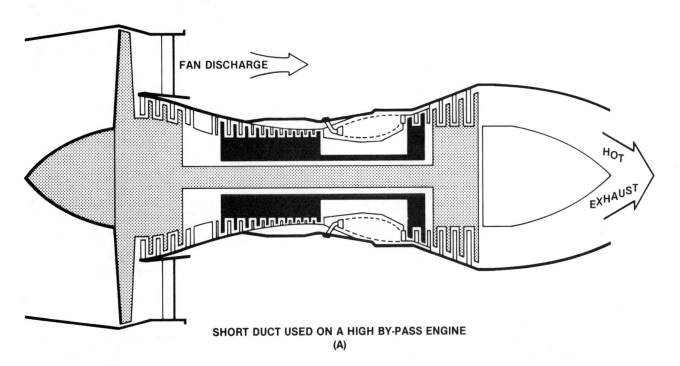

SHORT DUCT USED ON A HIGH BY-PASS ENGINE
(A)

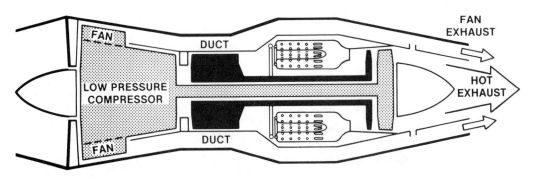

LONG DUCT USED ON LOW AND MEDIUM BY-PASS ENGINES
(B)

Fig. 2A-2   Induction systems for a turbofan engine

15

In order to vary the geometry, or the shape, of the inlet duct, a movable restrictor is sometimes used to give this duct a convergent-divergent, or C-D, shape. A C-D-shaped duct is necessary to reduce the supersonic airflow to a subsonic speed. This speed reduction is important, because at subsonic flow rates, air acts as though it were an incompressible fluid; but at supersonic flow rates, air can be compressed, and when it is, it creates shock waves.

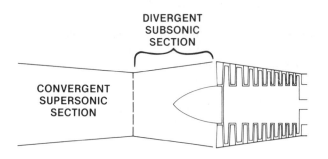

Fig. 2A-3 Convergent-divergent inlet duct used on supersonic airplanes to slow the air entering the compressor to a speed below Mach one

Fig. 2A-4A shows a fixed C-D duct in which the supersonic airflow is slowed by the formation of both oblique and normal shock waves. Once the speed of the inlet air is reduced to Mach one, it enters the subsonic diffuser section where velocity is further reduced and its pressure is increased before it enters the compressor.

The supersonic diffuser-type inlet shown creates a shock wave across the inlet and then directs the air into a variable convergent-divergent portion of the duct whose shape can be changed to accommodate the various flight conditions from takeoff to cruise. Fig. 2A-4 B and C shows a variable geometry inlet in both its supersonic and its subsonic position.

3. Bellmouth compressor inlets

Bellmouth inlets are converging in shape and are found on helicopters and other slower moving aircraft which generally fly below ram-recovery speed. This type of inlet produces a great deal of drag, but this is outweighed by their high degree of aerodynamic efficency.

When turbine engines are calibrated in test stands they use a bellmouth fitted with an anti-ingestion screen. Duct loss is so slight with this

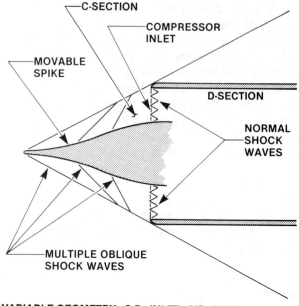

VARIABLE-GEOMETRY C-D INLET AIR DUCT USING A MOVABLE SPIKE INLET

(A)

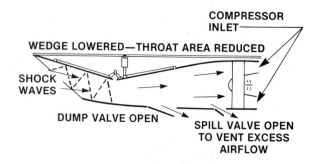

MOVABLE-WEDGE-TYPE INLET DUCT IN ITS SUPERSONIC CONDITION

(B)

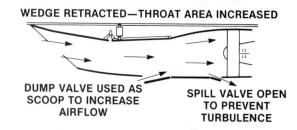

MOVABLE-WEDGE-TYPE INLET DUCT IN ITS SUBSONIC CONDITION

(C)

Fig. 2A-4 Supersonic air inlet ducts

design that it is considered to be zero. Engine performance data are collected when the engine is fitted with a bellmouth compressor inlet.

Fig. 2A-5 Bellmouth inlet ducts with anti-ingestion screens are used in engine test cells for calibrating gas turbine engines.

4. Compressor inlet screens

The use of compressor inlet screens is usually limited to rotorcraft, turboprop, and ground turbine installations. Inlet screens are seldom used on turbojet- or turbofan-powered aircraft, because icing and screen fatigue failures have caused so many maintenance problems. Many helicopter engines and some turboprop engines, however, are fitted with inlet screens to protect them against foreign object ingestion.

Some aircraft are also fitted with sand or ice separators as well as inlet screens, many of which are removable and are used only when operating conditions require them.

B. Variable Compressor Stator Vane System

The variable vane actuating system is incorporated on many gas turbine engines, especially on engines with high compression or where the compressor may have inherent compressor stall problems during acceleration or deceleration at low or intermediate speeds. The variable vane

system automatically varies the geometry (area and shape) of the compressor gas path to exclude unwanted air and maintain the proper relationship between compressor speed and airflow in the front stages. At low RPM the variable stator vanes are partially closed. As compressor rotor speed increases, the vanes open to allow more and more air to flow through the compressor. In effect, varying the vane angle schedules the correct angle of attack relationship between airflow and the rotor blade leading edge and allows for a smooth and rapid acceleration from low to high RPM.

Another way of viewing this situation is that the pre-whirl of airflow imposed on the airstream by varying vane angles slows the airstream's axial velocity before it reaches the rotor blades. Thus the low RPM of the rotor blade and the low axial velocity of the airstream are matched.

1. System operation (General Electric CF-6)

The example system in Fig. 2A-6 shows the inlet guide vanes and the six stator stages of the N_2 compressor as being variable. To permit rotation, the vanes are fitted with Teflon sockets at both ends. The remaining vane stages are conventionally installed stationary vanes.

This system utilizes fuel control discharge pressure to position hydraulic actuators located on the compressor case. The actuators move the vanes open by a beam arrangement which attaches to vane actuating rings and the stator vanes at their outermost point outside the engine.

The compressor inlet temperature sensor shown in Fig. 2A-6A is a heat sensitive, gas filled bulb which receives a fuel signal at constant pressure and modifies that pressure to become return pressure of a regulated value as per the temperature in the inlet. The CIT sensor contains a fuel metering orifice to accomplish this. The fuel control receives the return signal and uses it to schedule vane position through the rod end and head end fuel actuating lines. Even when the power lever is in a fixed position the CIT sensor will function to control compressor vane angles.

As the power lever is moved forward, fuel pressure increases at the head end of the actuator to open the vanes. The mechanical feedback cable signals the fuel control to cut off fuel pressure and allows the vane position to stabilize at the

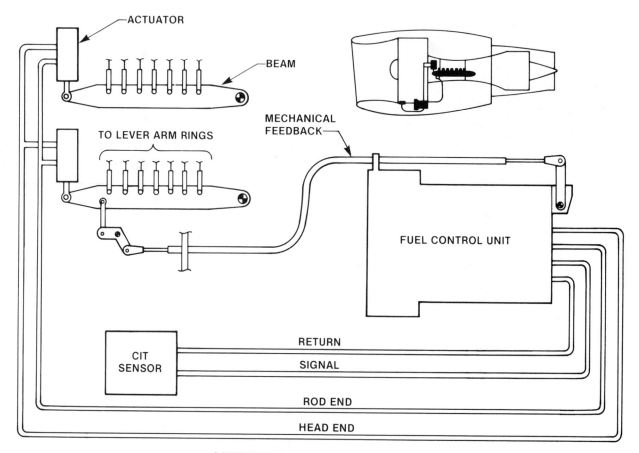

ACTUATOR

BEAM

TO LEVER ARM RINGS

MECHANICAL FEEDBACK

FUEL CONTROL UNIT

CIT SENSOR

RETURN

SIGNAL

ROD END

HEAD END

SCHEMATIC OF THE CONTROL SYSTEM
(A)

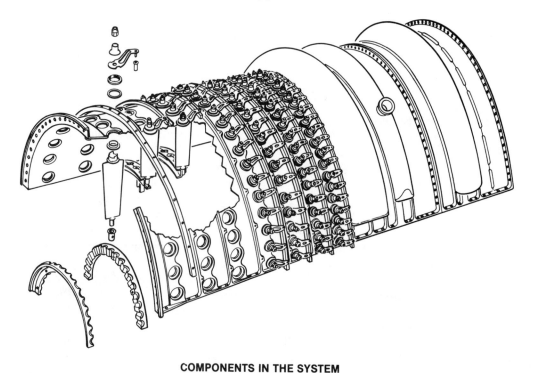

COMPONENTS IN THE SYSTEM
(B)

Fig. 2A-6   Typical variable inlet guide vane and stator vane system for a gas turbine engine

correct angle. The compressor inlet temperature sensor is installed to reset the vane-open schedule to match ambient temperature changes which affect mass airflow. At low CIT values the air will have increased density and the vane system will start to open at a lower compressor speed to increase airflow. This is necessary because cool air has a tendency to move more slowly than warm air. In this way the air velocity and compressor speed will remain in the proper relationship maintaining a correct angle-of-attack. In other words, the CIT sensor functions to keep the same compression ratio at a given compressor speed if ambient temperature changes.

Another way to view this situation is that if CIT decreases, compression increases in the rear stages which do not have variable vanes. Added compression in the rear stages starts to slow airflow in the rear stages. When the CIT sensor schedules the vanes more open, increased airflow (mass and velocity) occurs in the front stages to push air in the rear stages back up to its former velocity. Again this keeps the RPM vs. velocity in proper relationship to maintain the angle-of-attack. If a temperature shift downward occurs when the variable vanes are fully open, they obviously cannot open wider to compensate for the reduced velocity brought about by increased compression ratio. Therefore, another sensor in the fuel control, the burner pressure sensor, functions to force a reduction in fuel flow and compressor speed. This then brings compressor speed back down into line with air velocity. This situation, of course, is opposite to what was previously discussed, when the air velocity was brought back up into line with compressor speed. Here compressor speed is brought down to coincide with air velocity.

2. Variable vane schedule

Another aspect of the variable system is its operating schedule. In illustration 2A-7 we see

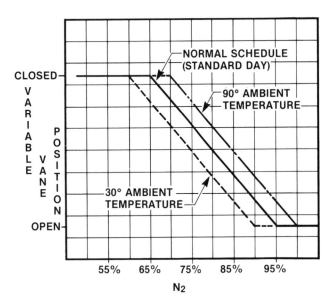

Fig. 2A-7 *Operational schedule chart for variable inlet guide vane system*

that when on normal schedule, the vanes will remain closed up to 65% N_2 speed, at which time they will start to open, and by 95% they will be fully open.

When inlet temperature increases or decreases from standard-day conditions the vane schedule also changes. If for instance, ambient temperature drops to 30° F, the vanes will start to open at approximately 60% N_2 speed instead of 65% and they would be fully closed at 90% instead of 95%. This of course accomplishes the same compressor vs. air velocity change as in the previous discussion about temperature shifts and angle-of-attack. If the CIT sensor fails or the linkages to the variable vanes are incorrectly adjusted, the correct vane schedule will not be met and a compressor stall is likely to occur.

An engine mounted protractor is installed to give the technician a means of checking the vane position as the engine is being operated to check its schedule of operation.

3. Troubleshooting the variable vane system

1. Compressor stall on acceleration or deceleration:			
Possible Cause:	**Check For:**		**Remedy:**
a. Variable vane system	(1) Out-of-track condition or binding		Re-rig system
	(2) Feedback cable out of adjustment or binding		Adjust or clean
b. Compressor inlet sensor	Gas leak from sensor tube		Replace
c. Fuel control	Correct fuel pressure to vane system		Replace control
2. Engine unable to attain full power:			
Possible Cause:	**Check For:**		**Remedy:**
Variable vane system	Vanes not fully open		Re-rig system

C. Compressor Anti-Surge Bleed System

The compressor anti-surge bleed system, as with the variable vane system, is installed on some gas turbine engines to minimize compressor acceleration and deceleration stall problems at low and intermediate speeds. Rather than exclude unwanted air, as is the case with the variable vane system, the compressor bleed system automatically dumps unwanted air overboard from a particular stage. Some high compression engines utilize both systems to increase their compressor stall margin.

Except at normal operation RPM and higher, the compressor cannot handle the amount of air passing through the engine. Another way of describing this situation is that at low and intermediate speeds, a relationship between compressor rotor RPM and airflow cannot be maintained to give the rotating airfoils the correct effective angle of attack to the oncoming airstream. At high rotational speeds, the compressor is designed to handle maximum airflow without aerodynamic disturbance so the bleed system is scheduled closed.

In comparing the two systems, the variable vane system and the compressor bleed system:

1. At the low end of the RPM range the variable vane system maintains axial velocity of airflow in the front stages by allowing less air to enter. This in turn keeps compression low and prevents piling up of air molecules in the rear stages as velocity of flow tends to increase.

2. At the low end of the RPM range the compressor bleed system maintains axial velocity of airflow in the front stages by bleeding away the excess of air molecules in the rear stages, which is in effect accomplishing the very same results.

On larger engines one or more bleed valves are used to dump unwanted air either into the fan duct or directly overboard. On smaller engines it is more convenient to use a sliding band which uncovers bleed ports which bleed away unwanted air.

Also on large engines a combination of bleeds and variable vanes is often used. The higher the compression ratio, the greater the need for systems which control the stall margin.

1. System operation (Avco-Lycoming T-53)

The example seen in Fig. 2A-8 is incorporated in some models of the Lycoming T-53 turboshaft and turboprop engines. Components of the system include a bleed band assembly and a control valve. The bleed band is an air operated unit which covers and uncovers ports in the compressor case to dump axial compressor discharge air overboard. The control valve is a fuel control operated unit which functions to permit control (P_3) air to open or close the bleed band.

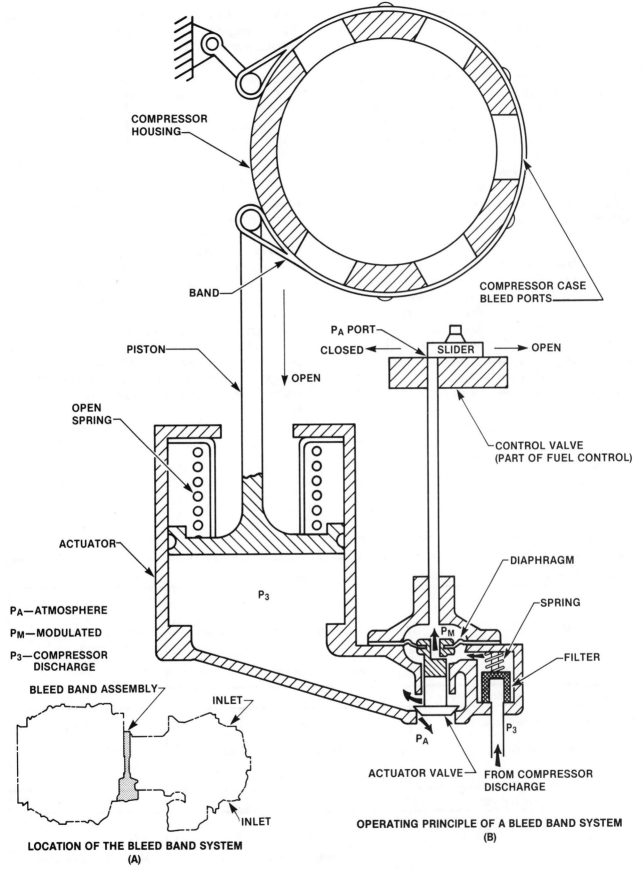

COMPRESSOR
HOUSING

COMPRESSOR CASE
BLEED PORTS

BAND

P_A PORT

CLOSED ← SLIDER → OPEN

PISTON

OPEN

OPEN
SPRING

CONTROL VALVE
(PART OF FUEL CONTROL)

ACTUATOR

DIAPHRAGM

SPRING

P_3

P_M

P_A—ATMOSPHERE

P_M—MODULATED

P_3—COMPRESSOR
DISCHARGE

FILTER

BLEED BAND ASSEMBLY

INLET

INLET

P_A

P_3

ACTUATOR VALVE

FROM COMPRESSOR
DISCHARGE

OPERATING PRINCIPLE OF A BLEED BAND SYSTEM
(B)

LOCATION OF THE BLEED BAND SYSTEM
(A)

Fig. 2A-8   Bleed band system for a gas turbine engine

21

427

In the schematic, we see the P_3, air which is tapped from the last stages of compression, pressurizes the actuator cavity, holding the piston up. Observe also that P_m air is locked in by the slider, and with the same pressure on both sides of the P_m diaphragm, the actuator valve is designed to close. This operational mode occurs when the fuel control is set at some predetermined high power setting and after the engine has had a stall free acceleration from a lower power setting. When the power lever is moved rearward, the fuel control will schedule the control valve slider to the right uncovering the P_a port, and P_m control air will bleed to atmospheric.

This causes a pressure drop on the P_m side of the diaphragm, and an oil-canning type movement upwards takes place which opens the actuator valve. This action allows P_3 air to dump from the actuator cavity and allows the open-spring to push the piston downward which, in turn, slackens the band, uncovering the bleed ports. In this operational mode, the bleeds remove a portion of the pressurized air in the compressor to the atmosphere and cause the axial compressor velocity of flow in the front stages to increase to match compressor RPM for angle-of-attack controlling purposes. This system is needed on many engines, because at lower compressor speeds the high pressure ratio in the rear stages tends to slow airflow in the front stages.

Operationally, when the bleed band opens and closes, cockpit instruments such as EPR and RPM will make a noticeable shift.

To bias system operation to ambient temperature the slider will open and close at the direction of a sensor in the fuel control. At cooler ambient temperatures the slider opens earlier on the RPM scale so that the heavier, slower moving airflow will be speeded up in the front compressor stages. This maintains the correct angle-of-attack relationship within the compressor.

There are several compressor bleed air systems, both engine and non-engine. The bleed band is not a compressor air bleed source for aircraft air conditioning or fuel tank pressurization, which is more properly referred to as customer bleed air.

2. Troubleshooting compressor bleed systems

1. Compressor stall on acceleration from low or intermediate speeds:		
Possible Cause:	**Check For:**	**Remedy:**
Bleed band system	(1) Early closing of bleed band ports	Reset schedule or replace control device
	(2) Binding mechanism	Adjust
2. Engine unable to attain full power:		
Possible Cause:	**Check For:**	**Remedy:**
Bleed band system	(1) Airbleed ports not fully closed	Re-rig system
	(2) Binding mechanism	Adjust
3. Fluctuating RPM and EGT:		
Possible Cause:	**Check For:**	**Remedy:**
Bleed band system	Modulating bleed band	Adjust or replace as necessary

D. Anti-Icing Systems

On many aircraft the compressor inlet case, inlet guide vanes, nose dome, and nose cowling are configured with internal passages which allow the circulation of hot air for anti-icing purposes.

On some aircraft, icing is not a problem for the engine because ice does not form in sufficient quantity and no anti-icing provisions are necessary. On some turboprops the oil reservoir is located within the propeller reduction gearbox, providing some anti-ice capability, and only minimum hot air flow is required to anti-ice the inlet area.

Engine bleed air is extracted from a stage of compression providing air at the correct temperature. This air, when directed radially inward, heats the surfaces so ice will not form in the inlet. Unlike certain deicing systems on wing leading edges and propellers, this system does not allow ice to form. If this system is turned on after ice accumulates, severe compressor damage can result which could seriously damage the engine.

Icing conditions are most prevalent when operating the engine at high speeds on the ground. Ice can form in the inlet up to 40° F ambient temperature due to the chilling effect of high inlet velocities.

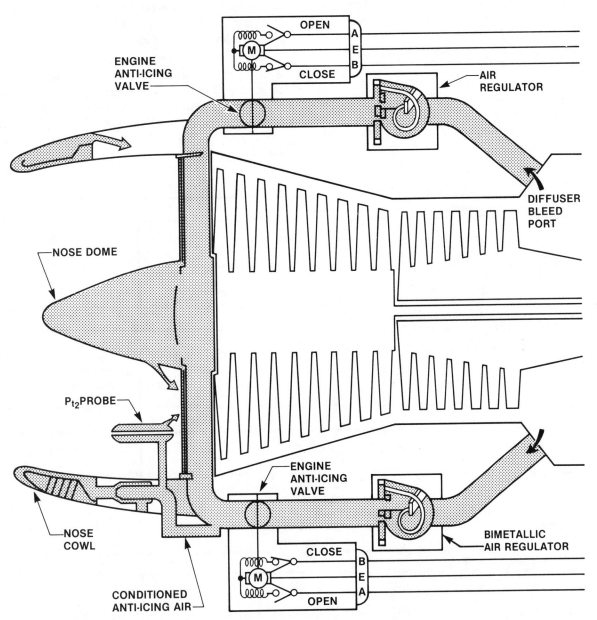

Fig. 2A-9 Typical inlet anti-icing system for a two-spool gas turbine engine

During flight, anti-icing is turned on only when visible moisture is present in the form of clouds or precipitation. The ambient temperature is below freezing at all cruise altitudes for a gas turbine powered aircraft, and ram pressure will not raise inlet temperature sufficiently above freezing. However, most of the flight time will be above cloud level and anti-icing will not be required. The usual method of initiating anti-icing is to select one engine, then watch the engine parameters stabilize, after which the remaining engine(s) are selected in a similar manner.

On takeoff, climbout, descent, and landing, the pilot will have to carefully assess the need for anti-icing according to the prevailing weather conditions. To prevent engine malfunction or damage, the technician will have to make the same assessment when running the engine on the ground.

1. System operation

The system illustrated in Fig. 2A-9 contains two air shutoff valves which are opened simul-taneously by actuation of a cockpit switch. Once the valves are opened, bimetallic coils inside the air regulator valves control the amount of airflow by the temperature of the air. Air that is too hot can affect the material strength of inlet components and also affects engine performance as the anti-ice air is ingested into the compressor. The idea here is that for this engine, the changes in temperature of compressor bleed air with changes in power settings would adversely affect engine performance, so the flow is regulated. Some aircraft have no need for regulator valves because the change in temperature has a negligible effect on either performance or material strength. On some large fan engines, only the flight inlet is anti-iced because the absence of inlet guide vanes and the slinging action of the fan eliminates engine inlet ice formation.

An indicator light turns on in the cockpit and a slight rise in exhaust gas temperature occurs, approximately 10° C, to indicate proper operation of the anti-icing system. Other engine instruments such as EPR and RPM might also shift noticeably at this time due to the momentary change in compression to the combustor.

2. Troubleshooting the anti-icing system

1. Ice forms in the inlet with the anti-ice system turned on:		
Possible Cause:	**Check For:**	**Remedy:**
Anti-ice valves	(1) Correct input voltage	Correct as necessary
	(2) Proper operation of valves	Replace valve(s)

2. Compressor stalls at high power setting with anti-icing system off:		
Possible Cause:	**Check For:**	**Remedy:**
Inlet icing	Ambient conditions	Shut down and remove ice, continue run with anti-ice on

3. Low power on takeoff:		
Possible Cause:	**Check For:**	**Remedy:**
Anti-ice shutoff valves (system off)	Valve(s) stuck open; carfully feel forward side of valve with hand for air leakage heating	Replace valve(s)
Air regulator(s) (system on)	Malfunctioning bimetallic coil (stuck full open)	Replace or bench check

4. Fluctuating EGT and RPM:		
Possible Cause:	**Check For:**	**Remedy:**
Anti-ice valves (system off)	Modulating valve motor	Adjust microswitch or replace motor

Section 7-II

Engine Cooling Systems

I. RECIPROCATING ENGINE COOLING SYSTEMS

An aircraft engine is a form of heat engine, which means that it converts heat energy into mechanical energy. But it is an extremely inefficient converter, as it changes only about one third of the energy into useful work. Every gallon of aviation gasoline we burn in the engine releases about 80,000 Btu of heat energy that does not produce any useful work, and all of this heat must be disposed of or it will destroy the engine.

About one-half of this heat is carried out of the engine through the exhaust system, and most of the rest of it is absorbed by the oil or by the metal of the engine and is transferred into the air through the cooling system.

Many of the early aircraft engines were adaptations of automobile engines and were cooled by passing water through jackets around the cylinders to absorb the heat from the cylinder walls and the cylinder heads. The hot water was then carried outside of the engine into a radiator where air passing through the coils absorbed the heat.

Water-cooled engines have been the standard for automotive and industrial engines for years, but since aircraft operate at altitudes where the air pressure is low, water boils at a low temperature and the systems had to be sealed to hold the water under pressure so the engines could have an efficient operating temperature. Pressurizing

these lightweight vibration-plagued systems led to many leaks and in-flight failures.

There have been several forms of radiators used to cool the water from the engines. Some airplanes had radiators in front of the engine with the propeller passing through the radiator core, while others had the radiator mounted either below the engine or in the cabane struts near the wing center section. And some high-speed racing airplanes used skin radiators in which the cooling water passed between thin sheets of brass that were installed over the wing surfaces.

Many of the problems of liquid-cooled engines were solved by substituting ethylene glycol for the water as a cooling agent. Ethylene glycol, often called by the trade name "Prestone," is a thick, liquid alcohol that has a much lower freezing point and a higher boiling point than water. When it is used in a sealed cooling system, it requires a much smaller volume than a water system that could remove an equivalent amount of heat.

The U.S. Navy decided against using liquid-cooled engines as early as the late 1920's, but the U.S. Air Corps and later the U.S. Air Force continued to use them up through World War II. Two ethylene glycol-cooled Allison V-12 engines powered the Lockheed P-38, which was one of the most effective American fighter planes in World War II, and the British RAF used several high-powered liquid-cooled engines in their first-line fighters and bombers.

Fig. 1B-1 The Lockheed Lightning of World War II fame was powered by two ethylene glycol-cooled V-12 turbosupercharged engines.

Fig. 1B-2 This small Szekely engine transferred the heat from the cylinders directly into the air as the air flowed through the fins on the cylinder barrels and the cylinder heads.

Rather than transferring heat from the engine into a liquid coolant and then carrying this coolant to an external radiator where the heat could be transferred into the air, some of the early engine designers put fins on the outside of the cylinder barrels and heads and let them stick out into the air. The air flowing over the fins absorbed the heat and effectively cooled the cylinders.

Radial engines lend themselves naturally to direct air cooling since all of the cylinders are exposed to the direct flow of air, and the most successful engine of its time, the Wright J5-C Whirlwind that powered the Spirit of St. Louis when Charles Lindbergh made his historic flight in 1927, was air cooled, with the cylinder heads and part of the barrels sticking directly into the flow of air.

Fig. 1B-3 The cylinder heads and part of the cylinder barrels of this Wright J5C engine stuck out in the airstream to remove heat.

Air cooling is good, but the penalty imposed by the increased drag is a high price to pay, and as the speed of the airplanes continued up beyond the 120 mile per hour mark, some method of reducing the drag was needed, and the Townend ring, or speed ring, cowling was developed that enclosed the cylinder heads in a streamlined cover and minimized the drag caused by turbulent airflow over the cylinder heads.

Fig. 1B-4 The Townend ring used on the Boeing P26A fighter reduced the drag by smoothing out the airflow over the engine cylinder heads.

As both speed and power increased, the need for more efficient cooling with less air drag became apparent, and by the beginning of the 1930's, the NACA cowling brought in pressure cooling for radial engines. The engine is enclosed in a streamlined cowling that completely covers all portions of the powerplant, and baffles seal the area between the cylinder heads and the cowling and all of the spaces between the cylinders.

Fig. 1B-5 The full NACA cowling used on the Laird Super Solution provided efficient cooling for the high-powered radial engine and at the same time produced a minimum amount of aerodynamic drag.

Intercylinder baffles cover part of the rear side of the cylinders so air flowing through the fins is forced into the area at the back center of the cylinders, an area that is insufficiently cooled on installations without baffles. The speed of the aircraft forces air into the forward side of the engine and through the baffles to provide effective cooling and low drag for radial engines. Even

the two-row and four-row radial engines are effectively cooled by pressure cooling.

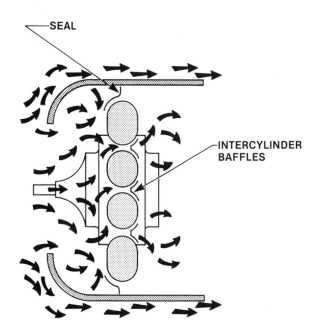

Fig. 1B-6 The cylinder head-to-cowling seal and intercylinder baffles force cooling air from the forward side of the engine through the cylinder fins in an engine housed in an NACA cowling to remove the maximum amount of heat.

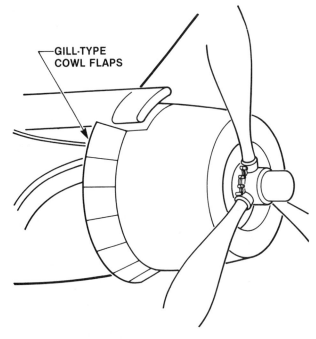

Fig. 1B-7 The amount of air flowing through the cooling fins on an engine housed in an NACA cowling is controlled by the opening of gill-type cowl flaps at the rear of the cowling.

In order to control the amount of cooling, gill-type cowl flaps may be installed at the rear of the cowling. When these flaps are open, the air flowing past them creates a low-pressure area behind the cylinders and increases the pressure drop across the cylinders which in turn increases the air flowing through the fins. When the airspeed is high and the need for cooling is decreased, the cowl flaps can be closed to decrease the airflow through the cylinders, and to streamline the engine installation.

Pressure cooling is used not only on radial engines, but also for in-line engines and horizontally opposed engines. The cowling for in-line engines brings the air in on one side of the engine, and it passes through the cylinder fins where it is held by intercylinder baffles. Then it is drawn out into the low pressure area created by the cowl flaps on the opposite side of the engine.

Horizontally opposed engines, such as the one seen in Fig. 1B-8, make effective use of pressure cooling. Air enters the front of the engine

COURTESY OF CESSNA

Fig. 1B-8   This horizontally opposed engine is efficiently cooled by pressure cooling. Air flows into the upper portion of the cowling from the front of the aircraft and out through a flanged opening at the rear of the lower cowling.

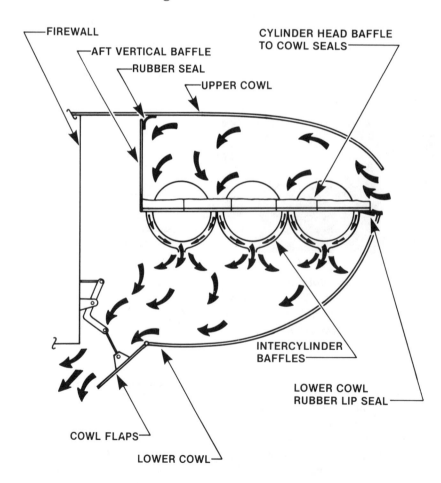

Fig. 1B-9   Cooling airflow in a pressure cooling system for a six-cylinder horizontally opposed engine

through openings in the cowling. Here the ram air pressurizes the area above the engine, and then it flows through the cylinders to the space below the engine and out through the cowl flaps at the aft end of the lower cowl. Small aircraft may not use cowl flaps, but rather a lip on the lower cowl creates a pressure drop to assist the air flowing through the cylinders.

The intercylinder baffles and the aft vertical baffle all have plastic, rubber, or leather strips to provide an air seal between the baffles and the cowling to prevent an air loss in these critical areas, and it is vital that when installing the cowling, to pay special attention to these seals to assure that they are all in good condition and that they all point in the direction shown in the aircraft manufacturer's service manual.

Augmenter tubes are used on some aircraft to augment, or increase, the airflow through the cylinders. The exhaust from the cylinders on each side, flow through a collector and discharge into the inlet of a stainless steel augmenter tube. This flow of high-velocity gas creates a low pressure and draws air from above the engine through the cylinder fins.

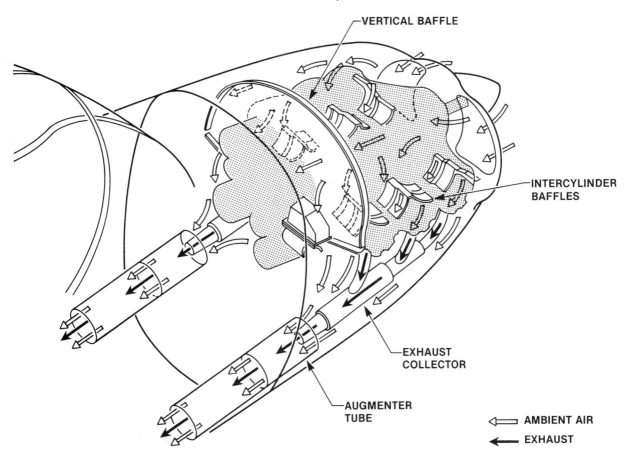

Fig. 1B-10 Augmenter tube installation for a six-cylinder horizontally opposed aircraft engine

II. TURBINE ENGINE COOLING SYSTEMS

Turbine engines, like reciprocating engines, are heat engines that release heat energy from the fuel and convert it into useful work. And, like the reciprocating engine, they are terribly inefficient. Much of the heat that is released must be disposed of before it can damage the lightweight components of the engine.

In a four-stroke reciprocating engine, heat is released only every other time the piston moves inward in the cylinder, and during the rest of the cyle, this heat is removed and the cylinder is filled with a cool fuel-air mixture. In the turbine engine, the flame burns continually, and if some provi-

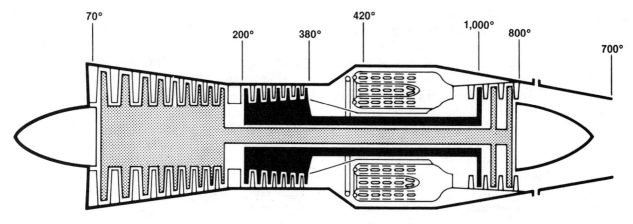

Fig. 2B-1   *Typical temperatures in a two-spool turbojet aircraft engine*

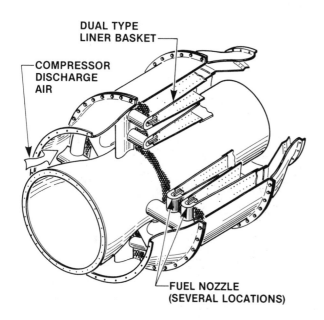

PHYSICAL APPEARANCE OF THE ANNULAR COMBUSTOR
(A)

sion is not made to remove a great deal of this heat, the engine will be destroyed.

The burner cans, or combustors, where the energy is released, are made of thin alloy steel that cannot withstand the continual high temperature of the burning gases, so a film of high-velocity cooling air is directed along the inner walls of the combustors, and the fuel nozzles are adjusted so they spray the fuel down the combustor away from the walls.

The hottest point within the engine is at the nozzle guide vanes for the first stage turbine. On many engines, high-pressure air from the N_2 compressor flows through the hollow nozzle guide vanes and the hollow turbine blades and removes some of their heat, and then this air discharges into the exhaust.

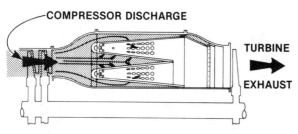

THE SECONDARY AIRFLOW THROUGH THE COMBUSTOR PROTECTS THE WALLS FROM THE INTENSE HEAT OF THE FLAME.

(B)

Fig. 2B-2   *Annular-type combustor for a turbojet engine*

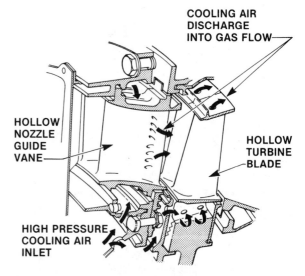

Fig. 2B-3   *The hollow inlet guide vanes and hollow turbine blades are cooled by a flow of high-pressure compressed air.*

436

30

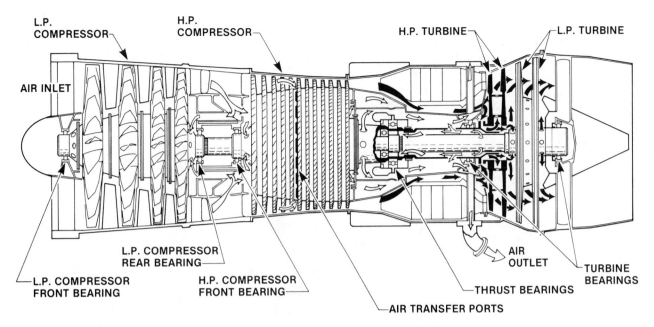

L.P. COMPRESSOR

H.P. COMPRESSOR

H.P. TURBINE

L.P. TURBINE

AIR INLET

L.P. COMPRESSOR REAR BEARING

L.P. COMPRESSOR FRONT BEARING

H.P. COMPRESSOR FRONT BEARING

AIR TRANSFER PORTS

AIR OUTLET

THRUST BEARINGS

TURBINE BEARINGS

Fig. 2B-4   Cooling airflow through the hot section of a typical gas turbine engine

In addition to the cooling air passing through the hollow nozzle guide vanes and the hollow turbine blades, some of the high-pressure air also flows up the face of the turbine disks and is then discharged into the exhaust system. And air from the low-pressure compressor flows through the hollow turbine shaft and removes heat from the compressor and turbine bearings. This air, after performing its cooling function, is discharged to the outside air through an air outlet. (Fig. 2B-6).

Cooling air is taken in from outside the engine and directed around the outside of the engine case to remove excess heat that could damage the aircraft structure. This air is discharged into the surrounding air so the heat can be carried away from the structure.

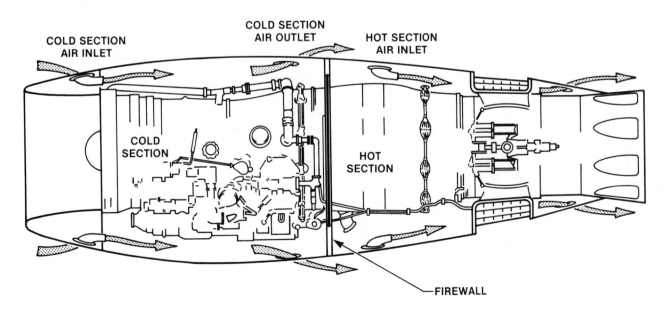

COLD SECTION AIR INLET

COLD SECTION AIR OUTLET

HOT SECTION AIR INLET

COLD SECTION

HOT SECTION

FIREWALL

Fig. 2B-5   Cooling air taken in from outside the engine pod flows through the pod and around the outside of the engine to remove excessive heat so it cannot cause structural damage.

31

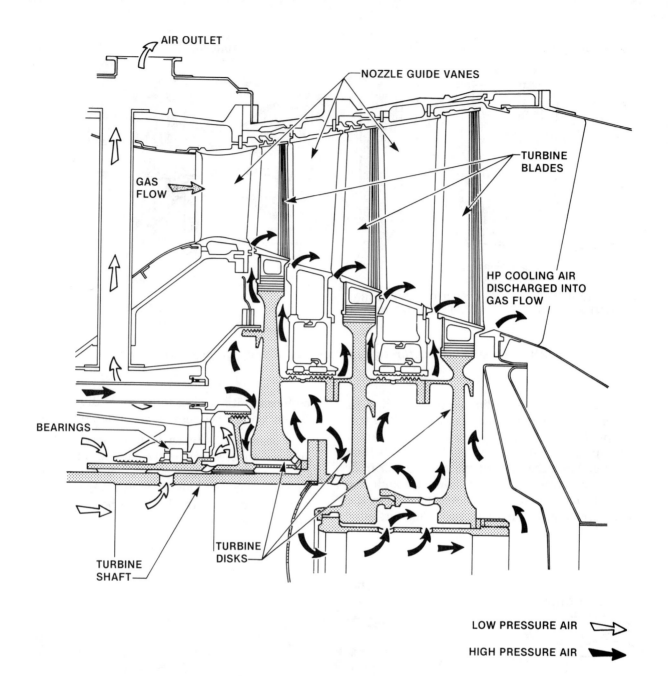

Fig. 2B-6 *High pressure air flows up the face of the disks in this three-stage turbine section to remove heat. Low-pressure air flows around the turbine bearings to remove heat.*

Section 7-III

Engine Exhaust Systems

I. RECIPROCATING ENGINE EXHAUST SYSTEMS

We have seen that a large part of the total energy in the fuel burned in an aircraft engine must be disposed of through the engine exhaust system. And these systems have undergone as many evolutionary changes as any other portion of an aircraft.

The very simplest exhaust system was that used by the rotary radial engine that was popular in World War I. The cylinders of these engines actually revolved around the stationary crankshaft as they ran, and as the cylinders passed below the fuselage, the exhaust valve opened and the gases were forced out of the cylinders.

In-line and V-engines of the same era often used straight stacks which were simply short sections of steel tubing welded to a flange and bolted to the cylinder head at the exhaust port. These straight stacks were effective in getting the exhaust out of the engine compartment, but they had no silencing capability, and when the aircraft was side slipped, cold air could flow into these stacks and warp the exhaust valves.

Short stacks for V-engines were soon replaced with collectors that were welded to all of the stacks on each side of the engine, and the exhaust gas was carried away from the engine and discharged over the side through a tail pipe on each side of the aircraft.

It was discovered that if the ends of the collectors were cut at a taper and the exhaust discharged through a relatively narrow slot rather than through the straight open pipe, the noise

was reduced, and the air flowing over this bayonnet, as it was called, created a slightly low pressure that helped decrease the exhaust back pressure. Bayonnet stacks were used for radial engines as well as for in-line and V-engines.

STRAIGHT STACK
(A)

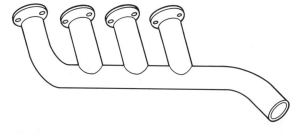

COLLECTOR STACKS USING A STRAIGHT EXHAUST DISCHARGE

(B)

BAYONNET-TYPE DISCHARGE OF AN EXHAUST STACK
(C)

Fig. 1C-1 Exhaust stacks used on early V-type engines

33

439

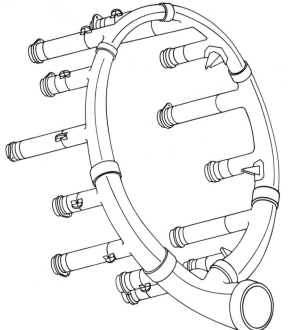

Radial engines use an exhaust manifold made up of a number of pieces of tubing that are fitted together with loose slip fits. When the engine is running, the components expand and fit together tightly, so there is no leakage.

Turbocharged engines require a tight-fitting, leak-proof exhaust system to prevent loss of manifold pressure when the turbocharger must put out all of the air it can. In Fig. 1C-3, we have the exhaust system of a turbocharged six-cylinder horizontally opposed engine. A crossover tube connects the exhaust stacks on the left side of the engine with the stacks on the right side, and at each location where expansion and contraction must be provided for, there are bellows that allow for the change in physical dimensions

Fig. 1C-2 Typical exhaust collector ring for a 14-cylinder twin-row radial engine

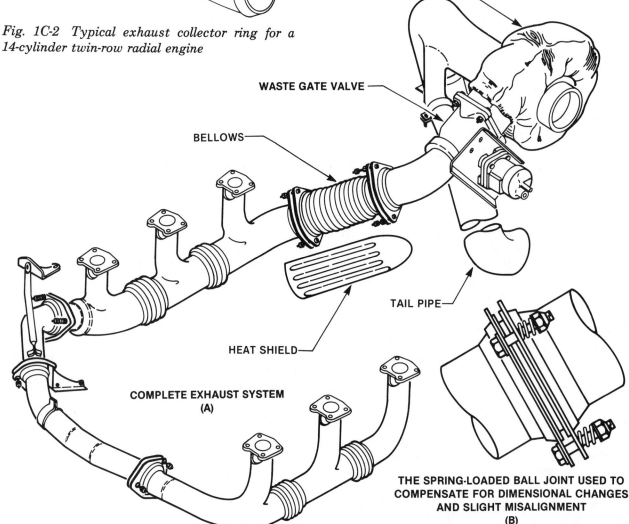

Fig. 1C-3 Typical exhaust system for a six-cylinder horizontally opposed turbocharged engine

without any leakage. The waste gate valve is hydraulically opened to allow exhaust gases to pass directly out the tail pipe, or closed to force these gases out through the turbocharger.

The sections of the exhaust system are joined with spring-loaded ball joints that allow movement without leakage and allow for slight misalignment of the parts.

So much energy is left in the exhaust as it leaves the engine that several systems have been devised to recover it. The one system that is most successful is the turbocharger, which uses the exhaust to drive a turbine which in turn drives an air compressor to increase the pressure of the air entering the engine. This turbine, naturally, produces some back pressure and robs the engine of some power, but the power increase provided by the air compression more than compensates for the loss.

Some of the large radial engine, the Wright R-3350, for example, uses three velocity-type power recovery turbines, commonly called PRT's. The exhaust from three groups of six cylinders

each are manifolded together and each group drives one of the turbines. The output shafts of the three turbines are geared to the engine crankshaft through fluid couplings.

The power recovered by the PRT's is appreciable, and the amount of back pressure caused by the velocity turbines is not excessive; but this method of power recovery has, up to this time, been limited to large radial engines. Study is presently being made, however, into the use of this type of device for the smaller engines.

Noise is a problem with aviation engines, and studies have been made to find practical ways of increasing the frequency and reducing the intensity of the noise. Propellers produce a large portion of the total noise, but the energy in the exhaust also accounts for an appreciable amount.

Bayonnet stacks were an early approach to reducing the amount of exhaust noise, and in the last thirty years, mufflers similar to those used so effectively on automobiles have been used to reduce the engine noise to a tolerable level.

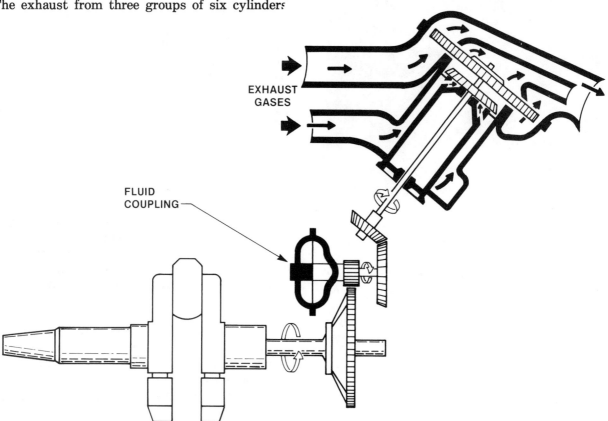

Fig. 1C-4   Turbocompound engines use velocity-type power recovery turbines (PRT's) coupled through a fluid coupling to the engine crankshaft. This recovers part of the energy that is passed out of the engine through the exhaust system.

The muffler in Fig. 1C-5 receives the exhaust from two cylinders and passes it through a series of baffles to break up the sound energy, and then this exhaust is passed out the single tail pipe. The flanges around the ends of the muffler support a sheet metal shroud that holds cabin air or carburetor air in contact with the outside of the muffler skin so it can absorb heat from the muffler.

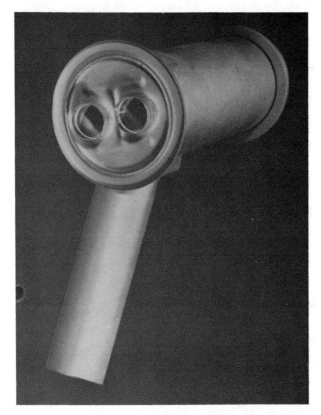

Fig. 1C-5 Typical muffler for use on one side of a four-cylinder horizontally opposed aircraft engine

The corrosion-resistant steel of which exhaust systems are made is thin, and the systems operate at high temperatures. These difficult conditions, coupled with the fact that exhaust system failure almost always causes a fire hazard as well as allowing noxious fumes to enter the aircraft cabin, make inspection of the exhaust system extremely important.

Cracks are the more common problem with an exhaust system, and on all inspections the entire system should be carefully inspected for any indication of cracks. A crack will usually allow exhaust gas to escape, and this often shows up as a white feather-line indication on the outside of the stack or muffler. Weld areas are specially subject to cracks, as the expansion and contraction of the

thin material of which the system is made produces enough stresses to cause cracks to form.

One effective method of checking an exhaust system for cracks is to remove all of the heater shrouds and connect the output of the discharge side of a vacuum cleaner to the exhaust tail pipe and then paint a soap and water solution over all of the joints and welds. Cracks will allow enough air to pass through that soap bubbles will form.

Any repair made to any exhaust system component must follow the aircraft manufacturer's recommendations in detail.

II. TURBINE ENGINE EXHAUST SYSTEMS

A. *Exhaust Cone, Tail Cone, and Tail Pipe*

The exhaust section is located directly behind the turbine section and consists of a convergent exhaust cone and an inner tail cone. The exhaust cone, sometimes referred to as the exhaust collector, collects the exhaust gases discharged from the turbine wheel and gradually converts them into a solid jet. This process is accomplished by the tail cone and its radial support struts. The tail cone keeps the air from becoming turbulent in back of the turbine wheel and the struts return the airflow to an axial direction. The exhaust cone is the terminating component of most engines, as delivered by the engine manufacturer.

On some very early model engines the tail cone was movable. In order to increase exhaust velocity and thrust, the tail cone was mechanically moved rearward, decreasing the effective size of the jet nozzle. Today, smooth and rapid acceleration is accomplished by more sophisticated fuel scheduling techniques using a fixed area tail pipe.

On some late model smaller engines, the exhaust section loses its traditional location and shape. The engine shown in Fig. 2C-1 uses turning vanes rather than a tail come and the strut arrangement to flow the gases in the correct direction.

The tail pipe is an airframe part, used to adapt an engine to a particular aircraft installation. The tail pipe is a convergent duct and is also referred to as the jet pipe or exhaust duct. Its con-

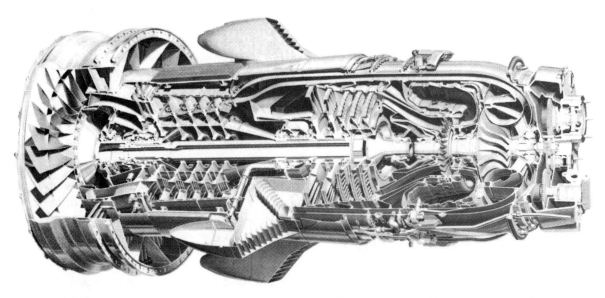

Fig. 2C-1   The exhaust on this Garrett ATF-3 turbofan engine exits through the turning vanes near the center of the engine.

vergent shape causes the gases to accelerate to the design speed necessary for producing the required thrust. The gases can never travel faster than the speed of sound with this shape of nozzle. This convergent tail pipe is for use on most subsonic aircraft. The shape is generally of fixed geometry and cannot be altered. Although some tail pipes are manufactured in both standard and other sizes which can be used to regain lagging engine performance.

On some older model engines small tabs called "mice," could be fitted to change the effective open area of the jet nozzle to recover some small amount of lost performance but these tabs are rarely seen today.

The convergent shaped exhaust ducts as described here can accelerate the exhaust gases to Mach 1 (the speed of sound) and no faster. The jet nozzle being an orifice forces the gas molecules to pile up as the mass airflow increases with airspeed, fuel, etc. When the gas exits the orifice it accelerates radially (spreads out) faster than it accelerates axially and this axial velocity is fixed at Mach 1. (The complete explanation of this phenomenon involves the understanding of molecular motion in thermodynamics, which is beyond the scope of this text.)

If more fuel is added after Mach 1 gas velocity is reached, engine speed, compression and mass airflow would increase and pressure (pileup) in the tail pipe would increase. The additional jet nozzle pressure would give a small increase in thrust but would soon be uneconomical for fuel consumption purposes. Also temperatures within the engine would elevate significantly. If supersonic exhaust nozzle velocities are needed for supersonic flight, a convergent-divergent nozzle is required rather than a simple convergent-type jet nozzle.

The area of the jet nozzle opening is calculated by the manufacturer and must be maintained within his limits; otherwise the back pressure it is designed to create will be altered. This in turn will affect the stall margin in terms of compression ratio, RPM, and mass airflow and also effect EGT and thrust.

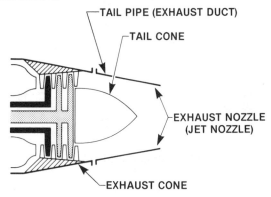

Fig. 2C-2   Exhaust system for a two-spool turbojet engine

1. Convergent-divergent tail pipe

Supersonic aircraft utilize the C-D type of tail pipe. The advantage of the convergent-divergent nozzle is greatest at high Mach numbers because of the high pressure ratio across the tail pipe resulting from supersonic inlet ram pressure.

To ensure that a constant weight or volume of a gas will flow past any given point after sonic velocity is reached, the rear portion of the duct is enlarged. This divergent section handles the gases, further increasing their velocity after they emerge from the throat and become supersonic.

Gas traveling at supersonic speeds expands outward faster than it accelerates rearward. The C-D nozzle takes advantage of this principle to create the thrust necessary to propel the aircraft at supersonic speeds. The C-D tail pipe is in fact an afterburner such as one would see on the Concorde-SST.

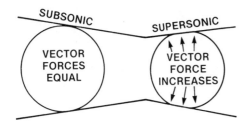

PRODUCTION OF THRUST BY THE SUPERSONIC AIR-STREAM IN THE CONVERGENT PORTION OF THE DUCT
(A)

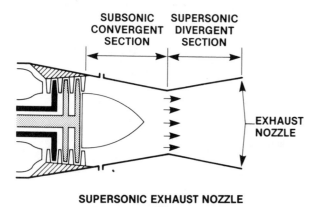

SUPERSONIC EXHAUST NOZZLE
(B)

Fig. 2C-3 Supersonic exhaust nozzle used on a two-spool turbojet engine

The forward convergent section causes pressure to build as the waist area chokes, creating a back pressure. The aft, divergent section allows this confinement of flow to reduce. The velocity then increases to the desired Mach number depending on the force being applied by combustion. If shaped properly, the C-D duct will effectively control gas expansion and produce the required thrust.

The shape of supersonic aircraft tail pipes is such that the rearmost section has a slight divergence in normal mode (non-afterburning). In this mode the supersonic aircraft is generally quite capable of supersonic flight speeds. When afterburner mode is selected, the mechanism is made such that the divergence of the jet nozzle increases to handle (accelerate properly) the new mass airflow created by afterburner fuel flow. In other words, the divergent section functions as a variable area jet nozzle. In non-afterburner the jet nozzle area is closed down and when afterburner is selected, the nozzle is scheduled to a more open position.

The addition of afterburner fuel and the expansion of gases it creates causes outward pressure forces on the walls of the afterburner rear duct and the forces are vectored forward by the diverging shape. Scientific testing indicates that fully expanded supersonic flow results in the best thrust augmentation. That is to say, pressure across the jet nozzle returns to ambient value.

Supersonic flow causes the familar shock wave phenomena to occur but a properly designed afterburner will have no shock distortion to airflow within the duct; only shock rings which are visible in the jet exhaust stream. The balloon analogy in Fig. 2C-3 demonstrates the C-D principle, showing that the expansion outward creates thrust vectors forward.

2. Afterburning

The addition of an afterburner to a gas turbine engine is made possible by the fact that an excess of oxygen is available. The products of combustion in the tail pipe contain unburned oxygen. The 60-80% of compressor discharge air used for combustor cooling mixes with combusted air and flows downstream to the tail pipe. A set of afterburner fuel nozzles called spraybars is fitted into the tail pipe entrance along with a suitable ignition system. The unburned oxygen, when ignited here, is turned into additional propulsive power as the mass airflow is further accelerated.

Along with fuel and ignition components, another device called a flame holder is required for good combustion. It is a tubular grid or spoke-shaped obstruction placed downstream of the fuel nozzles. As gases impinge on the flame holder, turbulence is created which enhances fuel-air mixing. This promotes complete and stable combustion in a very fast moving airstream.

In effect, the afterburner is a form of ramjet attached to the rear of a gas turbine engine. The only types of gas turbines, however, that utilize afterburning are the turbojet and the turbofans with mixed exhausts. That is, turbofans in which the fan and core engine gases mix and issue from one jet nozzle.

Afterburning is used primarily for takeoff with heavy aircraft loading and for rapid climb-

out speeds, the fuel used in afterburning generally being less than what would be burned to reach cruise altitude at slower non-afterburning speeds. Some military aircraft whose design is more for speed and maneuverability than conservation of fuel, however, may have very high fuel consumption for quick reaction speeds during combat. These aircraft could have as much as 70% additional thrust in afterburning. On the other hand, some very modern aircraft have very powerful engines and require only limited help in the 15 to 20% range. In this case the tendency today is to refer to the C-D tail pipe as a thrust augmentor rather than an afterburner.

B. Thrust Reversers

Airliners powered by turbojets and turbofans and an increasing number of business jets are equipped with engine thrust reversers to aid in braking and to reduce brake maintenance costs. Reversers also provide an added safety factor

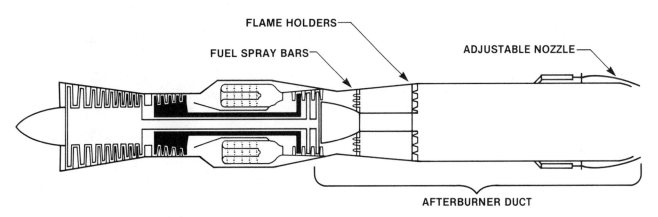

FLAME HOLDERS

FUEL SPRAY BARS

ADJUSTABLE NOZZLE

AFTERBURNER DUCT

DUAL AXIAL FLOW COMPRESSOR TURBOJET WITH AFTERBURNER
(A)

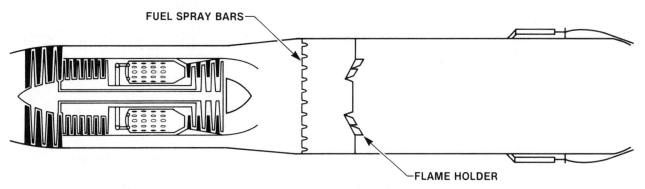

FUEL SPRAY BARS

FLAME HOLDER

DUAL AXIAL FLOW COMPRESSOR MIXED EXHAUST TURBOFAN ENGINE WITH AFTERBURNER
(B)

Fig. 2C-4 Afterburners

39

during emergency landings and rejected takeoffs. Some thrust reverse systems can also be used in flight as speed brakes to increase the aircraft's rate of descent.

The two types of thrust reversers in popular use are the aerodynamic blockage, sometimes referred to as cascade reversers, and the mechanical blockage, often referred to as clamshell reversers. The aerodynamic type consists of a set of cascade turning vanes in a pre-exit position in either the fan exhaust or hot exhaust. They turn the escaping gases to a forward direction which in turn causes a rearward thrust. The mechanical blockage type can be placed in either a pre-exit position or in a post-exit position as shown in Fig. 2C-5. This type reverser, when deployed, forms a solid blocking door in the jet exhaust path.

The gases hit the clamshell portion and are turned forward just enough to give reverse thrust but not enough to allow hot gas reingestion into the engine inlet. Thrust reversers are controlled by a cockpit lever at the command of the pilot.

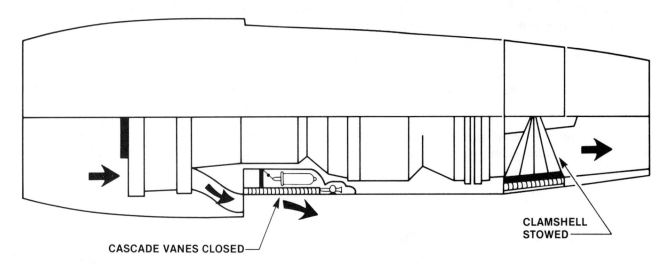

CASCADE VANES CLOSED—

CLAMSHELL
STOWED—

IN THE FORWARD-THRUST POSITION, THE REVERSERS ARE STOWED.
(A)

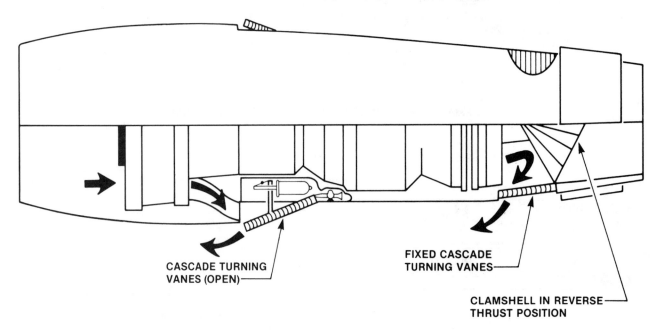

CASCADE TURNING
VANES (OPEN)—

FIXED CASCADE
TURNING VANES—

CLAMSHELL IN REVERSE—
THRUST POSITION

IN THE REVERSE-THRUST POSITION, THE REVERSERS ARE DEPLOYED.
(B)

Fig. 2C-5   Fanjet engine thrust reverser

After thrust reverse is selected, the pilot can move the reverse lever from idle select-position up to takeoff position as required by landing conditions.

Reversers are especially helpful when landing on wet or icy runways and provide approximately 40 to 50% reverse thrust. The usual operation of the reverser system is to apply reverse as soon as the aircraft is firmly on the runway and then to apply as much reverse power as is needed for existing conditions of wetness, ice, etc. Then, as the aircraft slows to approximately 80 knots, power is reduced back to reverse-idle and then to forward thrust as soon as practical.

Operating in reverse at low ground speeds can cause reingestion of hot gases and compressor stalls. It can also cause ingestion of fine sand and debris that can abrade gas path components and even find its way through main bearing air-oil seals into oil sumps.

A mathematical definition or explanation of thrust reverse can be arrived at if we recall the thrust distribution example in the turbine engine theory portion of this Integrated Training Program. The jet nozzle had -722.2 pounds of thrust with a rearward velocity of 1,900 ft/sec. If the velocity direction (vector) is changed to an angle in the forward direction, this velocity component changes from a plus to a minus value. But because the angle is not straight forward for reasons already explained, the resultant velocity vector will drop to perhaps 60% of the former value of 1,900 ft/sec.

EXAMPLE:

Consider the jet nozzle is now unchoked, and: F_g at the jet nozzle is:

where:
$$M_s = 29 \text{ lb/sec}$$
$$V = .60 \times (-1,900) = -1,140$$
$$I = 2,958.4 \text{ lb.}$$
$$g = 32.2 \text{ ft/sec}^2$$

$$F_g = \left[\frac{M_s(V)}{g} \right] - I$$

$$F_g = \left[\frac{29\,(-1,140)}{32.2} \right] - 2,958.4$$

$$F_g = \frac{-33,060}{32.2} - 2,958.4$$

$$F_g = -1,026.7 - 2,958.4$$

$$F_g = -3,985.1$$

Now, if the sum of forward and rearward thrust forces is:

	Positive Thrust
Compressor	+ 3,660.0
Diffuser	+ 360.2
Combustor	+ 5,250.0
Exhaust Cone	+ 542.9
	+ 9,813.1

	Negative Thrust
Turbine	− 6,854.7
Jet Nozzle	− 3,985.1
	−10,839.8

	Resultant Thrust
Total Reverse (F_g)	−10,839.8
Total Fwd (F_g)	+ 9,813.1
Total Reverse (F_g)	− 1,026.7 lb.

The figure 1,026.7 lb. represents 45.9% reverse thrust of the former 2,236.2 lb. forward thrust.

Turboprop aircraft may also have a form of thrust reversing, namely a full reversing propeller. This system works so well that it may be used to back the aircraft into ground parking spaces.

The latest design in thrust reversing for turbofan engines involves reversing the pitch of the fan blades. Presently it is in limited use, but will probably be seen on aircraft in the future.

C. Noise Suppression

Noise is best defined for gas turbine engine purposes as unwanted sound because it can be both irritating and harmful. The sound level of the average business jet or airliner during takeoff, as heard by persons on the airport or off the end of the runway, would probably be in the range of 95 to 100 decibels. This noise level would be similar to a subway train noise as heard from the boarding platform. Right at the aircraft the noise level could be as high as 160 decibels and painful to the ears.

Even the lower level of noise mentioned is felt by many people to be excessive and harmful. The industry has reacted to this by continually improving noise reduction techniques on every new generation of engine and aircraft to satisfy the public's need for more effective noise abatement.

Noise suppressor units as such, are not generally utilized on business jets or airliners today, but FAR Part 36 requires their use on many older commercial jet aircraft. Newer aircraft have inlets and tail pipes lined with noise attenuating materials to keep sound emission within the established effective perceived noise decibel (epndb). However, one can see the old style noise suppressor being fitted to some newer engines to meet the new noise standards.

Noise generated as the exhaust gases leave the engine is at a low frequency level, such as from a ship's foghorn, and in the same way carries for long distances. It is this low frequency noise that tends to bother people who live close to airports.

The noise generated by a turbofan engine is much less than that generated by a turbojet. This is principally because the turbofan will generally employ more turbine wheels to drive the compressor and the fan. This, in turn, causes the hot exhaust velocity and noise level to be lessened. Fig. 2C-6 shows a hot stream sound suppressor with an increased perimeter design and also with a cold pre-mixing capablity.

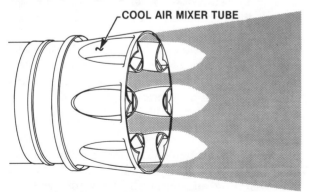

COOL AIR MIXER TUBE

Fig. 2C-6 Premixture-type noise suppressor. Cool air mixes with the hot exhaust gases to decrease the noise production by a turbojet engine.

Some new generation fully ducted turbofan engines are designed with what is termed exhaust mixing to blend the fan and hot airstreams more effectively and lessen the sound emission issuing from the exhaust duct. On these engines the sound from the inlet is likely to be louder than from the tail pipe. This is also the case today with the high by-pass fan engines which draw so much energy from the hot gases to drive the fan, compressor and accessories that the fan emits the greatest noise.

Because of the characteristic of low frequency noise to linger at relatively high volume, noise reduction is achieved by raising the frequency. This is accomplished by increasing the perimeter of the exhaust stream, allowing more cold and hot air mixing. This reduces the tendency of hot and cold air molecules to shear against each other and to break up the large turbulences in the jet wake which produces the low frequency (loud) noise.

In other words, reducing turbulence also changes the frequency of noise to a higher state which is more readily absorbed by the atmosphere. The noise then is lessened for any given distance from the noise source and the effective perceived noise reading on a decibel meter will be lower.

Note: Epndb is a standard measure of loudness combined with frequency of sound and is used specifically for aircraft noises in the atmosphere. Epndb can also be an estimated value where atmospheric absorption prevents completely accurate measurement—such as an aircraft flying overhead where wind, temperature, moisture, etc., should interfere with accuracy.

As stated previously, some older model Boeing 707 and DC-8 aircraft with turbojet powerplants have the type suppressors shown in Fig. 2C-6. Also some newer military aircraft which are designed for a quiet approach in combat situations have similar devices attached to their exhausts.

I. RECIPROCATING ENGINE STARTERS

The earliest aircraft engines were started by hand "propping" them. The mechanics swung the propeller by hand to pull the engine through one or two compression strokes, and if everything was as it should be, the engine started. Cold weather made the oil heavy, and the engines were hard to start. As the engines grew larger and the propellers longer, the procedure for starting became more difficult, and several methods were devised to replace the hand-cranking procedure.

The bungee starter used a long piece of rubber shock cord and a leather boot. The engine was primed and the boot placed over one end of a propeller blade, and the bungee cord was stretched across the propeller. One mechanic held the propeller still, while two or three other pulled on the bungee cord.

When everyone was ready, the mechanic holding the propeller gave it a flip, and the bungee cord pulled the propeller through about one revolution.

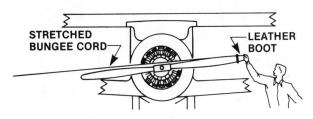

STRETCHED BUNGEE CORD — LEATHER BOOT

Fig. 1D-1 A bungee starter system was used on some of the early engines as an improvement over the hand-propping method of cranking an aircraft engine.

Compressed air was used for starting some of the engines. The engine was primed and the piston in one cylinder placed just beyond top center, and compressed air was directed into the cylinder to force the piston down.

These methods were primitive, at best, and only when the electric starter was developed did starting an engine cease to become one of the more dangerous parts of engine operation.

One of the first types of starters coupled to the crankshaft was the hand inertia starter. In this type of starter, a relatively heavy flywheel is spun with a hand crank through a step-up gear arrangement. When the flywheel is spinning at a high rate of speed with lots of kinetic energy stored in it, the crank is removed, and the engage handle is pulled to extend a ratchet-type jaw out of the starter to engage with a similar ratchet on the rear end of the crankshaft. A torque-overload clutch is used between the flywheel and the jaws to prevent the sudden engagement damaging either the engine or the starter. With the jaws engaged, the energy stored in the flywheel turns the engine over enough to get it started.

Hand inertia starters were soon improved by mounting an electric motor on the back of the starter to spin the flywheel. When the flywheel was up to speed, the motor was de-energized and the engage handle pulled to engage the flywheel to the engine. In case the aircraft battery was run down, there were provisions on these starters to use a hand crank to drive the flywheel. This combination electric and hand inertia starter was pretty much the standard starter for large engines up until well into World War II.

When electric motors were improved so that they had a high torque with a light weight, the direct-cranking electric starter, similar to the one in Fig. 1D-2, was developed. This starter uses a series-wound electric motor that produces a very high starting torque.

When the starter switch is placed in the Engage position, the ratcheting jaws extend, and the motor begins to turn the engine over for starting. When the engine starts before the starter is disengaged, the ratchet mechanism prevents the starter being damaged, and there is a clutch between the motor and the jaws that prevents the sudden application of torque damaging the starter.

Fig. 1D-2 Direct-cranking electric starter for a large aircraft engine

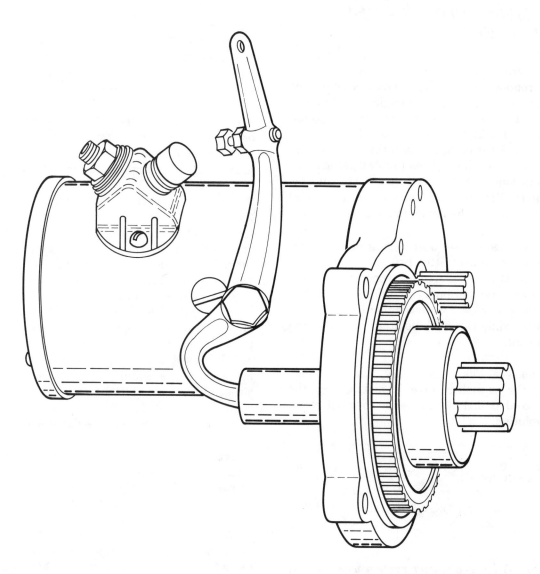

Fig. 1D-3 Direct-cranking starter for a small reciprocating aircraft engine. This starter uses a reduction gearing for increasing the torque and an over-running clutch to prevent damage to the starter or to the engine when the engine starts.

44

Many of the small horizontally opposed aircraft engines use a form of direct-cranking electric starter similar to the one in Fig. 1D-3. This starter uses a small series-wound electric motor with a small gear on the end of its shaft. This small gear meshes with a large gear that is part of an over-running clutch. These starters may be engaged either by a hand-pulled cable or with a solenoid that operates the shift lever, Fig. 1D-4.

When the shift lever is pulled, it compresses the meshing spring and forces the pinion to mesh with the starter gear inside the accessory case of the engine. When the pinion and the starter gear are fully meshed, further movement of the shift lever closes the starter motor switch, and the starter cranks the engine. When the engine starts, the over-running clutch allows the pinion to spin without doing any damage, and as soon as the toggle is released or the solenoid is de-energized, the return spring pulls the pinion away from the starter gear.

The popular series of Avco-Lycoming horizontally opposed engines that have the large diameter starter gear just behind the propeller use a direct-cranking electric starter similar to the ones used on automobile engines. These starters are driven through a Bendix drive. See Fig. 1D-5.

When the starter switch is closed, the series-wound electric motor spins the pinion through the Bendix drive spring. The drive pinion fits loosely over the drive shaft, and as the armature spins, the pinion moves forward on the helical splines, and it engages the teeth on the periphery of the engine starter gear. The starter then cranks the engine. As soon as the engine starts and the starter gear begins to spin the Bendix

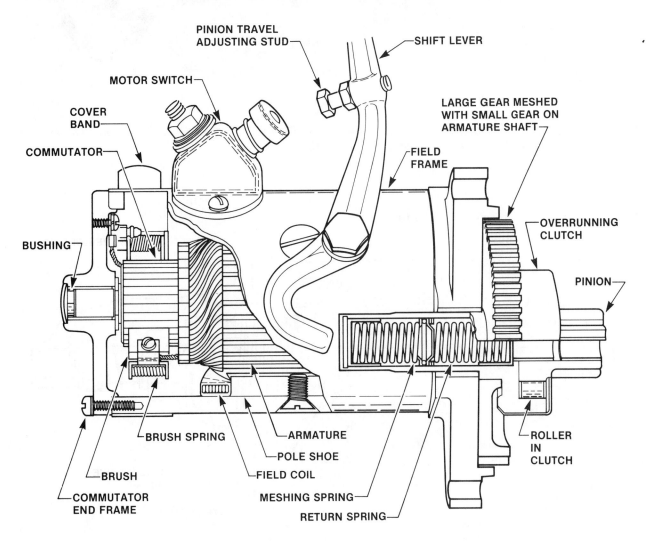

Fig. 1D-4 Cutaway view of a direct-cranking starter using an over-running clutch.

45

451

COURTESY OF AVCO-LYCOMING

Fig. 1D-5 *A direct-cranking starter using a Bendix drive is used on the Lycoming engine. The starter drive gear meshes with the large gear on the front of the engine.*

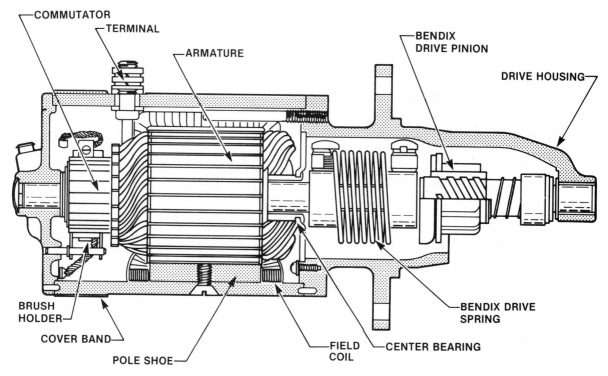

Fig. 1D-6 *Cutaway view of an aircraft starter using a Bendix drive*

drive pinion, the pinion is forced back along the helical splines, and it disengages from the starter gear. See Fig. 1D-6.

COURTESY OF PRESTOLITE

Fig. 1D-7 Aircraft starter using a Bendix drive

Some of the new generation of Teledyne-Continental aircraft engines use a starter mounted on the side of the accessory case, and it cranks the engine through a right-angle worm-gear-type adaptor.

In Fig. 1D-9, we see the way this system works. The series-wound electric motor drives the

Fig. 1D-8 Starter adapter used on a Teledyne-Continental aircraft engine

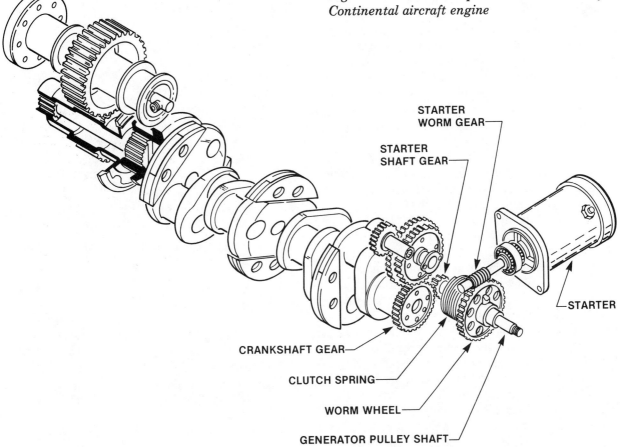

STARTER WORM GEAR

STARTER SHAFT GEAR

STARTER

CRANKSHAFT GEAR

CLUTCH SPRING

WORM WHEEL

GENERATOR PULLEY SHAFT

Fig. 1D-9 Direct-cranking starting system such as is used on some of the modern Teledyne-Continental aircraft engines

starter worm gear, which meshes with the worm wheel. The clutch spring is attached to the worm wheel, and when the worm wheel turns, the spring tightens around a knurled drum on the starter shaft gear, and the starter gear, which is meshed with the crankshaft gear, cranks the engine. As soon as the engine starts, the starter shaft turns faster than the worm wheel, and the spring releases the knurled drum. The generator drive pulley is mounted on the end of the starter gear shaft, and with the clutch spring disengaged, the shaft serves as the generator drive shaft.

II. TURBINE ENGINE STARTER SYSTEMS

Gas turbine engines are generally started by a starter power input to the main gearbox which in turn rotates the compressor. On the dual axial compressor gas turbine, the starter rotates the high speed compressor system only. On free turbine, turboprop, or turboshaft engines, again the compressor rotor system only is rotated by the starter through the accessory gearbox. The usual starting sequence is to energize the starter and then at 5 to 10% rotor speed, to energize ignition and open the fuel lever. A normal lightoff will usually occur in 20 seconds or less. If lightoff does not occur within 20 seconds, the start would generally be aborted to investigate the malfunction. Such problems as low starting power, weak ignition, or air in the fuel lines can cause starting problems.

Compressor rotation by the starter provides the engine with sufficient air for combustion and also aids the engine in self-accelerating to idle speed after combustion occurs. Neither the starter nor the turbine wheel have sufficient power on their own to bring the engine from rest to idle speed, but when used in combination, the process takes place smoothly in approximately 30 seconds on the typical engine. The start is often automatically terminated by a speed sensor device 5 to 10 percent RPM after self-accelerating speed is reached; at this point turbine power is sufficient to take the engine up to idle. If the engine is not assisted to the correct speed, a hung (stagnated) start may occur. That is, the engine stabilizes at or near the point of starter cutoff. To remedy this situation, the engine must be shut down for investigation of the problem. Any attempt to accelerate by adding fuel will quite often result in a hot start as well as a hung start,

because the engine is operating with insufficient airflow to support further combustion.

Turboprop and turboshaft engines are started either in low pitch to reduce drag on the rotor and provide more speed and airflow or they are configured with a free-turbine driving the propeller, which allows for a low drag acceleration in that the compressor rotor system only is being turned by the starter.

A. Electric Starters

Electric starters are not in wide use on flight engines because the combination starter-generator is more feasible for small engines. Also, large engines require such high starting power that starter weight becomes a problem. However, electric starters are widely used on auxiliary and ground power units and some small flight engines.

Most electric starters contain an automatic release clutch mechanism to disengage the

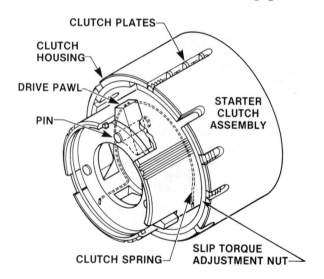

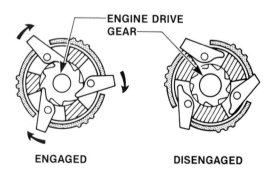

ENGAGED   DISENGAGED

Fig. 2D-1 Over-running starter clutch mecha-used in an electric starter for a gas turbine engine

starter drive from the engine drive. Fig. 2D-1 shows a clutch assembly that performs two functions. Its first function is to prevent the starter from applying excessive torque to the engine drive. At approximately 130 inch-pounds of torque, small clutch plates within the clutch housing slip around and act as a friction clutch. This setting is adjustable. During starting, the friction clutch is designed to slip until engine speed and starter speed increase to develop less than the slip torque setting.

The second function of the clutch assembly is to act as an overrunning clutch. This pawl and ratchet-type mechanism contains three pawls which are spring loaded to the retracted position. When the starter is rotated, inertia causes the pawls to move inward to engage a ratchet-type engine drive gear. This occurs because the pawl cage assembly, which floats within the pawl clutch housing, tries to remain stationary when the armature starts to drive the clutch housing around. However, the pawl clutch housing quickly forces the pawls inward by a bumping action, overcoming the retracting spring force. When engine speed approaches idle, its speed exceeds that of the starter and the pawls slip out of the tapered slots of the engine drive gear and throw outward, via force of the retracting spring. This overrunning feature prevents the engine driving the starter to burst speed.

B. Starter-Generator

The combination starter-generator is most widely utilized on corporate size jet aircraft due to the weight saving feature of one engine accessory taking the place of two. Because of its dual purpose, the drive mechanism differs from the electric starter. The starter-generator has a drive spline which stays permanently engaged to the engine.

An analysis of the example diagram will be made easier by tracing the circuit in the following steps: See Fig. 2D-3.

1. Master switch closed (up) allows either battery power or external power to reach the fuel valve, the throttle relay coil, the fuel pumps, and the ignition relay contactor.

2. Battery and start switch closed (up) allows bus power to illuminate the cockpit light, to close the ignition relay, and to close the motor relay. Ignition occurs at this time.

Fig. 2D-2   DC aircraft starter-generator used on a gas turbine engine

3. Closing the motor relay in turn allows the undercurrent relay to close and the starter to operate.

4. The battery and start switch can be released and current will continue to flow to the relays via the emergency stop lead. However, as engine speed increases and less than 200 amperes of current is flowing, the undercurrent relay opens to shut down the starter and ignition circuit operation.

5. Pulling the emergency stop button will open the ignition circuit at the same point as the undercurrent relay. This button can be used if a malfunction occurs and the relay contacts stick closed or if continuous high ampere draw from a false start prevents normal cutout.

6. An external power receptacle door microswitch prevents both external power and battery power on the bus at the same time. Some aircraft installations provide no battery start capability or require that battery starting only be used in emergencies to increase service life of the battery.

C.   Air Turbine Starter

The air turbine starter was developed as a high power-to-weight ratio device. It has only one-fifth the weight of a comparable electric starter. This starter is used almost exclusively on commercial aircraft and is also becoming an optional accessory on some corporate jets.

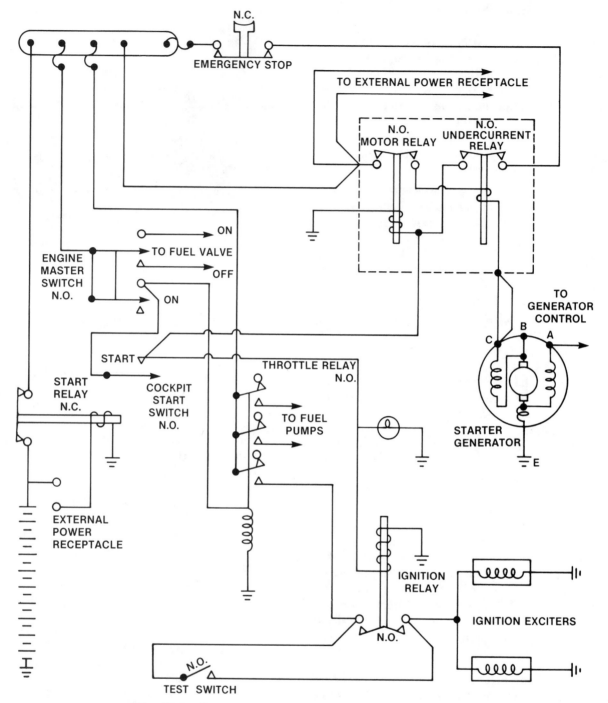

Fig. 2D-3 Circuit for a typical aircraft starter-generator

A low pressure, high volume air source of approximately 40 psig at 50-100 lb/min is supplied to this starter from an on board auxiliary power unit (APU), a ground power unit, or from the cross bleed air source of an operating main engine. Air enters the starter inlet and passes through a set of turbine nozzle vanes to change pressure to velocity and impinge at high kinetic energy levels on the turbine blades. The exhaust air exits overboard through a cowl fairing. The air supply is shut off automatically by a centrifugal cutout switch which closes the inlet air supply valve.

The turbine rotates upwards of 60 to 80 thousand RPM and is geared down 20 to 30 times. An

456

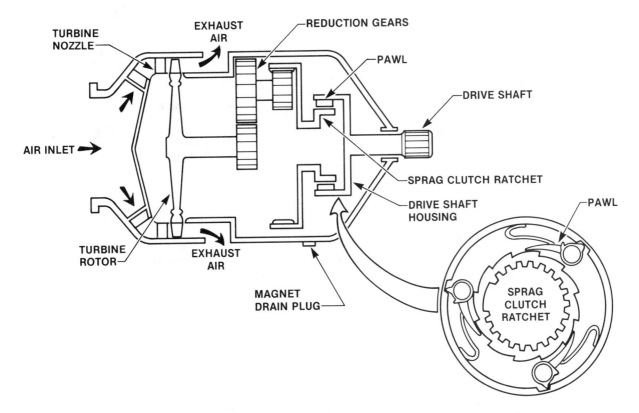

Fig. 2D-4 Schematic of an air turbine starter for a gas turbine engine

integral oil supply provides lubrication of the gear train. The oil level and magnetic drain plug are frequent inspection items for the maintenance technician.

On shutdown, at approximately 30% engine speed, the pawls of the overrunning clutch on this type starter re-engage sufficiently for re-starting if it should be required. This procedure can damage some starters and must only be accomplished within the manufacturer's recommended limits.

Safety features usually incorporated in this starter include a drive shaft shear point, which will break at a predetermined gear train induced torque force if the clutch does not release and the engine drives the starter above design speed. Another safety feature to prevent the starter from reaching burst speed if inlet air does not terminate on schedule, is that the stator nozzle airflow becomes choked and turbine wheel speed stabilizes in an overspeed condition. After either of these malfunctions occurs, a special inspection of the magnetic drain plug is generally required.

The illustration shows that the overrunning clutch, called a sprag clutch assembly, is different from the one previously mentioned for electric

starters. The clutch in this installation is in a driveshaft housing which stays permanently engaged to the engine gearbox drive. The pawls are forced inward by small leaf springs to engage the sprag clutch ratchet. At a preset engine speed, the pawls experience sufficient G force to throw outward, disengaging the drive shaft assembly from the sprag clutch ratchet. The sprag clutch ratchet and starter gear-train coast to a halt and the drive shaft housing containing the pawls will continue to rotate at engine gearbox speed. A clicking sound heard during coastdown on this type starter is not a malfunction, but rather the result of the pawls riding on the ratchet.

1. Starter pressure-regulating and shutoff valve

The starter air valve is installed in the inlet air line to the starter. It consists of a control head and a butterfly valve and is powered open electrically by means of a cockpit switch.

In Fig. 2D-5, the solenoid is energized upward by the cockpit start switch, and the following events occur:

1. The control crank rotates counterclockwise, pushing the control rod to the right and ex-

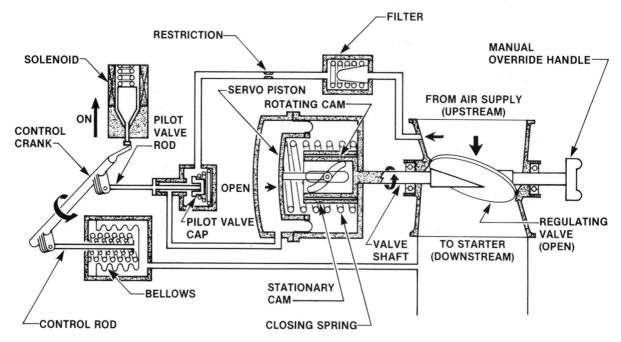

Fig. 2D-5   Pressure regulating and shutoff for an air turbine starter for aircraft gas turbine engines

tending the bellows fully. The butterfly type regulating valve being closed permits this because no pressure is present in the sensing line at this time.

2. The control crank also forces the pilot valve rod and cap to the right against a spring tension.

3. Air that had been blocked in the filtered inlet line flows past the cap to the servo piston and opens the butterfly valve.

4. As pressure builds in the downstream supply line, the sensing line takeoff directs air to partially compress the bellows. As this occurs, the pilot valve rod offseats, allowing servo piston air to vent to atmosphere.

5. When downstream air pressure reaches a preset value, the amount of air flowing to the servo piston through the restrictor equals the amount of air being bled to atmospheric, and the system is in a state of equilibrium. This feature is present to protect the starter if inlet air pressure is set too high.

6. When a predetermined starter drive speed is reached, the centrifugal cutout flyweight switch de-energizes the solenoid and the butterfly returns to a closed position.

7. The manual override handle is present to manually rotate the butterfly open and closed if corrosion or icing is causing excessive friction within the system. After freeing the restriction to movement the valve must operate normally or be replaced.

2. *Air turbine starting system*

The auxiliary power unit is generally mounted in the rear of the airplane, and an air manifold which runs through the entire ship interconnects the APU, the ground service connection, the engine bleed air ports and the starter inlets. A common procedure with this system is to start one engine from a ground or onboard starting unit and then to start the remaining engine(s) from the bleed air source of the operating engine.

The starting procedure of the typical dual-spool engine would be to:

1. Place cockpit start switch to "on."

2. Watch for slight drop in starter air manifold pressure on the cockpit gage.

3. See N_2 speed start to increase on cockpit tachometer indicator.

4. See N_1 speed start to increase on cockpit tachometer indicator.

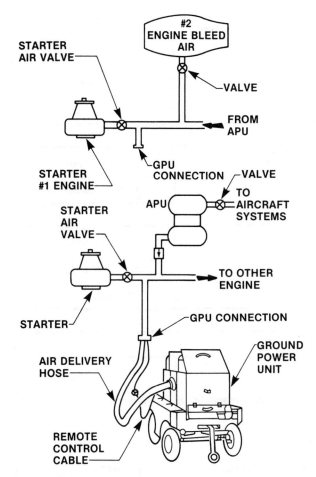

Fig. 2D-6 Air turbine starting system for aircraft gas turbine engines

5. Place ignition/fuel lever to "on."

6. Watch for EGT rise and oil pressure indication.

7. Monitor all engine instruments as engine spools up to idle speed.

8. Follow prescribed checklist for remainder of start sequence.

D. Other Starting Systems

Many other starting systems have been developed in the past for military and commercial engines. They are not in common use in either general or commercial aviation.

1. Cartridge-Pneumatic Starter — An accessory gearbox mounted starter which can use either an explosive charge or a low pressure, high volume air source similar to the air turbine starter.

2. Fuel-Air Combustion Starter — An accessory gearbox mounted starter which utilizes a high pressure 3,000 psi air source and a combustion process. It is very similar to a small gas turbine engine.

3. Impingement Starting — A low pressure, high volume air source of 40 psig at 200 to 300 lb/min is directed onto the engine turbine wheel. The air source terminates after self-accelerating speed is reached. No accessory is required in this system; only an inlet air port.

4. Hydraulic Starter — An accessory gearbox mounted hydraulic starter motor. It is driven by fluid from an APU mounted hydraulic pump.

E. Troubleshooting Starter System

STARTER-GENERATOR		
1. Engine does not rotate when the starter switch is closed during a battery start.		
Possible Cause:	Check For:	Remedy:
a. External power receptacle door	door open or faulty microswitch	close door or repair switch
b. Battery relay	power at DC start bus	repair as necessary
c. Motor relay	proper closing	repair or replace
d. Starter-generator	(1) correct input voltage	charge battery
	(2) sheared shaft	replace unit

2. Engine does not rotate when the start switch is closed during external power start.

Possible Cause:	Check For:	Remedy:
a. External power source	correct connection to aircraft	make connection
b. 1a, 1b, 1c, 1d, above		

3. Starting terminates when start is released.

Possible Cause:	Check For:	Remedy:
Undercurrent relay	proper closing	replace as necessary

4. Engine starts but will not self-accelerate. False or hung start.

Possible Cause:	Check For:	Remedy:
a. Power supply	low voltage	charge or change battery
b. Starter-generator	internal malfunction	change unit

AIR TURBINE STARTER

5. Engine does not rotate when the start switch is closed.

Possible Cause	Check For:	Remedy:
a. APU or GPU	presence of air in starting manifold (cockpit gage)	~~cprrect~~ correct as necessary
b. Starter air valve	(1) solenoid operation	repair or replace
	(2) ice preventing valve operation	apply heat
c. Starter	shaft sheared	replace starter

6. Starter does not rotate to normal cutoff speed.

Possible Cause:	Check For:	Remedy:
a. APU or GPU	low air supply	repair or replace air unit
b. Starter	centrifugal switch cutout setting	adjust setting or replace starter
c. Air valve	internal malfunction	replace valve

7. Starter does not cut off.

Possible Cause:	Check For:	Remedy:
5b or 5c above (flyweight switch inoperative)		

8. Metal particles on magnetic drain plug.

Possible Cause:	Check For:	Remedy:
Internal starter malfunction	(1) small fuzzy particles	normal
	(2) chips or slivers	replace starter

I. RECIPROCATING ENGINE LUBRICATING SYSTEMS

A. Functions of the Lubricating System

1. Reduces friction

If we could microscopically examine the surfaces of the metal parts inside an aircraft engine, we would find that they are not perfectly smooth, but, rather, they are made up of peaks and valleys. When two parts with surfaces such as these are rubbed together, there is friction that soon wears the metal away.

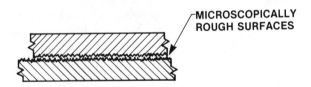

EVEN SURFACES THAT APPEAR SMOOTH HAVE MICROSCOPICALLY ROUGH SURFACES THAT LOCK WITH OTHER SURFACES TO CAUSE FRICTION AND WEAR.

(A)

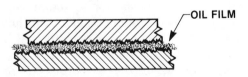

LUBRICATING OIL WETS THE SURFACES AND SEPARATES THEM SO THE PARTS SLIDE OVER EACH OTHER ON A FILM OF OIL RATHER THAN THE METAL-TO-METAL CONTACT. THIS REDUCES FRICTION AND MINIMIZES WEAR.

(B)

Fig. 1A-1 Functions of a lubricating oil

In order to reduce this friction, a film of oil is placed between the moving parts. This oil wets the surfaces, fills in the valleys, and holds the metal surfaces apart. The movement is now between layers of the oil rather than the metal surfaces, and the oil slides over itself with very little friction.

The viscosity of an oil is a measure of its fluid friction, and the amount of clearance between the moving parts determines the viscosity of the oil that is needed to prevent the film breaking away and allowing the metal-to-metal contact that causes wear.

2. Provides cooling

The lubricating oil is in intimate contact with the moving parts of an aircraft engine, and it absorbs some of the heat from the combustion process. This heated oil then flows through the system into the oil cooler where the heat is given up to the outside air as it passes through the core of the cooler. (See Fig. 1A-2.)

3. Seals and cushions

The viscous nature of oil—that is, its ability to wet the surface it contacts—makes oil a good sealing agent between moving parts. The oil film on the cylinder walls and around the piston forms a tight seal in the cylinder, and the thin film of oil between the rocker arm and its bushing takes up much of the hammering shock from the valve action.

4. Protects against corrosion

When metal remains uncovered in the presence of moisture or some of the chemicals that

Fig. 1A-2 Modern high-performance aircraft engine using an external oil cooler to transfer heat from the engine oil to the outside air.

contaminate the air, rust and other forms of surface corrosion will attack the metal. This is especially true of metal surfaces such as cylinder walls and crankshafts which have been hardened by the nitriding process. A film of oil covering these surfaces will prevent oxygen reaching the metal and causing it to oxidize or corrode.

5. Cleans

Dirt, dust, carbon, and water all enter the oil, and the ability of an oil to hold these contaminants until they can be trapped in the filter helps keep the inside of the engine clean.

B. Types of Aircraft Engine Lubricating Oil

1. Straight mineral oil

MIL-L-6028B is a straight mineral oil that has been used for many years as the chief lubricant for aircraft engines. It has one main limitation, that being its tendency to oxidize when it is exposed to elevated temperatures or when it is aerated.

Large amounts of carbonaceous deposits form in turbochargers because of the heat from the exhaust gases that are used to spin the turbine. When the engine is shut down, the turbocharger housing acts as a heat sink and cooks the oil in the bearings, forming carbon. Sludge also forms in this oil at the relatively low temperature of 150° F or lower. It forms from such combustion products as partly burned fuel, water vapor, and lead compounds, and these particles unite to form a loosely linked mass which clogs oil screens and scores engine bearings.

2. Metallic-ash detergent oil

Certain metallic-ash-forming additives have been used with mineral oil to increase its oxidation stability. These additives were chosen to have the minimum effect in the combustion chamber and to minimize spark plug fouling and preignition. But their cleaning action loosens any carbon deposits and sludge that have formed in the engine and, once loosened, they will pass through the engine where they are prone to clog oil passages and filters.

Detergent oils were given only a limited approval by some of the engine manufacturers and because of their limitations, they are no longer considered a suitable oil for use in aircraft engines.

3. Ashless-dispersant oil

By far the most important oil in use today is ashless-dispersant, or AD, oil. It does not have the carbon forming restrictions of straight mineral oil nor does it form ash deposits, as the detergent oil did.

Meeting the specification of MIL-L-22851, AD oil is approved by both Lycoming and Continental for use in their engines and is the only oil used by the military services in their aircraft reciprocating engines.

Ashless-dispersant oil does not contain any ash-forming additives, but rather uses additives of the dispersant type which, instead of allowing the sludge-forming materials to join together, causes them to repel each other and stay in suspension until they can be picked up by the filters.

It has been argued that these contaminants, held in suspension, will act as a liquid hone and accelerate the wear of the engine parts, but this has been proven untrue. It is interesting to note that most engine manufacturers recommend their new engines to be operated on straight mineral oil for the first fifty hours, or at least until oil consumption stabilizes, and then switched to AD oil. The reason for this is that the AD oil has so much better lubricating properties that it does not allow enough wear to properly seat the piston rings.

4. Synthetic oil

The higher operating temperatures of modern reciprocating engines, and the lower temperature environment in which turbine engines operate, have in the past few years caused synthetic oil to be produced which is proving superior to mineral oils for lubrication. At the present time, this oil has only limited approval for reciprocating engine use, not because of any inherent problems, but because of lack of service experience.

The low-temperature operating characteristics of synthetic oil make it admirable for cold weather and high altitude operation. Its viscosity at −20° F is about the same as ashless-dispersant mineral oil at 0° F, and engines operated on synthetic oil have been successfully started and run at ambient temperatures as low as −40° F. This characteristic has the potential advantage in cold weather operation of eliminat-ing the extensive engine preheat and the practice of draining the oil when arriving in a warm climate, making it truly an all-weather lubricant.

Additive synthetic oils appear to have the advantage over mineral oils with respect to engine cleanliness. Oil oxidation at high temperatures, with its resultant deposits, has proven in laboratory tests to be less than that produced with either straight mineral oil or ashless-dispersant oil. Synthetic oils have replaced mineral oils for turbine operation since their high power and high temperature operation is a limiting factor. These oils inhibit oxidation and thermal decomposition at temperatures far higher than mineral oils can tolerate. In the 150-hour certification tests, they have demonstrated their ability to operate satisfactorily with bulk oil temperatures as high as 245° F.

The wear characteristics of synthetic oil appears to be about the same as those of ashless-dispersant oil and superior to straight mineral oil.

One of the problems with synthetic oil is its more pronounced softening effects on rubber products and resins. One manufacturer requires a more frequent replacement of the intercylinder drain lines when synthetic oil is used, and pleated paper oil filters must be examined more closely to be sure that the oil does not dissolve the resins and allow the filter to collapse.

Synthetic oil is considerably more expensive than mineral oils, but the extended drain period for synthetic oil appears to compensate for its higher cost. Any decision to use synthetic oil, taking into consideration cost factors, must be an individual matter based on specific operating conditions to determine whether or not the advantage of the oil justifies its increased cost.

The concept of synthetic oil for reciprocating engines is relatively new, and engine manufacturers have not as yet resolved all of the problems, nor have they issued a blanket approval for them. This does not infer any condemnation; it merely indicates that sufficient data has not yet been obtained, nor has field evaluation been completed.

C. Compatibility of Oils

Contrary to popular opinion, oils within their basic categories are compatible. All straight mineral oils are physically compatible with each other, and it is extremely doubtful if engine performance or cleanliness would be compromised by mixing several brands of oil.

All ashless-dispersant mineral oils meeting MIL-L-22851 specifications are physically compatible, and one brand may be added to an engine operating on another brand without destroying the overall cleanliness of the engine. This was proven by the military when no performance nor cleanliness deterioration resulted from operating between suppliers having different brands.

Ashless-dispersant oils are also compatible with straight mineral oil. The airlines proved this while converting to the ashless-dispersant type. In the transition, various amounts of straight mineral oil and ashless-dispersant oil were mixed, and it was found that there were no adverse side effects other than a reduction in the degree of the advantages from operating with all AD oil.

At the present time, because of lack of experience, it is not advisable to add synthetic oil to engines operating on straight mineral oil or on ashless-dispersant oil. Before changing from either of the mineral oils to synthetic oil, be sure to follow in detail the procedures for flushing and draining established by the manufacturer of the synthetic oil.

D. Engine Oil Rating

1. Viscosity

The viscosity, or fluid friction, of an engine oil is one of its more important ratings, and it is measured by a rather elaborate laboratory instrument known as a Saybolt Universal Viscosimeter. The number of seconds required for 60 cubic centimeters of the oil to flow through an extremely accurately calibrated orifice at a specific temperature is known as the S.S.U., or Saybolt Seconds Universal viscosity.

	AN 1065 AVIATION 65 SAE 30	AN 1080 AVIATION 80 SAE 40	AN 1100 AVIATION 100 SAE 50	AN 1120 AVIATION 120 SAE 60
VISCOSITY S.S.U. @ 100° F S.S.U. @ 130° F S.S.U. @ 210° F	443 215 65.4	676 310 79.2	1,124 480 103.0	1,530 630 123.2
VISCOSITY INDEX	116	112	108	107
GRAVITY API	29.0	27.5	27.4	27.1
COLOR ASTM	1.5	4.5	4.5	5.5
POUR POINT °F	− 20	− 15	− 10	− 10
POUR POINT DILUTED °F	− 70	− 70	− 70	− 50
FLASH POINT °F	450	465	515	520
CARBON RESIDUE % W	0.11	0.23	0.23	0.40

Fig. 1A-3 Characteristics of aircraft engine lubricating oil

A convenient grading of engine oil for aviation is done by rounding off the S.S.U. numbers for 210° F, as we see in Fig. 1A-3. The Society of Automotive Engineers, the SAE, and the military have different numbering systems, but the numbers all relate to the S.S.U. viscosity of the oil. In civilian aviation we nearly always use the SAE designation system.

2. Viscosity index

Viscosity index is a measure of the change in viscosity of an oil with a given change in temperature. The index itself is based on the viscosity changes with temperature of two reference oils. One oil is rated at 100, and the other is rated at zero. The smaller the change in the viscosity for a given temperature change, the higher the viscosity index.

3. Gravity API

The American Petroleum Institute has formulated a measurement of the specific gravity of petroleum products which is an expansion of the regular specific gravity scale. A conversion chart must be used to relate any API number to its corresponding specific gravity.

4. Color

The color of an oil is rated by comparing its color with an ASTM (American Society of Testing and Materials) color chart. The reference 1.00 on the chart is pure white, and 8.00 is a red, darker than claret wine.

5. Pour point

The pour point of an oil is the lowest temperature at which the oil will pour without disturbance.

6. Flash point

The flash point is that temperature to which oil must be raised before it will momentarily flash but not sustain combustion when a small flame is passed above its surface.

7. Carbon residue

A given amount of oil is placed in a stainless steel receptacle and is heated to an accurately controlled high temperature until it is evaporated. The container is weighed before and after the test, and the amount of carbon residue left in the container is expressed in this rating as a percent of the weight of the sampled material.

E. Lubrication Systems

1. Types of lubrication systems

a. Dry sump systems

Radial engines are not built in such a way that they can carry their lubrication oil inside the engine itself, and it is for this reason that the dry sump lubrication system was designed. The oil is carried outside the engine in an oil tank usually mounted in the engine compartment. This type of system proved so successful that it was used as the standard system for many years and was adapted to some of the horizontally opposed engines. (See Fig. 1A-4.)

The oil flows from the tank into the inlet of the engine pressure pump by gravity, and after it has passed through the pump, it flows through a pressure screen such as the one we see in Fig. 1A-5. These screens are usually made of two different sizes of wire mesh. The larger mesh provides the mechanical shape and strength, and the fine mesh does the actual straining.

Some strainers have a by-pass valve built into their element that will open and allow the oil to bypass the strainer if it becomes clogged with contaminants. Other installations use an external valve. The check valve inside the strainer shown in Fig. 1A-5 is used with radial engine installations to prevent oil from the tank draining into the engine and flooding the lower cylinders when the engine is not running. As soon as the engine starts, the pump will build up enough pressure to force the valve off its seat so the oil can flow into the engine with a minimum of opposition.

The oil temperature may be measured by a temperature bulb in the line just before it enters the pressure pump, or it can be inserted into the oil strainer. In either location it measures the temperature of the oil as it enters the engine.

After the oil leaves the strainer, it flows through drilled passages inside the engine to lubricate the crankshaft and cam bearings and through the hollow propeller shaft to furnish oil

465

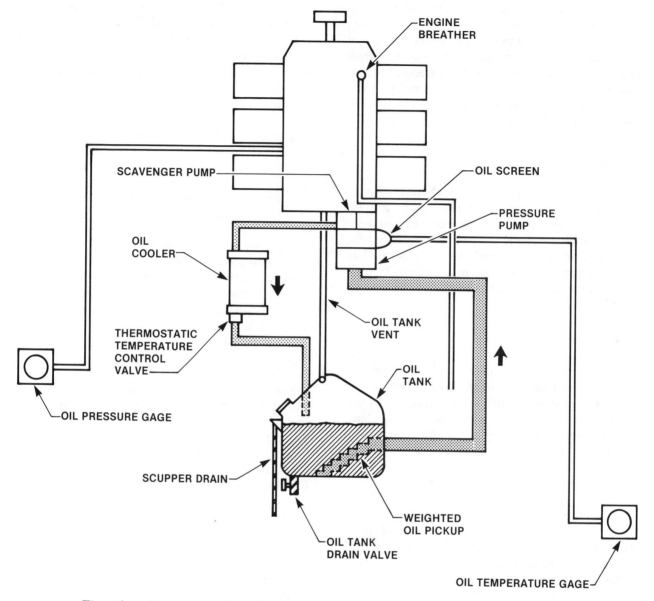

Fig. 1A-4 Dry sump lubrication system for a horizontally opposed aircraft engine.

for hydraulic propeller control. A spring-loaded oil pressure relief valve dumps the oil back into the inlet of the pump when the pressure exceeds that for which it is set.

Oil to lubricate the valve mechanism flows into the rocker boxes through hollow push rods and drains back either through intercylinder drain lines or the push rod housing.

After the oil has circulated through all of the engine, it drains by gravity and collects in one or more sumps, which are small compartments connected by oil pickup tubes to the scavenger

pump. A scavenger pump is similar to a pressure pump, but it is considerably larger because the oil being scavenged is hot and has air entrapped in it, causing its volume to be greater than that delivered by the pressure pump.

From the scavenger pump, the oil leaves the engine and is directed through the oil cooler back into the tank. (See Fig. 1A-5.)

b. Wet sump systems

Most modern aircraft engines use a wet sump system in which all of the oil is carried in a sump

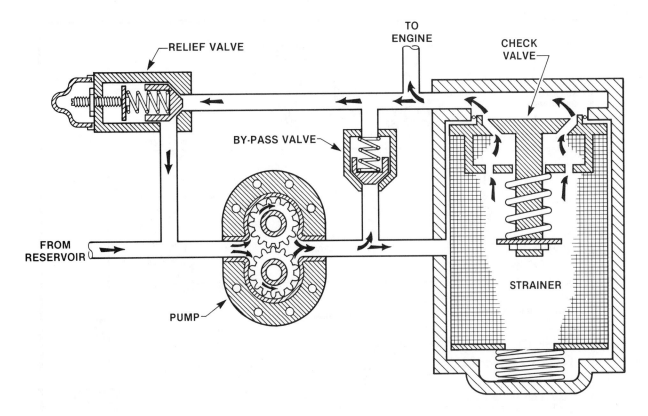

Fig. 1A-5   Typical oil pressure system in an aircraft engine

which is part of the engine itself. The oil is picked up through an oil suction screen by the gear-type oil pump. From the pump exit, the oil is directed to the oil filter chamber, and if the filter should become clogged, a spring-loaded by-pass valve will open, permitting oil to flow directly from the pump to the oil filter outlet. From here, a passage leads to the adjustable oil pressure relief valve, and any time the pressure exceeds that for which the valve is set, the excess will bypass back into the sump. (See Fig. 1A-6.)

If the temperature of the oil is below that for which the oil cooler by-pass valve is set, the oil goes directly into the oil gallery in the engine crankcase; but if it is hotter than it should be, it must pass through the core of the cooler. This valve keeps the oil temperature within the range specified for the engine operation.

The oil flows through drilled passages to the front of the engine, providing lubrication for the crankshaft, camshaft, and propeller shaft bearings, the valve mechanism, and the propeller governor.

The gear-type pump in the propeller governor boosts the engine oil pressure and directs it back into the crankcase through drilled passages, through the hollow propeller shaft, and into the propeller to control the pitch.

The accessory drives are lubricated by oil fed into their bearings through drilled passages, and the gears are cooled and lubricated by a spray of oil in the accessory case at the rear of the engine.

The cylinder walls are lubricated, and the pistons are cooled by oil from around the crankshaft main bearings which is squirted in a continuous stream against the inner dome of the piston. This oil, after it has lubricated the pistons and cylinders, drains down into the sump.

When a turbocharger is installed, it is lubricated by oil from the crankcase gallery, which, after it has lubricated these critical turbocharger bearings, drains into an oil separator, which removes the air, and then the oil is picked up and returned into the engine sump by a gear-type scavenger pump.

467

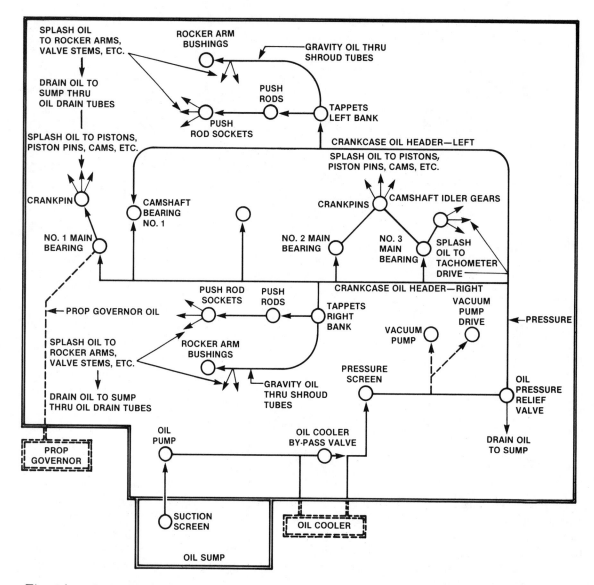

SPLASH OIL
TO ROCKER ARMS,
VALVE STEMS, ETC.

ROCKER ARM
BUSHINGS

GRAVITY OIL THRU
SHROUD TUBES

DRAIN OIL TO
SUMP THRU
OIL DRAIN TUBES

PUSH
RODS

TAPPETS
LEFT BANK

SPLASH OIL TO PISTONS,
PISTON PINS, CAMS, ETC.

PUSH
ROD SOCKETS

CRANKCASE OIL HEADER—LEFT

SPLASH OIL TO PISTONS,
PISTON PINS, CAMS, ETC.

CRANKPIN

CAMSHAFT
BEARING
NO. 1

CRANKPINS

CAMSHAFT IDLER GEARS

NO. 1 MAIN
BEARING

NO. 2 MAIN
BEARING

NO. 3
MAIN
BEARING

SPLASH
OIL TO
TACHOMETER
DRIVE

CRANKCASE OIL HEADER—RIGHT

PROP GOVERNOR OIL

PUSH ROD
SOCKETS

PUSH
RODS

TAPPETS
RIGHT
BANK

VACUUM
PUMP
DRIVE

PRESSURE

SPLASH OIL TO
ROCKER ARMS,
VALVE STEMS, ETC.

ROCKER ARM
BUSHINGS

VACUUM
PUMP

PRESSURE
SCREEN

OIL
PRESSURE
RELIEF
VALVE

DRAIN OIL TO SUMP
THRU OIL DRAIN TUBES

GRAVITY OIL
THRU SHROUD
TUBES

PROP
GOVERNOR

OIL
PUMP

OIL COOLER
BY-PASS VALVE

DRAIN OIL
TO SUMP

SUCTION
SCREEN

OIL COOLER

OIL SUMP

Fig. 1A-6   Lubrication system for a modern horizontally opposed wet sump aircraft engine

2.   Lubrication system components

a.   Pumps

(1)   Gear-type

Two spur gears, one driven by the engine, and the other by the drive gear, rotate in a close fitting housing. As the teeth unmesh on the inlet side of the pump, the volume of the cavity increases, lowering the pressure and drawing oil into the pump. This oil is trapped between the teeth and the housing and is carried around the outside of the gears to the outlet side. As the teeth of the gears mesh, the volume of the cavity decreases, and the oil is forced out of the pump into the drilled passages in the engine crankcase.

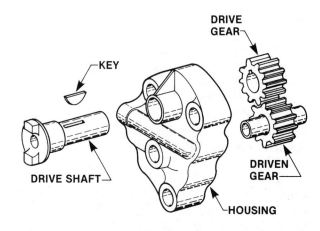

DRIVE
GEAR

KEY

DRIVE SHAFT

DRIVEN
GEAR

HOUSING

Fig. 1A-7   Gear-type oil pump

A gear pump is a constant-displacement pump, meaning that a specific amount of fluid is moved each time the pump rotates, and provision must be made to relieve the excess pressure so the system pressure can remain constant as the pump speed varies.

(2) Gerotor-type

Another form of constant-displacement pump used for moving lubricating oil through a reciprocating engine is the gerotor pump, a special form of gear pump. In the illustration of Fig. 1A-8, we see a six-tooth spur gear driven from an engine accessory drive. This gear rides inside a rotor which rotates freely in the housing. A seven-tooth internal gear is cut inside the rotor, and as the drive gear is turned, it turns the rotor. In view A, the two marked teeth are meshed and there is a minimum of space between them. As the two gears rotate, the volume between the teeth increase as you follow the marked teeth in

views A, B, and C. A plate with two kidney-shaped openings covers the gears, forming a seal for their ends. As the gears rotate beneath the inlet port, the volume between the teeth continually increases, and as they rotate beneath the outlet port, as seen in views D and E, the volume decreases, moving the fluid out into the system. Gerotor pumps may be designed to pump relatively large volumes of fluid without their having to be excessively thick.

b. Relief valves

(1) Simple relief valve

Almost all pumps used in aircraft engine lubrication systems are of the constant-displacement type, and provision must be made to relieve some of the oil back to the inlet side of the pump, to keep the pressure constant as the engine RPM changes. In Fig. 1A-9 we see the operating principle of a simple pressure relief valve. Oil

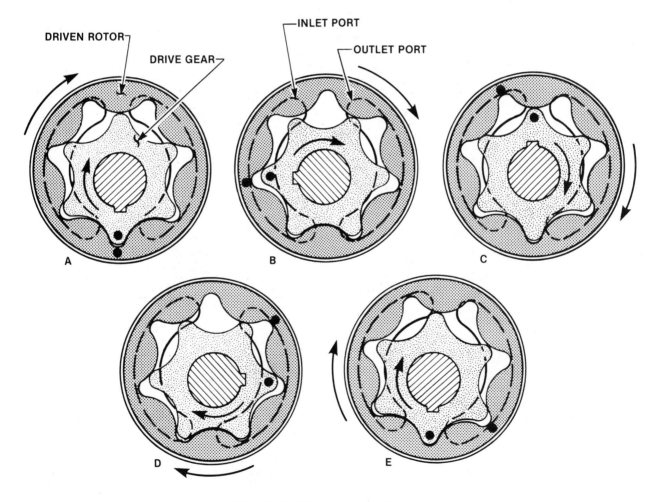

Fig. 1A-8   Gerotor-type oil pump

9

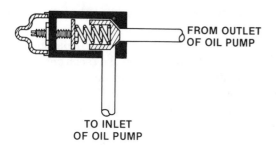

FROM OUTLET
OF OIL PUMP

TO INLET
OF OIL PUMP

Fig. 1A-9 Simple oil pressure relief valve

from the discharge side of the pump flows into the lubricating passage, which is connected by a spring-loaded valve back into the inlet side of the pump. As long as the pressure is below that for which the relief valve is set, the valve remains on its seat; but when the pressure rises, the valve will move off its seat, and oil will return to the inlet side of the pump, holding the system pressure constant. An adjustment screw, locked with a jam nut and covered with a protective cap, is used to change the oil pressure. Loosening the jam nut and turning the adjustment screw clockwise raises the oil pressure.

(2) Compensated relief valve

Some of the larger engines require a very high oil pressure to force cold oil through all of the bearings for starting. But when the oil warms up and becomes thinner, this high pressure would cause excessive oil consumption, so a compensated oil pressure relief valve is used to hold the pressure high until the oil warms up and then automatically lowers it to the normal operating range.

This is done by using two springs to hold the valve on its seat when the oil is cold, and when the oil warms up, a thermostatically operated valve opens a passage and allows oil to flow be-

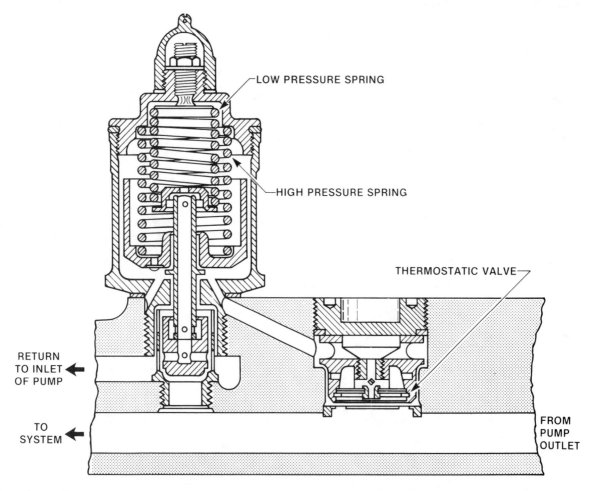

LOW PRESSURE SPRING

HIGH PRESSURE SPRING

THERMOSTATIC VALVE

RETURN
TO INLET
OF PUMP

TO
SYSTEM

FROM
PUMP
OUTLET

Fig. 1A-10 Compensated oil pressure relief valve. This valve allows the oil pressure to be high enough to force cold oil through all of the passages inside the engine, but decreases the pressure to the proper operating range when the oil is warm.

470

neath a piston, relieving the force on the high-pressure spring. Normal operating pressure is maintained by the force of the low-pressure spring alone.

c. Oil coolers

Much of the heat that is released from the fuel is picked up by the oil, so some provision must be made to transfer part of it into the air. This transfer is done in a radiator, usually called an oil cooler.

Depending on the design of the lubricating system, the cooler may be either between the oil filter and the bearings or in the return line between the scavenger pump and the oil tank. In either case, a thermostatic control valve allows the oil to bypass the core of the cooler when it is cold, and as it warms up, the valve forces the oil through the core so the excess heat can be picked up by the air. (See Fig. 1A-11.)

d. Oil reservoir

Dry sump engines carry their oil in an external reservoir which must have a capacity compatible with the amount of fuel carried, plus a margin that will provide for adequate circulation and cooling.

Engines operated in cold weather must have some provision for thinning the oil for starting. In extreme cases, the oil may be drained while it is still hot from the last flight of the day and warmed the next morning before it is put back into the oil tank; but a much simpler method for thinning the oil is by oil dilution.

Before the engine is shut down and while the oil is still hot, fuel from the carburetor is directed into the engine oil pump inlet, where it mixes with the oil and dilutes it. The oil is diluted for the proper length of time for the temperature expected for starting, and when the engine is started, the diluted oil flows through the engine and assures proper lubrication. When the oil warms up, the gasoline evaporates from it and leaves the engine through the crankcase breather, and the oil returns to its original condition.

Oil tanks used with engines equipped with Hydromatic propellers feed the pressure pump through a standpipe which sticks into the tank for a few inches. The tank outlet to the propeller

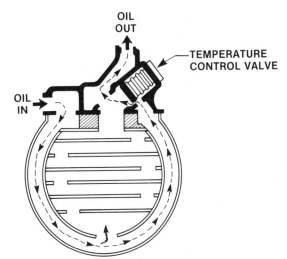

WHEN THE OIL IS COLD, IT FLOWS AROUND THE CORE OF THE COOLER.

(A)

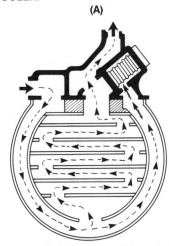

WHEN THE OIL IS HOT, THE TEMPERATURE CONTROL VALVE SHUTS OFF THE PASSAGE IN THE SHELL OF THE COOLER, AND THE OIL MUST PASS THROUGH THE CORE, WHERE IT TRANSFERS ITS HEAT TO THE AIR PASSING THROUGH THE COOLER.

(B)

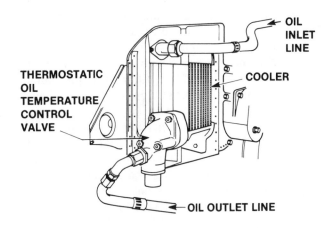

TYPICAL OIL COOLER INSTALLATION
(C)

Fig. 1A-11 Engine oil cooler

feathering pump is taken from the bottom of the tank, and with this arrangement, if an oil line should break and all of the oil be pumped overboard by the engine pump, there will still be enough oil in the tank to feather the propeller.

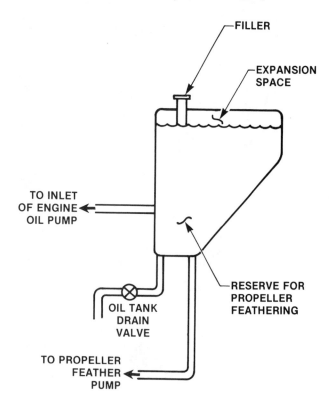

Fig. 1A-12 Typical dry sump oil tank for an engine equipped with a Hydromatic feathering propeller.

e. Filters

Solid contaminants and sludge pumped through an aircraft engine lubricating system can clog the oil passages and damage the bearing surfaces, so provision must be made to remove as much of it as possible. There are two ways this may be done: by a full-flow filter, through which all of the engine oil must flow each time it circulates through the engine, or with a by-pass filter that filters only a portion of the oil each time it circulates, but which eventually gets all of the oil. By-pass filters can be made much finer because, if they should clog and prevent oil flowing through them, they will not starve the engine of oil. (See Fig. 1A-13 and 1A-14.)

The majority of filters used in modern aircraft engines are of the full-flow type, and there are four methods of filtration used.

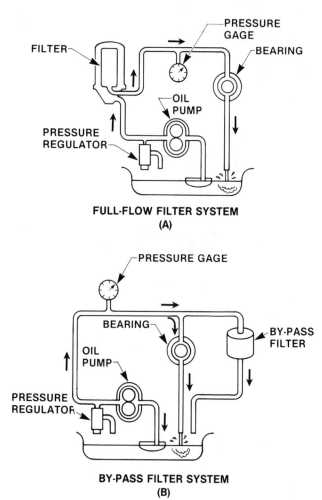

Fig. 1A-13 Oil filter systems

(1) Edge filtration

A long, flat spiral with a wedge-shaped cross section forms the filtering element for an edge-type filter. The oil flows from the outside of the element into its center, and any contaminants collect on the outside where the slots are the smallest. Since the cross section of the spiral is wedge-shaped, if a contaminant should be small enough to pass through, it will not clog the filter. Some edge-type filters are cleaned with a built-in blade that is rotated to scrape off the contaminants that collect on the outside of the element. (See Fig. 1A-15.)

(2) Depth filtration

Depth filters consist of a matrix of fibers that are closely packed to a depth of about one inch. The oil flows through this mat and the contaminants are trapped in the fibers. Depth-type filters

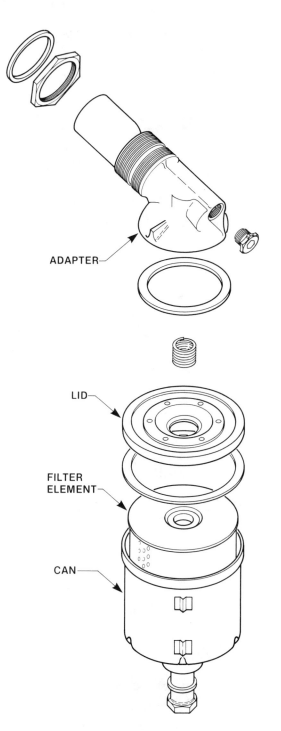

ADAPTER

LID

FILTER
ELEMENT

CAN

Fig. 1A-14 Typical installation of a full-flow-type oil filter.

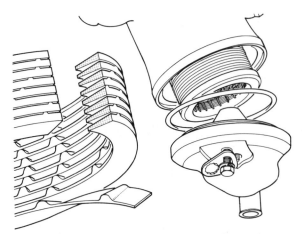

Fig. 1A-15 Edge filtration-type oil filter

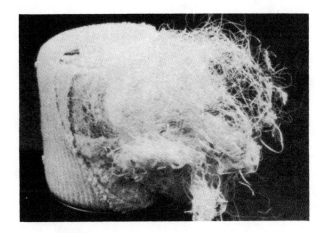

Fig. 1A-16 Depth-type oil filter

are more efficient than edge type because they can hold so much more contamination, but because of the non-uniformity of the matted material, it is possible for the high pressure oil flowing through it to find a weak spot and channel through, causing the filter to lose a great deal of its effectiveness. (See Fig. 1A-16.)

(3) Surface filtration

Standard equipment in almost all aircraft engines is a woven wire-mesh oil filter that is useful for trapping some of the larger contaminants that flow through the engine. Particles of contamination larger than the size of the screen openings are trapped on the surface, and the smaller particles pass through. To get better filtering, the size of wire mesh must be decreased, but when this is done, the cleaning process becomes quite complex and the advantage of simplicity is lost. As a result, this type of filter is often supplemented by a more efficient type. (See Fig. 1A-17.)

(4) Semi-depth filtration

The most popular filter in common use for general aviation aircraft is a disposable, semi-

ENLARGED VIEW OF THE FIBERS IN THE FILTER ELEMENT
(A)

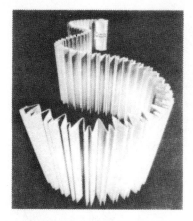

PLEATED FILTER ELEMENT
(B)

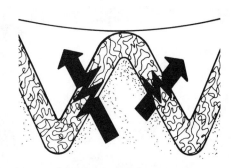

METHOD OF TRAPPING SEDIMENT IN THE FILTER ELEMENT
(C)

THE PLEATED FILTER ELEMENT IS ASSEMBLED AROUND A SHEET STEEL CORE
(D)

FILTER ELEMENT
(E)

SPIN-ON-TYPE OIL FILTER
(F)

METHOD OF OPENING A SPIN-ON-TYPE OIL FILTER FOR INSPECTION OF CONTAMINANTS
(G)

Fig. 1A-18   Semi-depth-type oil filter

Fig. 1A-17 Screen-type surface filtration oil strainer

depth filter made up of resin-impregnated fibers, formed into a long sheet and folded in a pleated fashion. All of the oil that flows in the system must pass through the filter element, and in doing so the contaminants are trapped within its fibers. There is so much uniform surface area that there is little tendency for the oil to channel through. The pleated material is assembled around a sheet steel core, and either fitted in a heavy sheet steel case which is an integeral part of the filter or may be made so it can be installed inside a housing which is part of the engine.

When this type of filter is removed from the engine at inspection time, it is cut open and inspected for the presence of any metal particles that might indicate an impending engine failure. Sealed, spin-on-type filters are opened with a special roller-type can cutter that removed the top of the container without introducing any metal particles that might make the examination more difficult. (See Fig. 1A-18.)

f. Instrumentation

(1) Pressure measurement

Oil pressure is measured at the outlet of the engine-driven pump and is indicated on a gage on the pilot's instrument panel. To prevent gage fluctuations and keep the loss of oil to a minimum if the line to the instrument should break, a small orifice, about a number 60-drill size, is installed in

the fitting where the gage line attaches to the oil pressure gallery in the engine.

(2) Temperature measurement

The oil temperature measurement read on the instrument panel is the temperature of the oil as it enters the engine. On dry sump engines, the temperature pickup bulb is often located in a special fitting between the oil tank and the pump, and on wet sump engines, it is usually installed inside the oil screen immediately after the pump. Temperature may be measured electrically, by measuring the change in resistance of a special temperature probe, or mechanically by measuring the pressure of a gas sealed inside a bulb held in the oil. The pressure of the gas varies in proportion to its temperature.

II. TURBINE ENGINE LUBRICATING SYSTEMS

A. Introduction

The lubrication system supplies oil to the various moving parts within the engine which are subjected to friction loads and heating from the gas path. The oil is supplied under pressure along the main rotor shaft and to the gearboxes to reduce friction, to cool, and to clean. It is then returned by a scavenging system to the oil storage tank to be used again and again. Oil consumption is low in gas turbine engines as compared to piston engines, and this accounts for the relatively small bulk oil storage tanks used. They can be as small as 3 to 5 quart capacity on business jet size engines and 20 to 30 quarts on large commercial type engines. The oil is not exposed to great quantities of combustion products and is kept fairly clean by filtration. Heat, however, is a problem which can cause rapid oil decomposition, and for this reason temperature is carefully controlled by automatic cooling devices and is carefully monitored by the engine operators.

1. Principles of turbine engine lubrication

In theory, lubricating fluids fill all surface irregularities and prevent metal-to-metal contact. The primary purpose of a lubricant is to reduce friction between moving parts. As long as this oil film remains unbroken, metallic friction is replaced by internal fluid friction.

In addition to reducing friction, this oil film acts as a cushion between metal parts, absorbs heat, and collects foreign matter as it circulates through the engine.

2. *Requirements of turbine engine lubricants*

Gas turbine engine lubricating oil must have a high enough viscosity for good load carrying ability, but it must also be of sufficiently low viscosity to provide good flow ability.

Synthetic, rather than petroleum base lubricants, are used in turbine engines. The desirable characteristics of synthetic lubricants are as follows:

1. Low volatility — to prevent evaporation at high altitudes

2. Anti-foaming quality — for more positive lubrication

3. Low lacquer and coke deposits — keeps solid particle formation to a minimum

4. High flash point — the temperature at which oil when heated gives off flammable vapors that will ignite if near a flame source

5. Low pour point — the lowest temperature at which oil flows by gravity

6. Excellent film strength qualities of cohesion and adhesion—a characteristic of oil molecules that allows them to stick together under compression loads and stick to surfaces under centrifugal loads

7. Wide temperature range — in the order of −60° F to +400° F; preheat not required to approximately −40° F

8. High viscosity index — meaning the oil will tend to retain its viscosity when heated to its operating temperature

a. *FAA requirements*

The following is a list of minimum FAA requirements from Part 23 and Part 25:

1. The word "oil" and the system capacity must be stenciled in the area of the filler opening.

2. A means of visually checking the oil level on the ground must be provided with a sight gage, dip stick, visual filler opening, or similar device.

3. Expansion space of 10% or 0.5 gallon, whichever is greater, must be provided.

4. The scavenge subsystem must be at least double the capacity of the pressure subsystem to accommodate increased volume caused by entrained air.

b. *Viscosity*

An SAE rating for petroleum base lubricants is determined by heating 60 milliliters (cubic centimeters) of oil to a specific temperature and measuring the flow time as the oil is poured through an appropriate orifice.

Many automotive and some aviation oils are now classified as multi-grade; for example, SAE 5W-20. This means that the oil when cold will have the viscosity of SAE-5 and when at its normal operating temperature will thin out no more than SAE-20. Viscosity, then, is a measure of an oil's pourability, and multi-viscosity oils are designed to pour or flow sufficiently for quick lubrication at low temperatures and remain thick enough for good lubrication at higher temperatures.

SAE numbers are not used to identify turbine oils, but they make synthetic oil viscosity more understandable by comparison.

c. *Viscosity index*

Viscosity index is a means of testing the viscosity change when a liquid lubricant is heated to two different temperatures. The quality of synthetic lubricants is determined in this way.

The centistoke value (metric viscosity measurement) can be seen on some container labels of synthetic lubricants. A rough equivalent to SAE values is: 3cSt. oils are approximately equal to SAE-5, 5cSt. oils are approximately equal to SAE 5W-10 multiviscosity oils, 7cSt. oils are approximately equal to SAE 5W-20 multiviscosity oils. It follows then that it would be more common to see a 7cSt. oil in use in a turboprop where high gear loading is present than a lower cSt oil.

3. *Oil sampling*

After shutdown, and just prior to servicing, many air carriers require ground personnel to take an oil sample from a sediment-free location in the main oil tank. From this, one can analyze particles or contaminants that are suspended in the oil. These particles are a good indicator of engine wear when counted in a device known as an oil spectrometer. The procedure of reading parts per million of contaminants is referred to as spectrometric oil analysis. With a spectrometer, contaminant levels are automatically registered by analyzing the color and measuring the intensity of brightness that occurs when the metal particles are burned in a certain light spectrum. Many private companies offer this service to customers, who in turn use the information to plot trend analyses of internal engine wear. This allows the operator to take appropriate action to avert costly repair or loss of equipment.

4. Synthetic lubricants

Synthetic lubricants are by their makeup multiviscosity, similar to automotive grades SAE-5 to SAE-20.

They are a blend of certain diesters, which are themselves man-made (synthesized) extracts of mineral, vegetable, and animal oils.

The blending of these diesters with suitable chemicals in different amounts produces a lubricant which meets a prescribed specification of the petroleum industry and the aviation industry. Synthetic oils are not compatible with and cannot be mixed with mineral base oils. In addition most manufacturers recommend that different brands or types of synthetic oils not be mixed or mixed only within strict guidelines of same-type and certain compatible brands.

There are two different types of synthetic lubricants being used in turbine engines today: Type-1 (MIL-L-7808) and Type-2 (MIL-L-23699). Type-2 is the most recent synthetic to be developed and is used in most of the more modern engines. It is designed to meet current engine requirements and does not necessarily have the same chemical composition as Type-1. Engines originally designed to use Type-1 oil are still using this oil. Continuous modifications to the chemical structure of Type-1 have kept it essentially the same quality as Type-2. Type-2 was developed to withstand higher operating temperatures and to have improved anti-coking characteristics. It does not, however, have the low temperature range of Type-1.

Type-2 has a low temperature range of −40° F, whereas Type-1 is −65° F. Changing oils from Type-1 to Type-2 is not generally recommended as the latter may have higher detergent quality which would be detrimental to older engines.

Two common reminders one sees in oil company materials concerning synthetic lubricants are as follows:

WARNING: Synthetic turbine lubricants contain additives which are readily absorbed through the skin and are considered highly toxic. Excessive and/or prolonged exposure to the skin should be avoided.

CAUTION: Silicone based grease, such as is sometimes used to hold O-rings in place during assembly, can cause silicone contamination to the lube system. This contamination can cause engine oil to foam and result in oil loss through oil tank vents and also lead to engine damage from oil pump cavitation and insufficient lubrication.

5. Lubrication system servicing

Before servicing an engine's oil system, the technician should refer to the engine or aircraft type certificate data sheets or operations manual for the correct oil.

A partial list of typical synthetic lubricants the maintenance technician is likely to see is as follows:

Type-1 (MIL-L-7808)	Type-2 (MIL-L-23699)
Aeroshell 300	Aeroshell 500 or 700
Mobil Jet I	Mobil Jet II
Stauffer I	Stauffer II
Castrol 3c	Castrol 205
Enco 15	Enco 2380
Exxon 15	Exxon 25
Exxon 2389	Exxon 2380
Caltex 15	Caltex 2380
Shell 307	Texaco 7388, Starjet-5
	Caltex Starjet-5
	Chevron jet-5
	Sinclair type-2

From this list it can easily be seen that no standard identification system is currently in use. In fact, not all oil companies include the type number or Mil Spec on the oil can label. If needed, the technician would have to refer to oil company literature for these specifications.

Synthetic oil for turbine engines is usually supplied in one quart containers to minimize the chance of contaminants entering the lubrication system. Ground personnel should pay careful attention to cleanliness during servicing to maintain the integrity of the lubricant. In addition, use of a clean service station type oil spout is recommended instead of can openers which tend to deposit metal slivers in the oil.

If bulk oil is used, rather than quart containers, filtering with a 10 micron filter or smaller is generally required.

In the event of inadvertent mixing of incompatible lubricants, many manufacturers require the oil system to be drained and flushed before refilling. Also when changing to another approved oil, a system drain and flush would more likely be required if the oils are not compatible.

Draining is usually accomplished at the oil tank, the accessory gearbox sump, the main oil filter, and other low points in the lube system. Flushing generally means reservicing and draining after motoring the engine over with the starter and no ignition.

After final reservice, the engine will generally be run for a short period of time to resupply the lines, sumps, etc., with the residual oil normally held within the system.

If a new oil has been used, the placard stencil near the filler opening or metal oil identification tag, whichever is used, should be changed accordingly.

Another important consideration when servicing the oil system is to insure that servicing is accomplished within a short time after engine shutdown. Manufacturers normally require this in order to prevent overservicing. Overservicing may occur on some engines which have the tendency to allow oil in the storage tank to seep into lower portions of the engine after periods of inactivity.

When the oil level is checked later than the prescribed time after shutdown, a typical procedure is as follows:

1. If the oil level is low, but still visible on the dipstick, motor the engine over with the starter for 20-30 seconds; then recheck the oil level.

2. If the oil level is not visible on the dipstick, add oil until an indication appears on the dipstick; then motor the engine for 20-30 seconds and recheck the oil level.

An important consideration after oil servicing is recording the amount of oil serviced. A steady oil consumption within allowable limits provides a valuable trend analysis of engine wear at main bearing seal locations.

Oil consumption-oil change

Oil consumption of turbine engines is very low. Many business jet sized engines require only one quart of oil replenishment per 200 to 300 flight hours. A typical oil change interval is 300 to 400 operating hours or 6 months on a calendar interval.

On larger engines—the engines of a wide-bodied jet, for instance—one could expect to service no more than 0.2 quart per hour. By comparison, an 18-cylinder radial engine could consume as much as 20 quarts per operating hour and be considered airworthy.

Many airlines do not establish oil change intervals. The reason being that in the average 20 to 30 quart capacity oil tank, normal replenishment automatically changes the oil at regular 50 to 100 hour intervals.

6. *Wet sump lubrication systems*

The wet sump system is the oldest design but rarely seen in modern flight engines. Components of a wet sump system are similar to a dry sump system, except for the location of the oil supply. The dry sump carries its oil in a separate tank, whereas the wet sump oil is contained integrally in an engine sump.

Fig. 2A-1 shows an engine with a wet sump lubrication system and the oil contained in its accessory gearbox. The bearings and drive gears within are lubricated by a splash system. The remaining points of lubrication receive oil from a gear-type pressure pump, being directed from oil jets in a liquid stream.

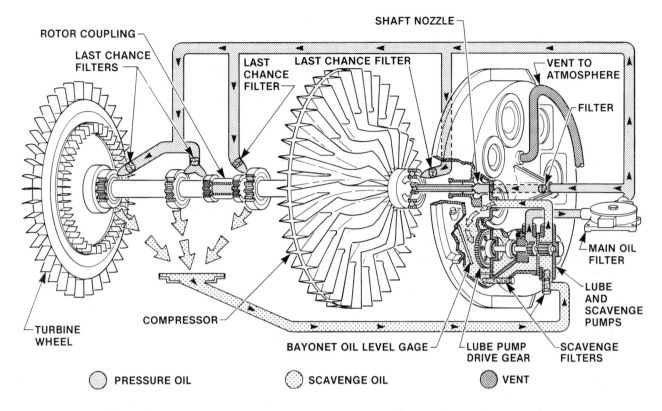

Fig. 2A-1   Wet sump lubrication system for an Allison J33 turbojet engine

Most wet sump engines do not incorporate a pressure relief valve and are known as variable pressure systems. With this system the pump output pressure depends directly on engine RPM.

Scavenged oil is returned to the sump by a combination of gravity flow from the bearings and suction created by a gear-type scavenge pump located within the pump housing.

The vent line is present to prevent over pressurization of the gearbox from air seeping past main bearing seals and finding its way to the gearbox via the scavenge system.

Wet sump systems of this type are seen on some types of auxiliary and ground power units, but rarely on flight engines.

7. Dry sump lubrication systems

Most turbine engines of the axial flow configuration use a dry sump lubrication system consisting of pressure, scavenge, and breather-vent subsystems.

The main oil supply is carried either in a tank mounted integrally within the engine or external-ly on the engine, or in the aircraft. A smaller supply is usually contained in a gearbox sump which also houses the oil pressure pump, oil scavenge pump, oil filter, and other lube system components. Another small amount of oil is residual within the oil system lines, sumps, and components.

a. System components

(1) Oil tank

The oil supply reservoir is usually constructed of sheet aluminum or stainless steel and is designed to furnish a constant supply of oil to the engine during all authorized flight attitudes. In most tanks, a pressure build-up is desired to assure a positive flow of oil to the oil pump inlet and suppress foaming in the tank which in turn prevents pump cavitation. This build-up is accomplished by running the tank overboard vent line through an adjustable relief valve to maintain a positive pressure of approximately 3 to 6 psig. Some dry sump oil tanks are of the integral type. Whereas the sheet metal type is a separate assembly located outside the engine, the integral oil tank is formed by space provided within the engine. Sometimes it is the propeller reduction

479

gearbox that houses the oil, and sometimes it is a cavity between major engine cases.

The distinction between the wet sump and dry sump is that the wet sump is located in the main gearbox at the lowest point within the engine, facilitating splash lubrication. The dry sump is seldom located at the low point on the engine and usually gravity flows oil to the main oil pump inlet.

Fig. 2A-2 shows an illustration of a dwell chamber, sometimes referred to as a deaerator, which provides a means of separating entrained air from the scavenge oil. The tank shown is of typical capacity for a business jet, approximately 4 quarts, 3 of which are usable. The location of the outlet in this example tank keeps 1 quart as residual oil and provides a low point for sediment to collect until drained. Other tanks would have a similar arrangement or perhaps a standpipe.

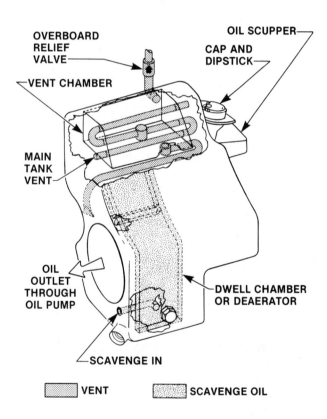

Fig. 2A-2 External dry sump oil tank

Today many oil tanks are configured with a pressure remote fill capability. An oil pumping cart can be attached to the tank and the oil hand pumped into the tank until full. The oil filler cap is usually removed during this operation to pre-

vent over-servicing. The hand filling method is, however, still the most common. The scupper shown on the illustraiton is present to catch oil that is spilled during servicing and to route this spillage through a tube drain location at the bottom of the engine.

In place of a dipstick some oil tanks incorporate a sight gage to satisfy the FAR requirement for a visual means of checking oil level. However, these glass indicators tend to cloud over after prolonged use and many operators have gone back to the dipstick.

(2) Oil pumps

The function of the oil pump is to supply oil under pressure to the parts of the engine that require lubrication. Many oil pumps consist not only of a pressure element, but one or more scavenge elements as well—all in one housing.

The three most common types of oil pumps are: the vane, gerotor, and gear types. All are classed as positive displacement pumps, as they deposit a fixed quantity of oil in the outlet, each revolution.

These pumps are also referred to as constant-displacement types because they displace a constant volume each revolution.

(a) Vane pump

The pump in Fig. 2A-3 could be a single element or one element of a multiple pump. Multiple pumps of this type generally contain one pressure element and one or more scavenge elements, all of which are mounted on a common shaft. The drive shaft mounts to an accessory gearbox drive pad and all pumping elements rotate together. Pumping action takes place as rotor drive shaft and eccentric rotor, which are actually one rotating piece, drive the sliding vanes around. The space between each vane pair floods with oil as it passes the oil inlet opening and carries this oil to the oil outlet. As the spaces diminishes to a zero clearance, the oil is forced to leave the pump. The downstream resistance to flow will determine the pump output pressure.

Vane pumps are considered to be more tolerant of debris in the scavenge oil. They are lighter in weight than the gerotor or gear pumps.

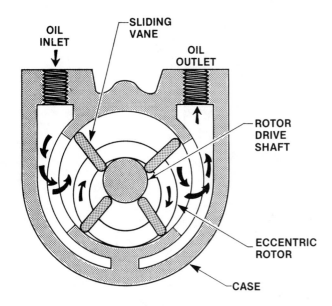

Fig. 2A-3   Sliding-vane-type oil pump

(b)   Gerotor pump

The gerotor pump, sometimes referred to as gear-rotor, utilizes a principle similar to the vane pump. The gerotor uses a lobe-shaped gear in an elliptically-shaped cavity to displace oil from the inlet to the outlet port.

Refer to Fig. 1A-8 and its accompanying text for the operation of the gerotor pump.

(c)   Gear pump

The gear type pump takes in inlet oil and rotates in a direction which allows oil to move between the gear teeth and the pump inner case until the oil is deposited in the outlet. The idler gear seals the inlet from the outlet preventing fluid backup and also doubles the capacity per revolution. This pump also incorporates a system relief valve in its housing which returns unwanted oil to the pump inlet. Fig. 2A-5 shows a dual pump with a pressure and a scavenge element.

(3)   Oil filters

Oil filters are an important part of the lubrication system since they remove foreign particles that collect in the oil.

The contaminants which are seen in filter bowls or on filter screens are always a matter of concern to the maintenance technician. Usually the determination as to whether the engine re-quires maintenance or whether it is airworthy is a matter of professional judgment after having seen cases of normal and abnormal levels of contamination. If a spectrometric oil analysis is available, a read-out of the various metal particles suspended in the oil can be used in the decision making process of whether or not there is sufficient cause to perform maintenance on the system.

Another common observation of engine oil is to see it turn dark brown or even blackish but with little or no contaminants. This is a chemical reaction to excessive heat causing oil decomposition. The cause of the overheat could be anything from low oil service to serious engine overheating and must be thoroughly analyzed.

The more common types of main system filters are: (1) disposable laminated paper, or fiber, (2) cleanable screen, and (3) screen and spacer types.

All three have micronic ratings in that they are designed to prevent passage of micronic-size contaminant particles into the system.

Fig. 2A-6A is an illustration of an in-line bowl type filter which could be either disposable or cleanable. A typical rating for this filter is 20 microns. This means it will filter out particles bigger than 20 microns in diameter. (A micron is a metric linear measurement equal to one millionth of a meter or 1/25,000th of an inch.)

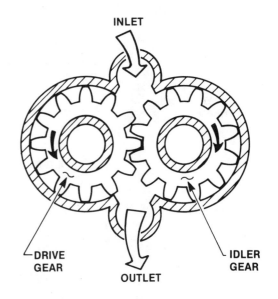

Fig. 2A-4   Typical gear-type oil pump

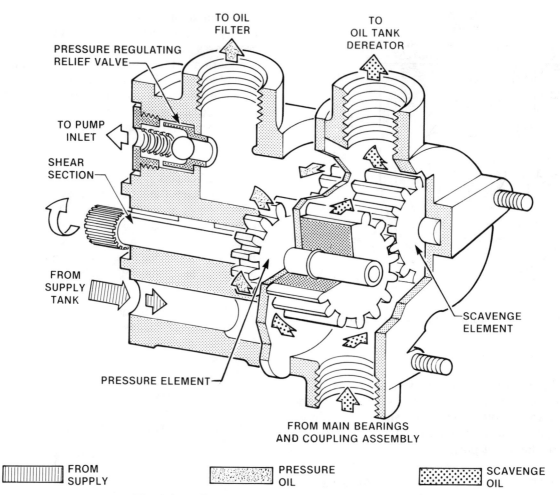

TO OIL
FILTER

TO
OIL TANK
DEREATOR

PRESSURE REGULATING
RELIEF VALVE

TO PUMP
INLET

SHEAR
SECTION

FROM
SUPPLY
TANK

PRESSURE ELEMENT

SCAVENGE
ELEMENT

FROM MAIN BEARINGS
AND COUPLING ASSEMBLY

FROM
SUPPLY

PRESSURE
OIL

SCAVENGE
OIL

Fig. 2A-5   Cutaway view of a gear-type oil pump

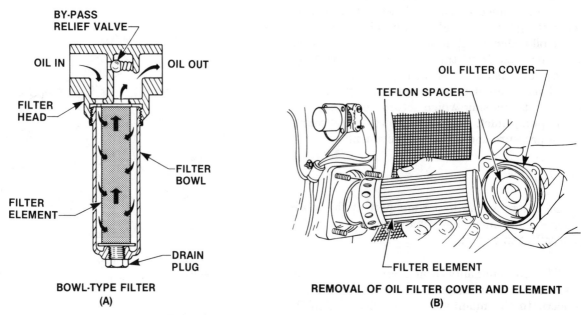

BY-PASS
RELIEF VALVE

OIL IN

OIL OUT

FILTER
HEAD

FILTER
BOWL

FILTER
ELEMENT

DRAIN
PLUG

BOWL-TYPE FILTER
(A)

OIL FILTER COVER

TEFLON SPACER

FILTER ELEMENT

REMOVAL OF OIL FILTER COVER AND ELEMENT
(B)

Fig. 2A-6   Turbine engine oil filters

Observe that oil fills the bowl, then forces its way through the filtering element to the core; exiting at the port near the spring side of the bypass relief valve. On a cold morning when oil is highly viscous, or if filter clogging restricts oil flow through the element, the bypass will open, allowing unfiltered oil to flow out to the engine.

An observation of the paper disposable and screen mesh filters would reveal that most are heavily pleated. This is to provide a maximum surface area for filtration. The screen type has an actual micronic size measurable in microns. The paper, or fiber, type has an equivalent micronic rating.

Sediment becomes trapped in the filter bowl and in the pores of the filtering element. This provides the technician with visible evidence of oil-wetted engine part serviceability. Generally it takes an experienced technician to properly analyze contaminant levels in turbine engine filter elements and sediment bowls.

Fig. 2A-6B is a filter which fits into a gearbox annulus and provides the exact same service as the bowl type.

The screen and spacer type filter, more common to Pratt & Whitney engines, is one that can be disassembled for inspection and cleaning. This filter usually fits into an annulus provided in the main accessory gearbox. The filter configuration is a series of wafer thin screens between spacers which allows oil to flow in the inward direction.

The filters illustrated are generally thought of as being located in the pressure subsystem of the engine. Some engines also provide filtration for the scavenge subsystems which route oil from the engine back to the supply tank.

Filter cleaning

Traditional methods of hand cleaning filters in solvent are still commonly used and acceptable. However, several cleaning devices that induce high frequency vibrations into the fluid, such as the one shown in Fig. 2A-8, are also available and will do a more complete job of removing all the contaminants from the filtering elements.

A written record of findings during filter inspections is a widely acceptable procedure. This provides a trend analysis of contamination build-up during subsequent filter inspections.

It is normal to find particles of metal on filter surfaces. If it is found that this contamination level is in excess of manufacturer's limits, or if any large metal chips are found, the source of the contamination must be located and the problem corrected. Then the engine should be drained, re-

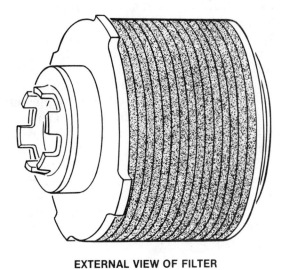

EXTERNAL VIEW OF FILTER
(A)

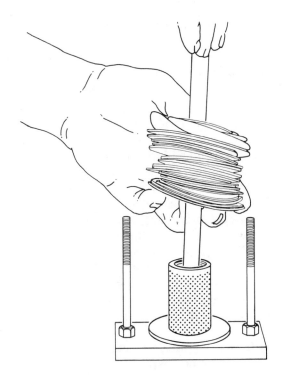

REMOVING THE OIL FILTER SCREENS
(B)

Fig. 2A-7 Screen and spacer-type oil filter

483

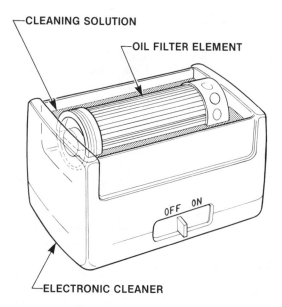

CLEANING SOLUTION

OIL FILTER ELEMENT

OFF ON

ELECTRONIC CLEANER

Fig. 2A-8   Field cleaning of oil filter screens

serviced and run for a short period of time. Sometimes several cycles of this procedure are needed to clear the system to the point where the filter remains clean.

8.   Typical lubrication system

a.   Oil pressure subsystem (General Electric CJ610)

Some statements in this discussion are specific to the CJ610 and some are general in nature.

In the schematic diagram in Fig. 2A-9, oil flows from a 0.75 gallon capacity tank to the pressure subsystem pumping element. The pump moves oil at the rate of 2.5 gallons per minute to the relief valve. This valve is adjustable and in this case, it is set to relieve at 115 to 125 psig. It relieves the system by routing oil back to the oil tank whenever fluid pressure reaches a pre-set value usually referred to as the cracking pressure. In this system the relief valve is referred to as a cold-start relief valve, because it only cracks when back-pressure caused by low ambient temperature and high oil viscosity forces it open. This pressure subsystem operates with varying oil pressure from idle power, 5 psig, to takeoff power, 60 psig, at normal oil operating temperature.

On other engines, the relief valve would be referred to as a regulating relief valve and the oil pressure would be held in a much narrower range

of say 45± 5 psig. The relief valve would be cracked open slightly at idle speed and much more open at takeoff power. Fig. 2A-12 has this type of relief valve.

After passing the relief valve, oil flows to the fuel-oil cooler. If congealed oil or other restrictions to flow occur, the pressure downstream of the by-pass valve which backs up the tensioning spring, will diminish. When a differential pressure (Δp) of 26 to 34 psid exists, the bypass will open to protect the system. The actual point at which the by-pass valve opens depends on the temperature of the oil and the tension of the spring. The spring causes a pressure of 34 psi when the oil is cold and 26 psi when the oil is at normal operating temperature.

In Fig. 2A-9 observe that an anti-static leak check valve is installed at the filter inlet to prevent oil seepage from the tank to the sump during periods of engine inactivity.

After the oil is cooled it is filtered. A similar cold starting and filter clogging by-pass valve is present, set to open at a pressure differential of 33-37 psid.

To see how this bypass operates, imagine the engine operating at normal oil temperature with a normal 5 psid across a clean filter and an oil pressure at this time, upstream of the filter, 60 psig and downstream of the filter, 55 psig. Holding the by-pass valve closed is 55 psig downstream oil pressure plus 33 psi spring pressure. This value is obviously much higher than the pressure on the open, upstream, side (60 psig). If the filter starts to clog, the pressure upstream begins to rise and the pressure downstream starts to drop. When the differential reaches 33 psi or slightly greater, the bypass will open to maintain sufficient lubrication. In some systems a differential pressure switch is placed across the filter inlet to outlet to show a warning light in the cockpit if bypassing is about to occur.

Once past the relief valve, oil under pump pressure flows downstream to the oil jets. The entire oil supply circulates approximately three times per minute.

(1)   Oil cooler

The oil cooler shown is a liquid-to-liquid heat exchanger. It contains numerous soda-straw type

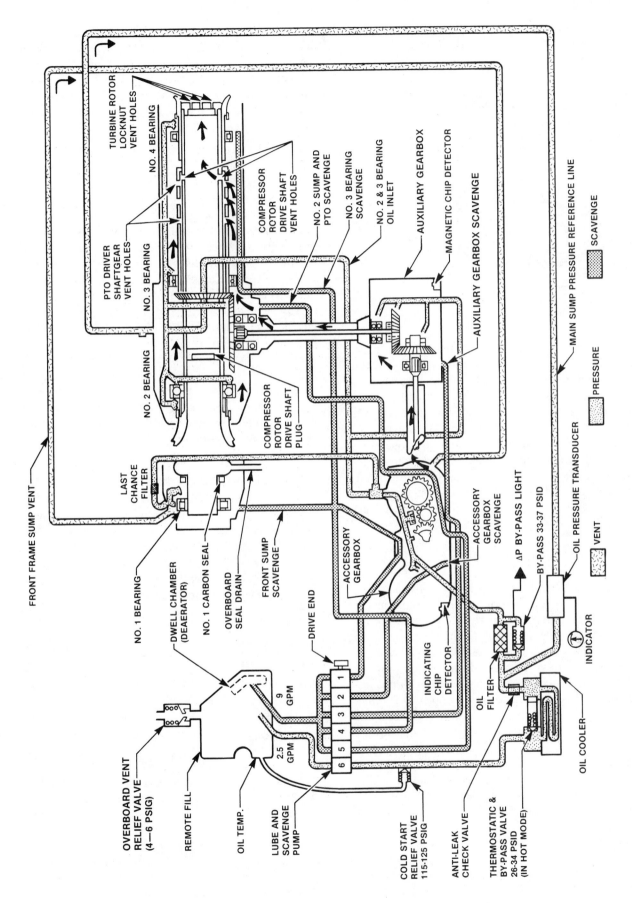

FRONT FRAME SUMP VENT

OVERBOARD VENT RELIEF VALVE (4—6 PSIG)

REMOTE FILL

OIL TEMP.

LUBE AND SCAVENGE PUMP

COLD START RELIEF VALVE 115-125 PSIG

ANTI-LEAK CHECK VALVE

THERMOSTATIC & BY-PASS VALVE 26-34 PSID (IN HOT MODE)

OIL COOLER

INDICATING CHIP DETECTOR

OIL FILTER

INDICATOR

OIL PRESSURE TRANSDUCER

ΔP BY-PASS LIGHT

BY-PASS 33-37 PSID

ACCESSORY GEARBOX SCAVENGE

ACCESSORY GEARBOX

DRIVE END

FRONT SUMP SCAVENGE

OVERBOARD SEAL DRAIN

NO. 1 CARBON SEAL

DWELL CHAMBER (DEAERATOR)

NO. 1 BEARING

LAST CHANCE FILTER

9 GPM

2.5 GPM

1 2 3 4 5 6

NO. 2 BEARING

COMPRESSOR ROTOR DRIVE SHAFT PLUG

PTO DRIVER SHAFTGEAR VENT HOLES

NO. 3 BEARING

TURBINE ROTOR LOCKNUT VENT HOLES

NO. 4 BEARING

COMPRESSOR ROTOR DRIVE SHAFT VENT HOLES

NO. 2 SUMP AND PTO SCAVENGE

NO. 3 BEARING SCAVENGE

NO. 2 & 3 BEARING OIL INLET

AUXILIARY GEARBOX

MAGNETIC CHIP DETECTOR

AUXILIARY GEARBOX SCAVENGE

MAIN SUMP PRESSURE REFERENCE LINE

PRESSURE

SCAVENGE

VENT

Fig. 2A-9 Hot-tank lubrication system for a General Electric CJ610 turbojet engine

25

485

passageways for fuel on its way to the combustor to flow through while the oil circulates around the straws. This allows an exchange of heat to occur between the fuel and the oil.

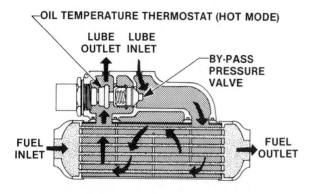

OIL TEMPERATURE THERMOSTAT (HOT MODE)

LUBE LUBE
OUTLET INLET

BY-PASS
PRESSURE
VALVE

FUEL
INLET

FUEL
OUTLET

Fig. 2A-10 Fuel-oil heat exchanger

The usual arrangement is to install a combination pressure and thermostatic by-pass valve at the cooler. When the oil is cold, the thermo valve opens, allowing oil to take the path of least resistance, bypassing the cooling chamber and flowing directly to the system. When this oil heats up, the thermo valve expands to close, forcing the oil to flow through the cooler. If a restriction occurs from cooler clogging, pressure build-up offseats the by-pass valve, and oil flows, uncooled and at a slightly reduced pressure, to the system.

A typical oil cooler operation might be as follows: oil cooler thermostatic valve starts to close at 165° F and is fully closed at 185° F with normal engine oil temperature stabilizing at 210° F. This indicates that the cooler capacity for fuel flow and oil flow regulate the operational oil temperature rather than the thermo valve. The thermostatic valve is present to quickly bypass oil to the lube points on a cold start. It is not uncommon, however, to see an oil cooler without a thermostatic device if the lubrication can be distributed rapidly enough in a particular engine.

One of the checks on an oil cooler is to see a correct oil temperature rise on engine deceleration and drop on acceleration as a function of fuel flow. From this, the technician can conclude that the thermo valve is not struck in a transient position and is not bypassing.

The fuel-oil cooler is used almost exclusively on larger engines and on many smaller engines,

but another type, the air-oil cooler, is in popular use in general aviation aircraft. The air-oil cooler and thermostatic valve arrangement is quite similar in principle and design to the fuel-oil cooler shown, and the heat exchanger section looks similar to the small radiator type used on reciprocating engines.

(2) Oil jets

Oil jets, or nozzles, as they are sometimes called, are located at the various places within the engine that are to be lubricated and are the terminating point of the pressure subsystem. They deliver either an atomized spray or a fluid stream of oil to bearings, gears and other parts. The fluid stream method is the most common. In most cases this stream of oil is directed onto the bearing surfaces from what is termed a direct lubrication oil jet. Another less common method is called a mist and vapor lubrication oil jet where the oil stream, sometimes an air-oil stream, is aimed at a splash pan and slinger ring device. This allows for a wider area of lubrication from a single spray jet and is utilized in some larger engines.

Oil jets can be checked for size and cleanliness with a new numbered drill shank. It is very important that the drill shank be free from any nicks or burrs. Another method of checking for restrictions to flow is the smoke check. Smoke or shop air is directed into the oil nozzle inlet port and a check is made of discharge rate through the orifice. A comparison is usually made to a known good or new oil jet.

A flow tester is also available in most larger repair facilities. This device can measure rate of flow accurately in gallons per minute with the oil jet installed in the engine.

(3) Last-chance filters

Quite often, last-chance filters are installed in oil lines to prevent plugging of the oil jets. Observe point A in Fig. 2A-11. This is a last-chance filter which is not readily cleaned because it is often accessible only during engine overhaul. To prevent engine damage from clogged screens or plugged oil jets, the main filters are inspected at frequent intervals by ground personnel.

486

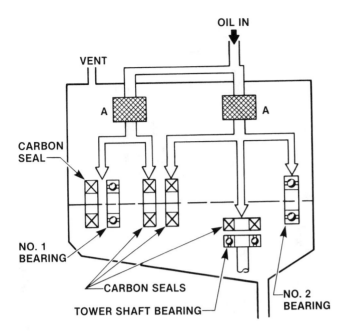

VENT

OIL IN

A   A

CARBON
SEAL

NO. 1
BEARING

CARBON SEALS

TOWER SHAFT BEARING

NO. 2
BEARING

Fig. 2A-11 Location of last-chance filter elements and oil jets

b. Scavenge subsystem

The scavenge subsystem Fig. 2A-9 removes oil from the bearing compartments and gearboxes by suction from the five scavenge return elements in the oil pump. All five scavenge elements route oil to one return line which enters the dwell chamber in the oil tank. The dwell chamber acts as an air-oil separator. The total return oil capacity is 9 gallons per minute. The entrained air that accumulates in the oil increases its volume and necessitates the use of a much higher capacity scavenge subsystem.

Many scavenge systems contain magnetic chip detectors at low points to attract ferrous metals—metals which would otherwise circulate back to the oil tank and engine, possibly causing wear or damage. These chip detectors are a point of frequent inspection for the technician.

As a general rule, the presence of gray metallic paste is considered satisfactory and the results of normal wear, but metallic chips or flakes are an indication of serious internal wear or malfunction. Fig. 2A-12 (insert) shows an indicating-type magnetic chip detector, with a warning circuit feature. When debris bridges the gap between the magnetic positive electrode in the center and the ground electrode (shell), a warning light shows in the cockpit. The flight crew will take whatever action is warranted, such as inflight shutdown, continued operation at flight idle, or continued operation at normal cruise, depending on the other engine instruments' readings.

c. Vent subsystem

The presence of pressurized air in the bearing cavities results from sealing air leaking across carbon and labyrinth type oil seals. This air pressure facilitates oil return by putting a head of pressure on the scavenge oil at the bearing sumps, but at the same time, some of this air is vented overboard so that an undesirable buildup does not occur. On some engines a separate subsystem is installed to vent this seal leakage air overboard, while other remove this seal leakage air with the scavenge oil.

A common problem associated with the vent system is coking. Vent air is oil-laden and over a period of time the heat to which this mixture is subjected causes some of the oil particles to decompose and turn to a solid form referred to as coke. The buildup of coke can slow or even block airflow through some of the smaller passageways to the extent that excessive pressures occur in portions of the vent subsystem. Problems such as low oil flow to bearings and high oil temperatures can result from this restriction to normal venting. The usual procedure when troubleshooting for this malfunction is to isolate the vent system at various points on the engine, measure the pressure, and compare this pressure against the standards in the maintenance manual.

In the CJ610 turbine engine, vent air is deposited in the gearbox along with scavenge oil and then routed to the mid-frame. From there it is channeled internally down the compressor shaft and out at the turbine into the gas path. A system of seals controls the rate of leakage and prevents turbine station air from backing up into the vent system.

Another function of pressurized air in the bearing cavities is to ensure a proper oil spray from the oil jet. If the amount of back pressure can be regulated, the quantity of oil flow from the oil jet is also regulated.

In some larger engines, greater vent system airflow results from higher compression and higher gas path pressure. These engines utilize a vent pressurizing valve and a centrifuge type rotary air-oil separator to assist the vent system.

The rotary separator (see Fig. 2A-12), is an impeller or centrifuge-like device located in the

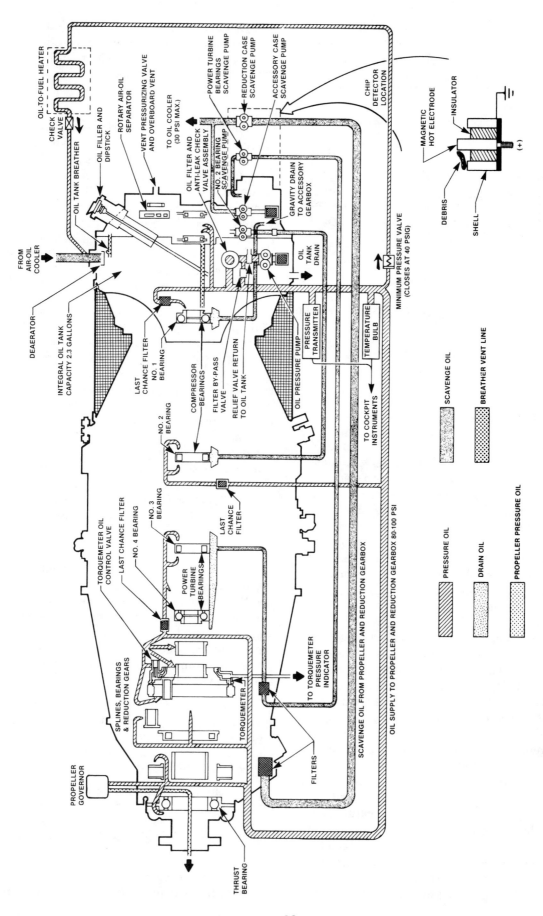

Fig. 2A-12 Cold-tank lubrication system for a Pratt & Whitney PT6 turboprop engine

gearbox near the vent outlet. As the oil-laden vent air enters the rotating slinger chamber, centrifugal action throws the oil outward to drain into the sump, while clean vent air is routed to the vent pressurizing valve and subsequently overboard.

The vent pressurizing valve shown in Fig. 2A-13 consists of an aneroid-operated spring and bellows and a spring-loaded blowoff valve located in the overboard line of the vent subsystem. At sea level, the bellows valve is open, but it closes with increasing altitude in order to maintain vent pressure sufficient to assure oil nozzle flows similar to those at sea level.

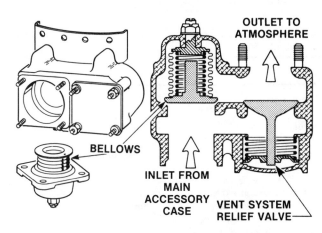

Fig. 2A-13 Vent pressurizing valve

9. PT6 lubrication system

The PT6 gives the appearance of having a wet sump lubrication system but it has in actuality, an integral dry sump oil tank. Tank capacity is 2 gallons, with 1.5 gallons usable and .5 gallon expansion space.

Pressure subsystem sequence of flow is:

1. Oil tank — pressurized 3 to 6 psig

2. Oil pump — a gear type pump, capacity 4 gpm

3. Regulating relief valve — sets oil pressure 80-100 psig

4. Main oil filter — bypass setting 25-30 psig.

5. No. 1 bearing and accessory gearbox

6. Bearings 2, 3, 4 and propeller gearbox — oil pressure and temperature are taken in this line

7. Fuel heater — the check valve closes at 40 psig. If engine is shut down in flight, flow terminates even though engine may be windmilling

Scavenge subsystem sequence of flow is:

1. No. 1 bearing sump — drains directly to accessory gearbox

2. No. 2 bearing sump — oil pumped to accessory gearbox by No. 2 scavenge pump

3. No. 3 and 4 bearing sump — oil pumped to accessory gearbox by free power turbine scavenge pump

4. Propeller shaft area — oil pumped to supply tank by reduction gearbox scavenge pump through air-oil cooler

5. Accessory gearbox — oil pumped to air-oil cooler by accessory gearbox scavenge pump

Vent subsystem flow is:

1. Propeller gearbox, bearing sumps 2, 3, 4 — vent via internal passageways into No. 1 bearing sump

2. No. 1 bearing sump — vents to accessory gearbox case

3. Oil tank — vents to accessory gearbox case

4. Accessory case — vents to atmosphere through rotary air-oil separator

10. Hot tank or cold tank systems

Whether an engine has a hot or cold oil tank system, as shown in the two typical lubrication systems, is either a matter of necessity or merely convenience of location for the manufacturer.

In the hot tank system the oil cooler is located in the pressure subsystem.

The oil has less entrained air and maximum heat exchange occurs. This allows a smaller cooler to be used as well.

In the cold tank system the oil cooler is located in the scavenge subsystem which allows the oil to return to the supply tank in a cooled condition. When oil enters the cooler, it is still aerated from the action of the rotating parts within the engine and a reduced heat exchange is said to occur, requiring a larger cooler.

The PT6 engine, Fig. 2A-12, has no convenient means of placing the oil cooler in the pressure subsystem because some of the oil from the oil pump is routed through internal passageways within the engine to lubrication points. This necessitates the use of the cold tank system.

11. Troubleshooting procedure

The following information on oil system problems, the probable causes, checking procedures, and remedies are general in nature and intended only to acquaint the reader with the concepts of troubleshooting. This information in no way takes precedence over the manufacturer's recommended procedures.

The philosophy the technician should adopt in troubleshooting is to start at the obvious or most likely cause, and work toward remote possibilities. To protect the interest of the company or the customer, the technician should check all possibilities and replace components only when reasonably sure the malfunction has been located.

TROUBLESHOOTING OIL SYSTEMS		
1. No Engine Oil Pressure (no oil leaks)		
Possible Cause:	Check For:	Remedy:
a. Low oil level	Check oil level	Add oil
b. Circuit breaker tripped	Check for location if installed	Reset or check circuit wiring
c. Defective indicator	1. Check power input and/or exchange indicator from another engine	Repair circuit or replace indicator
	2. Slave in another indicator or bench check	
d. Defective transmitter	1. Check power input	Repair circuit or replace transmitter
	2. Slave in another transmitter or bench check	
e. Obstruction in oil tank	Remove line at pump inlet and check flow rate	Remove obstruction or replace tank
f. Defective oil pump	1. Motor engine with outlet line removed and check flow rate	Replace pump
	2. Check for leaks between elements or sheared shaft	
2. Low Oil Pressure (no oil leaks)		
Possible Cause	Check For	Remedy
a. Same as 1a, 1b, 1c, 1d, 1e, 1f	Check as necessary	
b. Improper regulating relief valve setting	Check security of valve and install test gages	Reset or replace as necessary
c. High vent pressure	Check for high pressure by installing test gages	Possible engine teardown

3. High Oil Pressure

Possible Cause	Check For	Remedy
a. Same as item 1c, 1d, 2b	Check as necessary	
b. Oil by-pass line obstructed	Check line, relief valve to tank	Repair or replace

4. Fluctuating Oil Pressure

Possible Cause	Check For	Remedy
a. Loose electrical connection	Check circuit	Tighten as necessary
b. Defective relief valve	Check for sticking components	Clean or replace
c. Defective transmitter	Bench check or slave in new transmitter	Repair or replace

5. Excessive Oil Consumption

Possible Cause	Check For	Remedy
a. External oil leaks	Visually check entire engine	Tighten lines, replace gaskets
b. Gas path oil leaks	Check inlet and exhaust, refer to manual for limits	Possible teardown
c. Overboard vent discharging oil	Check for high vent subsystem pressure from possible damaged carbon or labyrinth oil seal	Possible teardown
d. Damaged main bearing oil seal	Check overboard vent for oil discharge and check vent pressure	Vent pressure over limits usually requires engine teardown
e. Overboard accessory seal drain discharging excessive oil	Check drainage quantity against allowable limits	Replace leaking gearbox seal
f. Vent pressurizing valve sticking open at altitude	Check for evidence of oil on cowling at vent opening	Bench check breather valve

6. Increasing Oil Quantity

Possible Cause	Check For	Remedy
Oil cooler core leak (fuel intrusion)	Requires bench check	Replace as necessary

7. Excessive Oil in Gas Path

Possible Cause	Check For	Remedy
a. Overservicing	Check servicing procedure, service only during prescribed period after engine shutdown	Remove excess oil and run engine to dry out
b. Inoperative scavenge pump	Check output with direct pressure gage	Replace if accessible

8. Oil in Tail Pipe Overnight

Possible Cause	Check For	Remedy
Pressure subsystem anti-static-leak check valve	Check valve for contamination or worn seals	Clean or replace seals, run engine check for reappearance of oil leak

9. Oil Tank Rupture		
Possible Cause	Check For	Remedy
Oil tank check valve	Check for valve sticking closed	Clean or replace

10. Oil Pressure Indication Follows Power Lever Movement		
Possible Cause	Check For	Remedy
Regulating relief valve	Check for sticking	Clean or replace

11. Oil Temperature High		
Possible Cause	Check For	Remedy
a. Vent subsystem coking	Check vent pressure	Clean or possible engine teardown
b. Oil cooler thermostat stuck open	Sticking, by bench-checking unit	Replace thermostat
c. Main bearing overheating	Clogged oil jet	Possible engine teardown

12. Oil Filter Screen Collapsed—Yet Clean		
Possible Cause	Check For	Remedy
Filter by-pass valve	Check for sticking (cold weather starts)	Clean or change

Section 8-II
Propellers

I. INTRODUCTION

Throughout the development of controlled flight as we know it, every aircraft required some kind of device to convert engine power to a usable form of thrust. With few exceptions, nearly all of the early practical aircraft designs used propellers to create necessary thrust, and during the latter part of the 19th century, many unusual and innovative designs for propellers made their debut on the early flying machines. These ranged from simple fabric-covered wood paddles to elaborate multi-bladed wire-braced designs, and some of these designs were used successfully as a means of propelling the early dirigibles and heavier-than-air designs.

As the science of aeronautics progressed, propeller designs improved from flat boards which merely pushed air backward, to actual airfoil shapes that produced lift, as do wings, to pull the aircraft forward by aerodynamic action. By the time the Wright brothers began their first powered flights, propeller designs had evolved into the standard two-bladed style similar in appearance to those used on today's modern light aircraft.

World War I brought about an increase in aircraft size, speed, and engine horsepower and further improvements in propeller designs were required. The most widely used improvement during the war was a four-bladed wood propeller. Other design improvements which were developed during the war included an aluminum fixed-pitch propeller and the two-position controllable propeller. But these improvements did not come into wide usage until the 1920's.

As aircraft designs improved, propellers were developed which used thinner airfoil sections and had greater strength, and because of its structural strength, these improvements brought the aluminum alloy propeller into wide usage.

The advantage of being able to change the propeller blade angle in flight led to wide acceptance of the two-position propeller and, later, the constant-speed propeller system.

Refinements of propeller designs and systems from the 1930's through World War II included the feathering propeller for multi-engine applications, reversing propellers, which allowed shorter landing runs and improved ground maneuverability, and many special auxiliary systems such as ice elimination, simultaneous control systems, and automatic feathering systems. Moreover, with the development of the turbine engines, propeller systems were adapted for use with these engines to allow their efficient use at low altitudes and low airspeeds.

Today, propeller designs continue to be improved by the use of new airfoil shapes, composite materials, and multi-blade configurations. Some recent improvements include the use of laminar flow airfoils, composite materials, and gull wing propeller desings.

Fig. 1B-1 Three-blade propeller used on a modern airplane

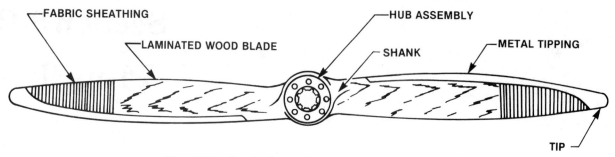

Fig. 1B-2 *Typical wood aircraft propeller*

A. Nomenclature

Before starting any discussion about propellers, it is necessary to define some basic terms to avoid confusion and misunderstanding.

A propeller is a rotating airfoil that consists of two or more blades attached to a central hub which is mounted on an engine crankshaft. The function of the propeller is to convert engine horsepower to useful thrust. The blades have a leading edge, trailing edge, a tip, a shank, a face, and a back as shown in Figs. 1B-2 and 1B-3.

A term that will be used throughout our study of propellers is blade angle. This is the angle between the propeller plane of rotation and the chord line of a propeller airfoil section.

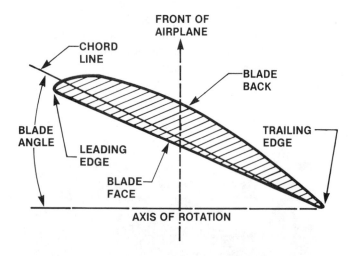

Fig. 1B-3 *Typical airfoil section of a propeller*

Blade station is a reference position on a blade that is a specified distance from the center of the hub.

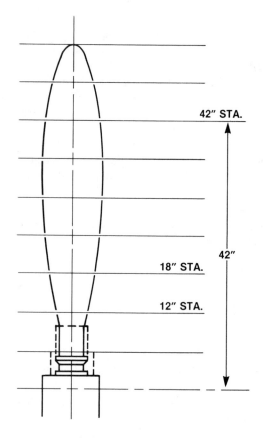

Fig. 1B-4 *Propeller blade stations*

Pitch is the distance, in inches, that a propeller section will move forward in one revolution.

Pitch distribution is the gradual twist in the propeller blade from shank to tip.

Fixed-pitch propellers are simple propellers whose blade angle cannot be changed in normal operation. They are usually made of wood or aluminum alloy and are usually found on light, single-engine airplanes.

494

34

Ground-adjustable propellers are similar to fixed-pitch propellers in that their blade angles cannot be changed in flight, but the propeller is made so that the blade angles can be changed on the ground. The pitch, or blade angle, can be adjusted to give the desired flight characteristics: that is, low blade angle if the airplane is used for operation from short airstrips or high blade angle if high-speed cruise flight is of most importance. This type of propeller was widely used on aircraft built between the 1920's and the 1940's.

A two-position controllable propeller allows the pilot to select one of two blade angles while in flight. This allows him to use a low blade angle for takeoff and then change to a high blade angle for cruise. This is something like a two-speed transmission in an automobile. Two-position controllable propellers were used on some of the more sophisticated airplanes in the 1920's and the 1930's.

Controllable-pitch propellers, those that allowed the pilot to set the blades at any angle in the propeller's range while in flight, gave him more control over the propeller than the two-position propeller, and this style of propeller was the forerunner of the modern constant-speed propeller.

Most medium- and high-performance airplanes produced today are equipped with constant-speed propeller systems. This propeller system uses a controllable propeller which the pilot controls indirectly by adjusting a constant-speed control unit, commonly called the governor. The propeller blade angle is adjusted by the governor to maintain the engine speed (RPM) which the pilot calls for. Because of this controllability, coupled with light weight and the relatively low cost of this constant-speed system, many of the earlier propeller designs have become less common.

Most multi-engine aircraft are equipped with constant-speed propeller systems which also have the capability of being feathered. When a propeller is feathered, its blades are rotated so they present their leading edge to the wind, eliminating the drag associated with a windmilling propeller.

Reversing propeller systems are refinements of the constant-speed feathering systems. The propeller blades can be rotated to a negative angle to produce reverse thrust, which forces air forward instead of backwards and permits a shorter landing roll and improved ground maneuvering.

B. Propeller Requirements

The Federal Aviation Administration has furnished guidelines regarding propeller system designs and the maintenance required for propeller systems. These are given to us in the Federal Aviation Regulations, the FAR's.

FAR 23, "Airworthiness Standards: Normal, Utility, and Acrobatic Aircraft," and FAR, Part 25, "Airworthiness Standards, Transport Category Aircraft," outline the requirements for propellers and their control systems for aircraft certification. Because there is very little difference in the wording of Parts 23 and 25, we can consider them to be identical for the purposes of this discussion.

Part 43 of the FAR's defines the different classes of maintenance for the propeller system and the minimum requirements for 100-hour and annual inspections.

The information that is required to be permanently affixed to a propeller is discussed briefly in FAR 45, "Identification and Registration Markings."

The licenses required to perform or supervise the maintenance or repair of a propeller and related systems are covered by FAR, Part 65 "Certification: Airmen Other Than Flight Crewmembers." This section distinguishes between the authority of a powerplant mechanic, a propeller repairman, and an authorized inspector.

1. Propeller requirement for aircraft certification

Aircraft propellers must be certificated under FAR, Part 35, and the following information must be clearly displayed on the hub or butt of the propeller blade: The manufacturer's name, model designation, serial number, and production certificate number. (FAR 45.13).

a. Static RPM

An aircraft which uses a fixed-pitch propeller will not operate at maximum RPM (tachometer red line) on the ground in a no-wind condition when the engine is producing maximum allowable horsepower. It is designed this way so that the propeller will limit the engine RPM to the maximum allowable when the engine is operating at full power and the aircraft is flying at its best rate-of-climb speed. This prevents damage to the engine from overspeeding.

The propeller must also prevent the engine from exceeding its rated RPM by more than 10% in a closed-throttle dive at the aircraft's never-exceed speed (FAR 23/25.33). As the airspeed or wind speed increases, the engine RPM will increase because it is easier for the propeller to rotate. All two-position and controllable-pitch propellers must comply with this requirement at their low blade angle setting.

A constant-speed propeller system must limit engine speed to the rated RPM at all times when the system is operating normally. If the governor should fail, the system must be designed to allow a static RPM of no more that 103% of rated RPM (FAR 23/25.33). This is accomplished by using the correct low blade angle setting; the greater the blade angle, the lower the static RPM.

b. *Cockpit controls and instruments*

Propeller control levers in the cockpit must be arranged to allow easy operation of all controls at the same time, but not to restrict the movement of individual controls (FAR 23/25.1149).

The propeller controls must be rigged so that an increase in RPM is obtained by moving the controls forward, and a decrease in RPM is caused by moving the controls aft. The throttles must be arranged so that forward thrust is increased by forward movement of the control, and reverse thrust is increased by aft movement of the throttle (FAR 23/25.779).

Cockpit powerplant controls must be arranged to prevent confusion as to which engine they control. Recent regulation changes require the control knobs to be distinguished by their shape and color (FAR 23/25.781). We see these shapes in Fig. 1B-5.

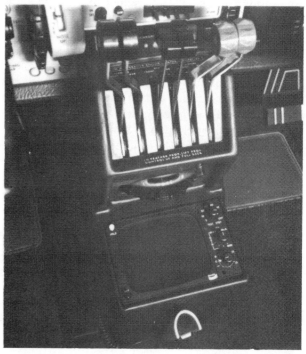

Fig. 1B-6 *Powerplant control console in a modern twin-engine airplane*

Cockpit instruments such as tachometers and manifold pressure gages must be marked with a green arc to indicate the normal operating range; a yellow arc for takeoff and precautionary range, a red arc for critical vibration range, and a red radial line for maximum operating limit (FAR 23/25.1549). (See Fig. 1B-7.)

c. *Minimum terrain and structural clearances*

The minimum ground clearance (distance from the level ground to the edge of the propeller disc) for a tail wheel airplane in the takeoff attitude is nine inches. For a tricycle-geared aircraft in its most nose-low attitude, with normal tire and strut inflation, the minimum ground clearance is seven inches. If the tire and strut are deflated, there need be only a positive ground clearance; that is, the propeller disc must not touch the ground.

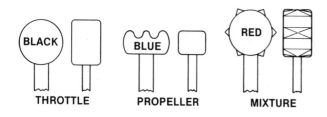

Fig. 1B-5 *Powerplant control knobs*

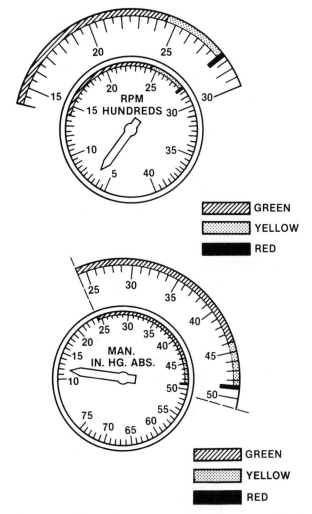

GREEN
YELLOW
RED

GREEN
YELLOW
RED

Fig. 1B-7   Typical range marking for powerplant instruments

For a seaplane, there must be at least 18 inches clearance between the water and the propeller disc.

All airplanes must be designed so that the edge of the propeller disc comes no closer than one inch to the airframe. This is known as radial clearance. (See Fig. 1B-8.)

FAR 23/25.925 gives full details of the propeller clearance requirements. (See Fig. 1B-8.)

d. Feathering system requirements

If a propeller can be feathered, there must be some means of unfeathering it in flight (FAR 23/25.1153).

If a propeller system uses oil to feather the propeller, a supply of oil must be reserved for feathering use only, and a provision must be made in this system to prevent sludge or foreign matter from affecting the feathering oil supply (FAR 23/25.1027). These requirements are normally met by using a standpipe in the engine oil tank with an outlet below the standpipe accessible only to the propeller feathering system.

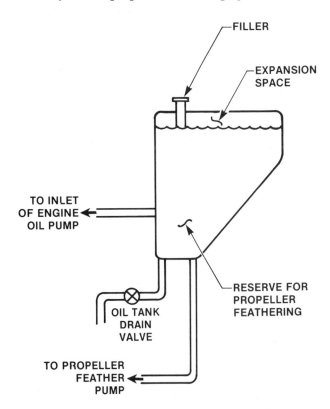

FILLER

EXPANSION SPACE

TO INLET OF ENGINE OIL PUMP

OIL TANK DRAIN VALVE

TO PROPELLER FEATHER PUMP

RESERVE FOR PROPELLER FEATHERING

Fig. 1B-9   Oil tank for use on an aircraft engine equipped with a Hydromatic feathering propeller. The propeller feathering reserve oil is below the tank outlet to the engine oil pump.

A separate feathering control is required for each propeller, and it must be so designed that it prevents accidental operation (FAR 23/25.1153). This may be done by using a separate feathering control such as a feathering button or by requiring an extreme movement of the propeller control.

2.   Propeller maintenance regulations

a.   Authorized maintenance personnel

The inspection, adjustment, installation, and minor repair of a propeller and its related parts and appliances on the engine are the responsibility of the powerplant mechanic. The powerplant

497

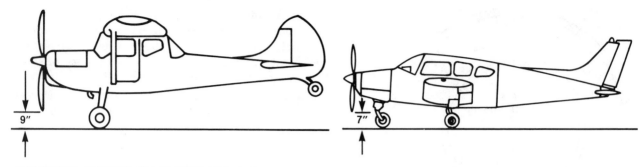

PROPELLER GROUND CLEARANCE FOR A TAILWHEEL
AIRPLANE
(A)

PROPELLER GROUND CLEARANCE FOR A TRICYCLE
LANDING GEAR AIRPLANE
(B)

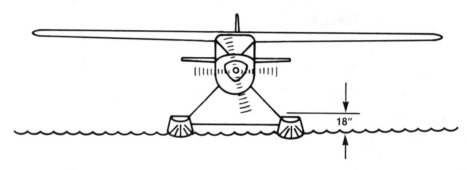

PROPELLER CLEARANCE FROM THE WATER FOR A FLOAT PLANE
(C)

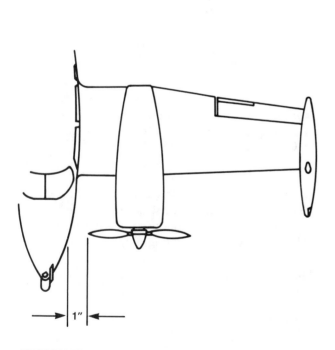

PROPELLER RADIAL CLEARANCE FROM AIRCRAFT
STRUCTURE
(D)

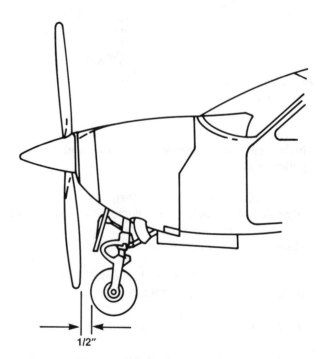

PROPELLER LONGITUDINAL CLEARANCE FROM AIR-
CRAFT STRUCTURE
(E)

Fig. 1B-8 Propeller clearance requirement

mechanic may also perform the 100-hour inspection of the propeller and its related components (FAR 65.87).

A propeller repairman may perform or supervise the major overhaul and repair of propellers and related parts and appliances for which he is certificated, but the repair and overhaul must be performed in connection with the operation of a certificated repair station or, within limits, by the holder of a commercial operator or air carrier certificate. (FAR 65.103).

An A&P mechanic who holds an Inspection Authorization may perform the annual inspection of a propeller, but he may not approve for return to service major repairs and alterations to propellers or their related parts and appliances. Only an appropriately rated facility, such as a propeller repair station, may return a propeller or accessory to service after a major repair or alteration (FAR 65.81 and .91).

b. Preventive maintenance

The following are types of preventive maintenance that may be associated with propellers and their systems: replacing defective safety wiring or cotter keys; lubrication not requiring disassembly other than removal of nonstructural items such as cover plates, cowlings, and fairings; applying preservative or protective material (paint, wax, etc.) to components when no disassembly is required and where the coating is not prohibited or contrary to good practice. (FAR 43, Appendix A(c)).

c. Major alterations and repairs

The following are major propeller alterations when not authorized in the FAA propeller specifications: a change to the blade or hub design; a change in the governor or control design; installation of a governor or feathering system; installation of a propeller deicing system; installation of parts not approved for the propeller.

Propeller major repairs are classified as any repair to or straightening of steel blades; repairing or machining of steel hubs; shortening of blades; retipping of wood propellers; replacement of outer laminations on fixed-pitch wood propellers; repairing elongated bolt holes in the hub of fixed-pitch wood propeller; inlay work on wood blades; repairs to composition blades; replacement of tip fabric; replacement of plastic covering; repair of propeller governors; overhaul of controllable-pitch propellers; repairs to deep dents, cuts, scars, nicks, etc., and straightening of aluminum blades; the repair or replacement of internal blade elements. (FAR 43 Appendix A(a)(3)and (b)(3)).

Major repairs and alterations to propellers and control devices are normally performed by the manufacturer or a certified repair station.

When a propeller or control device is overhauled by a repair facility, a maintenance release tag must be attached to the item to certify that the item is approved for return to service. This tag takes the place of a FAA Form 337 and should be included in the maintenance record (FAR 43 Appendix B (b)).

d. Annual and 100-hour inspections

When performing a 100-hour or annual inspection, Appendix D of FAR 43 specifies that the following areas relating to propellers and their controls must be inspected: engine controls for defects, improper travel, and improper safety; lines, hoses, and clamps for leaks, improper condition, and looseness; accessories for apparent defects in security of mounting; all systems for improper installation, poor general condition, defects, and insecure attachment; propeller assembly for cracks, nicks, binds, and oil leakage; bolts for improper torquing and lack of safetying; anti-icing devices for improper operation and obvious defects; control mechanisms for improper operation, insecure mounting, and restricted travel.

These inspection are the minimum required by regulation, and you should always refer to the manufacturer's manuals for specific inspection procedures.

II. PROPELLER THEORY

As a propeller rotates, it produces lift and causes an aircraft to move forward. The amount of lift produced depends on variables such as engine RPM, propeller airfoil shape, and aircraft speed.

A. Propeller Lift and Angle of Attack

Because a propeller blade is a rotating airfoil, it produces lift by aerodynamic action and pulls an aircraft forward. Before discussing ways of varying the amount of lift produced by a propeller blade, we must understand some of the propeller design characteristics.

Starting from the centerline of the hub of a propeller, each blade can be marked off in one-inch segments known as blade stations. If the blade angle is measured at each of these stations, the blade angle nearest the center of the propeller will be highest, with the blade angle decreasing toward the tip. This decrease in blade angle from the hub to the tip is called pitch distribution. A cross section of each blade station will show that low-speed airfoils are used near the hub and high-speed airfoils toward the tip. The pitch distribution and the change in airfoil shape along the

length of the blade are necessary, because each section moves through the air at a different velocity, with the slowest speeds near the hub and the highest speeds near the tip.

To illustrate the difference in the speed of airfoil sections at a fixed RPM, consider three airfoil sections on a propeller blade. If a propeller is rotating at 1800 RPM, the 18-inch station will travel 9.42 feet per revolution (192.7 miles per hour), while the 36-inch station will travel 18.84 feet per revolution or 385.4 miles per hour. And the 48-inch station will move 25.13 feet per revolution, or 514 miles per hour. The airfoil that gives the best lift at 193 miles per hour is inefficient at 514 MPH. Thus the airfoil is changed gradually throughout the length of the blade as we see in Fig. 2B-1.

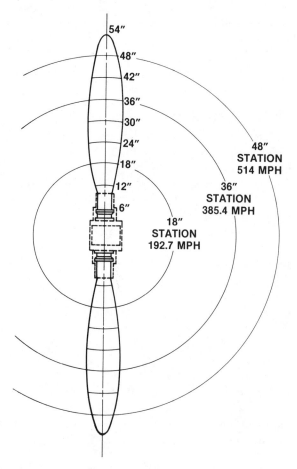

Fig. 2B-2 *Comparative velocities at three blade stations of a typical propeller*

Fig. 2B-1 Pitch distribution in a typical aircraft propeller

A look at one blade section will illustrate how the angle of attack on the blade of a fixed-pitch propeller changes with different flight conditions. The angle of attack is the angle between the air-

500

foil chord line and the relative wind, and the direction of the relative wind is the resultant of the combined velocities of rotational speed (RPM) and airspeed.

If the aircraft is stationary with no wind flowing past it, and the engine is turning at 1200 RPM, the propeller blade angle of 20 degrees at the 20-inch station will have an angle of attack of

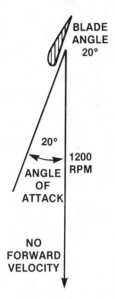

Fig. 2B-3 With no forward velocity, the angle of attack of a propeller blade is the same as the blade angle.

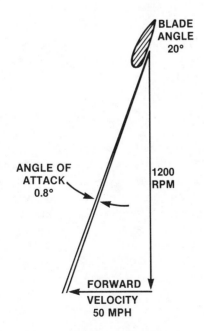

Fig. 2B-4 As the forward velocity of the aircraft increases, the angle of attack of the propeller blade decreases.

20 degrees. This is because the direction of the relative wind is opposite to the movement of the propeller (Fig. 2B-3).

When this airplane is moving forward at 50 miles per hour, the relative wind now causes an angle of attack of 0.8 degrees (Fig. 2B-4).

Now if we increase the propeller speed to 1500 RPM, the relative wind will cause the angle of attack to be 4.4 degrees.

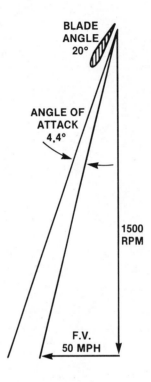

Fig. 2B-5 As the engine RPM increases for a given forward velocity, the angle of attack increases.

The most effective angle of attack is between two and four degrees, and any angle above 15 degrees is ineffective because of the possibility of a stall. Fixed-pitch propellers may be selected to give this two- to four-degree angle of attack at either climb or cruise airspeeds and RPM, depending upon the desired flight characteristics.

B. Forces Acting on the Propeller

When a propeller rotates, many forces interact and cause tension, twisting, and bending stresses within the propeller.

1. Centrifugal force

The force which causes the greatest stress on a propeller is centrifugal force. Centrifugal force can best be described as the force which tries to pull the blades out of the hub. The amount of stress created by centrifugal force may be greater than 7,500 times the weight of the propeller blade.

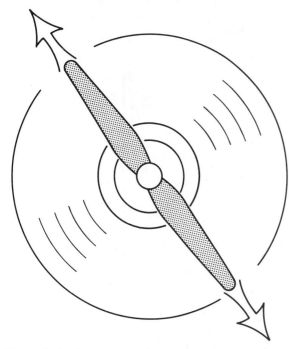

Fig. 2B-6   Centrifugal force tries to pull the propeller blades out of the hub.

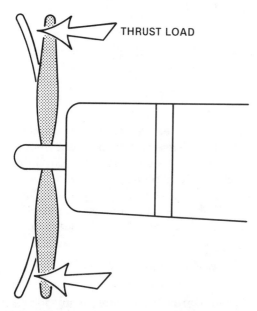

Fig. 2B-7   Thrust bending forces try to bend the propeller blade tips forward.

2. Thrust bending force

Thrust bending force attempts to bend the propeller blades forward at the tips, because the lift toward the tip of the blade flexes the thin blade sections forward. Thrust bending force opposes centrifugal force to some degree (Fig. 2B-7).

3. Torque bending force

Torque bending forces try to bend the propeller blade back in the direction opposite the direction of rotation.

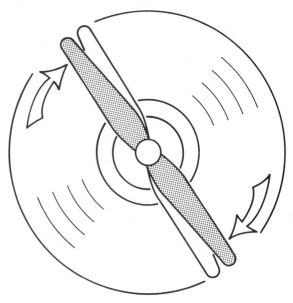

Fig. 2B-8   Torque bending forces try to bend the blade in the plane of rotation opposite the direction of rotation.

4. Aerodynamic twisting moment

Aerodynamic twisting moment tries to twist a blade to a higher angle. This force is produced because the axis of rotation of the blade is at the midpoint of the chord line, while the center of the lift of the blade is forward of this axis. This force tries to increase the blade angle. Aerodynamic twisting moment is used in some designs to help feather the propeller (Fig. 2B-9).

5. Centrifugal twisting moment

Centrifugal twisting moment tries to decrease the blade angle, and opposes aerodynamic twisting moment. This tendency to decrease the blade angle is produced since all the parts of a

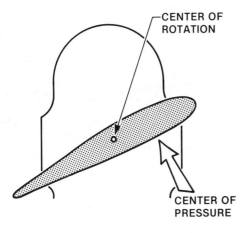

Fig. 2B-9 The aerodynamic twisting moment tries to increase the blade angle.

rotating propeller try to move in the same plane of rotation as the blade centerline. This force is greater than the aerodynamic twisting moment at operational RPM and is used in some designs to decrease the blade angle.

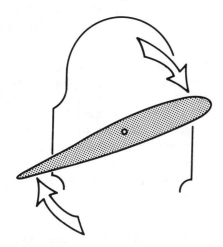

Fig. 2B-10 The centrifugal twisting moment tries to decrease the blade angle.

6. Vibrational force and critical range

When a propeller produces thrust, aerodynamic and mechanical forces are present which cause the blade to vibrate. If this is not compensated for in the design, this vibration may cause excessive flexing and work-hardening of the metal and may even result in sections of the propeller blade breaking off in flight.

Aerodynamic forces cause vibrations at the tip of a blade where the effects of transonic speeds cause buffeting and vibration.

Mechanical vibrations are caused by the power pulses in a piston engine and are considered to be more destructive in their effect than aerodynamic vibration. These power pulses cause a propeller blade to vibrate and set up standing wave patterns that cause metal fatigue and failure. The location and number of stress points change with different RPM settings, but the most critical location for these stress concentrations is about six inches in from the tip of the blades.

Most airframe-engine-propeller combinations have eliminated the detrimental effects of these vibrational stresses by careful design, but some combinations are sensitive to certain RPM's, and this critical range is indicated on the tachometer by a red arc. The engine should not be operated in the critical range except as necessary to pass through it to set a higher or lower RPM. If the engine is operated in the critical range, there is a possibility of structural failure in the aircraft because of the vibrational stresses set up.

C. Propeller Pitch

The geometric pitch of a propeller is defined as the distance, in inches, that a propeller will move forward in one revolution, and this is based on the propeller blade angle at the 75% blade station. Geometric pitch is theoretical because it does not take into account any losses due to inefficiency.

Effective pitch is the distance the aircraft actually moves forward in one revolution of the propeller. It may vary from zero, when the aircraft is stationary on the ground, to about 90% of the geometric pitch during the most efficient flight conditions.

The difference between geometric pitch and effective pitch is called slip (Fig. 2B-11).

If a propeller has a pitch of 50 inches, in theory, it should move forward 50 inches in one revolution. But, if the aircraft actually moves forward only 35 inches in one revolution, the effective pitch is 35 inches and the propeller is 70% efficient in pitch.

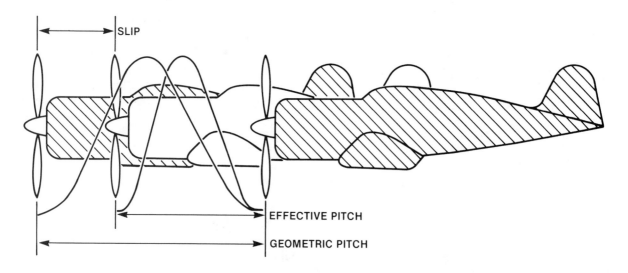

Fig. 2B-11   *Effective and geometric pitch of a propeller*

III. FIXED-PITCH PROPELLERS

A. *Wood Propellers*

Wood propellers are often found on older single-engine airplanes. Most have a natural wood finish, with some models having a black or gray plastic coating.

1. *Construction*

Wood propellers are made of several layers of wood bonded together with a waterproof resin glue, and except for very few instances, they are made of birch. Each layer of a propeller is normally of the same thickness and type of wood, with a minimum of five layers of wood being used. When the planks of wood are glued together, they form what is called a propeller blank.

During fabrication, the blank is roughed to shape and is allowed to set for a week or so to allow the moisture to be distributed equally through all of the layers.

The white, as the rough-shaped blank is called, is finished to the exact airfoil and pitch dimensions required for the desired performance characteristics, and during this process, the center bore and bolt holes are drilled.

At this stage of its manufacture, the tip fabric is applied to the propeller. Cotton fabric is glued to the last 12 to 15 inches of the propeller

blade, where it reinforces the thin sections of the tip. The fabric is doped to prevent deterioration by weather and by the rays of the sun. The propeller is then finished with clear varnish to protect the wood surface (Fig. 3B-1).

Monel, brass, or stainless steel tipping is applied to the leading edge of the propeller to prevent damage from small stones during ground operations. The metal is shaped to the leading edge contour and is attached to the blade with countersunk screws in the thick blade sections and with copper rivets in the thin sections near the tip. The screws and rivets are safetied into place with solder (Fig. 3B-2).

Three number 60 holes are drilled 3/16-inch deep into the tip of each blade to release moisture from the propeller and allow the wood to breathe.

Wood blades for controllable-pitch propellers are constructed in the same manner as fixed-pitch propellers, except that the shank of the blade is held in a metal sleeve with lag screws.

2. *Inspection, maintenance, and repair*

Wood propellers have many components and require a close inspection of each part to assure proper operation and to prevent failures.

Defects that may occur in the wood include separation of laminations, dents or bruises on the surface—especially on the face of the blade, scars across the blade surface, broken sections, warp-

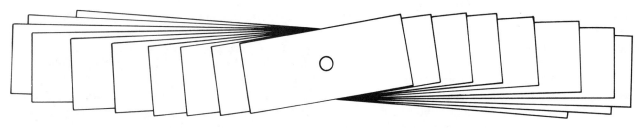

PLANKS ARE GLUED TOGETHER TO FORM THE PROPELLER BLANK.
(A)

THE ROUGH-SHAPED PROPELLER BLANK IS CALLED A "WHITE."
(B)

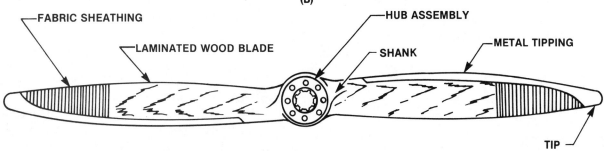

THE FINISHED WOOD PROPELLER WITH THE FABRIC SHEATHING, METAL TIPPING, AND STEEL HUB INSTALLED.
(C)

Fig. 3B-1   Stages in the production of a wood propeller

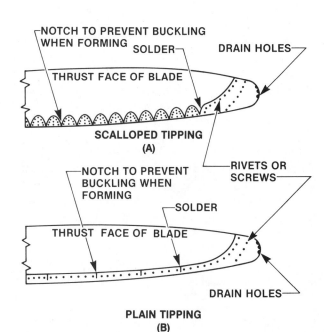

Fig. 3B-2   Types of metal tipping applied to a wood propeller

ing, and worn or oversize center bore and bolt holes.

Separated laminations are not repairable unless they occur in the outside lamination of fixed-pitch propellers. Delamination of the outer layers may be repaired by a repair station.

Inspect dents, bruises, and scars on the blade surfaces with a magnifying glass while flexing the blade to expose any cracks. Some cracks may be repaired by an inlay at a repair facility.

Small cracks parallel to the grain may be repaired by working resin glue into the crack. When the glue is dry, sand the area and refinish with varnish. Small cuts are treated in the same manner.

Broken sections may be repaired by a repair facility, depending on the location and severity of the break.

505

Check the tip fabric for cracks or bubbles in the material, chipping of the paint, and wrinkles that appear when the tip is twisted or flexed.

If the tip fabric has surface defects of three-quarters of an inch or less, and it does not indicate a breakdown in the wood structure, the defect may be filled with several coats of lacquer until the defect blends in with the fabric surface. Defects larger than three-quarters of an inch should be referred to a repair facility.

When inspecting the metal tipping, look for looseness or slipping, loose screws or rivets, cracks in the solder joints, damage to the metal surface, and cracks in the metal, especially on the leading edge.

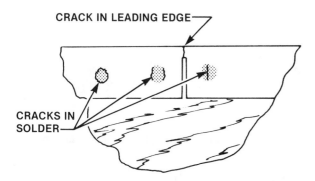

Fig. 3B-3 Inspection of metal tipping. Check for cracks in the solder that is used to safety the screw heads.

Cracks in the solder joint near the blade tip may be indications of wood deterioration. Inspect the area closely while flexing the blade tip. If no defects are found, the joint may be resoldered, but inspect the area closely at each opportunity for evidence of recurrence.

On a controllable-pitch blade made of wood, check the metal sleeve and the wood next to the sleeve for cracks. This may indicate loose or broken lag screws and should be referred to an overhaul facility for correction.

If the varnish should begin to chip or peel, sand the surface lightly to feather in the edges of the irregularity and then apply a fresh coat of varnish to the area.

The following defects are not repairable and are reasons for considering a propeller unairworthy: a crack or deep cut across the grain; a split blade; separated laminations, except for the out-

side laminations of a fixed-pitch propeller; unused screw or rivet holes; any appreciable warp; an appreciable portion of wood missing; cracks, cuts, or damage to the metal sleeve of a changeable-pitch propeller; an oversized crankshaft bore in a fixed-pitch propeller; cracks between crankshaft hole and bolt holes; cracked internal laminations; and excessively elongated bolt holes.

When a wood propeller is stored, place it in a horizontal position to keep the moisture evenly distributed throughout the wood. The storage area should be cool, dark, dry, and well ventilated. Do not wrap the propeller in any material that will seal off the propeller from the surrounding airflow, or the wood will rot.

B. Aluminum Alloy Propellers and Blades

Aluminum propellers are the most widely used type of propellers in aviation. Aluminum propellers are more desirable than wood propellers because thinner, more efficient airfoils may be used without sacrificing structural strength. Better engine cooling is also achieved by carrying the airfoil sections close to the hub and directing more air over the engine. These propellers require much less maintenance than wood propellers, thereby reducing the operating cost.

1. Construction

Aluminum propellers are made of high-strength aluminum alloys and are finished to the desired airfoil shape by machine and manual grinding. The final pitch is set by twisting the blades to the desired angles. As the propeller is being finished by grinding, its balance is checked and adjusted by removing metal from the tip of the blade to adjust horizontal balance and from the boss or leading and trailing edges of the blades to adjust vertical balance. Some fixed-pitch propeller designs have their horizontal balance adjusted by placing lead wool in balance holes near the boss, and their vertical balance corrected by attaching balance weights to the side of the boss.

Once the propeller is ground to the desired contours and the balance is adjusted, the surfaces are finished by anodizing and painting.

2. Inspection, maintenance, and repair

As we have mentioned, an advantage of aluminum propellers is the low cost of maintenance.

This is because of their one-piece construction and the hardness of the metal from which the propellers are made. However, when damage does occur, it is usually critical and may result in blade failure. For this reason, the blades must be carefully inspected and any damage must be repaired before further flight.

Before inspecting a propeller, clean it with a solution of mild soap and water to remove all of the dirt, grease, and grass stains. Then inspect the blades for pitting, nicks, dents, cracks, and corrosion, especially on the leading edges and face. A four-power magnifying glass will aid in these inspections, and dye-penetrant inspection should be performed when cracks are suspected.

A majority of the surface defects that occur on the blades can be repaired by the powerplant mechanic. Defects on the leading and trailing edge of a blade may be dressed out by using round and half-round files. The repair should blend in smoothly with the edge and should not leave any sharp edges or angles. The approximate maximum allowable size of a repaired edge defect is 1/8-inch deep and no more than 1-1/2 inch in length. Repairs to the face and back of a blade are performed with a spoon-like riffle file which is used to dish out the damaged area. The maximum allowable repair size of a surface defect is 1/16-inch deep, 3/8-inch wide, and 1 inch long. All repairs are finished by polishing with very fine sandpaper, moving the paper in a direction parallel to the length of the blade, and then treating the surface with Alodine, paint, or some other appropriate protective coating.

Inspect the hub boss for damage and corrosion inside the center bore and on the surfaces which mount on the crankshaft. The bolt holes should be inspected for damage, security and dimensions.

Light corrosion in the boss can be cleaned with sandpaper and then painted or treated to prevent the recurrence of corrosion. Propellers with damage, dimensional wear, or heavy corrosion in the boss area should be referred to a repair station for appropriate repairs.

Damage in the shank area of a propeller blade should be referred to an overhaul facility for corrective action. Since all forces acting on the propeller are concentrated on the shank, any damage in this area is critical.

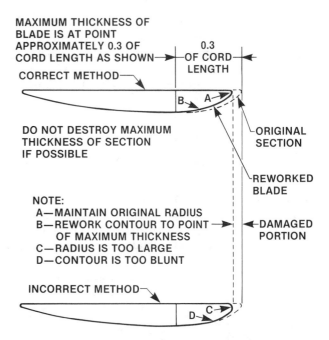

METHODS OF REPAIRING LEADING EDGE DAMAGE
(A)

BEFORE-AND-AFTER ILLUSTRATIONS OF TYPICAL DEFECTS
(B)

Fig. 3B-4   Allowable repairs to a metal propeller blade

If a blade has been bent, measure the angle of the bend and the blade station of the bend center and, by using the proper chart, determine the repairability of the blade. To make this decision, find the center of the bend, and measure from the center of the hub to determine the blade station of the bend center. Next, mark the blade one inch on each side of the center of the bend and measure the degree of bend by using a protractor similar to the one in Fig. 3B-5. Be sure the protractor is tangent to the one-inch lines when measuring the angle. Use a chart approved by the propeller manufacturer, similar to the one in Fig. 3B-6, to determine if the bend is repairable. When reading

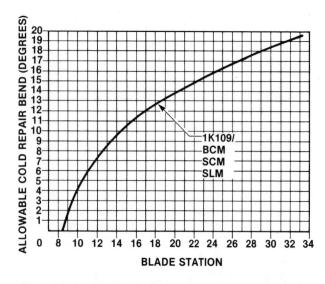

ANGLE OF BEND

THICKNESS OF BLADE

MEASURE AT A POINT OF TANGENCY TAKEN ONE INCH EACH SIDE ₵ OF BEND

Fig. 3B-5 Method of measuring a propeller blade bend using a protractor.

Fig. 3B-6 Typical chart for determining the allowable amount of bend in a metal propeller blade.

the chart, any bend above the graph line is not repairable. If the proper chart is not available, take the measurements and contact an overhaul facility for a decision before sending the propeller to them for straightening.

After the propeller has been repaired, repaint the surfaces. Paint the face of each blade with one coat of zinc chromate primer and two coats of flat black lacquer from the six-inch station to the tip. The back of the blade should have the last four inches of the tip painted with one coat of zinc chromate primer and two coats of a high visibility color such as red, yellow, or orange. The color scheme on the back of the blade on some aircraft differs from that described here, so the original color scheme may be duplicated if desired.

C. Fixed-pitch Propeller Designation System

Two propeller designation systems are discussed here, so the mechanic will be able to recognize the systems and understand the difference in propeller designs by their designation. The McCauley and Sensenich systems are representative of those presently in use.

1. McCauley designation system

A McCauley propeller designated as 1B90/CM7246 has a basic design designation of 1B90. The CM component of the designation indicates the type of crankshaft the propeller will fit, its blade tip contour, the adapter used, and provides other information pertaining to a specific aircraft installation. The 72 indicates the diameter of the propeller in inches, and the 46 indicates the pitch of the propeller at the 75% station.

2. Sensenich designation system

The Sensenich designation 76DM6S5-2-54 indicates a propeller with a designed diameter of 76 inches. The D designates the blade design, and the M6 indicates the hub design and mounting information, such as bolt hole size, dowl pin location, etc. The S5 designates the thickness of the spacer to be used when the propeller is installed. The 2 indicates that the diameter has been reduced two inches from the designed diameter, meaning that the propeller has an actual diameter of 74 inches. The 54 designates the pitch in inches, at the 75% station.

48

In either designation system, a change in pitch will be indicated by the pitch stamping on the hub being restamped to indicate the new pitch setting.

Other propeller manufacturers use designation systems that are similar to the McCauley or the Sensenich system.

IV. PROPELLER INSTALLATIONS

A. Flanged Shaft Installation

Flanged propeller shafts are used on most horizontally opposed and some turboprop engines. The front of the crankshaft is formed into a flange four to eight inches across and perpendicular to the crankshaft centerline. Mounting bolt holes and dowel pin holes are machined into the flange and, on some flanges, threaded inserts are pressed into the bolt holes.

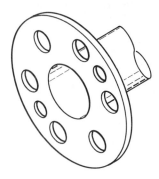

FLANGE WITH DOWEL PIN HOLES
(A)

FLANGE WITH THREADED INSERTS
(B)

Fig. 4B-1 Typical crankshaft flanges for mounting a propeller

1. Preparation for installation

Before installing the propeller, inspect the flange for corrosion, nicks, burrs, and other surface defects. Any defects found should be repaired in accordance with the engine manufacturer's recommendations. Light corrosion can be removed with very fine sandpaper, and if a bent flange is suspected, a run-out inspection should be performed. The bolt holes and threaded inserts should be clean and in good condition.

When the flange area is clean and smooth, apply a light coat of oil or anti-seize compound to prevent corrosion and make removal of the propeller easy.

Inspect the mounting surface of the propeller and prepare it in the same way as you did the flange.

The attaching bolts should be in good condition and inspected for cracks with either a dye-penetrant or magnetic particle inspection process. Washers and nuts should also be inspected, and new fiber lock nuts used if they are required in the installation.

2. Installation

The propeller is now ready to mount on the crankshaft. If dowel pins are used, the propeller will fit on the shaft in only one position, but if there is no dowel, install the propeller in the position specified in the aircraft or engine maintenance manual, as propeller position is critical for maximum engine life in some installations. If no position is specified on a four-cylinder horizontally opposed engine, the propeller should be installed so that the blades are at the 10 o'clock and 4 o'clock position when the engine is stopped. This reduces vibration in many instances and puts the propeller in position for hand propping the engine. Other engine configurations, if not doweled, are not normally affected by installation position (Fig. 4B-2.)

The bolts, washers, and nuts are installed next, according to the particular installation. Tighten all of the bolts slightly, then use an alternating torquing sequence to tighten the bolts to the desired value, which is usually 35 foot-pounds or higher for metal propellers and about 25 foot-pounds for wood propellers (Fig. 4B-3).

Fig. 4B-2 When installing a propeller on a four-cylinder horizontally opposed engine, one of the blades should come to a rest at the ten o'clock position.

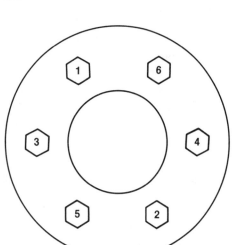

Fig. 4B-3 Typical torque sequence for tightening propeller retaining bolts.

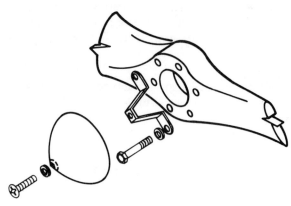

Fig. 4B-4 Skull-cap-type spinner installation

When a "skull cap" spinner is used, the mounting bracket is installed with two of the propeller mounting bolts. And if a full spinner is used, a rear bulkhead is slipped on the flange before the propeller is installed, and a front bulkhead is installed, on the front of the boss before the bolts are slipped through the boss. After the bolts are tightened and safetied, the spinner is installed with machine screws through the spinner into nut plates on the bulkheads.

When a wood propeller is installed, a face-plate is normally placed on the front of the pro-

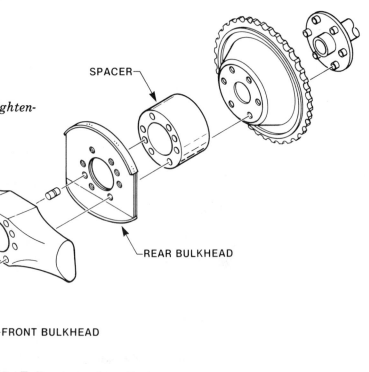

SPACER

REAR BULKHEAD

FRONT BULKHEAD

Fig. 4B-5 Full spinner installation

510

peller boss before installing the bolts. This faceplate distributes the compression load of the bolts over the entire surface of the boss.

After the bolts are installed and properly torqued, the propeller is tracked and safetied.

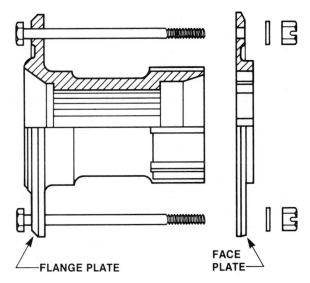

FLANGE PLATE

FACE PLATE

Fig. 4B-6 Propeller hub used to mount a wood propeller on a tapered or splined-shaft engine

B. Tapered Shaft Installations

Tapered shaft crankshafts are found on older model engines of low horsepower. This type of crankshaft requires a hub to adapt the propeller to the shaft.

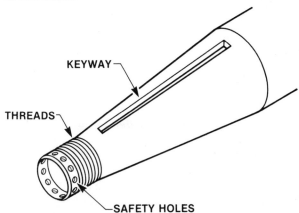

KEYWAY

THREADS

SAFETY HOLES

Fig. 4B-7 Tapered shaft used on some of the smaller aircraft engines

1. Preinstallation checks

Before installing the propeller, carefully inspect the taper of the shaft for corrosion, thread

condition, cracks, and wear in the areas of the keyway. The keyway is critical, since cracks can develop in the corners of the keyway and result in the crankshaft breaking. A dye-penetrant inspection of the keyway area is advisable each time the propeller is removed.

If surface irregularities are found, dress or polish out the defects as the engine manufacturer recommends.

Inspect the hub components and mounting hardware for wear, cracks, corrosion, and warpage. Inspect the hub and bolts with dye penetrant or magnetic particle inspection methods and correct any defects that are found.

The fit of the hub on the crankshaft should be checked by the use of a liquid transfer ink such as Prussian Blue. The Prussian Blue is applied in a thin, even coating on the tapered area of the crankshaft. Then, with the key installed in the keyway, install the hub on the shaft and tighten the retaining nut to the recommended installation torque. Remove the hub and note the amount of dye transferred from the crankshaft to the hub.

The dye transfer should indicate that there is a minimum contact area of 70%. If less than 70% contact area is indicated, the hub and crankshaft should be checked for surface irregularities such as dirt, wear, and corrosion. The surfaces may be lapped to fit by removing the key from the crankshaft and lapping the hub to the crankshaft, using a fine-grit polishing compound, until a minimum of 70% contact area is achieved. (Check the engine manufacturer's instructions for specific information about this lapping proedure.)

This inspection and corrective action may be done with the propeller installed on the hub, and when sufficient contact area is obtained, clean the hub and shaft of the dye and polishing compound.

2. Installation

Apply a very light coat of oil or anti-seize compound to the crankshaft, making sure that the key is installed properly, and then place the propeller and hub assembly on the shaft. Be sure that the threads on the shaft and nut are clean and dry, then install the retaining nut and torque the nut to the proper value. Install the puller snap ring and track and safety the propeller.

3. Removal

To remove the propeller from the tapered shaft, remove the safety and back the retaining nut off with a bar to pull the propeller from the shaft. A snap ring is installed inside the hub so the retaining nut can act as a puller to loosen the hub from the shaft as the nut is unscrewed. If no snap ring is installed, hub removal may be very difficult.

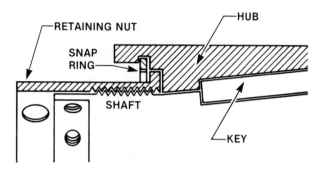

Fig. 4B-8 The snap ring is installed inside the hub of a propeller mounted on a splined or tapered shaft to aid in pulling the propeller from the shaft.

C. Splined Shaft Installations

Splined crankshafts are found on most radial engines and some horizontally opposed, in-line, and even turboprop engines. The splined shaft has grooves and splines of equal dimensions, and a master, or double-width, spline so that a hub will fit on the shaft in only one position.

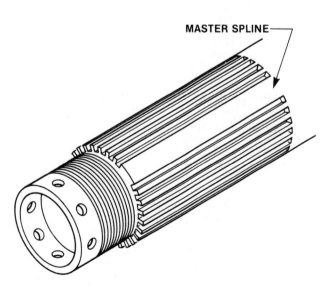

Fig. 4B-9 Splined propeller shaft showing the master spline.

1. Preinstallation checks

Inspect the crankshaft for cracks, surface defects, and corrosion, and repair any defects in accordance with the engine manufacturer's instructions.

Inspect the splines on the crankshaft and on the hub for wear by using a go-no go gage which is 0.002-inch larger than the maximum space allowed between the splines. The crankshaft or spline is serviceable if the gage cannot be inserted between the splines for more than 20% of the spline length. If it will go in more than 20% of the way, the hub or the crankshaft is worn excessively and should be replaced.

The cones that are used to center the hub on the crankshaft should be inspected for general condition. The rear cone is made of bronze and is split to allow flexibility during installation and to assure a tight fit when it is installed. The front cone is made in two pieces and is a matched set. The two halves are marked with a serial number to identify the mates in a set.

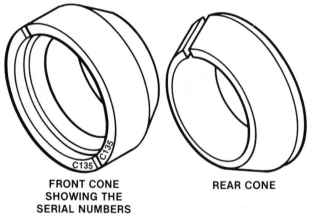

Fig. 4B-10 Front and rear cone for mounting a propeller on a splined propeller shaft

2. Trial Installation

Slip the rear cone and, in some installations, a bronze spacer on the crankshaft, and push them all the way back on the shaft. Apply a thin coat of Prussian Blue to the rear cone, and slide the hub on the shaft, taking care to align the hub on the master spline. Push the hub back against the rear cone. Coat the front cone halves with Prussian Blue and place them around the lip of the retaining nut. Install the nut in the hub and tighten it to the proper torque.

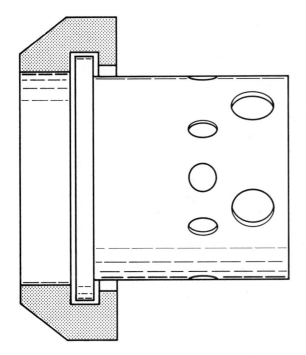

Fig. 4B-11 Propeller retaining nut and the front cone used to secure a propeller to a splined shaft

Remove the retaining nut and the front cone and note the amount of Prussian Blue transferred to the hub. A minimum of 70% contact is required. Remove the hub from the crankshaft and note the transfer of dye from the rear cone. Again, a minimum of 70% contact is required. If contact is insufficient, lap the hub to the cones, using special lapping fixtures.

If no dye is transferred from the rear cone during the transfer check, a condition known as rear cone bottoming exists. This happens when the apex, or point, of the rear cone contacts the land on the rear seat of the hub before the hub can seat on the rear cone. Correct rear cone bottoming by removing up to 1/16-inch from the apex of the cone with sandpaper on a surface plate.

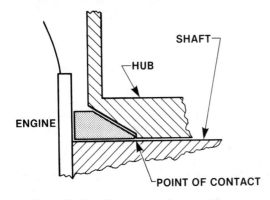

Fig. 4B-12 Rear cone bottoming

Front cone bottoming occurs when the apex of the front cone bottoms on the splines of the crankshaft before it seats on the hub. Front cone bottoming is indicated when the hub is loose on the shaft after the retaining nut has been torqued and there is no transfer of Prussian Blue to the front hub seat. Correct front cone bottoming by using a spacer of no more than 1/8-inch thickness behind the rear cone. This moves the hub forward so that the front cone can seat properly.

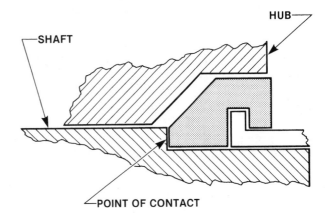

Fig. 4B-13 Front cone bottoming

3. *Installation*

Install the propeller on the hub in the same manner as is used for a tapered shaft installation. The position of the propeller on the hub in relation to the master spline is critical. Some installations require that one blade align with the master spline while other installations require that the blades be perpendicular to the master spline position; so be sure to consult the engine maintenance manual for the requirements of a particular installation.

Once the propeller is mounted on the hub, coat the crankshaft with oil or an anti-seize compound and slide the propeller and hub assembly in place on the shaft. Install the front cone and the retaining nut. Track the propeller and then safety the installation.

Propeller removal is done in the same way as is done with a tapered shaft installation.

4. *Tracking the propeller*

Once the propeller is installed and torqued, check the track. The track of the propeller is defined as the path which the tips of the blades follow as they rotate with the aircraft stationary. For light aircraft with propellers of up to approximately six feet diameter, metal propellers can be out of track no more than 1/16-inch, and the track of a wood propeller may not be out more than 1/8-inch.

Before the propeller can be tracked, the aircraft must be made stationary by chocking the wheels so that the aircraft will not move. Next, place a fixed reference point within 1/4-inch of the propeller arc. This may be done by placing a board on blocks under the propeller arc and taping a piece of paper to the board so the track of each blade can be marked. Rotate the propeller by hand until one blade is pointing down at the paper, and mark this position on the paper. Now turn the propeller so that the track of the next blade can be marked on the paper, and repeat this for each blade. The maximum difference in track for all of the blades should not exceed the limits established for the installation.

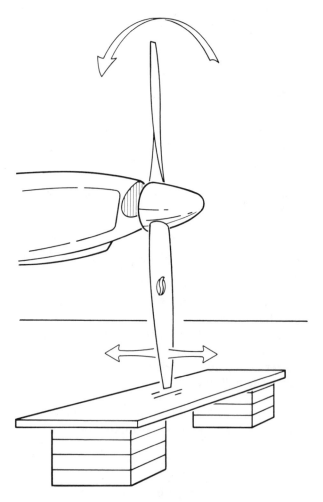

Fig. 4B-14   Method of tracking a propeller, using a reference board

If the propeller track is off more than is allowed, and presuming that the propeller, hub, and crankshaft are all within permissible dimensional tolerances, the track may be corrected by the use of shims. If allowed by a metal propeller's manufacturer, use a shim shaped to fit halfway around the face of the hub or flange and install it between the flange and the propeller on the side of the more rearward tracking blade, so that the blade will be moved forward and the forward blade will be moved rearward. The shim material may be thin brass shim stock of about 0.002 to 0.004 inch, as necessary to correct the track. If a wood propeller is used, install a shim as just described, but use another shim of the same thickness between the faceplate and the hub boss on the side of the more forward blade.

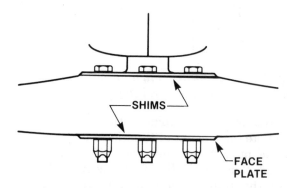

Fig. 4B-15   Adjusting propeller track by the use of shims

Once the shims are installed, reassemble the propeller and hub assembly and install the propeller on the crankshaft. Torque the nuts and recheck the track. If the track is within limits, safety the installation.

5.   *Safetying the propeller*

Once a propeller is properly tracked and torqued, it can be safetied. There is no one correct way to safety a propeller installation because of the many different types of installations, and for this reason only the more commonly used safeties will be discussed.

A flanged shaft installation has the largest variety of safety methods because of its many variations. If the flange has threaded inserts installed, the propeller is held on by bolts screwed into the inserts. The bolt heads are drilled and safetied with 0.041-inch stainless steel safety wire, using standard safety wire procedures.

If threaded inserts are not pressed into the flange, bolts and nuts are used. Some installations use fiber lock nuts which require no safetying, but the nuts should be replaced each time the propeller is removed. Other installations use castellated nuts and drilled bolts and the nuts are safetied to the bolts with cotter pins.

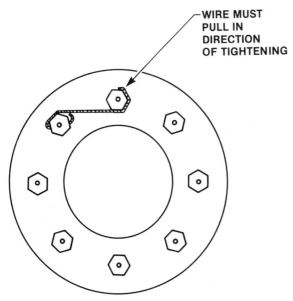

USE OF SAFETY WIRE TO SECURE PROPELLER BOLT HEADS
(A)

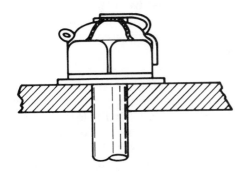

CORRECT METHOD OF SAFETYING CASTELLATED NUTS ON PROPELLER BOLTS
(B)

Fig. 4B-16   Method of safetying a propeller

The retaining nuts for tapered and splined shaft installations are safetied in the same way. A clevis pin is installed through the safety holes in the retaining nut and crankshaft. Position this pin with the head toward the center of the crankshaft so that the centrifugal force will hold the pin in the hole.

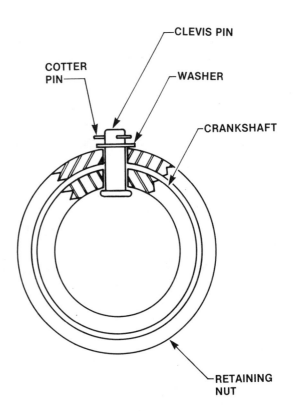

Fig. 4B-17   Method of safetying a retaining nut on tapered or splined propeller shaft

6.   Troubleshooting

Troubleshooting the installation of a fixed-pitch propeller usually involves determining the cause of vibration. When investigating vibration, it is important to know the recent history of the propeller. Accidents, repairs, and the type of operation may give clues to the cause of vibration.

Metal propellers vibrate if a repair has thrown the propeller out of balance, and this vibration would appear immediately after the repair. The propeller will have to be rebalanced.

If the propeller has been involved in a ground strike, both the balance and aerodynamic characteristics may have been altered. If a ground strike is suspected, check with the pilot, as not all ground strikes cause observable damage. If a ground strike has occurred, send the propeller to an overhaul facility for repair.

Some people use the propeller as a handle to pull the aircraft around on the ground. This may result in pulling the blades out of track, and the propeller must be overhauled if it is bent.

Vibration associated with wood propellers is often related to wood damage or moisture in the wood. If a wood propeller is stored improperly, or if the airplane has been idle for a period of time, moisture may be concentrated in one of the blades, causing the propeller to be out of balance. But the moisture will redistribute itself if the propeller is placed in a horizontal position for several days.

Wood propellers may warp, resulting in a change in the aerodynamic characteristics of the blade and causing aerodynamic imbalance. Warped propellers should be replaced.

Reasons for vibrations that are common to both wood and metal propellers include improper overhaul, uneven torquing on the mounting bolts, improper tracking, loose retaining nut, front or rear cone bottoming, and improper installation position relative to the crankshaft.

V. GROUND-ADJUSTABLE PROPELLERS

Ground-adjustable propellers are designed so that their blade angles can be adjusted on the ground to give the desired performance characteristics for various operational conditions that may be encountered.

A. Propeller Construction

A ground-adjustable propeller is designed so that its blades can be rotated in the hub to change the blade angles. The hub is made in two halves that must be separated slightly to loosen the blades so they can be rotated. And the hub is held together with clamps or bolts to prevent the blades from rotating during operation.

The propeller blades may be of wood, aluminum, or steel with shoulders machined on to the root to hold the blades in the hub against the centrifugal operating loads.

The hub of the propeller is made of aluminum or steel, with the two halves machined as a matching pair. Grooves in the hub mate with the shoulders on the blades. When steel blades are used, the hub is usually held together with bolts. When wood or aluminum alloy blades are used, the hub halves are held together with bolts or clamp rings.

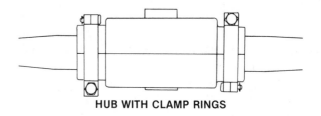

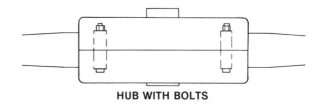

Fig. 5B-1 Typical propeller hubs for ground-adjustable propellers

B. Installation

Ground-adjustable propellers are made that fit flanged, tapered, or splined crankshafts, and their installation is basically the same as it is for fixed-pitch propellers.

Blade angle adjustment

Before the blade angles are adjusted, the reference station and the correct blade angle must be determined by referring to the propeller or aircraft maintenance manual. The propeller blade angles can be adjusted on the aircraft or on a propeller bench.

Before loosening the retaining bolts or clamps, mark the relative position of the hub and blades with a red lead, a white lead, or grease pencil. *Do not use a graphite pencil, as it can cause corrosion.* This mark will allow you to watch the initial movement of the blade and will help as you move all of the blades toward their new blade angle.

Place the propeller in a horizontal position, and if it is on the engine, loosen the propeller retaining nut. Now, loosen the hub bolts or clamps, to free the blades to turn in the hub. Use a propeller paddle to turn the blades to the desired angle.

Check the new blade angles with a propeller protractor; then tighten the blade bolts or clamps and the propeller retaining nut. The blade angle may change during the tightening process, since the blades tend to hang down when the hub halves are loosened. If it does, determine the amount the angle has changed and loosen the bolts or clamps and the retaining nut. Reset the blade angles, allowing for the change that will occur during the tightening process, and retighten the hub and measure the blade angles again. They should now be within allowable angular tolerance.

When the blade angle is correct, torque, track and safety the propeller and remove all of the reference marks.

C. Inspection, Maintenance, and Repair

Inspection of a ground-adjustable propeller blade is the same as for any propeller, whether it is made of wood or metal. Special attention should be given to the area of the shank where the metal sleeve is used on wood blades and in the area of the retention shoulders and grooves on the

blades and the hub. A dye-penetrant inspection of these areas is recommended any time the propeller is disassembled for shipment, for local inspection, or repair.

During 100-hour and annual inspections, check all torques and safeties and check all nuts, bolts, clevis pins, etc., for condition, and replace any that are damaged or missing.

D. The Propeller Protractor

One of the most useful propeller tools is the universal propeller protractor. This device is used to measure the propeller blade angle at a specified blade station to determine if the propeller is properly adjusted. This blade angle is referenced from the plane of propeller rotation, which is perpendicular to the crankshaft centerline.

The frame of the protractor is made of aluminum alloy, and three sides of it are ninety

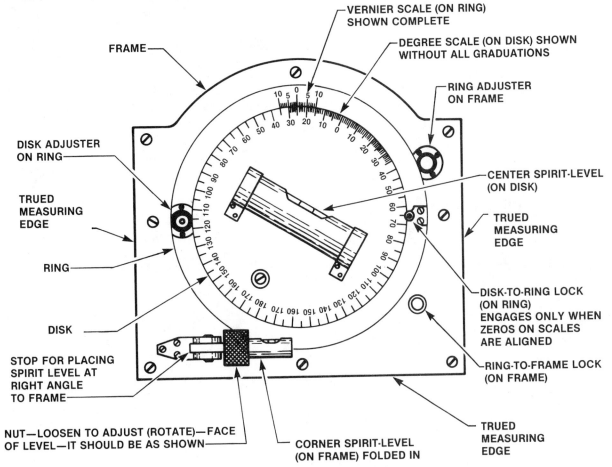

NOTE: PROTRACTOR HANDLE LOCATED ON OPPOSITE SIDE

Fig. 5B-2   Universal propeller protractor

57

degrees to each other. A bubble spirit level is mounted on one corner of the front of the frame, and this corner level swings out to indicate when the protractor is level. A movable ring is located inside the frame and is used to set the zero reference angle for blade angle measurements. The ring is engraved with vernier index marks, which allow readings as small as one tenth of a degree. A center disk is engraved with a degree scale from zero to 180 degrees, both positive and negative, and it contains a spirit level to indicate when the disk is level. We can see the locking and adjusting control in Fig. 5B-2.

When using the propeller protractor and before measuring the angle of a propeller blade, determine the reference blade station from the aircraft manufacturer's maintenance manual, and mark this reference station on the blade with chalk or with a grease pencil.

The next step is to establish the reference plane from the engine crankshaft centerline, rather than the airframe attitude, because some engines are canted in the aircraft. To zero the protractor, loosen the ring-to-frame lock, align the zeros on the disk and the ring, and then engage the disk-to-ring lock. Place one edge of the protractor on a flat surface on the propeller hub that is parallel to or perpendicular to the crankshaft centerline, and turn the ring adjuster until the spirit level in the center of the disk is level. The corner level should also be level. Now, tighten the ring-to-frame lock, and release the disk-to-frame lock. The protractor is now aligned with the engine crankshaft.

Place one blade of the propeller horizontal and move out to the reference station marked on the face of the blade to measure the blade angle. Stand on the same side of the airplane, facing in the same direction, as when you established the zero with the protractor; otherwise the measurement will be incorrect. Place the edge of the protractor on the face of the blade at the reference station and turn the disk adjuster until the spirit level centers. Now read the blade angle, using the zero line on the ring as the index. Tenths of degrees can be read from the vernier scale. Rotate each blade to the same horizontal position, and measure the angle.

If the face of the propeller blade is curved, use masking tape to attach a piece of 1/8-inch drill

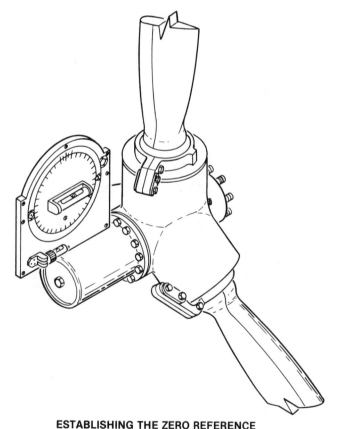

ESTABLISHING THE ZERO REFERENCE
(A)

METHOD OF MEASURING THE BLADE ANGLE AT THE REFERENCE STATION
(B)

Fig. 5B-3   Correct method of using a propeller protractor

518

rod, 1/2-inch from the leading and trailing edges and measure the angle with the protractor resting on the rods.

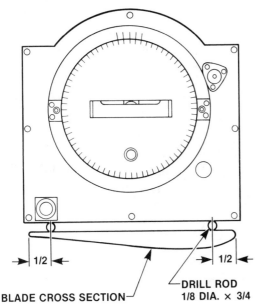

BLADE CROSS SECTION

DRILL ROD
1/8 DIA. × 3/4

TAPE RODS ON REFERENCE STATION
ON THRUST SIDE OF BLADE

Fig. 5B-4 Method of correcting for blade curvature when using a universal propeller protractor.

VI. TWO-POSITION AND CONSTANT-SPEED PROPELLERS

A. Hamilton-Standard Counterweight Propellers

The Hamilton-Standard counterweight propeller is one of the most successful controllable-pitch propellers that has been used on American designed and built airplanes. It was developed in the 1930's as a two-position propeller that allowed the pilot to use a low blade angle for takeoff and climb, so the engine could develop its full horsepower. Then when the airplane was at its cruising altitude, the propeller could be shifted to a high blade angle for efficient high-speed flight.

The two-position propeller uses a control valve to direct engine oil into the propeller to decrease the blade angle or drain the oil back into the engine so the blades will go to a high blade angle.

As the science of flight progressed, a fly-weight governor-controlled valve was developed to replace the two-position valve to direct oil into or out of the propeller to change the blade angle so it will hold the engine RPM constant with varying aerodynamic loads.

The propellers used for two-position operation and those used for constant speed operation are basically the same; the difference between the two systems is in the oil control mechanism.

The central component of the Hamilton-Standard counterweight propeller is the spider. The spider is designed for installation on a splined crankshaft and incorporates two or three arms on which the blades are mounted (Fig. 6B-1).

Aluminum alloy blades are used with these propellers, and counterweight brackets which are part of the pitch-change mechanism are mounted on the blade butts. The hollow ends of the blades fit over the arms of the spider (Fig. 6B-2).

The propeller barrel, which encloses the hub assembly, is made of two steel halves which are machined as a matched set, and the two halves are identified by identical serial numbers. The barrel halves, with required bearings and spacers, are placed around the spider after the blades are installed and are held together with bolts.

The cylinder fits between the counterweight brackets and is attached to the brackets by special Allen-head screws which act as follower pins during pitch changing operations.

The piston fits through the cylinder and doubles as the propeller retaining nut. The front cone, the snap ring, and a safety ring are attached to the piston, and leather seals used between the piston and cylinder are held in place by a seal nut.

The cylinder head and its copper-asbestos gaskets are installed in the forward end of the cylinder and safetied with a locking ring wire.

1. Two-position system operation

Two forces are used to cause the blade angle to change: engine oil pressure inside the propeller cylinder and centrifugal force acting on the counterweights. Other rotational and operational forces have a minimal effect on system operation.

59

519

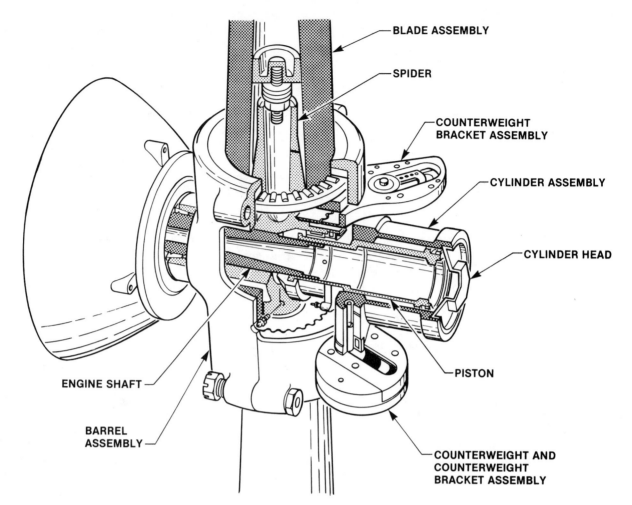

BLADE ASSEMBLY

SPIDER

COUNTERWEIGHT
BRACKET ASSEMBLY

CYLINDER ASSEMBLY

CYLINDER HEAD

PISTON

ENGINE SHAFT

BARREL
ASSEMBLY

COUNTERWEIGHT AND
COUNTERWEIGHT
BRACKET ASSEMBLY

Fig. 6B-1   Cutaway view of a Hamilton-Standard counterweight propeller

When the propeller control lever is moved forward to decrease the blade angle, the selector valve is rotated to direct engine oil pressure into the propeller cylinder.

As the oil moves the cylinder forward, it overcomes the centrifugal force on the counterweights and pulls them in toward the centerline of the propeller blades. The counterweights are attached to the blade butts so that as they move toward the blade centerline, the blades are rotated to a lower angle. The propeller movement continues until the follower pins contact the stops inside the counterweights (Fig. 6B-3).

To increase the blade angle, the cockpit control is moved rearward and the selector valve is rotated so that it releases the oil from the propeller. The centrifugal force on the counterweights is now greater than the force of the oil in the propeller cylinder, and the blades rotate to a higher blade angle. The oil is forced out of the

cylinder and returned to the engine sump as the cylinder is pulled inward by the action of the counterweights. The propeller movement continues until the follower pins contact the stop. The propeller is now held in high blade angle by centrifugal force acting on the counterweights.

When the engine is started, the propeller is at the high blade angle setting. This is to prevent oil from going into the propeller cylinder rather than to the engine bearings and causing unnecessary bearing wear. When the engine oil temperature and pressure are at the desired values, the propeller can be put into its low blade angle position.

When the engine is to be shut down, place the propeller in its high blade angle, so that the majority of the oil will be forced out of the propeller cylinder. This prepares the propeller for the next engine start, covers the piston surfaces with the cylinder to prevent corrosion and dirt accumulation on the piston, and prevents congealing of the

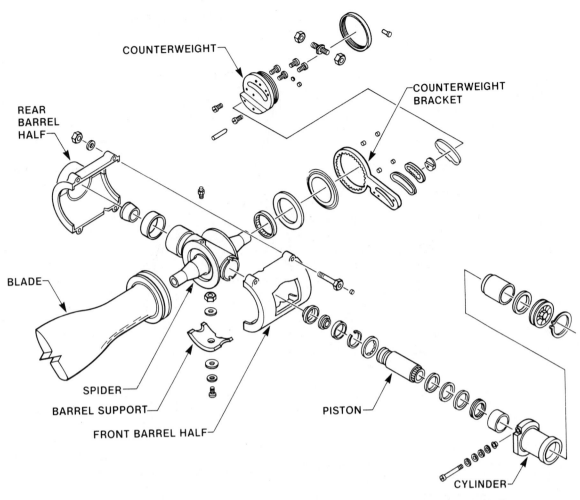

COUNTERWEIGHT

COUNTERWEIGHT BRACKET

REAR BARREL HALF

BLADE

SPIDER

BARREL SUPPORT

FRONT BARREL HALF

PISTON

CYLINDER

Fig. 6B-2   Exploded view of a Hamilton-Standard counterweight propeller

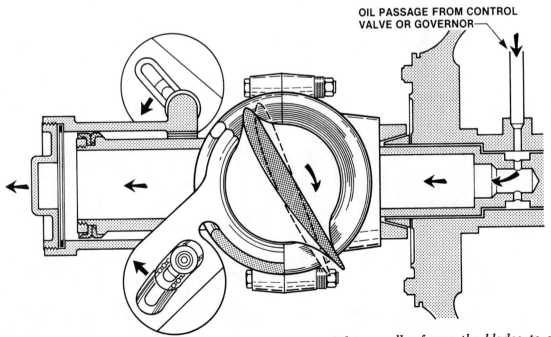

OIL PASSAGE FROM CONTROL VALVE OR GOVERNOR

Fig. 6B-3   Engine oil entering the cylinder of a counterweight propeller forces the blades to a lower angle.

61

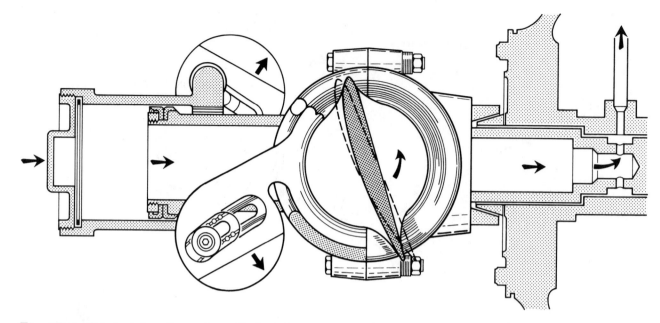

Fig. 6B-4 Oil draining from the cylinder of a counterweight propeller allows centrifugal force on the counterweights to move the blades to a higher angle.

oil in the cylinder when operating in cold climates.

2. Installation

Hamilton-Standard propellers are used on splined engine crankshafts. The standard checks are made for the proper front and rear cone contact and spline wear. The rear cone and propeller are placed on the shaft in the same manner we have previously discussed.

The propeller piston is not installed until the propeller is placed on the shaft. The cylinder is pulled forward and the piston is inserted through the cylinder. The piston lock ring and the snap ring are then placed on the portion of the piston which was inserted through the cylinder. The

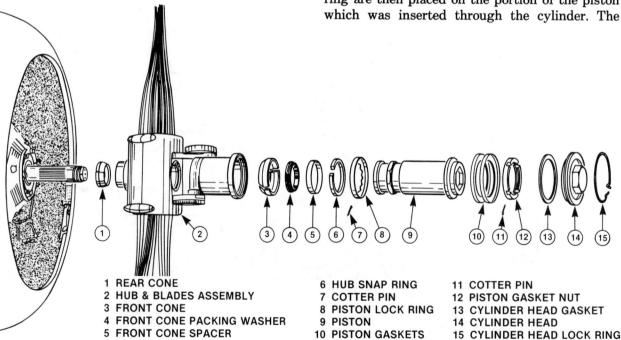

1 REAR CONE	6 HUB SNAP RING	11 COTTER PIN
2 HUB & BLADES ASSEMBLY	7 COTTER PIN	12 PISTON GASKET NUT
3 FRONT CONE	8 PISTON LOCK RING	13 CYLINDER HEAD GASKET
4 FRONT CONE PACKING WASHER	9 PISTON	14 CYLINDER HEAD
5 FRONT CONE SPACER	10 PISTON GASKETS	15 CYLINDER HEAD LOCK RING

Fig. 6B-5 Installation of a counterweight propeller.

front cone halves are placed on the piston and the piston is started on the threads of the crankshaft. The proper Hamilton-Standard tool is used to tighten the piston on the shaft.

If the threads are damaged or the cylinder is cocked, the piston may not start on the shaft. If the piston does not turn smoothly and easily onto the shaft, do not force it. Correct the problem, then reinstall the piston. If the threads on the crankshaft are damaged, the crankshaft may have to be replaced. Install all the seals that are required for a particular installation when mounting the propeller.

When torquing the piston, refer to the propeller or aircraft maintenance manual. A specific torque wrench reading is usually not given here. Rather, a procedure similar to the following may be specified: "Apply a force of 180 pounds to the end of a four-foot bar, and strike the bar twice with a 2-1/2 pound hammer while applying the torque."

The snap ring and piston lock ring are then installed and the lock ring is safetied to the spider with a cotter pin. The piston-to-cylinder leather seals are installed through the front of the cylinder along with the piston gasket nut, and the nut is torqued and safetied.

Install the copper-asbestos cylinder head gasket with the slit toward the cylinder and install and tighten the cylinder head. Safety the cylinder head with the wire lock ring.

The propeller installation is now complete. The propeller should be checked for proper track and, if it is within the allowable limits, the propeller blade angle should be set.

3. Propeller blade angle adjustments

The propeller blade angles are adjusted by means of the stop nuts on the index pin located under each counterweight cap.

The index pins are removed by first removing the clevis pin which safeties the counterweight cap and then unscrewing the cap. The index pin is now pulled out of its recess in the counterweight or pushed from behind the counterweight bracket with a small screwdriver.

Alongside the recess which held the index pin is a scale which is calibrated in degrees and half-

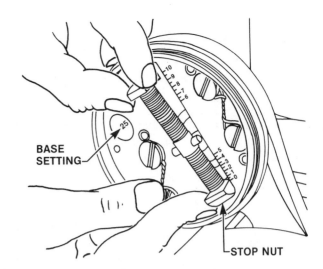

Fig. 6B-6 *The propeller blade angle of a counterweight propeller is adjusted by the position of the stop nuts on the index pins inside the counterweights.*

degrees with a numerical scale ranging from zero to ten. This scale is used to set the stop nuts on the index pin.

The propeller blade index number, also known as the base setting, should be stamped in a lead plug located near the index pin recess. This number indicates the maximum blade angle for which the propeller was adjusted during its last overhaul. This number is usually somewhere near 25 degrees and is used to calculate where the stop nuts on the index pin should be positioned.

If the blade index is, for example, 25 degrees, and the desired blade angles listed in the aircraft specifications are 17 and 22 degrees, the calculation is done in the following manner: $25 - 17 = 8$. And $25 - 22 = 3$. To set the low blade angle (17 degrees) the stop nut is positioned on the index pin so that the edge toward the center of the pin will align with the eight-degree mark on the scale beside the index pin recess. The other stop nut is positioned so that its edge lines up with the number three. Set the stop nuts for each counterweight in this manner. With these settings the propeller blades should have the approximate blade angles desired.

When the index pins are set, the counterweight caps are screwed on, and the blades are moved through their full range of travel once and then positioned for high blade angle. Measure the angle of each blade at the proper reference station, which is the 42-inch station for most models for these propellers. Next, move the blades to their low blade angle stop and check these angles. Make small adjustments as necessary for the stop nuts until the angles are within acceptable limits.

To complete the installation, the engine must be run up and the propeller checked for erratic operation. This may be caused by air trapped in the cylinder, and cycling the propeller through its pitch-change operation several times will purge all of the air from the system.

4. Constant-speed operation

A constant-speed propeller system is a system in which the propeller blade angle is varied by the action of a governor to maintain a constant engine RPM. The governor holds the RPM constant with changes in engine throttle setting and aircraft speeds.

Constant-speed systems are used on almost all modern medium- and high-performance airplanes.

a. The governor

The Hamilton-Standard governor is divided into three parts: the head, the body, and the base.

The head of the governor contains the flyweights and flyweight cup, the speeder spring, a speeder rack and pinion mechanism, and a control pulley. Cast on the side of the head is a flange for the pulley adjustment stop screw. Some head designs incorporate a balance spring above the speeder rack to set the governor to cruise RPM if the control cable breaks.

The body of the governor contains the propeller oil flow control mechanism, which is composed of the pilot valve, oil passages, and the pressure relief valve which is set for 180 to 200 psi.

The base contains the governor boost pump, the mounting surface for installation on the engine, and oil passages which direct engine oil to the pump and return oil from the propeller to the engine sump.

The head, body, and base are held together with studs and nuts. The governor drive shaft extends below the base to mate with the engine drive gear. The driveshaft passes up through the base where it drives the oil pump, through the body where it has oil ports so oil can flow to and from the propeller, and into the head, where it is attached to the flyweight cup and rotates the flyweights.

The governor designation system indicates the design of the head, body, and base used on a particular governor. For a governor model 1A3-B2H, the basic design of the head is indicated by the *1* with minor modifications to the head design indicated by the *B* following the dash. The body design is *A* with minor modification, *2*. The base is a *3* altered by an *H* minor modification.

The propeller governor is an RPM sensing device which responds to a change in system RPM by directing oil to or releasing oil from the propeller to change the blade angle and return the system RPM to the original value. The governor may be set up for a specific RPM by the cockpit propeller control.

The basic governor contains a drive shaft which is connected to the engine drive train, and

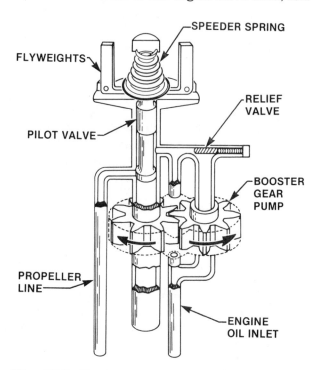

Fig. 6B-7 Basic configuration of a typical propeller governor.

it rotates at a speed that is proportional to the engine RPM. An oil pump drive gear is mounted on the drive shaft, and it meshes with an oil pump idler gear. These gears take oil at engine oil system pressure and boost it to the propeller operating pressure. Excess pressure built up by the booster pump is returned to the inlet side of the pump by a pressure relief valve.

The boosted oil is routed through passages in the governor to a pilot valve which fits in the center of the hollow drive shaft. This pilot valve can be moved up and down in the drive shaft, and it directs oil through ports in the drive shaft to or from the propeller to vary the blade angle.

The position of the pilot valve is determined by the action of the flyweights attached to the end of the drive shaft. The flyweights are designed to tilt outward when the RPM increases and inward when RPM decreases. When the flyweights tilt outward, they raise the pilot valve, and when they tilt inward, the pilot valve is lowered. The movement of the pilot valve in response to changes in RPM directs oil flow to adjust the blade angle to maintain the selected RPM.

The action of the flyweights is opposed by a speeder spring located above the flyweights and adjusted by the pilot through a control cable, pulley, and speeder rack. When a higher RPM is desired, the cockpit control is moved forward to compress the speeder spring. This increased speeder spring compression tilts the flyweights inward and the pilot valve is lowered. This causes the blade angle to decrease, and the RPM will increase until the centrifugal force on the flyweights overcomes the force of the speeder spring and returns the pilot valve to the neutral position.

The opposite action occurs if the cockpit control is moved aft: When the speeder spring compression is reduced, the flyweights tilt outward, the pilot valve is raised, and the blade angle increases until the engine slows down and the centrifugal force on the flyweights decreases. The pilot valve returns to its neutral position.

Whenever the flyweights tilt outward and the pilot valve is raised, the governor is said to be in an *overspeed* condition, with the RPM higher than the governor speeder spring setting calls for.

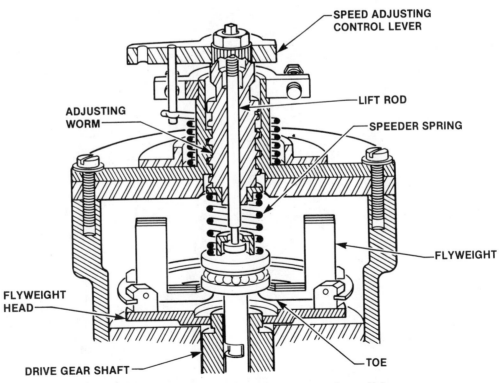

Fig. 6B-8 Propeller governor in the onspeed condition

65

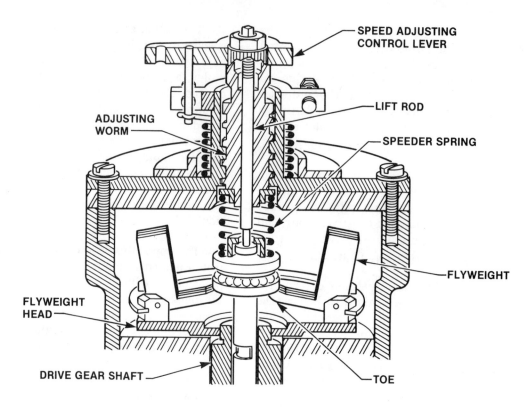

Fig. 6B-9   Propeller governor in the overspeed condition

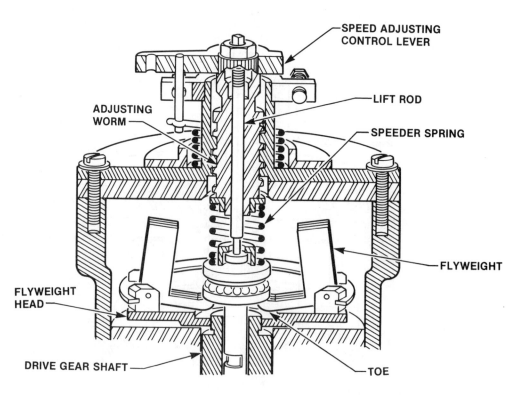

Fig. 6B-10   Propeller governor in the underspeed condition

When the flyweights tilt inward, the governor is in an *underspeed* condition, with the RPM lower than the speeder spring setting calls for. When the RPM is the same as the governor setting is calling for, the governor is in its *onspeed* condition.

The same governing action of the flyweights and pilot valve occurs with changing flight conditions. If the aircraft is in a cruise condition and the pilot begins a climb, airspeed will decrease and cause an increase in the angle of attack of the propeller blades. This increase produces more drag, and the RPM decreases. The governor senses this decrease in RPM by the reduced centrifugal force on the flyweights, and they tilt inward, lowering the pilot valve and producing an underspeed condition. When the pilot valve is lowered, the blade angle is reduced and the RPM increases to its original value. The system returns to the onspeed condition.

If the aircraft is placed in a dive from cruising flight, an overspeed condition is created, and the governor will cause an increase in blade angle to return the system to the onspeed condition.

A change in throttle setting will have the same effect as placing the aircraft in a climb or dive. Increasing the throttle will cause an increase in the blade angle to prevent an RPM increase. Retarding the throttle will result in a decrease in blade angle.

b. Instrument indications

An aircraft with a fixed-pitch propeller uses the tachometer to indicate the throttle setting, with the RPM increasing as the throttle is advanced and decreasing as the throttle is retarded. A constant-speed system uses a manifold pressure gage to indicate the throttle setting, and the tachometer is used to indicate the setting of the propeller control. However, there is some interaction between the propeller control and the manifold pressure gage when changing the RPM and holding the throttle setting fixed.

The throttle directly controls horsepower, so with a fixed throttle setting, the horsepower output of the engine is constant. For example, assume an engine is producing 200 brake horsepower at 23 inches of mercury manifold pressure and 2300 RPM. If the propeller control is advanced to 2400 RPM, the manifold pressure must decrease to some lower value; for example, 22 inches of mercury. If the RPM is reduced to 2200 RPM, the manifold pressure may increase to 24 inches. By these two examples we can see that when the RPM is adjusted, a change in manifold pressure will occur. The amount of change will vary, of course, with different engines and under different conditions, but if the throttle is not moved, the horsepower output does not change.

It is important to note that an engine can be seriously damaged by excessive manifold pressure for a given RPM, and the aircraft flight manual must always be considered when operating an aircraft engine.

With a fixed propeller control setting, the manifold pressure can be changed with the throttle. The RPM, however, will remain constant because the governor will adjust the blade angle to maintain a constant RPM.

When changing power settings, take care to prevent damage to the engine by creating too high a manifold pressure for a given RPM. When power is to be increased, it is best to increase the RPM to the desired setting and then advance the throttle to the desired manifold pressure. When decreasing power, pull the throttle back until the manifold pressure is about one inch below the desired setting and then reduce the RPM with the propeller governor control, and the manifold pressure will increase to the desired value.

B. McCauley Propellers

The McCauley constant-speed propeller system is one of the more popular constant-speed systems for light and medium size general aviation airplanes. This is the system used on most of the Cessna aircraft that require constant-speed propellers. It is also used on many other aircraft designs.

1. The propellers

Two series of propellers are currently being produced by McCauley, the threaded series and the threadless series. The threaded series propellers use a retention nut which screws into the propeller hub and holds the blades in the hub. The threadless design is the more modern of the two and uses a split retainer ring to hold the blades in the hub.

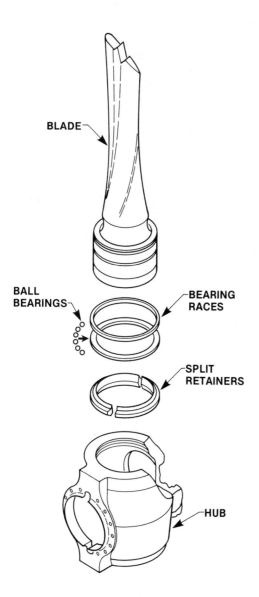

BLADE

BALL BEARINGS

BEARING RACES

SPLIT RETAINERS

HUB

Fig. 6B-11 *McCauley propeller with the thread-less blades*

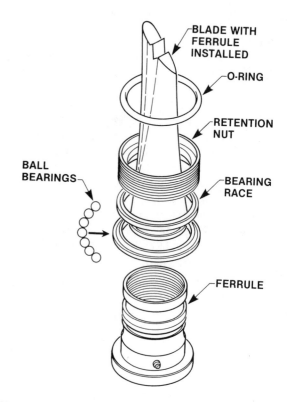

BLADE WITH FERRULE INSTALLED

O-RING

RETENTION NUT

BALL BEARINGS

BEARING RACE

FERRULE

Fig. 6B-12 *Threaded series of McCauley propel-ler blades*

McCauley propellers use oil pressure on an internal piston to cause an increase in the blade angle, and this opposes a spring inside the hub which is used to decrease the blade angle. The movement of the piston is transmitted by blade actuating links to the blade actuating pins that are located on the butt of the blades. All of the pitch changing mechanism is enclosed inside the hub.

The propeller blades, hub, and piston are made of aluminum. The propeller cylinder, blade actuating pins, piston rod, and spring are all made of steel, plated with chrome or cadmium, and the actuating links are made of a phenolic material.

The hollow piston rod through the center of the hub is used as an oil passage to direct oil from the engine crankshaft to the propeller piston. The pitch return spring is located around the piston rod and is compressed between the piston and the rear inside surface of the hub. O-ring seals are used to seal between the piston and the cylinder, the piston and the piston rod, and the piston rod and the hub.

All operating components of the propeller are lubricated at overhaul and receive no additional lubrication during operation.

Certain models of McCauley propellers have been modified to allow for an ongoing dye-penetrant type of inspection. The hub breather holes are sealed and the hub is partially filled with engine oil colored with a red dye. The red dye in the oil makes the location of cracks readily apparent and indicates that the propeller should be removed from service.

The McCauley propeller designation system is broken down as we see in Fig. 6B-14. The most important parts of the designation for the mechanic are the dowel pin location, the C-number,

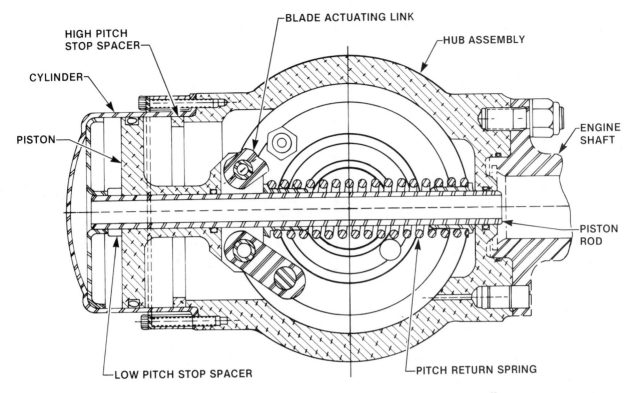

Fig. 6B-13 Pitch change mechanism in a McCauley propeller

and the modification or change letter after the C-number. The modification and change designation indicates the compliance with required or recommended alterations.

The blade designation is included with the propeller designation when determining which propeller will fit a specific aircraft. For example: a C203 propeller will fit a Cessna 180J aircraft, but the landplane version requires a 90DCA-8 blade of 82-inches diameter, while the seaplane version requires a 90DCA-2 blade of 88-inches diameter.

McCauley serial numbers on the propeller hub indicate the year in which the hub was manufactured.

2. The governors

McCauley governors use the same principles of operation as the Hamilton-Standard governors, except that oil is released from the propeller to decrease blade angle, directly opposite from the oil flow in the Hamilton-Standard system. The governor relief valve is set for an oil pressure of about 290 psi. The governor control lever is spring loaded to the high RPM setting. The overall construction of the governor is simpler than the Hamilton-Standard governor, being lighter

and smaller, and all governors incorporate a high RPM stop, while some governors also use an adjustable low RPM stop.

3. Installation and adjustment

McCauley constant-speed propellers are found only on flanged crankshafts and are installed following the basic procedures for fixed-pitch flanged shaft installations. Before the propeller is placed on the crankshaft, the O-ring in the rear of the hub is lubricated with a light coat of engine oil. When placing the propeller on the flange, take care to protect this O-ring from being damaged. The bolts or nuts should be torqued following a torquing sequence. Safety them as appropriate, and be sure to use new fiber lock nuts, if applicable.

The governor is installed using the procedure covered in the section on the Hamilton-Standard constant-speed governor. The McCauley governor uses a control arm instead of a pulley to connect the governor control shaft to the cockpit control cable. The push-pull cockpit control is adjustable in length through a limited range by an adjustable rod end, and the governor high RPM limit can be adjusted by the set screw on the head

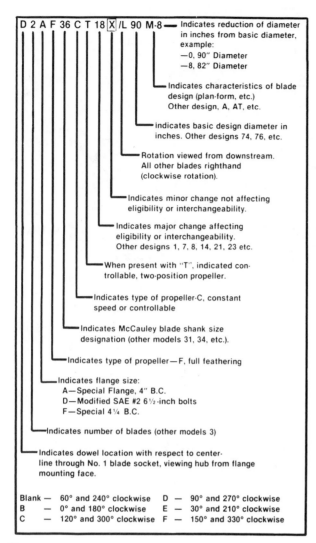

Fig. 6B-14 McCauley propeller designation system

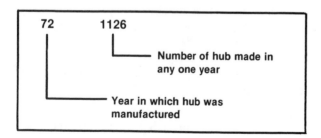

Fig. 6B-15 McCauley propeller and governor serial numbering system

of the governor, with one turn of the screw changing the RPM by 17, 20, or 25 RPM, depending upon the engine gear ratio and the governor.

When the propeller and governor are installed, the cockpit control is rigged for the proper cushion. Check the system for the proper operation on the ground, correcting adjustments and rigging as necessary to obtain the correct maximum RPM, the correct response to the cockpit control movement, and correct cushion. It should be noted that some systems cannot reach their rated RPM on the ground. A test flight should now be performed to check the operation and to check for oil leaks.

Install the propeller spinners and spacers in accordance with the applicable aircraft service manual.

4. Inspection, maintenance, and repair

The propeller should be inspected for surface defects on the blades and hub areas, security of the blades in the hub, proper safety installation, oil leaks, and security of mounting bolts and nuts.

Oil found coming from the hub breather holes indicates a defective internal O-ring. The piston-to-cylinder O-ring is often the cause of this leak. On some models, this can be replaced by a mechanic following the procedures outlined in the propeller or aircraft service manual. On other models, the propeller will have to be returned to a repair facility. If other seals are the cause of the leak, they will have to be replaced by a repair facility.

A dye-penetrant inspection of the blade retention areas of the hub and of the blade shanks is advisable at each 100-hour and annual inspection, but be sure to remove all of the residue after the inspection, as some of the substances used are corrosive.

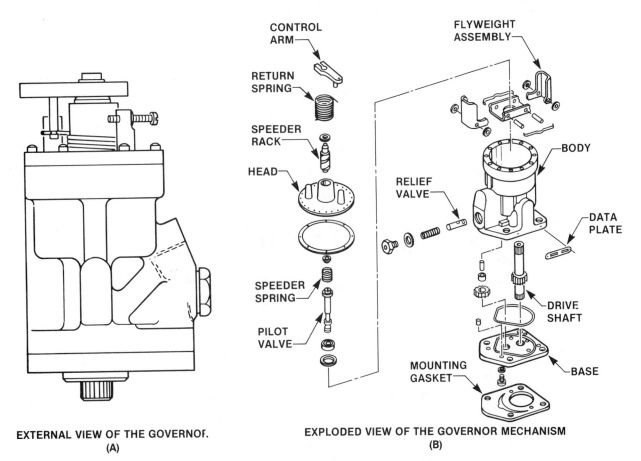

Fig. 6B-16   McCauley constant-speed governor

EXTERNAL VIEW OF THE GOVERNOR.
(A)

EXPLODED VIEW OF THE GOVERNOR MECHANISM
(B)

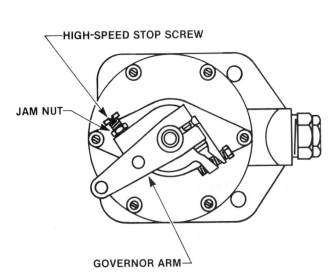

Fig. 6B-17   High-RPM stop screw adjustment on a McCauley propeller governor

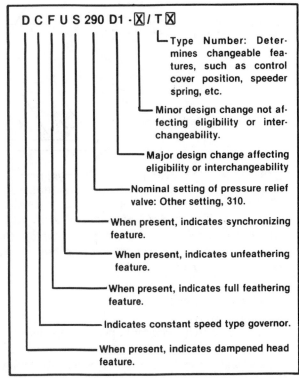

Fig. 6B-18 McCauley governor designation system

531

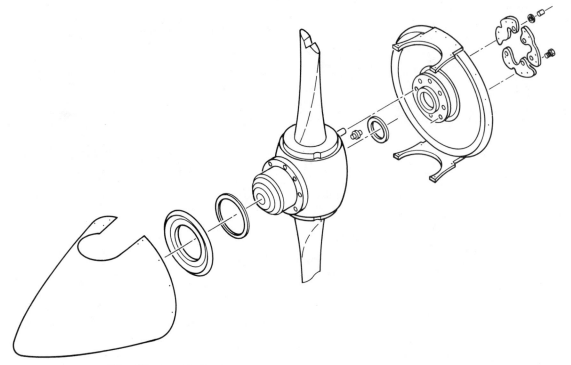

Fig. 6B-19 *McCauley constant-speed propeller installation*

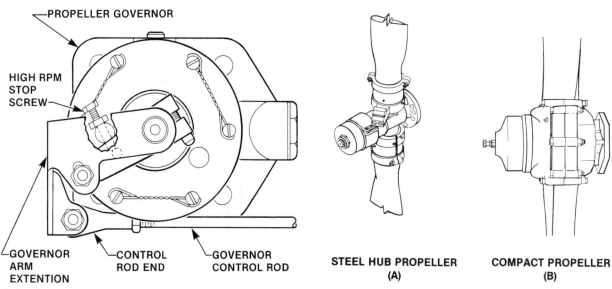

PROPELLER GOVERNOR

HIGH RPM
STOP
SCREW

GOVERNOR
ARM
EXTENTION

CONTROL
ROD END

GOVERNOR
CONTROL ROD

Fig. 6B-20 *The amount of travel of a propeller governor control arm may be regulated by the amount the governor control rod is screwed into the control rod end.*

STEEL HUB PROPELLER
(A)

COMPACT PROPELLER
(B)

Fig. 6B-21 *Hartzell constant-speed propellers*

C. Hartzell Propeller

The Hartzell constant-speed propeller systems are used in modern general aviation airplanes and share the market with McCauley. Hartzell systems are used extensively on Piper aircraft and on many other designs.

1. The propellers

Hartzell produces two styles of constant-speed propellers—a steel hub propeller and a "Compact" model. Steel hub propellers are identified by their exposed operating mechanism, while Compact models have the pitch changing mechanism inside the hub assembly.

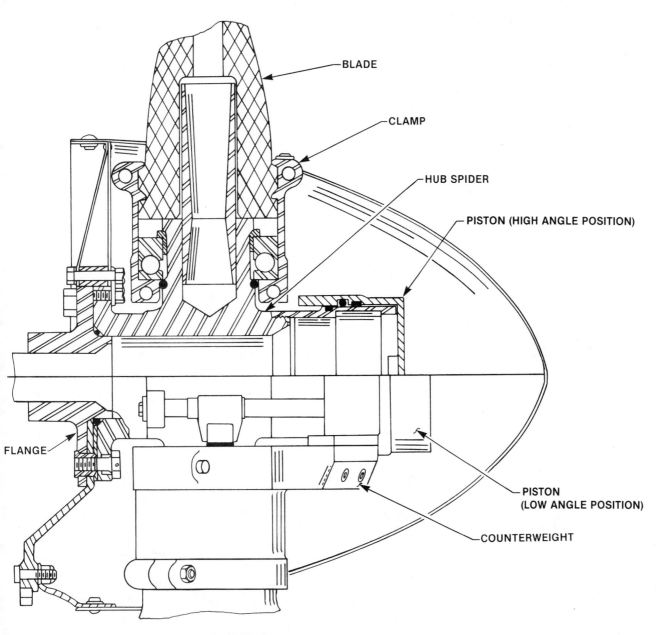

Fig. 6B-22   Hartzell steel hub propeller

Some models of the steel hub propellers use oil pressure to decrease blade angle and the centrifugal force on the counterweights to increase blade angle. Other models of the steel hub propellers have no counterweights and use centrifugal twisting moment to decrease the blade angle and oil pressure to increase blade angle.

Hartzell steel hub propellers use a steel spider as the central component. Bearing assemblies and aluminum blades are placed on the

spider arms and are held in place by two-piece steel clamps. A steel cylinder is screwed onto the front of the spider and an aluminum piston is placed over the cylinder. The piston is connected to the blade clamps by steel link rods. (Note that the terms "piston" and "cylinder" are reversed when compared to a Hamilton-Standard counterweight propeller.)

During operation, oil pressure is directed to the propeller piston through the engine crankshaft where it causes a change in blade angle.

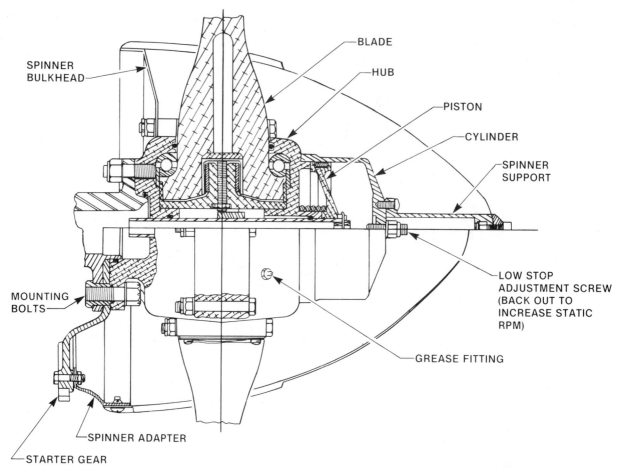

SPINNER
BULKHEAD

BLADE

HUB

PISTON

CYLINDER

SPINNER
SUPPORT

LOW STOP
ADJUSTMENT SCREW
(BACK OUT TO
INCREASE STATIC
RPM)

GREASE FITTING

MOUNTING
BOLTS

SPINNER ADAPTER

STARTER GEAR

Fig. 6B-23   Hartzell Compact propeller

Counterweight models use oil pressure to decrease the blade angle, and centrifugal force on the counterweights increases the angle. Noncounterweighted models use oil pressure to increase the blade angle and centrifugal twisting moment to decrease the angle.

Some steel hub propellers are made so they will mount on flanged crankshafts, and other mount on splined shafts. Compact designs are normally used only with flanged crankshafts.

Hartzell Compact propellers use aluminum blades mounted in an aluminum hub. The hub is held together with bolts and contains the pitch changing mechanism of the propeller. This mechanism consists of a piston, piston rod, and actuating links.

The Compact propeller uses governor oil pressure to increase blade angle and the centrifugal twisting moment acting on the propeller blades to decrease blade angle.

The Hartzell propeller designation system is the same for steel hub and Compact propeller models, as we see in the sample designations in Fig. 6B-24.

2.   *Governors*

Hartzell governors may be reworked Hamilton-Standard governors or Woodward governors. The Hartzell governor designation system is shown in Fig. 6B-25.

3. Installation and adjustments

Hartzell propellers are installed following the same basic practices as those previously discussed.

The steel hub propellers can be adjusted for the desired low blade angle by loosening the hub clamps and rotating the blades in the clamps until the desired blade angle is obtained. Then clamps are retorqued and safetied. This also changes the high blade angle, since the range between high and low blade angle is fixed by the range of the piston movement.

The low pitch setting of the Compact propellers can be adjusted by loosening the lock nut on the adjusting screw on the hub cylinder and rotating the screw in to increase the low blade angle or out to decrease the blade angle. When the desired angle is set, retighten the lock nut.

When changing the blade angles, always refer to the aircraft specifications and the propeller manufacturer's manual for instructions about specific propeller models.

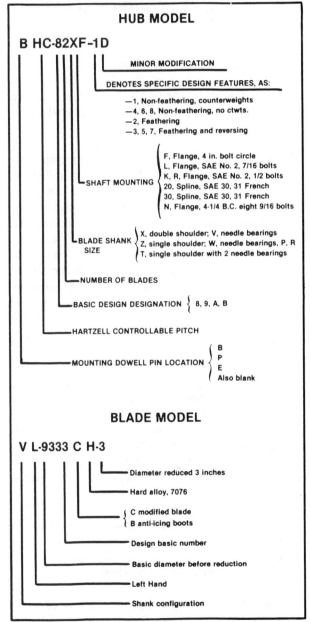

Fig. 6B-24 *Hartzell propeller and blade designation system*

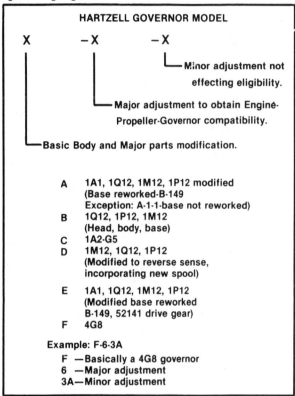

Fig. 6B-25 *Hartzell governor designation system*

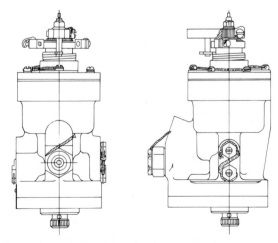

Fig. 6B-26 *Woodward governors for use with Hartzell constant-speed propellers*

4. Inspection, maintenance, and repair

The Hartzell constant-speed propeller system requires basically the same inspection, maintenance, and repair as for systems previously discussed. However, special care should be taken when lubricating the propeller blades to prevent damage to the blade seals.

Before lubricating a blade, remove one of the two zerk fittings for the blade and grease the blade through the remaining zerk fitting. This will prevent any pressure building up in the blade grease chamber and prevent damage to the blade seals. Some propeller models are serviced until grease comes out of the hole of the removed fitting. Other models require less grease, so check the service manual before lubricating the propeller. After lubrication, reinstall the zerk fitting, replace the protective cap, and safety it.

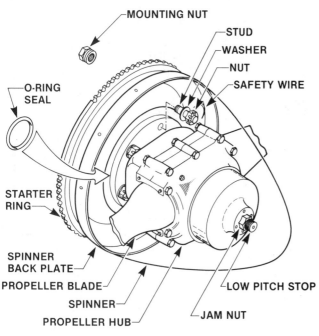

Fig. 6B-27 Hartzell Compact propeller installation

5. Troubleshooting

One of the principle troubleshooting practices on a Hartzell propeller is the determination of the source of a grease leak.

A grease leak is readily noticeable, and the cause should be determined and corrected as soon as possible. The most common causes of grease leakage are loose or missing zerk fittings, defec-

tive zerk fittings, loose blade clamps, defective blade clamp seals, and over-lubrication of the blades.

If the zerk fitting is loose, missing, or defective, it should be tightened or replaced as appropriate. Loose blade clamps should be torqued to the specified value for the particular model of propeller and resafetied. Be sure to check to make certain that the blade angle is not changed after retorquing.

VII. FEATHERING PROPELLER SYSTEMS

Feathering propellers are used on most modern multi-engine airplanes. The primary purpose of a feathering propeller is to eliminate the drag created by a windmilling propeller when an engine fails.

Feathering propeller systems are constant-speed systems with the additional capability of being able to feather the blades. This means that the blades can be rotated to an approximate 90 degree blade angle. The constant-speed controls and operations we have discussed in previous sections apply to the feathering propeller system, but the cockpit propeller control lever incorporates an additional range of movement to allow the propeller to feather, or else a separate cockpit control may be used to operate the feathering mechanism.

Feathering functions are independent of the constant-speed operation and can override the constant-speed operation to feather the propeller at any time. The engine does not have to be developing power, and in some systems the engine does not even have to be rotating for the propeller to feather. In short, propellers are feathered by forces which are totally independent of engine operation.

It should be noted that when the propeller is feathered, the engine stops rotating.

A. McCauley Feathering System

The McCauley feathering system is used on all current production piston-engined Cessna twins and on many twin-engined Beechcraft airplanes. The system incorporates a feathering pro-

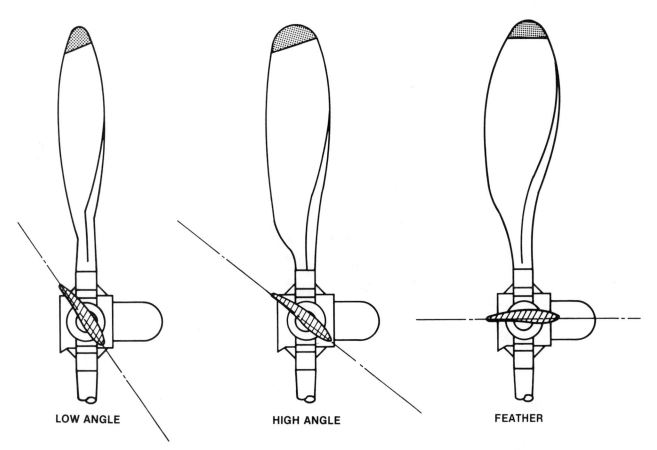

Fig. 7B-1 When a propeller is feathered, its blades are turned to an angle of approximately 90 degrees to the plane of propeller rotation.

LOW ANGLE   HIGH ANGLE   FEATHER

peller, governor and cockpit control levers which control both the constant-speed and feathering operations of the system. Unfeathering accumulators are available as optional equipment.

1.   Propellers

The outward appearance of a McCauley feathering propeller is similar to that of the constant-speed propellers except for the longer cylinder, which gives the propeller a greater blade angle range. These propellers also have counterweights. Feathering propellers use oil pressure from the governor to decrease blade angle and use the force of springs and counterweights to increase the blade angle and to feather the blades.

The propeller is spring-loaded and counterweighted to the feather position at all times so that if oil pressure is lost, the propeller will automatically feather. To prevent the propeller feathering when the engine is stopped on the ground, a spring-loaded latch mechanism engages at some low RPM—for example, 900 RPM. This prevents excessive load on the starter and engine system when starting the engine.

Three different latch mechanisms have been used with the McCauley feathering propellers: the inertial latch, the pressure latch, and the centrifugal latch. Since the centrifugal latch has proved to be the best system, most propellers either have been converted to this style or should be converted during the next overhaul.

The propeller designation system is the same as is used for the McCauley constant-speed propellers.

2.   Governors

Governors used with McCauley feathering propellers are basically the same as other constant-speed governors, except that the governor directs oil pressure to the propeller to decrease the blade angle and releases oil from the propeller to increase blade angle.

Feathering governors incorporate a lift rod connected to the speeder rack. This rod mechanically lifts the pilot valve and releases the oil from the propeller when the cockpit control lever is moved to the Feather position which is fully aft.

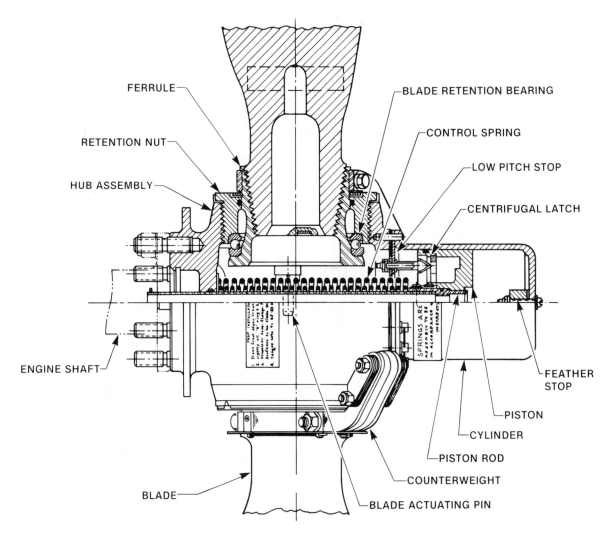

FERRULE

RETENTION NUT

HUB ASSEMBLY

ENGINE SHAFT

BLADE

BLADE RETENTION BEARING

CONTROL SPRING

LOW PITCH STOP

CENTRIFUGAL LATCH

FEATHER STOP

PISTON

CYLINDER

PISTON ROD

COUNTERWEIGHT

BLADE ACTUATING PIN

Fig. 7B-2   McCauley feathering propeller

This action may take place at any time and is independent of the position of the flyweights or the speeder spring.

If the feathering system includes an accumulator, a ball check valve is incorporated in the governor, so the accumulator can be charged during normal operation and kept from discharging during feathering. The ball check valve is released by a push rod when the propeller control is moved forward to the High RPM position. At this time, the stored oil pressure in the accumulator will be released to unfeather the propeller.

The governor designation system is the same as is used for the McCauley constant-speed governors.

3. Accumulator

The unfeathering accumulator may either be of the ball or cylinder type, and it contains a diaphragm or piston which is used to separate the air charge from the oil charge. An air preload charge of approximately 100 psi is used. The oil side is charged by the governor through a flexible line, and the accumulator stores oil at governor pressure of about 290 psi.

4. System operation

The constant-speed operation of the McCauley feathering propeller system is exactly the same as the McCauley constant-speed system, except for the change in oil flow direction.

Feathering of the McCauley system is done by moving the cockpit propeller control lever full aft. When this is done, the governor lever arm is pulled to the low RPM stop. The RPM lift rod in the governor raises the pilot valve and releases oil pressure from the propeller. Without oil pressure in the propeller, it is taken to the feather angle by a combination of spring force and centrifugal force on the blade counterweights. When the propeller blades are fully feathered, engine rotation stops. If an accumulator is used, the ball check valve in the governor holds a charge of oil pressure in the accumulator.

If an accumulator is used to unfeather the propeller, the governor ball check valve is opened by the push rod when the cockpit control is moved forward. This allows the oil pressure in the accumulator to flow through the governor to the propeller cylinder and force the blades to a lower angle.

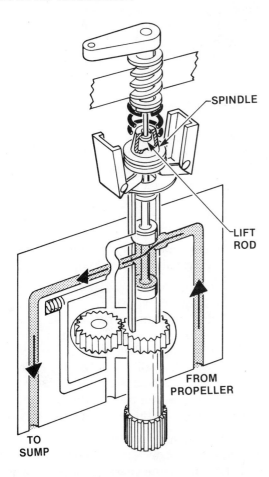

Fig. 7B-4 *Action of the governor lift rod during feathering*

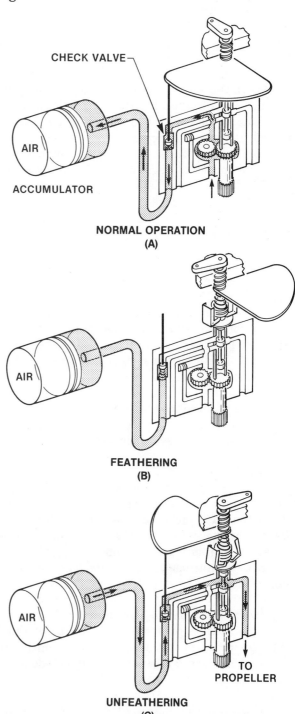

Fig. 7B-5 *Action of the governor push rod, ball check valve, and accumulator during the feathering and unfeathering operation of a propeller*

To unfeather the propeller without an accumulator, move the cockpit control forward, and turn the engine over with the starter. This allows governor oil pressure to build up and overcome the force of the springs and move the blades to a lower angle. As this blade angle decreases, the propeller will begin to windmill and help complete the unfeathering operation.

539

When the engine is stopped on the ground, it should be idled at about 800 RPM, with the propeller pitch control in the High RPM (full forward) position. When the blades are at this low blade angle and the engine is shut down, they will try to feather because of the loss of oil pressure. But feathering is prevented by a spring force which engages the latch mechanism, and the blades are held by latch plates a few degrees above the low blade angle as they move toward feather.

When the engine is started on the ground, the cockpit control should always be full forward. This causes the blade angle to decrease as governor oil pressure is generated, and the blades will move to the low blade stop from the latch angle. As RPM increases, these latch plates move outward to free the propeller blades to move through their full range of travel.

5. System maintenance

The installation and adjustment of a feathering propeller and governor are basically the same as those used in the constant-speed models. If an accumulator is used, it is normally installed in the engine compartment in accordance with standard airframe practices, and is connected to the governor with flexible hose similar to those used in hydraulic systems.

In addition to the troubleshooting procedures discussed for constant-speed systems, the feathering system may have some additional operating difficulties.

If the propeller will not feather in flight, the problem may be that the governor low RPM stop is set too high, preventing the lift rod from raising the pilot valve. It may also be caused by the cockpit controls being improperly rigged which restricts or prevents full movement of the governor lever arm.

If the propeller high blade angle is not correctly set, the propeller may continue to windmill after the propeller feathers. Refer the propeller to an overhaul facility to adjust the high blade angle stop.

If the propeller fails to unfeather in flight and an accumulator system is not used, the problem may be that the starter cannot generate sufficient RPM to restart the engine or develop sufficient

oil pressure in the governor to unfeather the propeller. This is a common problem in some older aircraft and involves a change in pilot technique and/or the addition of accumulators. Some aircraft can be unfeathered more easily if the aircraft is placed in a shallow dive to increase airspeed and the engine is then rotated with the starter.

If the propeller fails to unfeather with an accumulator installed, one of the following problems may be the cause: low air pressure in the accumulator; oil leaking from the accumulator or flex hose; or, a leak from the air side of the accumulator to the oil side. External air leaks can be located by pressurizing the system to normal operating pressure and checking for leaks with soapy water.

B. Hartzell Feathering System

Hartzell feathering systems are used on many current production Piper piston-driven twins, some Beechcraft and Aero Commander twins, and older Cessna twin-engine aircraft.

1. Propellers

Both compact and steel hub propeller designs are used for Hartzell feathering propellers.

Compact designs use governor oil pressure to decrease the blade angle. Air pressure in the propeller cylinder and, in some models counterweights and springs, are used to feather the propeller. A latch stop (called the automatic high pitch stop by Hartzell) is located inside the cylinder to hold the blades in a low blade angle when the engine is stopped on the ground. The latch mechanism is composed of springs and locking pins.

The Hartzell steel hub feathering propeller uses oil pressure to decrease blade angle and a combination of springs and counterweights to increase the propeller blade angle. An external latch mechanism is used to prevent the propeller from feathering when the engine is stopped on the ground.

2. Governors

Hartzell feathering systems may use either Hartzell or Woodward governors for operation. The governors may incorporate an internal mechanism with a lift rod and accumulator oil

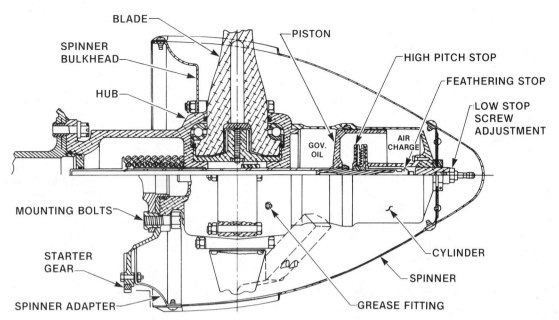

Fig. 7B-6   Hartzell Compact feathering propeller

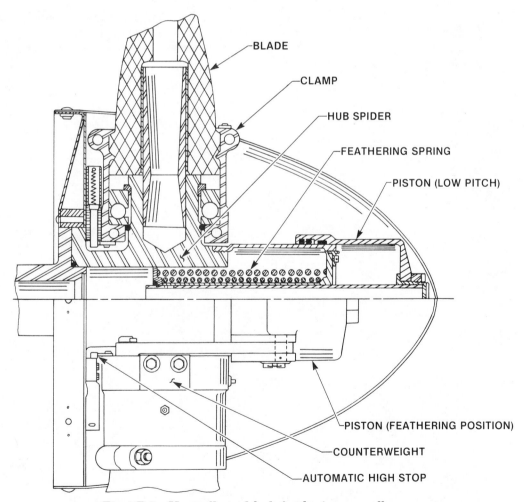

Fig. 7B-7   Hartzell steel hub feathering propeller

81

passages as well as valves. Or, they may have an external adapter which contains a shutoff valve linked to the governor control arm to control the accumulator operation.

3. System operation

The constant-speed operation of the Hartzell feathering propellers is the same as for the constant-speed models, except for the change in direction of oil flow in some models.

When the Hartzell propellers are feathered, the cockpit control is moved full aft and the governor pilot valve is raised by the lift rod to release oil from the propeller. With the oil pressure released, a steel hub propeller will move to the feather position by the force of the counterweights and/or springs. The Compact models will go to feather by the force of the air pressure in the cylinder and, in some models, the force of the counterwights and/or springs. The blades are held in feather by the spring force or by air pressure.

When unfeathering the propeller in flight, the system relies on engine rotation by the starter to initiate the unfeathering operation unless an ac-

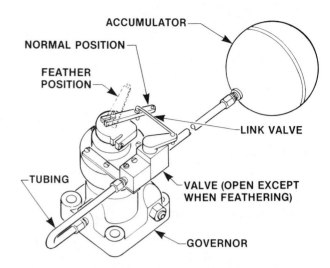

Fig. 7B-8 *Propeller governor with an external accumulator adapter to aid in unfeathering the propeller in flight.*

cumulator is used. The operation of the accumulator is the same as for the McCauley system.

When shutting the engine down after flight, the propeller cockpit control should be placed in full forward position while the engine is idling. This causes the spring in the latch mechanism to

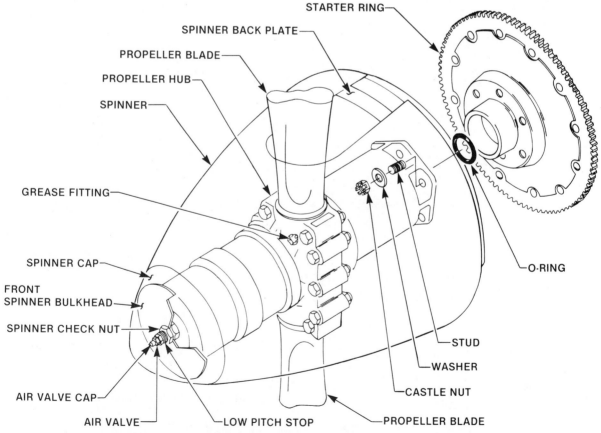

Fig. 7B-9 *Typical installation of a Hartzell Compact feathering propeller*

force the lock pin into a low-pitch lock position and to engage when the engine is shut down and the blades attempt to rotate toward feather.

4. System maintenance

Installation and adjustment of the propellers are the same as for the constant-speed models. If the blades are feathered, they can be rotated to the latch angle by placing a blade paddle on each blade and rotating the blades simultaneously to the latch angle. The air pressure in the Compact model propellers should be checked and serviced as necessary with the propeller at the latch angle.

Governor installation and adjustment is the same as for a constant-speed governor.

Accumulators are installed in accordance with the aircraft maintenance manual and are ser-viced with dry air or nitrogen to the value specified in the aircraft maintenance manual.

The inspection, maintenance, and repair of the Hartzell feathering propeller system are the same as described for other constant-speed and feathering systems. The Compact propeller air pressure should be checked at each 100-hour and annual inspection.

The system troubleshooting procedures are the same as described for other propeller systems.

If the air charge is too low in the Compact propeller, it may not feather or respond properly to constant-speed operation, and it may have a tendency to overspeed or surge. If the air charge is too great, the system may not reach full RPM and may feather when the engine is shutdown on the ground.

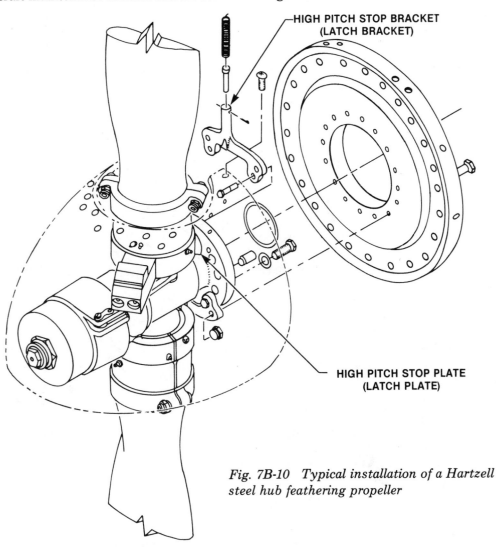

HIGH PITCH STOP BRACKET
(LATCH BRACKET)

HIGH PITCH STOP PLATE
(LATCH PLATE)

Fig. 7B-10 Typical installation of a Hartzell steel hub feathering propeller

C. Hamilton-Standard Hydromatic System

The Hamilton-Standard feathering system is used on many medium and large reciprocating engine transports. The Hamilton-Standard design goes by the trade name of *Hydromatic*, indicating that the principal operating forces are oil pressure.

1. Propellers

The Hamilton-Standard Hydromatic propeller is made up of three major assemblies—the hub or barrel assembly, the dome assembly, and the distributor valve.

The barrel assembly contains the spider, blades, blade gear segments, barrel halves, and necessary support blocks, spacers, and bearings. Front and rear cones, a retaining nut, and a lock ring are used to install the barrel assembly on the splined crankshaft.

The dome assembly contains the pitch-changing mechanism of the propeller and includes the dome shell, a piston, a rotating cam, a stationary cam, and two blade angle stop rings. The dome shell acts as the cylinder for the propeller piston.

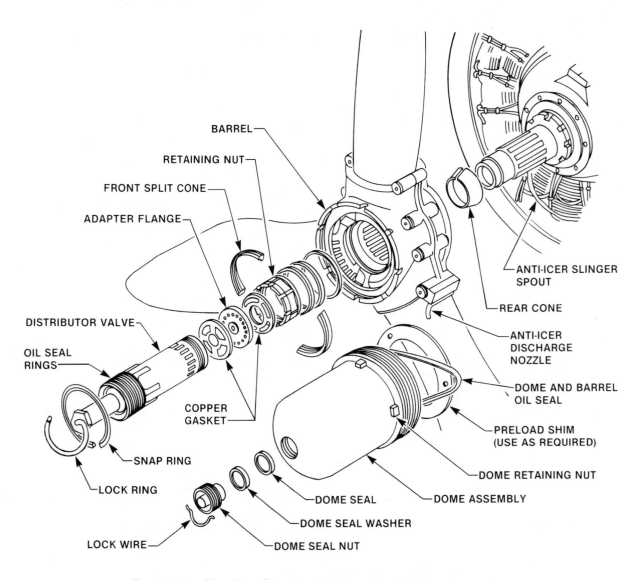

Fig. 7B-11   Hamilton-Standard Hydromatic feathering propeller

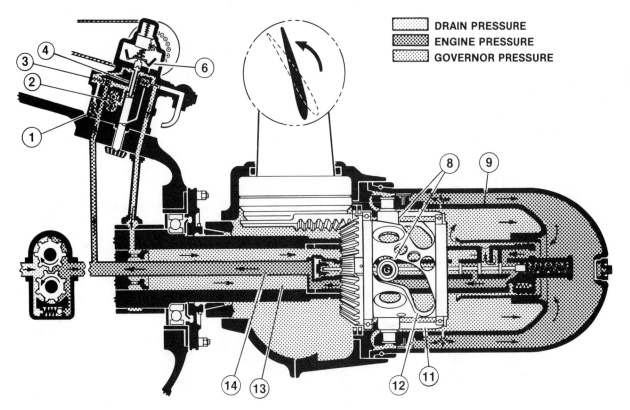

1	GOVERNOR DUMP VALVE	9	DOUBLE ACTING PISTON
2	GOVERNOR BOOSTER PUMP	11	FIXED CAM
3	GOVERNOR RELIEF VALVE	12	ROTATING CAM
4	PILOT VALVE	13	PROPELLER SHAFT GOVERNOR OIL PASSAGE
6	FLYWEIGHTS	14	PROPELLER SHAFT ENGINE OIL PASSAGE
8	CAM ROLLERS		

Fig. 7B-12 Hydromatic propeller installation in an overspeed condition

The distributor valve is used to direct oil from the crankshaft to the inboard and outboard side of the piston and is shifted during unfeathering to reverse the oil passages to the piston.

2. Governors

The feathering Hydromatic governor includes all of the basic governor components discussed for constant-speed governors. And, in addition, the Hydromatic governor contains a high-pressure transfer valve which is used to block the governor constant-speed mechanism out of the propeller control system when the propeller is feathered or unfeathered.

A pressure cut-out switch is located on the side of the governor and is used to automatically stop the feathering operation.

3. Feathering system components

The cockpit control for the feathering system is a push button approximately 1-1/4 inches in diameter and is used to feather and unfeather the propeller. The feathering button incorporates a holding coil to electrically hold the button in after it is pushed.

An electrically operated feathering pump is used to supply oil under high pressure of about 600 psi to the propeller when the feathering system is actuated. The pump takes its oil from the engine oil supply tank, at a level below the standpipe feeding the engine lubrication system.

4. System operation

The Hydromatic propeller uses governor oil pressure on one side of the propeller piston, opposed by engine oil pressure on the other side of

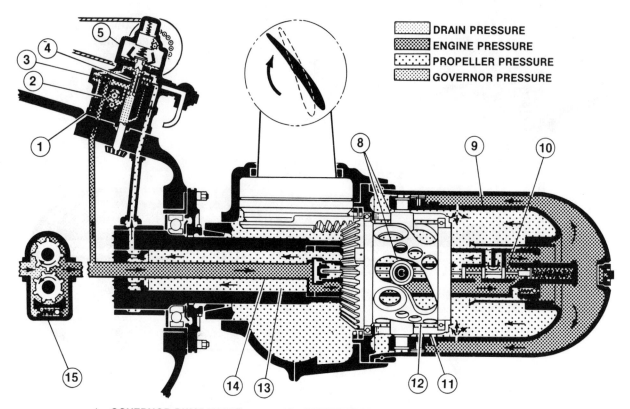

1	GOVERNOR DUMP VALVE	10	DISTRIBUTOR VALVE
2	GOVERNOR BOOSTER PUMP	11	FIXED CAM
3	GOVERNOR RELIEF VALVE	12	ROTATING CAM
4	PILOT VALVE	13	PROPELLER SHAFT GOVERNOR OIL PASSAGE
5	GOVERNOR	14	PROPELLER SHAFT ENGINE OIL PASSAGE
8	CAM ROLLERS	15	ENGINE OIL PUMP
9	DOUBLE ACTING PISTON		

Fig. 7B-13 Hydromatic propeller installation in an underspeed condition

the piston. Depending on the model of the propeller, governor oil pressure may be directed to the outboard side or inboard side of the piston.

For discussion purposes, consider that governor oil pressure is on the inboard side of the propeller piston and engine oil pressure is on the outboard side of the piston.

The Hydromatic propeller does not use any springs or counterweights for operation. The fixed force is the engine oil pressure, which is about 60 psi. And the governor oil pressure (200 or 300 psi, depending on the system) is controlled by the pilot valve during constant-speed operation.

When the system is in an overspeed condition, the pilot valve in the governor is raised and governor oil pressure flows to the inboard (rear) side of the propeller piston via the crankshaft transfer bearing and the distributor valve. The increase in pressure on the inboard side of the piston forces the piston outboard (forward). As the piston moves outboard, it rotates, following the slot in the stationary cam, and this causes the rotating cam to rotate. As the rotating cam turns, the gears on the bottom of the cam mesh with the gears on the blade segment and rotate the blade to a higher blade angle. With this increase in blade angle, the system RPM slows down and the governor returns to the onspeed condition. The oil in the outboard side of the piston is forced back into the engine lubrication system.

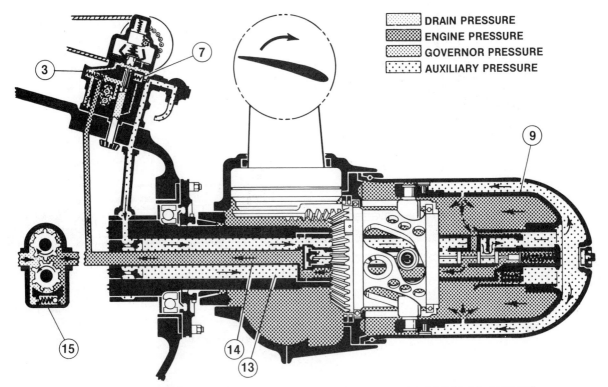

3	GOVERNOR RELIEF VALVE	13	PROPELLER SHAFT GOVERNOR OIL PASSAGE
7	GOVERNOR HIGH PRESSURE TRANSFER VALVE	14	PROPELLER SHAFT ENGINE OIL PASSAGE
9	DOUBLE ACTING PISTON	15	ENGINE OIL PUMP

Fig. 7B-14 Hydromatic propeller installation being feathered

When the system is underspeed, the pilot valve is lowered, and the governor oil pressure in the inboard side of the piston is released. This causes engine oil pressure on the outboard side of the piston to force the piston inboard. As the piston moves inboard, the rotation created by the piston and the cams causes the blades to rotate to a lower blade angle, allowing system RPM to increase to the onspeed condition.

To feather the propeller, the feather button in the cockpit is pushed. When this is done, electrical contacts close and energize the holding coil which holds the feather button in. Another set of feather button electrical contacts close at the same time in and cause the feathering relay to close.

The feathering relay completes the circuit from the battery to the auxiliary pump, and the high-pressure oil generated by the pump shifts the high-pressure transfer valve in the governor to block the governor constant-speed components out of the system. This high-pressure oil is then directed to the inboard side of the piston, and it moves the blades toward the feather angle.

When the rotating cam contacts the high-pitch stop, the piston stops moving and the blades have reached the feather angle. Since the piston cannot move any further, the pressure in the system starts to build rapidly. This increasing pressure is sensed by the pressure cutout switch on the governor, and it breaks the circuit to the feather button holding coil when the pressure reaches about 650 psi. This releases the feather relay and shuts off the auxiliary pump. With the engine stopped and the propeller in feather, all oil pressures drop to zero. The blades are held in their full-feather position by aerodynamic forces.

547

To unfeather the propeller, the feather button is pushed and held in to prevent the button popping back out when the pressure cutout switch opens. The auxiliary pump starts building pressure above the setting of the pressure cutout switch. This causes the distributor valve to shift and reverse the flow of oil to the piston. Auxiliary pump pressure is then directed to the outboard side of the piston, and engine oil lines are open to the inboard side of the piston. The piston moves inboard and causes the blades to rotate to a lower blade angle through the action of the cams. With this lower blade angle, the propeller starts to windmill, allowing the engine to be restarted.

At this point, the feather button is released and the system will return to constant-speed operation. If the feather button is not released, the dome relief valve in the distributor valve will off-seat and release excess oil pressure (above 750 psi) from the outboard side of the piston after the rotating cam contacts the low blade angle stop.

5. System maintenance

The installation of a Hydromatic propeller requires the barrel assembly to be installed first, following the basic procedure used to install a fixed-pitch propeller on a splined crankshaft. A hoist is required for most Hydromatic propellers because of their size and weight, and special tools are necessary to tighten the retaining nut. The standard checks for proper cone seating and spline wear are made when installing the barrel assembly.

The distributor valve gasket and distributor valve are installed in the crankshaft. The distributor valve is carefully screwed into the internal threads on the crankshaft, and it is then torqued to the value specified for the particular installation. The distributor valve and retaining nut are now safetied to the crankshaft with a special lock ring.

The dome assembly is prepared for installation on the barrel assembly. The pitch stop rings must first be installed so that the propeller will have the proper blade angle range. Refer to the appropriate service manual to set these rings.

Turn the rotating cam until the lugs on the high-pitch stop ring contact the dome stop lugs. This sets the dome in its feather position.

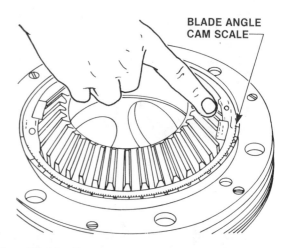

Fig. 7B-15 Stop rings inside the dome of a Hydromatic propeller

Rotate the blades in the barrel until the scales on the blade shanks indicate that each of the blades is in its feather position. Install the dome shim on the dome shelf in the barrel, and place the base gasket on the bottom of the dome. Install the dome on the barrel, following the procedure recommended by the manufacturer or the particular model of propeller. Torque and safety the dome and install and safety the dome plug.

Check the track of the propeller and the low- and high-blade angles of each blade.

The governor is installed and rigged in the same manner as for the Hamilton-Standard counterweight propeller. The only additional steps required are to connect the electrical plug to the pressure cutout switch and attach the oil line from the auxiliary pump.

If an electric head is used on the governor, governor rigging is simplified, since the electrical plug for the head needs only to be connected and the high and low RPM limit switches set.

The auxiliary pump, feather button, and other system components are installed and adjusted according to the particular aircraft service manual.

The propellers and governors are inspected and repaired in accordance with the procedure discussed in the previous sections of this text. The inspection primarily involves assuring proper operation, checking for oil leaks, and inspecting the external oil lines for signs of deterioration or abrasion.

548

Oil leaks in the propeller are normally caused by a faulty gasket or a loose nut or bolt. If oil covers all of the propeller, the dome plug is leaking. If oil appears on the barrel immediately behind the dome, the dome gasket is leaking or the dome nut is loose. The dome plug seal and the dome-to-barrel seal can be replaced in the field.

If oil comes from the blade shank area or from between the barrel halves, the hub bolts may be loose or the gaskets may be defective. If no irregularities are found, the bolts may be retorqued, but the gaskets must be replaced by an overhaul facility.

The propeller is lubricated by operating oil and it needs no other lubrication.

Troubleshooting procedures and solutions discussed for other systems are generally applicable to the feathering Hydromatic system.

If the propeller fails to respond to the cockpit propeller control lever, but can be feathered and unfeathered, the cause is most likely a failure of the governor or governor control system.

If the propeller fails to feather, check the system for electrical faults or for open wiring to the electrical components.

If the propeller fails to unfeather after feathering normally, the distributor valve is not shifting.

If the propeller feathers and immediately unfeathers, the problem may be a shorted line from the holding coil to the pressure cutout switch or a defective pressure cutout switch. The same thing will happen if the feather button is shorted internally.

Sluggish movement of the propeller may be the result of a buildup of sludge in the propeller dome or a worn out piston-to-dome seal inside the dome.

Erratic or jerky operation of the propeller is an indication of the wrong preload shim being used between the dome and the barrel assemblies, and the dome will have to be removed and the proper shim installed.

VIII. REVERSING PROPELLER SYSTEMS

A reversing propeller is usually a constant-speed, feathering propeller with the additional capability of producing a reverse thrust.

Reversing propeller systems are used on most modern multi-engine turboprop aircraft such as the Cessna Conquest, Beech King Air, Piper Cheyenne and on large transport aircraft such as the Douglas DC-7, and the turboprop conversions on such airplanes as the Convair 580. Some seaplanes and floatplanes use reversing propellers to improve their water maneuverability.

Reversing propellers have the advantage of decreasing the length of the landing roll, reducing brake wear, and increasing ground maneuverability. Some aircraft can use the reversing system to back the aircraft on the ground, while other designs allow the system to be used only to brake the aircraft during the landing roll.

The main disadvantages of the reversing system are the reduced engine cooling and increased blade damage caused by stones, sand, etc., when the propellers are in reverse.

When a propeller goes into reverse, the blades rotate below the low blade angle and into a negative angle of about −15 degrees. This forces the air forward and produces a reverse thrust.

The reversing operation is controlled in the cockpit by the throttles and is initiated by moving the throttles aft of the idle position. This reverses the blades, and the engine RPM and/or propeller blade angle is varied by moving the throttles within the reverse range to control the amount of reverse thrust. The farther aft the throttles are moved, the greater the reverse thrust produced.

Propellers cannot normally be reversed in flight, and the aircraft must often be below a specified airspeed on the landing roll before the reverse mechanism is engaged. Many systems require that the aircraft weight be on the landing gear before the throttles can be moved into the reverse range. This is controlled by a squat switch on the landing gear strut.

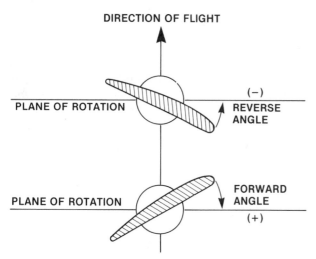

Fig. 8B-1 *When a propeller is put in reverse angle, the blades assume a negative angle.*

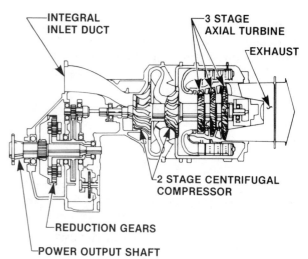

Fig. 8B-2 *Cutaway view of a Garrett-AiResearch TPE-331 turboprop engine*

In this section of the text, we will discuss the Hartzell reversing propellers used with the Garrett AiResearch TPE-331, and the Pratt and Whitney of Canada PT6 engines.

A. Hartzell Reversing Propeller System used on the Garrett AiResearch TPE-331 Engine

The Hartzell propeller on the TPE-331 is used on aircraft such as the Mitsubishi MU-2, the Short Skyvan, and the Aero Commander 690.

The TPE-331 engine is a fixed turbine engine that produces more than 600 horsepower when the engine is turning at about 40,000 RPM. A reduction gear assembly on the front of the engine couples the engine drive shaft to the propeller drive shaft and reduces the engine speed to about 2200 RPM at the propeller drive shaft.

The engine gear ratio and the installation configuration can be one of two basic designs. Some engines have the propeller drive shaft below the engine centerline, and other installations have the propeller above the engine centerline.

1. Propeller

The propeller commonly used on the TPE-331 is a three- or four-bladed Hartzell steel hub reversing propeller. The propeller is spring-loaded and counterweighted to the feather position and uses engine oil pressurized by the governor to decrease the blade angle. The propeller is flange-mounted on the drive shaft, and it locks in a flat angle of about two degrees when the engine is

shut down on the ground. This prevents excessive strain on the engine starter when the engine is being started.

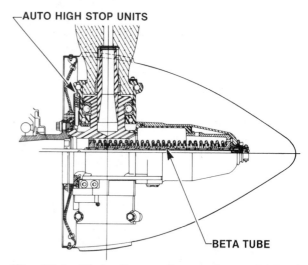

Fig. 8B-3 *Hartzell propeller used on a TPE-331 engine*

The propeller is constructed similar to the feathering steel hub designs. The principal additional component is the Beta tube, which passes through the center of the propeller and serves as an oil passage and follow-up device during propeller operation.

The same designation system is used for the reversing Hartzell propellers as is used for Hartzell feathering and constant-speed propellers.

The cockpit controls for the TPE-331 turboprop installation include a power lever which

550

controls the horsepower output of the engine, a speed lever which controls system RPM, a feather handle, and an unfeathering switch.

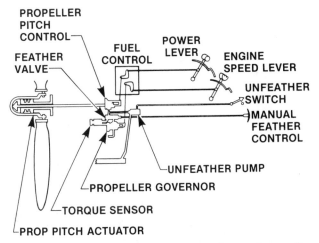

Fig. 8B-4 *System components located on the gear reduction assembly of a TPE-331 engine*

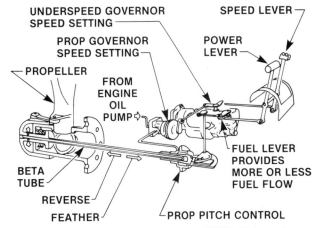

Fig. 8B-5 *Control system for a TPE-331 engine*

The power lever is similar to the reciprocating engine throttle in that it controls system horsepower. During ground operation, the power lever directly controls the propeller blade angle by positioning the propeller pitch control unit, and during flight operations, the power lever directly controls the engine fuel control unit.

The speed lever is similar to the propeller control lever in a reciprocating engine system in that it controls the system RPM. During ground operation, the speed lever adjusts the underspeed governor on the fuel control unit to vary the fuel flow and maintain a fixed RPM as the blade angle is changed by the power lever. During flight operations, the speed lever sets the RPM on the propeller governor which keeps the RPM constant by varying the blade angles as the engine power is changed with the power lever or when flight operations change.

Many aircraft use a feather handle connected to the feathering valve on the engine. Other aircraft connect the feathering valve to the speed lever so that full aft movement of the lever will cause the propeller to feather. When the feathering valve is moved by the cockpit control, oil is released from the propeller and the propeller feathers.

An unfeathering switch is used to control the electric unfeathering pump to unfeather the propeller.

2. System operation

The two basic operating modes of the TPE-331 system are the Beta mode, meaning any ground operation including start, taxi, and reverse operation, and the Alpha mode, which is any flight operation from takeoff to landing. Typically, Beta mode includes operation from 65% to 95% RPM, and Alpha mode includes operation from 95% to 100% of system rated RPM.

When the engine is started, the power lever is set at the ground idle position and the speed lever is in the start position. When the engine starts, the propeller latches are retracted by reversing the propeller with the power levers, and the propeller moves to a zero degree blade angle as the propeller pitch control is positioned by the power lever over the Beta tube. The Beta tube is attached to the propeller piston, and it moves forward with the piston as the propeller moves to the low blade angle. The propeller blade angle stops changing when the Beta tube moves forward to the neutral position in the propeller pitch control.

The speed lever is used to set the desired RPM through the underspeed governor during ground operation, and the power lever is used to vary the blade angle to cause the aircraft to move forward or rearward. If the power lever is moved forward, the propeller pitch control moves rearward, so that the oil ports on the end of the Beta tube are open to the gear reduction case and the oil in the propeller is forced out by the springs and counterweights. As the blade angle increases, the propeller piston moves inward until the Beta tube returns to its neutral position in the propeller pitch control unit. This causes a propor-

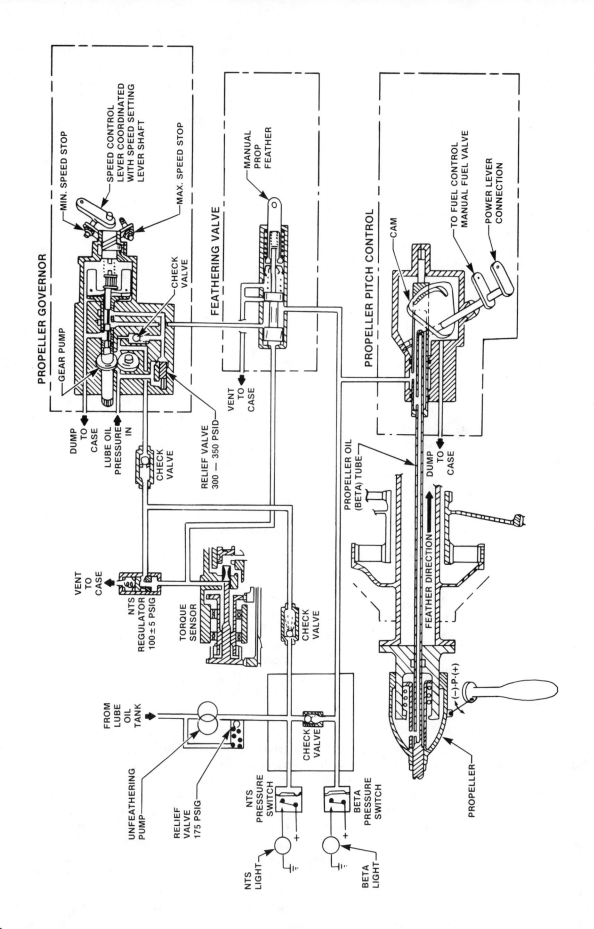

Fig. 8B-6 Schematic of the propeller control system for a TPE-331 engine

tional response of the propeller to the power lever movement.

With the increase in blade angle, the engine will start to slow down, but the underspeed governor, which is set by the speed lever, will adjust the fuel flow to the engine to maintain the selected RPM.

If the power lever is moved rearward, the propeller pitch control moves forward over the Beta tube, and governor oil pressure flows out to the propeller piston and causes a decrease in blade angle. As the piston moves outward, the Beta tube moves with it and returns to the neutral position as the blade angle changes. With this lower blade angle, the engine RPM will try to change, but the underspeed governor will reduce the fuel flow to maintain the selected RPM.

In the Alpha mode of operation, flight operation, the speed lever is moved to a high RPM setting of between 95% to 100%, and the power lever is moved to the flight idle position. When this is done, the underspeed governor is fully opened and no longer affects system operation. RPM control is now accomplished through the propeller governor.

When the power lever is moved to flight idle, the propeller pitch control moves forward so that the Beta tube is fully in the propeller pitch control and it no longer functions to adjust the blade angle. The power lever now controls the fuel flow through the engine fuel control unit.

With a fixed power lever setting in the Alpha mode, the propeller governor is adjusted by the speed lever to set the system RPM in the same manner as for any constant-speed system.

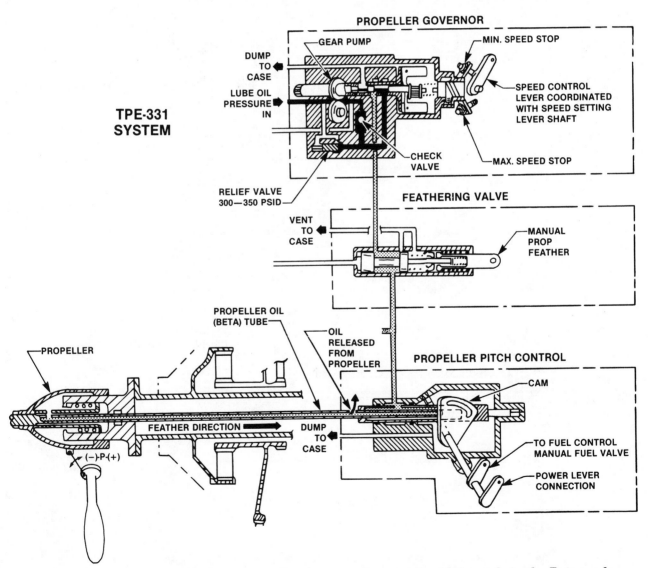

Fig. 8B-7 System components positioned to increase the propeller blade angle in the Beta mode   553

With a fixed speed lever setting in the Alpha mode, the power lever adjusts the fuel control unit to control the amount of fuel delivered to the engine. When the power lever is moved forward, fuel flow will increase and the propeller blade angle will be increased by the propeller governor to absorb the increased engine power and maintain the set RPM. When the power lever is moved aft, fuel flow will decrease and the propeller blade angle will decrease by the action of the propeller governor to maintain the selected RPM.

Whenever it is desired to feather the propeller, the feather handle is pulled or the speed lever is moved full aft, depending on the aircraft design. This action shifts the feathering valve, located on the rear of the gear reduction assembly, and releases the oil pressure from the propeller, returning the oil to the engine sump.

The springs and counterweights on the propeller force the oil out of the propeller, and the blades go into the feather angle.

To unfeather the propeller, the electric unfeathering pump is turned on with a toggle switch in the cockpit, and oil pressure is directed to the propeller to reduce the blade angle. This causes the propeller to start windmilling in flight and an air start can be accomplished. On the ground, the propeller can be unfeathered to the latch position with the unfeathering pump before starting the engine.

3. *Installation and adjustment*

The propeller is installed following the basic procedure used for the installation of other flange-shaft propellers. The Beta tube is installed through the propeller piston after the propeller is

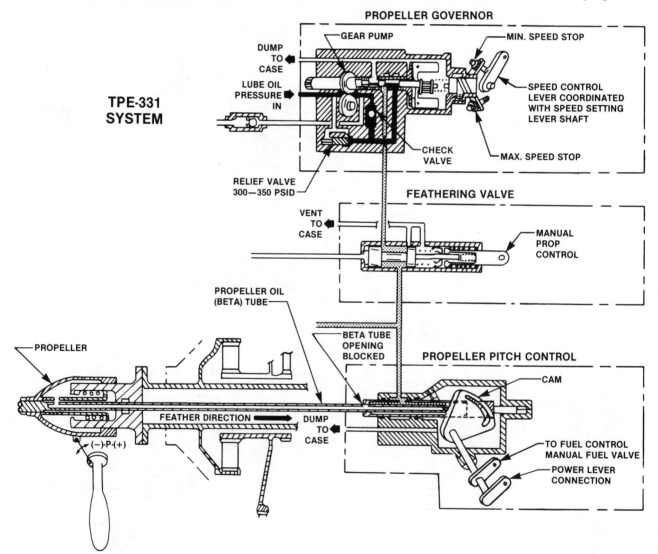

Fig. 8B-8   The Beta tube stops propeller blade angle change in the Beta mode by moving to the neutral position in the propeller pitch control unit.

installed and is bolted to the forward part of the piston.

To adjust the reverse blade angle, the Beta tube is adjusted in or out of the position to set the neutral position on the propeller pitch control unit. This angle must be adjusted according to the aircraft service manual during propeller installation.

The propeller governor, propeller pitch control, feather valve, and fuel control unit are mounted on the engine gear reduction assembly in accordance with the engine service manual.

The interconnection between the speed lever, the underspeed governor, and the propeller governor is rigged and adjusted according to the manual pertaining to the particular aircraft and engine model used. The same holds true for the interconnection between the power lever, the fuel control unit, and the propeller pitch control.

4. System maintenance

Inspect and repair the propeller following the basic procedures set forth for other versions of the Hartzell steel hub propeller. Take care when removing or installing the Beta tube to prevent damage to the tube surface. The Beta tube is trued for roundness and is machined to close tolerances.

Inspect the propeller control units for leaks, security, and damage. Check the linkages between these units for freedom of movement,

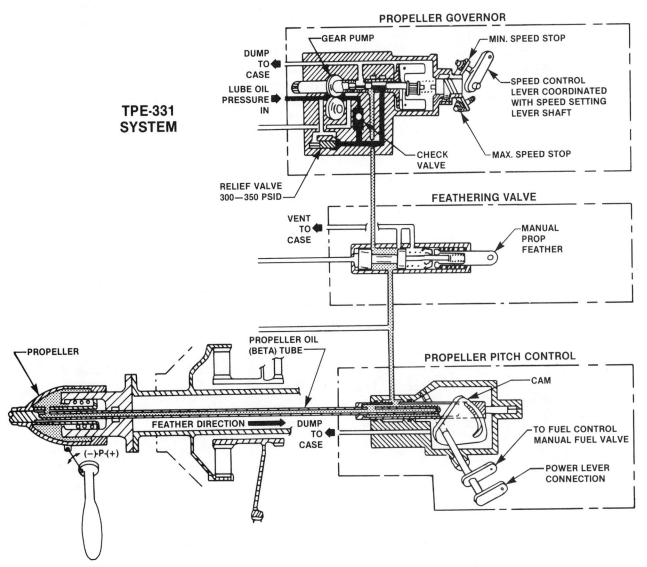

Fig. 8B-9   Decreasing the blade angle in the Beta mode

95

security, and damage. Replace any defective seals, adjust rigging, and secure all nuts and bolts as appropriate for the installation. Use the engine or aircraft maintenance manuals for specific instructions concerning each aircraft model.

Basic troubleshooting procedures as have been previously discussed apply to the Hartzell reversing propeller system. If the proper propeller response is not occurring, and there is no obvious defect, check the system for proper rigging.

In the Beta mode, if the RPM is not constant, investigate the underspeed governor on the fuel control unit. If the blade angle does not respond properly to power lever movement, check the propeller pitch control.

In the Alpha mode, if the RPM is not constant, check the propeller governor. If power does not change smoothly, check the fuel control unit.

B. Hartzell Reversing Propeller System on the Pratt & Whitney of Canada PT6 Engine

The Hartzell propeller on the PT6 engine is used on aircraft such as the Piper Cheyenne,

DeHavilland Twin Otter, and most models of the Beechcraft King Air series.

The PT6 engine is a free-turbine turboprop engine that produces more than 600 horsepower at 38,000 RPM. A geared reduction mechanism couples the engine power turbine to the propeller drive shaft with the propeller rotating at 2200 RPM at 100% RPM. The free-turbine design means that the power turbine is not mechanically connected to the engine compressor, but rather is air coupled. The hot gases generated by the gas generator section of the engine flow through the power turbine wheel and cause the power turbine and the propeller to rotate.

Another turbine section is mechanically linked to the compressor section and is used to drive the compressor. It is possible, during engine start, for the compressor and its turbine to be rotating while the propeller and the power turbine do not move. During engine start, the power turbine will eventually reach the speed of the compressor, but the starter motor is not under a load from the propeller and the power turbine during engine start. For this reason, the propeller can be shut down in feather and does not need a low blade angle latch mechanism for engine starting.

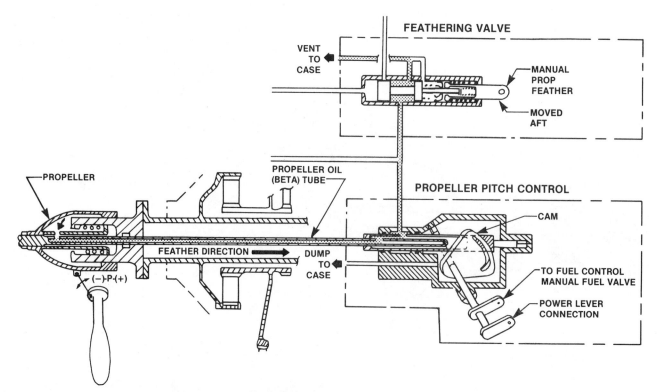

Fig. 8B-10 Feathering the propeller by the cockpit controls

1. Propeller

The propeller commonly used with the PT6 is a three- or four-bladed Hartzell steel hub reversing propeller. The propeller is flange-mounted on the engine and is spring-loaded and counterweighted to the feather position, with oil pressure being used to decrease the blade angle. A Beta slip ring assembly on the rear of the propeller serves as a followup mechanism in giving proportional propeller response to control inputs in the Beta mode.

2. Governor

The propeller governor used with the PT6 is basically the same as other governors discussed for constant-speed operation. It uses a speeder spring and flyweights to control a pilot valve which directs oil flow to and from the propeller. A lift rod is incorporated in the governor to allow the propeller to feather.

For Beta Mode operation, the governor contains a Beta control valve operated by the power lever linkage. It directs oil pressure generated by the governor boost pump to the propeller or relieves oil from the propeller to change the blade angle.

3. System components

A propeller overspeed governor is mounted on the gear reduction assembly and it releases oil from the propeller whenever the propeller RPM exceeds 100%. The release of oil pressure results in a higher blade angle and a reduction in RPM. The overspeed governor is adjusted by the overhaul facility, and it cannot be adjusted in flight. There are no cockpit controls to this governor except for a test mode in some aircraft.

A power turbine governor is installed on the gear reduction assembly as a safety backup in case the other propeller governing devices should

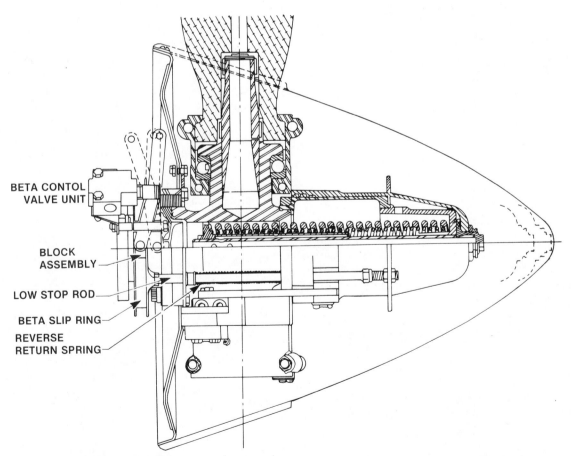

BETA CONTOL VALVE UNIT

BLOCK ASSEMBLY

LOW STOP ROD

BETA SLIP RING

REVERSE RETURN SPRING

Fig. 8B-11   Hartzell propeller for use on a Pratt & Whitney of Canada PT6 engine

fail. If the power turbine speed reaches about 105%, the power turbine governor will reduce the fuel flow to the engine. The power turbine governor is not controllable from the cockpit.

The engine fuel control unit is mounted on the rear of the engine and is linked through a cam assembly to the Beta control valve on the propeller governor and also to the Beta slip ring on the propeller. This interconnection with the fuel control unit is used during Beta mode operation.

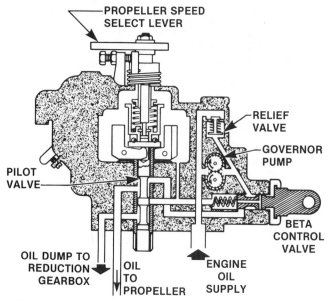

Fig. 8B-12 *Propeller governor for use on a PT6 installation*

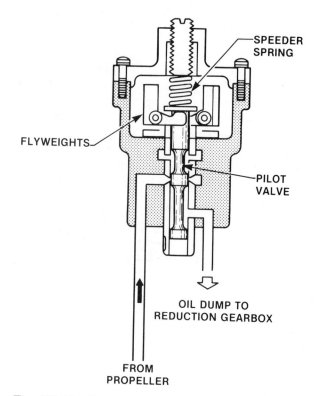

Fig. 8B-13 *Overspeed governor for use on a PT6 installation*

The propeller control lever adjusts system RPM in the Alpha mode through conventional governor operation. Full aft movement of the lever raises the lift rod in the governor and causes the propeller to feather.

The fuel cut-off lever turns the fuel to the engine on and off at the engine fuel control unit. Some designs have an intermediate position, called lo-idle, to limit system power while operating on the ground.

5. System operation

Beta mode operation is generally in the range of 50 to 85% RPM. In this range the power lever is used to control both fuel flow and propeller blade angle. When the power lever is moved forward, the cam assembly on the side of the engine causes the fuel flow to the engine to increase. At the same time, the linkage to the propeller governor moves the Beta control valve forward out of the governor body, and oil pressure in the propeller is released.

As the propeller cylinder moves rearward in response to the loss of oil, the slip ring on the rear of the cylinder moves rearward and, through the

4. Cockpit controls

The cockpit controls for the PT6 turboprop installation consist of a power lever controlling engine power output in all modes and propeller blade angle in the Beta mode. There is also a propeller control lever which adjusts the system RPM when in the Alpha mode and a fuel cutoff lever which turns the fuel at the fuel control on or off.

The power lever is linked to the cam assembly on the side of the engine and, from there, rearward to the fuel control unit and forward to the propeller governor Beta control valve. The power lever adjusts both engine fuel flow and propeller blade angle when operating in the Beta mode which is reverse to flight idle. But in the Alpha mode, the lever controls only fuel flow to the engine.

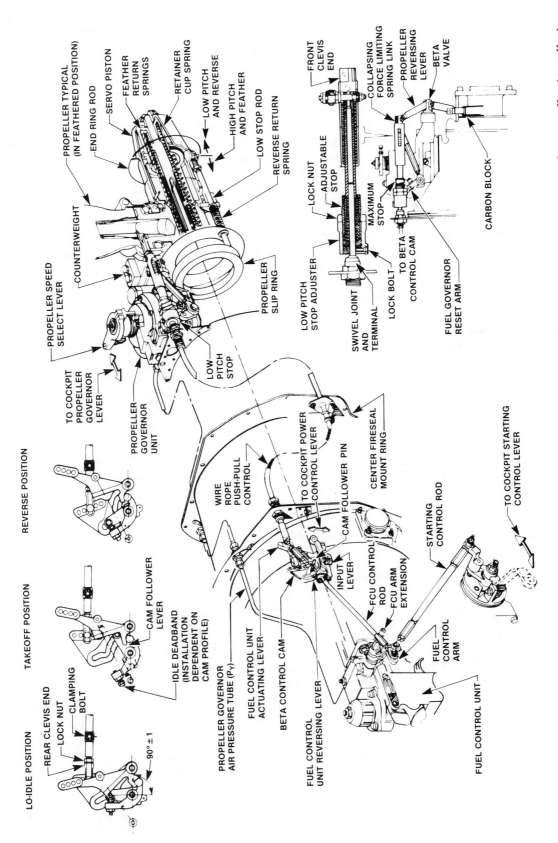

Fig. 8B-14 Side view of a PT6 engine showing the position of the fuel control, the cam mechanism, and the propeller installation

PROPELLER TYPICAL (IN FEATHERED POSITION)

END RING ROD

SERVO PISTON

FEATHER RETURN SPRINGS

RETAINER CUP SPRING

LOW PITCH AND REVERSE

HIGH PITCH AND FEATHER

LOW STOP ROD

REVERSE RETURN SPRING

COUNTERWEIGHT

PROPELLER SPEED SELECT LEVER

TO COCKPIT PROPELLER GOVERNOR LEVER

PROPELLER GOVERNOR UNIT

LOW PITCH STOP

PROPELLER SLIP RING

FRONT CLEVIS END

COLLAPSING FORCE LIMITING SPRING LINK

PROPELLER REVERSING LEVER

BETA VALVE

LOCK NUT

ADJUSTABLE STOP

LOW PITCH STOP ADJUSTER

SWIVEL JOINT AND TERMINAL

MAXIMUM STOP

LOCK BOLT

TO BETA CONTROL CAM

FUEL GOVERNOR RESET ARM

CARBON BLOCK

LO-IDLE POSITION

TAKEOFF POSITION

REVERSE POSITION

REAR CLEVIS END

LOCK NUT

CLAMPING BOLT

90° ± 1

CAM FOLLOWER LEVER

IDLE DEADBAND (INSTALLATION DEPENDENT ON CAM PROFILE)

PROPELLER GOVERNOR AIR PRESSURE TUBE (Py)

FUEL CONTROL UNIT ACTUATING LEVER

BETA CONTROL CAM

FUEL CONTROL UNIT REVERSING LEVER

WIRE ROPE PUSH-PULL CONTROL

TO COCKPIT POWER CONTROL LEVER

CAM FOLLOWER PIN

CENTER FIRESEAL MOUNT RING

STARTING CONTROL ROD

TO COCKPIT STARTING CONTROL LEVER

INPUT LEVER

FCU CONTROL ROD

FCU ARM EXTENSION

FUEL CONTROL ARM

FUEL CONTROL UNIT

carbon block and linkage, returns the Beta control valve to a neutral position. This gives a proportional movement to the propeller.

When the power lever is moved rearward, fuel flow is reduced and the Beta control valve moves into the governor body, directing oil pressure to the propeller to decrease the blade angle. And as the propeller cylinder moves forward, the Beta control valve returns to its neutral position by the action of the slip ring, carbon block, and linkage. This again gives a proportional response.

If the power lever is moved aft of the zero thrust position, fuel flow will increase and the blade angle goes negative to allow a variable reverse thrust. This change in fuel flow is caused by the cam mechanism on the side of the engine.

During operation in the Beta mode, the propeller governor constant-speed mechanism is underspeed and the pilot valve is lowered. The governor oil pump supplies the oil pressure for propeller operation in the Beta mode.

In the Alpha mode, the system RPM is high enough for the propeller governor to operate, and the system is in a constant-speed mode of operation. When the power lever is moved forward, more fuel flows to the engine to increase the horsepower, and the propeller governor causes an increase in propeller blade angle to absorb the power increase and maintain the selected system RPM. If the power lever is moved aft, the blade angle will be decreased by the governor to maintain the selected RPM.

To feather the propeller, move the propeller control lever full aft. This raises the pilot valve in the governor by a lift rod, and releases all of the oil pressure in the propeller. The springs and counterweights in the propeller will take it to feather.

To unfeather the propeller, start the engine. As it begins to rotate, the power turbine will rotate, and the governor or Beta control valve will take the propeller to the selected blade angle or governor RPM setting. When the engine is restarted, the engine will be started before the propeller is rotating at the same proportional speed because of the free-turbine characteristic of the engine.

If the propeller RPM should exceed 100%, the propeller overspeed governor will raise its pilot valve and release oil from the propeller to increase blade angle and prevent overspeeding of the propeller. The overspeed governor is automatic and is not controllable in flight.

The power turbine governor prevents excessive overspeeding of the propeller by reducing fuel flow to the engine at approximately 105%

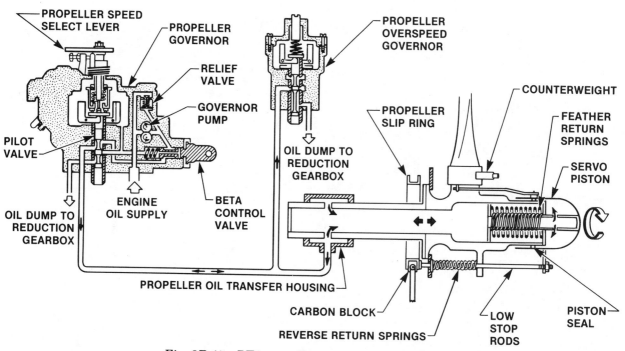

Fig. 8B-15   PT6 propeller system configuration

RPM. This governor is not controllable in flight and is automatic in operation.

6. *System installation and maintenance*

When installing the propeller, follow the basic procedures for flanged-shaft installations. The slip ring and carbon block arrangement must be installed following the procedures in the aircraft service manual for the specific model involved.

Adjust the governors and rig them with the fuel control unit and the cam mechanism, according to the appropriate maintenance manual.

Basic troubleshooting procedures as have been previously discussed are applicable to the Hartzell reversing system. If you do not get the proper propeller response, check the system for proper rigging before investigating individual units unless the defect is obvious.

In Beta mode operation, the interconnection should be checked between the power lever, the cam mechanism, the fuel control unit, the Beta control valve on the propeller governor, and the Beta slip ring.

In the Alpha mode, the propeller governor and linkage to it from the propeller control lever and the cam mechanism should be checked.

If the system maximum RPM is too low, the fault may be with the adjustment of the propeller overspeed governor or the power turbine governor. These components are not involved in the control linkage rigging, but do not forget them.

IX. PROPELLER AUXILIARY SYSTEMS

A. *Propeller Synchronizing Systems*

The propeller synchronization system is used to set all propellers at exactly the same RPM, thereby eliminating excess noise and vibration. It is used for all flight operations except takeoff and landing. A master engine is used to establish the RPM to which the other engine, called the slave engine, will adjust.

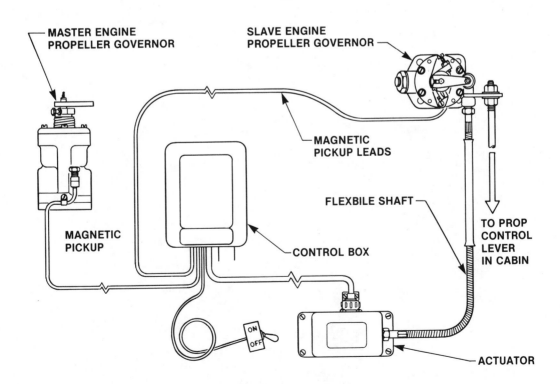

Fig. 9B-1 Woodward synchronization system for a light twin-engine airplane

561

A frequency generator built into the propeller governor generates a signal that is proportional to the RPM of the engine. A comparison circuit in the control box compares the RPM signal from the slave engine to the RPM signal from the master engine and sends a correcting signal to the slave engine governor control mechanism.

The comparison unit has a limited range of operation, and the slave engine must be within about 100 RPM of the master engine for synchronization to occur.

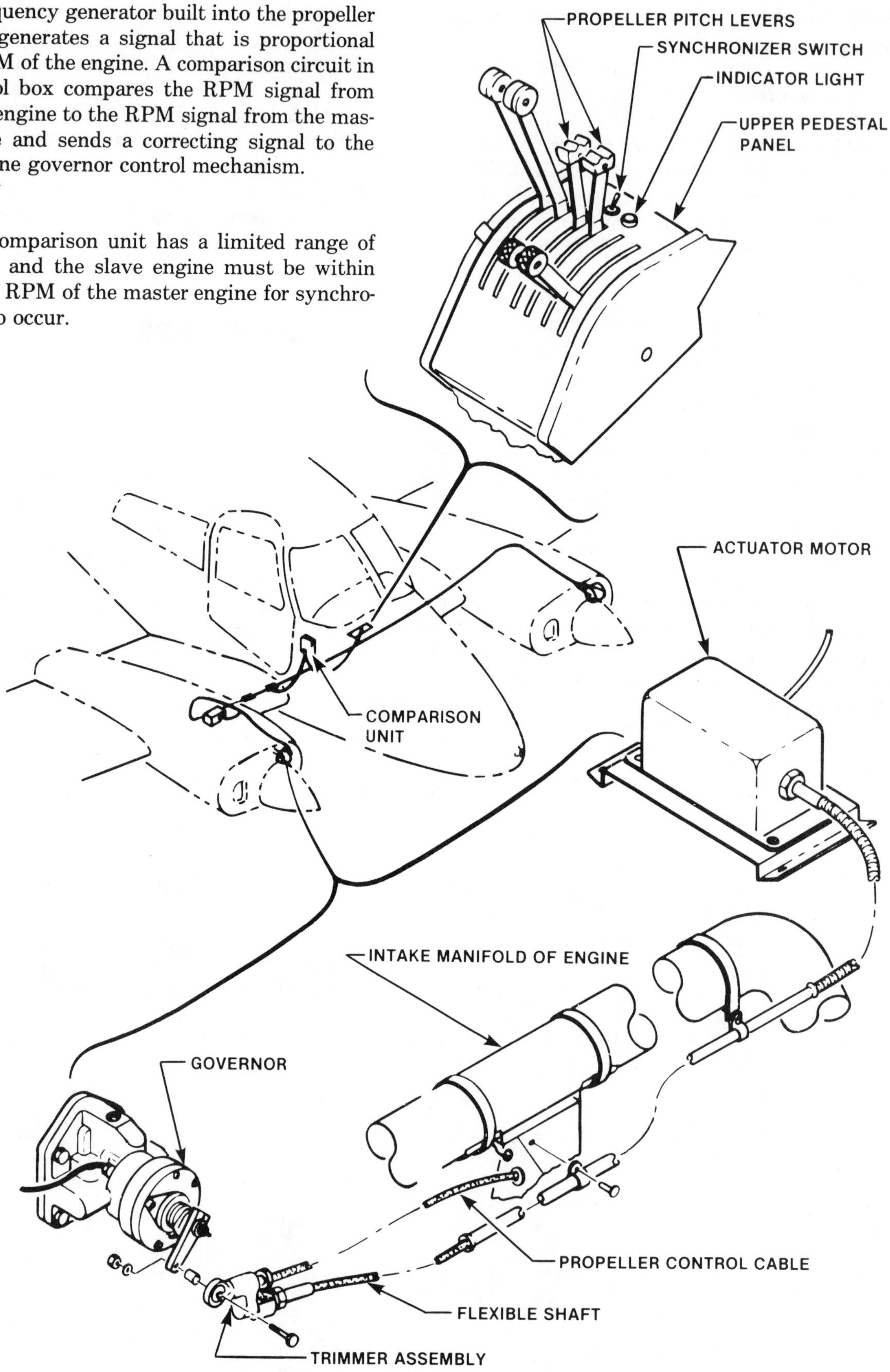

PROPELLER PITCH LEVERS
SYNCHRONIZER SWITCH
INDICATOR LIGHT
UPPER PEDESTAL PANEL

ACTUATOR MOTOR

COMPARISON UNIT

INTAKE MANIFOLD OF ENGINE

GOVERNOR

PROPELLER CONTROL CABLE

FLEXIBLE SHAFT

TRIMMER ASSEMBLY

Fig. 9B-2   Installation of a synchronization system in a light twin-engine airplane

562

102

B. Synchrophasing System

Synchrophasing is a refinement of synchronization which allows the pilot to set the angular difference in the plane of rotation between the blades of the slave engines and the blades of the master. Synchrophasing is used to reduce the noise and vibration created by the engines and propellers, and the synchrophase angle can be varied by the pilot to adjust for different flight conditions and still achieve a minimum noise level.

A pulse generator is keyed to the same blade of each propeller—blade number one, for example—and a signal is generated to determine if both number one blades are in the same relative position at any given instant. By comparison of the signals from the two engines, a signal is sent to the governor of the slave engine to cause it to establish the phase angle selected by the pilot.

A propeller manual phase control in the cockpit allows the pilot to select the phase angle which produces the minimum vibration and noise.

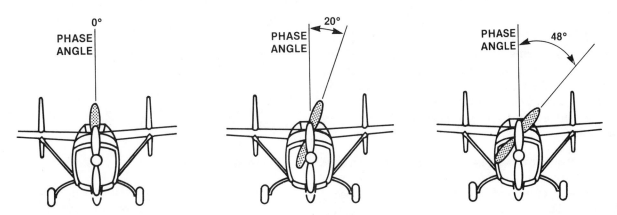

Fig. 9B-3 Synchrophasing allows the pilot to adjust the phase angle between the propellers on the various engines to reduce the noise and vibration to a minimum.

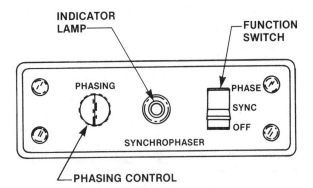

Fig. 9B-4 Synchrophasing control panel for a light twin-engine airplane

C. Automatic Feathering System

An automatic feathering system is used on some multi-engine aircraft to feather a propeller automatically if the engine fails. The system is normally armed for takeoff and landing, but is turned off during cruising flight.

1. System components

The system master switch is located on the pilot's or flight engineer's console and is normally covered by a guard. When the switch is turned on, an indicator light illuminates to indicate that the system is armed.

A throttle switch is used to arm the circuit further by closing a microswitch when the throttle is advanced to a specific position, for example, 75% of full throttle movement, depending on the aircraft. The circuit is open when the throttles are below this setting, and the system will not autofeather.

A torque pressure switch is used to sense the power output of the engine and will close a contact whenever engine power drops below a specific level. The amount of torque pressure loss required for the system to operate will vary with different aircraft, due to engine size and aircraft design.

A time delay unit is used in most circuits to prevent autofeathering if only a momentary interruption in power occurs. The power loss must exceed one to two seconds for the system to autofeather. This value also varies with aircraft designs.

The feather control is ultimately activated by the system when an engine fails. When the control is actuated by the autofeather systems, a red light in the cabin turns on to indicate to the pilot which propeller is feathered. The feather system can also be operated in a normal manner by the pilot at any time.

A blocking relay is used in the system to prevent more than one engine autofeathering. This relay may be electrically located between the master switch and the throttle switch or may be incorporated in some other part of the circuit. If one engine autofeathers, some systems can be reset to rearm the autofeather system in case another engine should fail. The pilot can feather any engine, at any time, regardless of whether or not a propeller has been autofeathered.

A test switch is used to bypass portions of the circuit so that the system operation can be checked on the ground without developing high power settings.

2. System operation

Before takeoff and landing, the system is armed by turning on the system master switch. As power is advanced for takeoff or during a missed landing approach, the throttle switch closes and the torque pressure switch is armed, but the torque pressure switch contacts are open. If a loss of engine power occurs, the torque pressure switch closes and, after the prescribed time interval, the time delay unit completes the circuit, energizing the feather control. At the same time, the blocking relay is actuated to break the circuit for the autofeather system on the other engines.

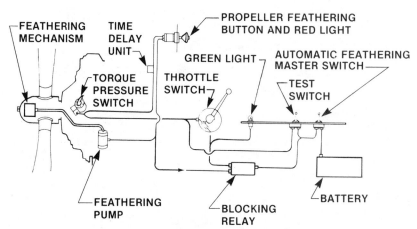

Fig. 9B-5 Basic automatic feather system diagram

3. System maintenance

Conducting an operational check of the system is the best way to check it for operation and possible defects. Start the engines and arm the system with the autofeather system master switch. Advance the throttles to develop the required torque to arm the torque pressure switch. Hold the test switch in the position for the engine being checked and retard the throttle to idle. This should cause the torque pressure switch and the time delay relay to close and start the feathering operation by activating the feather control and turning on the light in the cabin. Release the test switch and deactivate the feather control to prevent the propeller feathering. Note that with this check some components, for example the blocking relay and throttle switch, are not checked.

System components should be inspected and maintained in accordance with the aircraft service manual. Units can be removed and replaced as necessary to correct system operation.

If the system indicator light does not turn on when the system is armed, the bulb may be burned out, the system master switch may be open, or electrical power may not be getting to the system.

If the system operates properly during a ground test, but will not autofeather in flight, the components not in the test circuit may be open or incorrectly adjusted.

If the system will not operate during a test or in flight, but the armed light is turned on, the problem is most likely the components in the circuit when testing the autofeather system or the feathering system.

Index

A

Absolute pressure 5-21
AC
  motors 4-48
  systems 3-47–3-48
Acceleration 2-11
  pump 6-20
  well 6-19
Acceleration,
  float carburetor 6-19
  pressure carburetor 6-32
Accessories,
  fuel system components 6-74–6-82
  inspection 1-58
Accessory section, turbine
  engine 2-31
Additives, gasoline 6-11
Adjustment,
  blade angle 8-56, 8-63
  float carburetor 6-26
  pressure carburetor 6-35
  Teledyne-Continental 6-48
  turbine fuel control 6-61–6-70
Afterburning 7-38
Agents,
  fire extinguishing 5-56–5-57
  halogenated hydrocarbon 5-57
Air cooling 1-23
Air filters, induction 7-4
Air metering,
  Bendix 6-37
  pressure carburetor 6-28
Air turbine starting system 7-52
Airbleed enrichment 6-22
Airbleed, float carburetor 6-16
Airflow limiter 6-16
Airflow, interstage 2-39
Airspeed, and engine thrust 2-23
Alterations, propeller 8-39
Alternating current power
  systems 4-37–4-44
Alternator
  control unit 4-35
  controls, DC 4-30–4-35
  service, DC 4-35
Alternators, DC 4-26–4-35
Altitude performance 6-6
Altitude,
  and engine thrust 2-23
  density 6-4
Aluminum alloy propeller 8-46
Angle of attack 2-42
  propeller 8-40
Annual inspection 1-51–1-58
Annular combustor 2-46
Anti-icing systems 7-23–7-25
Anti-surge bleed, compressor 7-20–7-22
Armature reaction 4-12
Armature, generator 4-7–4-9

Aromatic series 6-9
Ashless-dispersant oil 8-3
Assembly, major overhaul 1-73
Attrition 1-48
Authorized personnel, propeller 8-37
Automatic feathering 8-104–8-105
Auxiliary
  power unit (APU) 6-59–6-60
  systems, propeller 8-101–8-105
Avco-Lycoming T-53 7-20–7-21
Axial flow compressors 2-35–2-39,
  2-43

B

Back-suction mixture control 6-17
Barrels,
  chrome-plated 1-28
  cylinder 1-27
  nitrided 1-29
Battery ignition system 3-1–3-2
Bearing
  distress terms 2-70
  installation, turbine engine 2-70
  seals, turbine engine 2-66
Bearings,
  crankshaft 1-39
  turbine engine 2-66
Bellmouth compresor inlets 2-28, 7-16
Bendix
  DP-L2 fuel control unit 6-54–6-56
  fuel injection 6-36–6-42
Bernoulli's Principle 2-12
Blade angle adjustment 8-56, 8-63
Blades 2-38
Bonding 5-14
Booster magnetos 3-11
Brake horsepower 1-10
Brush placement 4-12

C

Cam followers 1-41
Cams, compensated 3-17
Camshaft 1-40
Camshaft, major overhaul 1-71
Can-annular combustor 2-46
Carbon dioxide, fire extinguisher 5-56
Carbon residue 8-5
Carburetor
  air temp 6-4
  heat systems 7-2–7-3
Carburetors,
  float 6-3–6-28
  pressure 6-28–6-35
Centrifugal compressors 2-32–2-34,
  2-43
Certification, propeller 8-35–8-37
Choking, sonic 2-13
Chrome-plated barrels 1-28

Circuit
  breakers 5-17
  protection 5-16
Circuit,
  magnetic 3-7
  primary electrical 3-8
  secondary electrical 3-10
Clamping, wire bundle 5-13
Cleaning,
  compressor field 2-56
  float carburetor 6-23
  igniter plugs 3-50
  major overhaul 1-60
  oil filter 8-23
  spark plug -36
Clearances, propeller 8-36
Cockpit controls, Hartzell 8-98
Cold section maintenance 2-58–2-65
Cold tank, lubrication 8-30
Color, oil 8-5
Combination compressors 2-39
Combustion section,
  inspection 2-62
  turbine engine 2-44–2-47
Combustion, chemistry of 6-1–6-2
Combustor
  drain valve, turbine 6-81
  outlet, thrust 2-19
Combustor,
  annular 2-46
  can-annular 2-46
  multiple-can 2-46
  reverse-flow annular 2-47
Compensated cams 3-17
Compound-field motors 4-48
Compression
  ratio 1-13, 2-37, 6-4
  test 1-55
Compressor
  blade repair 2-61
  case repair 2-62
  field cleaning 2-56
  inlet screens 2-31
  outlet, thrust 2-18
  turbine engine 2-32–2-43
  stall 2-42
Compressor,
  anti-surge bleed 7-20–7-22
  axial flow 2-35–2-39, 2-43
  centrifugal 2-32–2-34, 2-43
  combination 2-39
  inlet screens 7-17
Computer, fuel flow 5-41
Computing, turbine 6-54
Connecting rods,
  horizontally opposed 1-34
  major overhaul 1-70
Connectors, MS 5-12
Constant-speed
  drive 4-38
  propellers 8-59–8-76

Construction,
    ignition harness 3-42
    spark plug 3-32

Contaminants,
    detection of 6-12
    fuel 6-12

Continuous loop fire detection 5-53

Control systems, turbocharger induction 7-8–7-13

Controls,
    Hartzell cockpit 8-98
    inspection 1-58

Converters, electronic 4-39

Cooler, oil 8-11, 8-24

Cooling system, inspection 1-51, 1-58

Cooling systems 1-22–1-23, 7-25–7-32
    reciprocating engine 7-25–7-28
    turbine engine 7-29–7-32

Corrosion 1-48

Crankcase,
    horizontally opposed 1-39
    major overhaul 1-65
    radial 1-44
    reassembly 1-73

Crankshaft bearings 1-39

Crankshaft,
    horizontally opposed 1-34
    major overhaul 1-68
    radial 1-43
    reassembly 1-72

Creep, inspection 2-64

Critical
    pressure and temperature 6-10
    range, propeller 8-43

Current limiter 4-19, 4-23

Curves, power 1-16

Cutout relay, reverse-current 4-14, 4-22

Cycloparaffin series 6-9

Cylinder
    arrangement 1-19–1-21
    head temperature indicator 5-30
    heads, horizontally opposed 1-29

Cylinders,
    inspection 1-51
    major overhaul 1-68
    radial 1-43
    reassembly 1-72

D

Dampers, dynamic 1-35

Danger zone, trim 6-68

Data plate speed check 6-65

DC
    motors 4-45-4-48
    systems 3-47–3-48

Density altitude 6-4

Depth filtration 8-12

Designation, fixed-pitch propeller 8-48

Detonation 6-5

Differential fuel pressure 5-25

Diffuser outlet,
    thrust 2-18
    section, turbine engine 2-44

Diffusers 2-13

Diffusion 2-13

Dimensional inspection 1-65

Direct current power system 4-1–4-36

Disassembly,
    float carburetor 6-23
    major overhaul 1-60

Displacement, piston 1-13

Double magnetos 3-5

Drive, constant-speed 4-38

Dry
    powder, fire extinguisher 5-56
    sump, reciprocating engine 8-5
    sump, turbine 8-19

Ducts, inlet 2-26–2-31

Dump valve, turbine 6-79–6-81

Dynamic dampers 1-35

E

Ear protection 6-68

Edge filtration 8-12

Efficiency,
    mechanical 1-13
    thermal 1-12, 2-22, 6-3
    turbofan engine 2-26
    volumetric 1-12

Electric motors 4-45–4-48

Electrical
    installation 5-1–5-19
    system requirements 5-16–5-19
    system, inspection 1-58
    systems 4-1–4-48

Electro-hydromechanical fuel controls 6-57–6-59

Electrodes, spark plug 3-31

Electromagnets 4-4

Electronic, converters 4-39

Emergency
    operating procedure 2-79
    shutdown procedure 2-80

Energy 2-11
    Energy release cycle 2-14
    transformation 1-5–1-16

Energy,
    conversion to power 6-1
    source 6-1
    transformation 6-1–6-8

Engine
    configurations 1-19–1-23
    construction, horizontally opposed 1-27–1-41
    construction, radial 1-42–1-44
    construction, V- 1-46
    identification 1-26
    performance charts 6-6–6-8
    power 6-4–6-6
    pressure ratio 5-25,6-62
    purging procedure 2-80
    requirements 1-18
    speed 1-15
    start 2-79

Enrichment,
    airbleed 6-22
    needle-type 6-20

Equivalent shaft horsepower (ESHP) 2-21

Exhaust
    back pressure 6-4
    cone 7-36–7-37
    cone outlet, thrust 2-19
    duct 2-50
    gas, temp gage reciprocating 5-31
    gas, temp system turbine 5-33

Exhaust (cont.)
    noise suppression 7-41–7-42
    section, inspection 2-63
    section, turbine engine 2-50–2-53
    system, inspection 1-50, 1-58

Exhaust systems 7-33–7-42
    reciprocating engine 7-33–7-35
    turbine engine 7-36

F

FAA engine power ratings 6-74

Fan engine thrust 2-20

Feathering
    components, Hamilton-Standard 8-85
    propeller 8-76–8-88
    system maintenance 8-80

Feathering,
    automatic 8-104–8-105
    Hartzell 8-80–8-83
    McCauley 8-76–8-80

Field connections 4-9

Field, generator 4-9

Fifty-hour inspection 1-50

Filter cleaning 8-23

Filters,
    induction air 7-4
    last-chance 8-26
    oil 8-12–8-14, 8-21

Filtration 8-12–8-13

Fire
    detection systems 5-51–5-55
    extinguishing 5-56–5-57
    protection 6-13
    protection systems 5-51–5-59
    zones 5-51

Fires, types 5-56

Fixed-pitch propeller 8-44–8-48

Flash point 8-5

Flat rating 6-70

Float carburetors 6-3–6-28
    service 6-22–6-28

Flow
    divider, Bendix 6-38
    matching, Teledyne-Continental 6-48
    meter, volume remote 5-40
    meter fluctuation, Teledyne-Continental 6-49

Fluorescent penetrant inspection 1-61

Force 2-10
    propeller centrifugal 8-42
    propeller thrust bending 8-42
    propeller torque bending 8-42
    propeller vibration 8-43

Foreign object damage (FOD) 2-55

Four-stroke-cycle reciprocating, energy transformation 1-8

Friction horsepower 1-11

Fuel
    additives, turbine 6-50
    consumption, specific 1-15, 6-3
    contamination 6-12, 6-13
    filters, turbine 6-76
    grades 6-11, 6-12
    handling 6-13
    heater, turbine 6-75
    inspection 1-50
    pressure, differential 5-25
    pressurizing, turbine 6-79
    requirements 6-9

Fuel control,
    Bendix DP-L2 6-54–6-56
    electro-hydromechanical 6-57–6-59
    Garrett-AiResearch ATF-3 6-57
    Rolls-Royce RB-211 6-57
    Teledyne-Continental 6-44
    turbine 6-50–6-70

Fuel flow
    computer 5-41
    instruments 5-40
    pressure 5-40

Fuel injection,
    Bendix 6-36–6-42
    Teledyne-Continental 6-43–6-49

Fuel metering
    forces, float carburetor 6-14–6-15
    systems, reciprocating engine 6-1–6-49
    systems, turbine 6-49–6-84

Fuel metering,
    Bendix 6-38
    pressure carburetor 6-28
    principles of 6-13
    turbine 6-52–6-53

Fuel nozzles,
    turbine atomizing 6-77
    turbine vaporizing 6-78

Fuel system
    components accessories 6-74–6-82
    inspection 1-57
    Pratt and Whitney JT-12 6-83

Fuel-air mixture ratio 6-6

Fuels,
    reciprocating engine 6-9–6-13
    turbine 6-49

Fuses 5-16

G

Gage pressure 5-24
Gage, manifold pressure 5-22
Gapping, spark plug 3-38
Garrett-AiResearch
    ATF-3 fuel control 6-57
    GTP-30, APU 6-60
    TPE-331, and Hartzell propeller 8-90

Gas turbine
    energy transformation 1-7
    engines 2-5
    operation 2-14–2-21
    performance 2-22–2-26

Gasoline
    additives 6-11, 6-12
    blends 6-9–6-10
    ratings 6-10

Gasoline, aviation 6-9–6-11
Gear pump 8-8, 8-21
General Electric
    CF-6 7-17–7-18
    CJ610, oil presssure subsystem 8-24

Generating system 5-16
    AC 4-38

Generator action 4-6–4-12
Generator controls,
    AC 4-39
    DC 4-13–4-23
    high-output 4-19
    low-output 4-14
    vibrator-type 4-14

Generator
    instrumentation, AC 4-43
    service, DC 4-23–4-25

Generators,
    AC 4-42
    high-output 4-11
    integrated drive 4-30

Gerotor pump 8-9, 8-21
Governor,
    Hamilton-Standard 8-64, 8-85
    Hartzell 8-74, 8-80
    Hartzell reversing 8-97
    McCauley 8-69, 8-77

Gravity API 8-5
Gross thrust 2-15
Ground
    adjustable propellers 8-56–8-58
    start 2-79–2-80

H

Halogenated hydrocarbon agents 5-57
Hamilton-Standard
    counterweight propellers 8-59–8-67
    hydromatic 8-84–8-88

Hartzell
    feathering 8-80–8-83
    propeller 8-72–8-76
    propeller, and Garrett AiResearch
    TPE-331 8-90
    reversing propeller and PT6 8-96–8-101

Heads, horizontally opposed cylinder 1-29
Heat
    range, spark plug 3-29
    systems, carburetor 7-1–7-3

Horizontally opposed engine
    construction 1-22–1-41
    cylinder arrangement 1-21
    firing order 1-25

Horsepower 1-9–1-11
    equivalent shaft (ESHP) 2-21
    thrust 2-20

Hot
    section inspection 2-62
    maintenance 2-58–2-65
    start 2-80
    tank, lubrication 8-29

Humidity 6-4
Hydrocarbon agents, halogenated 5-57
Hydromatic, Hamilton-Standard 8-84–8-88

I

Idle,
    Bendix 6-39
    float carburetor 6-18
    pressure carburetor 6-32

Igniter plugs 3-49–3-52
    cleaning 3-50
    inspection 3-50

Ignition harness 3-42–3-44
Ignition systems 3-1–3-52
    gas turbine 3-44–3-52
    inspection 1-50, 1-56
    "shower of sparks" 3-12

Ignition timing 1-14, 6-6
Impulse couplings 3-16, 3-21
In-line engines,
    cylinder arrangement 1-19
    firing order 1-24

Indicated horsepower 1-11
Induction systems 7-1–7-24
    inspection 1-57
    naturally aspirated 7-1
    reciprocating engine 7-1–7-13
    supercharged engine 7-4–7-13
    turbine engine 7-14–7-24

Induction vibrators 3-11
Injection
    nozzles, Bendix 6-39
    pump, Teledyne-Continental 6-43

Injector
    lines, Teledyne-Continental 6-46
    nozzles, Teledyne-Continental 6-46

Inlet ducts 2-26–2-31
    turbine engine 7-14

Inlet screens, compressor 2-32, 7-17
Inlets,
    bellmouth compressor 2-28,7-16
    subsonic 2-26, 7-14
    supersonic 2-28, 7-15

Input system,
    high voltage AC 3-47
    low voltage DC 3-46

Inspection,
    combustion section 2-62
    creep 2-64
    dimensional 1-65
    exhaust section 2-63
    float carburetor 6-23
    fluorescent penetrant 1-61
    ground-adjustable propeller 8-57
    Hartzell 8-76
    hot section 2-62
    igniter plugs 3-50
    magnetic particle 1-63
    magneto 3-23
    major overhaul 1-61–1-65
    McCauley 8-70
    powerplant 1-49–1-58
    propeller 8-39
    Teledyne-Continental 6-48
    case 2-63
    turbine engine 2-55–2-77
    turbine vanes 2-63
    turbine wheel 2-63
    untwist 2-64
    visual 1-61
    wood propeller 8-44

Installation,
    Bendix 6-41
    flanged shaft 8-49–8-50
    float carburetor 6-26
    ground-adjustable 8-56
    Hamilton-Standard propeller 8-62
    Hartzell 8-75
    Hartzell reversing 8-94, 8-101
    instrument 5-45–5-47
    magneto 3-17–3-24
    McCauley 8-69
    pressure carburetor 6-34
    propeller 8-49–8-55
    spark plug 3-40
    splined shaft 8-52–8-55
    tapered shaft 8-51
    wiring 5-6–5-8

Instrument
check 2-79
indications, Hamilton-Standard 8-67
installation 5-45–5-47
system, thermocouple 5-29
systems 5-20–5-50
Instrumentation,
AC generator 4-43
lubrication 8-15
Instruments,
fuel flow 5-40
powerplant evolution 5-20
pressure measuring 5-21–5-26
propeller 8-36
temperature measuring 5-26–5-36
Interstage airflow 2-39

J

Jets, oil 8-26
Joule ratings 3-45
Junction boxes 5-12

L

Last-chance filter 8-26
Lift, propeller 8-40
Limiter,
airflow 6-16
current 4-19, 4-23
Line maintenance, turbine engine 2-55–2-57
Liquid
cooling 1-22
nitrogen, fire extinguisher 5-57
Lockwiring 2-73
Lubricating
oil 8-2–8-3
system, functions of 8-1–8-2
systems 8-1–8-32
systems, reciprocating engine 8-1–8-15
systems, turbine engine 8-15–8-32
Lubrication
scavenge 8-27
system, inspection 1-50, 1-56
system, PT6 8-29
systems 1-23

M

Magnetic
circuit 3-7
particle inspection 1-63
Magnetism 4-1–4-5
check, turbine engine 2-70
Magneto ignition systems 3-3–3-4
Magnetos 3-3–3-24
booster 3-11
high tension 3-4
low tension 3-3
Magnets, permanent 4-2
Maintenance,
automatic feathering 8-105
feathering system 8-80
Hartzell reversing 8-95
improper 1-49
modular 2-73
propeller preventive 8-37
reciprocating engine 1-47–1-78
test cell 2-73
turbine engine 2-55–2-81

Major overhaul 1-59
Manifold
pressure gage 5-22
valve, Teledyne-Continental 6-46
Margin curve 2-42
Mass 2-11
Mass-flow, turbine engine 5-41
McCauley
feathering 8-76–8-80
propeller designation 8-48
propellers 8-67–8-72
Measurement,
mechanical 5-36–5-44
oil pressure 8-15
oil temperature 8-15
pressure 5-21–5-26
temperature 5-25, 5-26–5-36
torque 5-42–5-44
Mechanical
efficiency 1-13
measurement 5-36–5-44
Mesh size vs. micron rating 6-77
Metallic-ash detergent oil 8-2
Metering system, float carburetor 6-14
Micron rating vs. mesh size 6-77
Microorganisms, fuel 6-12
Mineral oil 8-2
Mixture control,
back-suction 6-17
Bendix automatic 6-41
Bendix manual 6-40
float carburetor 6-17
pressure carburetor 6-30
variable-orifice 6-17
Modular maintenance 2-73
Motors,
AC 4-48
compound-field 4-48
DC 4-45–4-48
electric 4-45–4-48
repulsion 4-48
series-field 4-47
shunt-field 4-46
Mounting, float carburetor 6-26
MS connectors 5-12
Multiple-can combustor 2-46

N

Naturally aspirated induction system 7-1
Needle-type enrichment 6-20
Net thrust 2-16
Newton's Laws 2-11
Nitrided barrels 1-29
Noise
suppression, exhaust 7-41–7-42
suppressors 2-53

O

Octane rating 6-10
Oil 8-1–8-32
change, turbine 8-18
filters 8-12–8-14, 8-21
pressure measurement 8-15
pressure subsystem General Electric CJ610 8-24

Oil (cont.)
rating, reciprocating engine 8-5–8-5
sampling 8-16
sealing 1-39
tank, turbine 8-19
temperature measurement 8-15
Oils,
compatibility 8-3
synthetic 8-3, 8-17
Olefin series 6-9
One-hundred-hour inspection 1-51–1-58
Operation, pressure carburetor 6-35
Overhaul,
float carburetor 6-23–6-26
major 1-59
powerplant 1-59–1-74
top 1-59

P

Paraffin series 6-9
Parts replacement, float carburetor 6-23
Performance
charts, engine 6-6–6-8
number, gasoline 6-11
Performance,
altitude 6-6
sea level 6-6
Phase rotation 4-41
Physics, principles of 2-10–2-13
Pins, piston 1-33
Piston
displacement 1-13
pins 1-33
rings 1-33
Pistons,
horizontally opposed 1-32
major overhaul 1-70
radial 1-43
reassembly 1-72
Pitch, propeller 8-43
Plugs, igniter 3-49–3-52
Polar-inductor magnetos 3-4
Pour point 8-5
Power 1-9, 2-10
check 2-79
curves 1-16
curves, full throttle vs. propeller load 6-8
enrichment, float carburetor 6-20–6-22
enrichment, pressure carburetor 6-33
ratings, FAA engine 6-74
systems, alternating current 4-37–4-44
systems, direct current 4-1–4-36
Power,
distribution of 1-16
engine 6-4–6-6
production 6-3
reactive 4-39
Powerplant
instruments, evolution 5-20
overhaul 1-59–1-74
removal, turbine engine 2-57
Pratt and Whitney JT-12 fuel system 6-83
Preflight inspection 1-49

Preignition 6-5

Preinstallation,
  splined shaft 8-52
  tapered shaft 8-51

Preservation, spark plug 3-40

Pressure
  carburetors 6-28–6-35
  drop, float carburetor 6-14–6-15
  ratio, engine 5-25
  switches 5-26

Pressure,
  absolute 5-21
  differential 5-25
  exhaust back 6-4
  fuel flow 5-40
  gage 5-24
  gage manifold 5-22

Pressure-regulating valve, starter 7-51

Pressure-type fire detection 5-55

Preventive maintenance,
  propeller 8-37

Primary electrical circuit 3-8

Propeller
  attachment 1-36
  auxiliary systems 8-101–8-105
  protractor 8-57
  reduction gearing 1-37, 1-44
  synchronizing 8-101–8-102
  synchrophasing 8-103
  theory 8-39–8-43

Propellers 8-33–8-105
  aluminum alloy 8-46
  constant-speed 8-59–8-76
  feathering 8-72–8-76
  fixed-pitch 8-44–8-48
  ground-adjustable 8-56–8-58
  Hamilton-Standard
    counterweight 8-59–8-67
  Hartzell 8-72–8-76
  inspection 1-58
  McCauley 8-67–8-72
  reversing 8-89–8-101
  two-position 8-59–8-76
  wood 8-44

Protection, circuit 5-16

Protractor, propeller 8-57

PT6
  and Hartzell reversing propeller 8-96–8-101
  JT-12 fuel system 6-83
  lubrication system 8-29

Pump,
  acceleration 6-20
  gear 8-8, 8-21
  gerotor 8-9, 8-21
  Teledyne-Continental injection 6-43
  vane 8-20

Purging procedure 2-80

Pushrods 1-40

R

Radial engine
  construction 1-42–144
  cylinder arrangement 1-21
  firing order 1-24

Ram recovery 2-28

Range marking 5-48

Ratings, gasoline 6-10

Ratio,
  compression 1-13, 2-37, 6-4
  engine pressure (EPR) 5-25, 6-62
  fuel-air mixture 6-6

Reach, spark plug 3-27

Reaction engines 2-3–2-7

Reactive power 4-39

Reassembly, float carburetor 6-25

Reciprocating engine
  cooling systems 7-25–7-28
  exhaust systems 7-33–7-35
  induction system 7-1–7-13
  lubricating systems 8-1–8-15
  maintenance 1-47–1-78
  removal 1-79
  replacement 1-80
  starters 7-43–7-47
  theory 1-1–1-46

Rectifier, DC alternator 4-28

Reduction gearing, propeller 1-37, 1-44

Regulator, voltage 4-16

Reid vapor pressure 6-10

Relays 5-19

Relief valves 8-9–8-10

Removal,
  reciprocating engine 1-79
  spark plug 3-33
  tapered shaft 8-52
  turbine engine 2-52

Repair,
  major overhaul 1-65–1-71
  propeller 8-39

**Replacement, reciprocating
  engine** 1-80

Repulsion motors 4-48

Requirements,
  engine 1-18
  propeller 8-35–8-39
  propeller feathering 8-37
  turbine engine lubricants 8-16

Reservoir, oil 8-11

Resistor, spark plug 3-30

Restrictions, trim 6-67

Resultant thrust 2-20

Retainers, horizontally opposed 1-31

Reverse-current cutout relay 4-14, 4-22

Reverse-flow annular combustor 2-47

Reversers, thrust 2-52, 7-39–7-40

Reversing propeller 8-89–8-101

Rings,
  major overhaul 1-70
  piston 1-33
  reassembly 1-72

Rocker arms 1-40

**Rods, horizontally opposed
  connecting** 1-34

Rolls-Royce RB-211 fuel control 6-57

Rotating-magnet magneto 3-5–3-17

Rotation, phase 4-41

Rotor, DC alterator 4-27

Run data 2-75

Run-up,
  post-inspection 1-58
  pre-inspection 1-51
  turbofan 2-78–2-80
  turbojet 2-78–2-80
  turboshaft 2-80

S

Safety 2-78

Safetying, splined shaft 8-54

Scavenge subsystem, lubrication 8-27

Sea level performance 6-6

Secondary electrical circuit 3-10

Semi-depth filtration 8-13

Sensenich propeller designation 8-48

Sensor-responder fire detection 5-55

Series-field motors 4-47

Service life 1-47

Service,
  DC alternator 4-35
  DC generator 4-23–4-25
  float carburetor 6-22–6-28
  spark plug 3-33–3-41
  turbine lubrication 8-17

Shielding 5-15
  spark plug 3-26

**Shop maintenance, turbine
  engine** 2-58

**"Shower of sparks" ignition
  system** 3-12

Shunt-field motors 4-46

Shutdown procedure 2-79
  emergency 2-80

Shutoff valve 7-51

Shuttle-type magnetos 3-4

Solenoids 5-19

Sonic choking 2-13

Spark plugs 3-25–3-41

Specific
  fuel consumption 1-15
  gravity, turbine 6-61

Speed check, data plate 6-65

Speed, engine 1-15

Stall, margin curve 2-42

Starter
  pressure-regulating valve 7-51
  systems, turbine engine 7-48–7-53

Starter-generator 4-36, 7-49

Starters,
  air turbine 7-48–7-53
  electric 7-48
  reciprocating engine 7-43–7-47

Starting systems 7-43–7-54

Starting,
  auxiliary systems 3-11–3-16
  Bendix 6-41
  Teledyne-Continental 6-47

Static RPM, propeller 8-35

Stator vane
  repair 2-61
  variable compressor 7-17–7-19

Stator, DC alternator 4-28

Storage, spark plug 3-40

Subsonic inlets 2-26

**Supercharged engine induction
  system** 7-4–7-13

Supercharging 6-4

Supersonic inlets 2-28

Suppressor, noise 2-53

Surface filtration 8-13

Surficants 6-12

Surge, margin curve 2-42
Switches 5-18
Switches, pressure 5-26
Synchronizing, propeller 8-101–8-102
Synchrophasing, propeller 8-103
Synchroscopes 5-39
Synthetic oil 8-3, 8-17

T

Tachometers 5-36–5-39
  AC electrical 5-37
  electrical 5-37
  electronic 5-37
  helicopter 5-38
  non-electrical 5-36
  reciprocating engine 5-37
  turbine engine 5-38
Tail cone 2-50, 7-36–7-37
Tail pipe 7-36–7-37
  convergent-divergent 7-38
  thrust 2-19
Taxi procedure 2-79, 2-81
Teledyne-Continental fuel injection 6-43–6-49
Temperature indicators, cylinder head 5-30
Temperature measurement 5-25
  electrical 5-28–5-35
  non-electrical 5-27
Temperature,
  and engine thrust 2-23
  carburetor air 6-4
Terminal strips 5-11
Terminals 5-9
Termination, wiring 5-9–5-12
Test
  cell maintenance 2-73
  facilities 1-73
  preparation 1-74
  run 1-74
Testing,
  ignition harness 3-44
  major overhaul 1-73
  spark plug 3-39
Thermal efficiency 1-12, 2-22, 6-3
Thermal-switch fire detection 5-51
Thermocouple instrument system 5-29
Thermocouple-type fire detection 5-53
Thermometers, resistance-type 5-28
Thrust
  calculations 2-15
  distribution 2-17–2-20
  horsepower 2-20
  reversers 2-52, 7-39–7-40
  specific fuel consumption (TSFC) 6-50
  with choked nozzle 2-16
Thrust,
  fan engine 2-20
  gross 2-15
  net 2-16
  resultant 2-20
Time between overhaul (TBO) 1-47, 2-76
Timing,
  dual magneto 3-21
  ignition 1-14, 6-6
  magneto 3-19–3-21

Top overhaul 1-59
Torque 2-71–2-72
  turbine engines 2-8
Torque, measurement 5-42–5-44
Tracking, splined shaft 8-53
Trim
  adjustments, turbine 6-61
  danger zones 6-68
  restrictions 6-67
Trim,
  EPR-rated engines 6-62
  part power 6-62
  speed-rated engine 6-65
Troubleshooting,
  anti-icing system 7-24
  compressor bleed system 7-22
  float carburetor 6-26
  fuel systems 6-83–6-84
  Hartzell 8-76
  lubrication 8-30
  powerplant 1-74
  splined shaft 8-55
  starter systems 7-53–7-54
  turbine engine 2-76
  turbine ignition systems 3-52
  variable vane system 7-20
Turbine
  case, inspection 2-63
  outlet, thrust 2-19
  section, turbine engine 2-47–2-50
  vanes, inspection 2-63
  wheel, inspection 2-63
Turbine engine
  cooling systems 7-29–7-32
  design 2-26–2-54
  exhaust system 7-36
  induction system 7-14–7-24
  lubricating systems 8-15–8-32
  maintenance 2-55–2-81
  operation 2-78–2-81
  RPM limits 2-24
  starters 7-48–7-53
  theory 2-1–2-54
Turbine fuels 6-49
  additives 6-50
  control adjustments 6-61–6-70
  metering systems 6-49–6-84
  systems, principles 6-49–6-50
Turbocharger,
  induction 7-5–7-7
  inspection 1-51, 1-58
Turbofan engines 2-6
  efficiency 2-26
Turbojet engines 2-5
Turboprop engines 2-8
Turboshaft engines 2-8
Twisting moment,
  propeller aerodynamic 8-42
  propeller centrifugal 8-43
Two-position propellers 8-59–8-76
Two-stroke-cycle reciprocating, energy transformation 1-7

U

Untwist, inspection 2-64

V

V-Engine
  construction 1-46
  cylinder arrangement 1-20
  firing order 1-24
Valve
  guides, horizontally opposed 1-31
  lifters, hydraulic 1-41
  lifters, solid 1-41
  operating mechanism, horizontally opposed 1-40
  operating mechanism, major overhaul 1-71
  operating mechanism, radial 1-44
  seats, horizontally opposed 1-31
  springs, horizontally opposed 1-31
Valves,
  horizontally opposed 1-30
  major overhaul 1-69
  relief 8-9–8-10
Vanes 2-38
  variable 2-39
Vapor pressure, Reid 6-10
Variable
  compressor stator vane 7-17–7-19
  vane schedule 7-19
  vanes 2-39
Variable-orifice mixture control 6-17
Velocity 2-11
Vent subsystem, lubrication 8-27
Vibrational force, propeller 8-44–8-48
Vibrators, induction 3-11
Viscosity
  index, reciprocating engine 8-5
  index, turbine 8-16
  reciprocating engine 8-4
  turbine 8-16
Visual inspection 1-61
  spark plug 3-34
Voltage control 4-10
Voltage regulator 4-16
  carbon pile 4-19
  transistor 4-32
  transistorized 4-31
Volumetric efficiency 1-12

W

Water injection, turbine 6-70–6-73
Water
  contamination 6-12
  fire extinguisher 5-56
Wear 1-48
Weight 2-11
Well, acceleration 6-19
Wet
  sump, reciprocating engine 8-6
  sump, turbine 8-18
Wire 5-1–5-13
  bundle routing 5-13
  identification 5-6
  size 5-2–5-5
  type 5-1
Wiring
  installation 5-6–5-8
  termination 5-9–5-12
Wood propeller 8-44
Work 1-9, 2-10